Electron and Phonon Spectrometrics

"Bond and nonbond relaxation by physical perturbation or transformation by chemical reaction, and the associated energetics, localization, entrapment, and polarization of electrons mediate the performance of substance accordingly."

—Chang Q Sun
Relaxation of the Chemical Bond
Springer Series in Chemical Physics 108, 2014

"O:H–O bond segmental cooperativity and specific-heat disparity form the soul dictating the extraordinary adaptivity, reactivity, recoverability, and sensitivity of water and ice subjecting to perturbation."

—Chang Q Sun and Yi Sun
The Attribute of Water: Single Notion, Multiple Myths
Springer Series in Chemical Physics 113, 2016

"Hydration of electrons, ions, lone pairs, protons, and dipoles mediates the O:H–O network and properties of a solution through O:H–O relaxation, H↔H and O:⇔:O repulsion, screened polarization, solute-solute interaction, and intrasolute bond contraction."

—Chang Q Sun
Solvation Dynamics: A Notion of Charge Injection
Springer Series in Chemical Physics 121, 2019

Chang Q Sun

Electron and Phonon Spectrometrics

 Springer

Chang Q Sun
School of Electrical and Electronic
Engineering
Nanyang Technological University
Singapore, Singapore

Yangtze Normal University
Chongqing, China

ISBN 978-981-15-3175-0   ISBN 978-981-15-3176-7   (eBook)
https://doi.org/10.1007/978-981-15-3176-7

This Springer imprint is published by the registered company Springer Nature Singapore Pte Ltd.
The registered company address is: 152 Beach Road, #21-01/04 Gateway East, Singapore 189721,
Singapore

Energy shift of electrons and phonons by physical perturbation or chemical reaction not only fingerprints the bonding, electronic, and molecular dynamics but also offers information for genomic engineering of matter and life.

In memory of and dedicated to my parents

Preface

This book was motivated by the following questions that arose during the Third International Conference on Scanning Tunnelling Microscopy and Spectroscopy (STM/S) held in Beijing in 1993. These questions are ubiquitous and have become increasingly important to the advancement of chemical and physical sciences:

(1) Can we probe the bonding network and the behaviour of electrons in various energy levels from sites surrounding a point defect or an impurity, and in the topmost atomic layer of a surface, to complement the STM/S signatures?

(2) How do electrons, atoms, and molecules perform in the energetic, spatial, and temporal domains cooperatively when the reaction takes places?

(3) How does the bond-electron-phonon-property relax under multifield perturbation—such as atomic irregular coordination, electrostatic polarization, mechanical activation, thermal excitation, and so on so forth?

(4) Can we distil the number fraction and stiffness of chemical bonds transiting from the vibration mode of the reference to the conditioned states under physical perturbation or chemical reaction?

As the basic means, diffraction crystallography, surface morphology, and electron and phonon spectroscopies characterize the performance of atoms, electrons, and molecules in the energetic, spatial, and temporal domains and their interactions. Spectral signatures are related to the intrinsic performance of chemical bonds and electrons in various energy levels or energy bands. Combined with the atomic-scale STM and the low energy electron or X-ray diffraction (LEED, XRD), spectroscopies of electrons and phonons reveal profound, quantitative information about the relaxation of the local bond-electron-phonon that fosters intrinsically the chemical and physical properties of the probed substance. However, it remains a challenge to fully tap the capacity of the spectroscopies in probing desired information. The conventionally well-adopted methods of spectral peak Gaussian-type decomposition of a certain peak into multiple components or an empirical simulation to the spectral peak evolution with external perturbation offer limited information, albeit hypothetic parameters involved and the lack of spectral decomposition constraints. Therefore, the presented electron and phonon spectrometrics for the intrinsic, local, dynamic, and quantitative information on the bond-electron-phonon-property correlation become inevitably desirable.

This book is devoted to reconciling spectrometrics of electron emission, electron diffraction, and phonon absorption and reflection from liquid and solid substance under various perturbations or reactions taken place. Focus is given on information distillation of the bond-electron-phonon dynamics, which is the reason to amplify the conventional spectroscopy to the presently featured spectrometrics. The entire work is grounded on principles of the bond order-length-strength (BOLS) correlation, nonbonding electron polarization (NEP), local bond average (LBA), multifield lattice oscillation and electron binding dynamics for systems under perturbation. The strategies include differential photoelectron/phonon spectrometrics (DPS) that distils transition of the length, energy, stiffness and the fraction of bonds upon being chemically or physically conditioned and the derived information on performance of electrons in various energy bands in terms of quantum entrapment and polarization.

The physical ground of the spectrometrics is the Hamiltonian perturbation through bond relaxation by physical perturbation and bond transformation by chemical reaction. One can establish the bond-electron-property correlation for a substance by shifting the phonon frequency and electron binding energy through perturbing its Hamiltonian by programmed electrostatic polarization, magnetization, mechanical and thermal excitation, atomic and molecular undercoordination, tetrahedral bond formation, charge injection by chemisorption or solvation, or a combination of these degrees of freedom. The mathematical foundation of the spectrometrics is the Fourier transformation that gathers electrons of the same binding energy or bonds vibrating in the same frequency into their characteristic spectral peaks, irrespective of their locations, bond orientations, or structural phases. One can thus focus on the behaviour of the representative bond or the representative electron for their own types in an examined specimen.

This volume contains three parts, with a focus on the principles, strategies, applications, and databases for the electron and phonon spectrometrics:

Part I features the coordination-resolved electron emission spectrometrics. Experimental technique includes STM/S, photoelectron spectroscopy (PES includes XPS and UPS), Auger electron spectroscopy (AES), and near-edge X-ray absorption/emission spectroscopy (NEXAS/XES). The XAS, XES and the AES and PES coincidence spectroscopy (APECS) could determine simultaneously energy shifts of the valence band and a core band, which resolves the effect of screen shielding of one level from the other and resolves the charge sharing between constituent reactants in reaction. The BOLS-NEP enabled the zone-resolved photoelectron spectrometrics (ZPS) that distils information on bond relaxation and associated energetics, charge localization, quantum entrapment and polarization pertaining to the irregularly coordinated atoms. Examination of the geometric registry and layer-number-order-resolved adatoms, point defects, solid and liquid skins, terrace edges, atomic clusters, nanocrystals, heterojunction interfaces and nanoalloys, and the confined (molecular undercoordination) water and salt solutions have led to rich genomic information. Findings clarified the following:

(1) Perturbation to the Hamiltonian by atomic undercoordination-induced bond contraction, the hetero-coordination-induced bond nature alteration, and the polarization of the nonbonding electrons dictate intrinsically the electron binding energy shift of the irregularly coordinated systems.

(2) Core electron entrapment and nonbonding electron polarization take place by atomic and molecular undercoordination, tetrahedral bond formation, and hetero-coordination. The local charge densification and quantum entrapment result in the globally positive core level shift, which is proportional to the local bond energy.

(3) Polarization of the nonbonding states pertained to the outermost shell of the even fewer coordinated edge or surface atoms by the locally densely entrapment bonding electrons negatively offsets the entrapped states and raises the local Fermi energy, or creates the polarized states above the E_F.

(4) Water molecular undercoordination and ionic electrostatic polarization have the same effect on O:H–O bond relaxation, O 1s electron entrapment, and valence electron polarization. Being similar to the APECS, the NEXAS pre-edge shift results from the combination of the core level and the valence band shifts, which fingerprints less sensitively the bond relaxation and polarization than the XPS.

The dominance of valence electron entrapment of Pt adatoms and Cu/Pd alloy ensures their acceptor-type catalysts, while the polarization dominance of Rh adatoms and Ag/Pd alloy endows their donor-type catalysts. It is the right entrapment or polarization dictates the single-atom catalysis. The isolation and polarization of the dangling σ bond electrons by the densely, locally entrapped bonding electrons surrounding the defects and at the zigzag edges create the graphitic Dirac-Fermi polaritons of graphene ribbons. Only one atomic neighbour loss transits the entrapment dominance of the smooth graphite skin to the polarization dominance of its point defect. It is suggested that the spin-coupled nonbonding electron polarization by atomic undercoordination and sp orbital hybridization dictate the skin dominance of monolayer high-T_C superconductivity and the topological insulator edge superconductivity.

Part II deals with the very low energy electron diffraction (VLEED), specifically by the O–Cu(001) surface reaction. Interplaying with STM/S and PES, VLEED probes the behaviour of atoms, bond geometry, and bonding dynamics in the outermost two atomic layers and their electrons in the valence band and above, which reveals dynamic information about the bond formation and relaxation, valence energy states, and potential barrier shape evolution. VLEED determines the variation of work function, atomic muffin-tin inner potential constant, the first two Brillouin zones, and the effective mass of electrons near the Brillouin zone boundaries. Most strikingly, the exposure-resolved VLEED clarifies the creation of four valence states of the bonding pairs, nonbonding lone pairs, ionic electronic holes, and antibonding dipoles associated with the four-stage bonding transition dynamics from the CuO_2 pairing pyramids to the Cu_2O_3 pairing tetrahedrons:

(1) Bond dissociation, formation, and relaxation and the valence electron and surface potential barrier evolution are involved simultaneously upon chemical reactions. O_2 dissociates into 2O atoms that bond to one Cu atom in the top layer to form a pair of off-centered pyramids, $[O^-Cu^{2+}O^-]$.

(2) Each of the O^- bonds to a Cu atom underneath to form the second Cu–O bond is associated with bond angle and length relaxation. The O^{2-} ionic polarization squeezes every fourth row of Cu atoms to evaporate from the $(\sqrt{2}\times2\sqrt{2})R45°$ surface, leading to the ordered missing-row vacancies.

(3) The sp^3 orbital hybridization takes place subsequently, and the nonbonding electron lone pairs polarize its Cu neighbours into paired dipoles across the missing-row vacancies.

(4) Further bond relaxation stabilizes the surface and turns the $\angle$Cu–O–Cu angle from 90° to 105° and the $\angle$Cu:O:Cu from 130° to 150° with ":" being the lone pair on oxygen, which expands the first layer spacing from 0.185 to 0.194 nm.

It ascertains that an O atom tends to form a tetrahedral structure in the solid phase and that one O cannot form two or more bonds with a specific host atom because of the bond geometry restriction. The bond formation dynamics are subject to the host lattice constant, geometry, and electronegativity. VLEED forms such a uniquely sophisticated means that integrates information on the bond-barrier-band energetics and dynamics which holds general for the oxidation of diamond and other metallic surfaces, and N and C absorption as well.

Part III is focused on the Raman and infrared phonon spectrometrics for information about bond length and bond energy under multifield perturbation. The BOLS-LBA derived DPS resolves the transition of abundance-stiffness-fluctuation of oscillating bonds upon perturbation or reaction. Exercises have quantified the local bond length and energy relaxation of group IV, III–V, II–VI nanocrystals, layered graphene ribbons and WX_2 flakes, and the hydrogen bond network of water ice and aqueous solutions. Practice has revealed the following:

(1) Molecular undercoordination-induced phonon frequency shift specifies the reference from which the phonon frequency shifts, bond nature index, as well as the manner and bond number in the specific mode of vibration. Atomic dimer vibration governs and stiffens the translational optical G mode of graphene, and the E_g mode of black phosphor, TiO_2, and WX_2, while the collective vibration of an atom with its nearest neighbours softens the longitudinal optical D modes of graphene, A_g mode of WX_2, and TiO_2 upon crystal size reduction.

(2) Pressure stiffened phonon frequency and elasticity enable the derivation of the binding energy density and the elasticity of a crystal, which is beyond the Grüneisen description. The uniaxial strain-induced phonon relaxation and phonon band splitting enable the derivation of the single-bond force constant and the relative direction between a specific bond and the applied strain in graphene and MoS_2.

(3) Temperature-dependent phonon relaxation, band gap, and elasticity follow the same Debye thermal decay, which gives rise to the atomic cohesive energy and the Debye temperature of a substance of interest.

(4) Molecular undercoordination, liquid heating, quasisolid cooling, electrification by a capacitor, or ionic injection upon the acid and salt solvation shorten the covalent part of the O:H–O bond, but the O:H nonbond responds to perturbation always contrastingly. Compression and base solvation relax the O:H–O bond contrastingly to the effect of salt solvation. The H–O and the O:H vibration frequencies and bonding energies determine the specific-heat curves and the critical temperatures for evaporation, liquidation, and crystallization for water ice and aqueous solutions. Aqueous solvation disperses the phase boundaries and modulates the critical temperatures for ice formation, melting, and evaporation.

Part II landmarked the beginning, and my career focus of *Coordination Bonding and Electronic Dynamics* and *Spectrometric Engineering* that enabled discoveries of orbital hybridization in chemisorption and undercoordination induced bond contraction. As an independent degree of freedom, atomic or molecular undercoordination should receive deserved attention, as it forms the foundation of defect physics, surface chemistry, nanoscience and nanotechnology. Orbital hybridization and bond relaxation have propelled considerable and systematic progress in dealing with physical perturbation and chemical reaction of solid and liquid substances and functional materials devising.

It is my great pleasure and obligation to share these personal thoughts and learnings with the community, though further refinement and improvement may be required—thus, critique from readers is most welcome. I hope that this volume, amplifying the capabilities of existing spectroscopy techniques, could inspire more analytically oriented approaches towards extracting bond-electron-phonon information and stimulate more research interest and activities towards predictive controlling of the coordination bonding and electronic dynamics. Directing effort to the engineering of bond and nobond, nonbonding electronics, materials genomics, aqueous science, and fine engineering of liquid and solid phases could be even more challenging, fascinating, promising, and rewarding.

I would like to express my gratitude to colleagues, friends, and peers for their encouragement, invaluable input, and support, to my students and collaborators for their contribution, and to my family, my wife Meng Chen and daughter Yi, for their assistance, patience, support, and understanding throughout this fruitful and pleasant journey.

Singapore   Chang Q Sun
February 2020

About This Book

This volume amplifies the traditional electron and phonon spectroscopies, including STM/S, XPS, UPS, AES, APECS, NEXAS, LEED, and Raman scattering and IR absorption, to their spectrometrics. This amplification allows for genomic information about the bonding, electronic, and molecular dynamics and energetics pertained to bond relaxation by physical multifield perturbation and bond transformation upon chemical reaction. Physical perturbation includes atomic and molecular undercoordination, mechanical compression, thermal excitation, electromagnetic radiation, etc. Chemical reaction includes CNHO chemisorption, atomic hetero-coordination, and aqueous solvation. Spectrometrics had led to atomic scale, dynamic, local and quantitative information on bond relaxation in length and energy, staged bond transforming dynamics, anti-hydrogen bond and super-hydrogen bond formation, valence band and potential barrier evolution, bonding electron quantum entrapment, nonbonding electron polarization, atomic cohesive energy, binding energy density, elastic modulus, Debye temperature, single-bond force constant, etc.

Targeted audience includes researchers, scientists, and engineers, in chemistry, physics, surface and interface science, and materials science and engineering.

Contents

Part III   Multifield Phonon Dynamics

About the Author

Chang Q Sun received his B.Sc. in 1982 from Wuhan University of Science and Technology and an M.Sc. in 1987 from Tianjin University and served on its faculty until 1992. He earned his Ph.D. in Physics at Murdoch University in 1997 and then joined Nanyang Technological University. He holds an honorary appointment at Yangtze Normal University, Chongqing, China.

Dr. Sun has been working on the theme of *Coordination Bonding and Electronic dynamics.* He has developed a comprehensive set of theoretical and spectral approaches, leading to a systematic understanding of the bonding, electronic and molecular dynamics and energetics pertained to atomic undercoordination, multifield perturbation to solid mechanics and water ice, aqueous solvation, and CNHO chemisorption. Major breakthroughs include: (i) reconciliation of the attributes of undercoordinated adatoms, defects, edges, surfaces, and the size dependency of nanostructures; (ii) resolution to anomalies of water and ice under perturbation; (iii) quantification of the CNHO chemisorption bonding and electronic dynamics; and, (iv) clarification of charge injection upon acid, base, salt, electrolyte and organic molecular solvation.

His contribution has been cultivated into four bilingual monographs, 30 perspectives/reviews, and over 400 peer-reviewed journal articles. He was bestowed with the inaugural Nanyang Award in 2005 and the first rank of the 25th Khwarizmi International Award in 2012. He is currently on the advisory boards for multiple journals.

Nomenclature

θ	Contact angle between the solution droplet and a certain substrate to feature the solution surface stress
γ	Grüneisen parameter
$\eta(C)$	Solute concentration C resolved solution viscosity
$\Theta_{DX}(\omega_X)$	Debye temperature (characteristic phonon frequency) of the x segment of the O:H–O bond
$\eta_x(T)$	Segmental specific heat featured by the Θ_{DX} and the E_x (specific-heat integral)
AES	Auger electron spectroscopy
APECS	Auger photoelectron coincidence spectroscopy
BE	Binding energy
BOLS	Bond order-length-strength correlation for undercoordinated atoms and molecules
BP	Black phosphorus
BZ	Brillouin zone
CLS	Core level shift
CN	Coordination number, z
DFT	Density functional theory
DOS	Density of states
DPS	Differential phonon spectrometrics
d_x/E_x	O:H–O bond x segmental length and bond energy (x = L for the O:H and H for the H–O bond)
E_ν	The νth electron energy level of an atom
E_{coh}	Atomic cohesive energy
E_{den}	Binding energy density
E_g	Bandgap; Raman translational vibration mode
$f(C)$	Solute concentration C resolved volume or number fraction of the hydrogen bonds transiting from the vibration mode of ordinary water to hydration states
FTIR	Fourier transformation of infrared spectroscopy

FWHM	Full width at half maximum
GNR	Graphene nanoribbon
H↔H	Anti-HB due to excessive H^+ injection serves as a point beaker fragilizing the HB network
KE	Kinetic energy
LBA	Local bond average
LFR	Low-frequency Raman mode
MD	Molecular dynamics
MR	Missing row
NEP	Nonbonding electron polarization
NEXAS	Near-edge X-ray absorption spectroscopy
O:⇔:O	Super-HB due to excessive lone pair injection serves as a point compressor
O:H–O	Hydrogen bond (HB) coupling inter- and intramolecular interactions
P_C/T_C	Critical pressure/temperature for phase transition
QS	Quasisolid phase bounded at $(-15, 4)$ °C due to O:H–O segmental specific-heat superposition, which disperses under external perturbation
SFG	Sum frequency generation spectroscopy
SPB	Surface potentials barrier
STM/S	Scanning tunnelling microscopy/spectroscopy
SWCNT	Single-walled carbon nanotube
TB	Tight binding
T_m/T_N	Melting/freezing temperature closing to the intersections of segmental specific-heat curves and dominated by the E_H/E_L
TO/LO	Translational/longitudinal optical wave
VLEED	Very low energy electron spectroscopy
X/UPS	X-ray/ultraviolet photoelectron spectroscopy
ZPS	Zone selective phonon/photoelectron spectroscopy

Part I
Electron Emission: Quantum Entrapment and Polarization

Electronic emission spectrometrics resolves atomistic, local, and quantitative information on the coordination-resolved bonding energetics and electronic densification, entrapment, and polarization as well as energy levels of an isolated atom.

Chapter 1
Introduction

Abstract Bonds and electrons perform differently at irregularly-coordinated atomic sites from they do inside large volumes of ideally fully-coordinated systems. The irregularly-coordinated atomic sites include adatoms, point defects, kink edges, grain boundaries, liquid and solid skins, nanostructures of various shapes, and the hetero-coordinated interfaces, alloys and compounds. These atoms and electrons offer novel phenomena and properties for basic science and practical applications including atomistic catalysts, adhesion, anticorrosion, friction, welding, high T_C super-conductors, nuclear radiation protection, topological insulators, etc. Fine-resolution detection and consistent understanding of the unusual behavior of bonds and electrons at such atomic sites become increasingly important. This part describes the practitioner's efforts and progress in this aspect.

Highlights

- Bond relaxation and the associated electron energetics matter the performance of a substance.
- Irregularly-coordination features the anomalies of defects, nanostructures, surface and interface.
- Resolving the atomistic local bonding, nonbonding, and electronic energetics is highly desired.
- Strategies for fine-resolution detection and consistently deeper insight still open for exploration.

1.1 Overview

1.1.1 Coordination Bonds and Energetic Electrons

Atomic undercoordination refers to atoms associated with grain boundaries, homogeneous adatoms, point defects, solid or liquid skins, terrace edges, and nanostructures of various sizes and dimensionalities such a monatomic chains and monatomic sheet. Atomic hetero-coordination means those associated with alloys, compounds, chemisorbed skins, dopants, impurities, and interfaces. Performance of bonds between the irregularly coordinated atoms and associated electrons in various bands are crucial to the properties of materials at these sites [1–8].

One can create a new substance with desired functionalities by breaking the old bonds and forming new kinds of bonds between different atoms. Harnessing the known properties of a substance by relaxing the bonds and the associated energetics and dynamics of electrons in localization, densification, polarization, and transportation becomes therefore increasingly important [1]. For instance, materials at sites surrounding irregularly-coordinated atoms perform differently from themselves in the bulk interior of elemental solids. Although they are traditionally unwanted, such irregularly-coordinated atoms are of key importance to the advancement of condensed matter physics, solid-state chemistry, materials sciences, and device technologies, in particular, at the nanometer scale [3].

Controllable relaxation of the coordination bonds and the energetic electrons at such irregularly-coordinated atomic sites provides profound impact to many areas of scientific and technological interest [4, 9]. These subject areas include, for instances, adhesion [10], adsorption [11], alloy formation [12, 13], catalytic reaction [14, 15], corrosion protection [16], decomposition [17], diffusion [18], doping [19] epitaxial growth [20, 21], hydrophobic lubrication [7], glass formation [22], dielectric modulation [23, 24], mechanical strength [25–28], thermal elasticity [8, 29], photon and electron emission and transportation [30], quantum friction [31], radiation protection [32], topological insulator conduction [33], superconductivity [34, 35], thermal stability [36, 37], wettability [38, 39], water and ice skin supersolidity [40, 41], etc.

Electrons associated with point defects [42, 43], homogeneous adatoms [44], adsorbates [45–47], terrace edges [48–50], monatomic chains and their ends [51, 52], and solid skins [53, 54] result in, for instance, new types of energy states that enhance tremendously the site-selective catalytic ability of a substance even though the bulk parent, like gold, is chemically inert [55]. For instance, the performance of a semiconductor could be promisingly improved if flaws that were previously thought irrelevant are reduced. However a specific defect impacts the ability of halide perovskite to hold energy derived from light in the form of electrons. Defects could be good or bad in semiconductors. Dislocations negatively impact the carrier dynamics of halide perovskite. Reducing dislocation densities by more than one order of magnitude is found to lead to an increase of electron lifetime by four times [56]. Defects rich MoS_2 ultrathin nanosheets with additional active edge sites could enhance electrocatalytic hydrogen evolution [57]. Atomic undercoordination laid the foundation

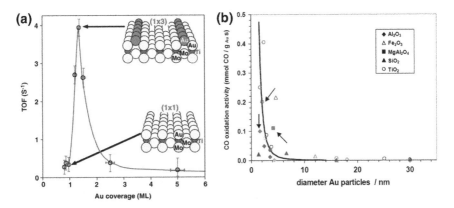

Fig. 1.1 a Atomic undercoordination enhanced catalytic reactivity of Au/TiO$_2$ for CO oxidation at room temperature and **b** CO oxidation activity of Au nanoparticles deposited on different oxides. Reprinted with permission from [62, 63]. Copyright 2004 American Association for the Advancement of Science and Elsevier

of porous structures or metal organic framed (MOFs) nanostructures that have many important application to chemical science and industry [58–61].

Figure 1.1a shows that the every third row of Au atoms added to a TiO$_2$ surface that has already been covered with a full-layer of Au atoms could improve the CO oxidation efficiency at room temperature by some 50 folds of the otherwise fully Au-covered surface [62]. Au particle size reduction raises its CO oxidation ability (Fig. 1.1b) and the ability of Guanine binding to the particle [64].

Likewise, the activation energy for N$_2$ dissociation is 1.5 eV lower at terrace edges than that at the flat Ru(0001) skin, yielding at 500 K a desorption rate that is at least 9 orders of magnitude higher on the terrace edges [65]. Similar attributes hold for NO decomposition on Ru(0001) skin, H$_2$ dissociation on Si(100) skin [66], and low-temperature nitridation of nano-structurally-roughened Fe skins [67].

Skin roughening with nanoscaled features forms an effective means improving the catalytic ability of a substance. The reactivity is three orders higher in magnitude for ammonia synthesis at the Re(11$\bar{2}$1) and the Re(11$\bar{2}$0) kink edges than at a smooth Re(0001) skin [68]. An addition of a certain kind of adsorbate roughens the skins of Ni(210), Ir(210) [69], Rh(553) and Re(12$\bar{3}$1) [70] to improve their catalytic efficiencies.

Undercoordinated atoms serve as the most active sites in reaction. A few percent of adatoms in a specimen could raise sufficiently the reactivity of the specimen in catalytic applications. The edged or faceted atoms account for ~70% of the total catalytic activity of the medium. The even-undercoordinated adatoms on Rh(111) skin favor the process of methane dehydrogenation more than atoms at steps or at the terrace edges [71, 72]. Adatoms deposited on oxides can activate the C–H bond scission [73], acetylene ciclomerization [74], and CO oxidation [75]. The catalytic efficiency of the undercoordinated atoms increases as their coordination numbers

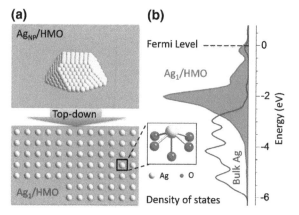

Fig. 1.2 **a** Atomically dispersed metal catalysts from supported Ag nanoparticles by a thermal diffusion process and **b** atomic undercoordination and oxygen support induced valence states polarization (reprinted with permission from [81])

(CNs) decrease. The efficiency further increases when the undercoordinated atoms grow hetero-coordinately on oxide supports.

Besides atomic undercoordination, strain generation by defect formation, implantation, or by substrate-particle interaction can also enhance the surface catalytic ability [76]. Argon plasma implantation into the Ru(0001) subsurface stretches the lattice, which promotes the adsorption efficiency of O and CO [77, 78] and enhances the NO dissociation probability in the stretched regions [79]. Adsorption of small clusters can induce a considerable strain in the skin and improve the catalytic ability of the clusters, demonstrating the joint effect of bond strain and atomic undercoordination [80].

Figure 1.2 examplifies the dispersion of Ag atoms in the homogeneous metal oxide for efficient catalysis and a comparison of the UPS spectra. Results show that the valence DOS goes up closing to the E_F upon atomic undercoordination [81]. Figure 1.3 compares the cluster size dependence of the Au 4f ionization energies. Similar to the supported clusters, the energy shift was interpreted as the effect of a large fraction of undercoordinated atoms.

The extremely high catalytic efficiency of undercoordinated atoms is indeed fascinating but the fundamental nature behind the efficiency elevation remained unclear. The following addressed the most advanced yet hypothetic mechanisms on the catalytic ability of gold adatoms [62, 83]:

(1) Gold adatoms have fewer nearest neighbors and a special yet unclear bonding geometry that creates more reactive orbits compared to the otherwise fully coordinated atoms.

(2) They exhibit the quantum size effects that may alter the energy band structure of nanoparticles.

(3) They may undergo electronic modification by interacting with the underlying oxide that causes partial electron donation to the atomic clusters.

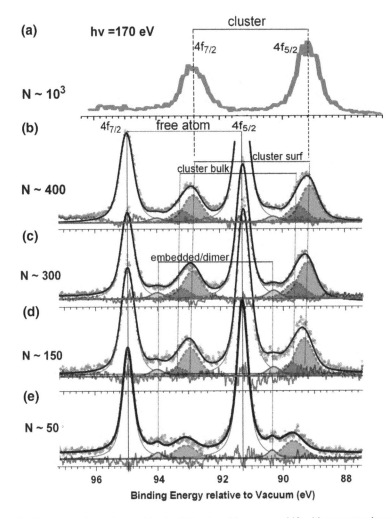

Fig. 1.3 Cluster size dependence of the Au $4f$ level positive energy shift with respect to the vacuum level. Reprinted with permission from [82]

Therefore, comprehension of the catalytic ability due to undercoordination from the perspective of local bond relaxation and the associated electronic energetics, i.e., entrapment or polarization, is of great importance.

Performance of the hetero-coordinated atoms for alloys, dopants, embedded nanocrystals, impurities, and interfaces is another subject of extensive study. Interfacial bond formation perturbs the local Hamiltonian, binding energy density, and atomic cohesive energy, associated with local charge entrapment or polarization, which determines the catalytic, electronic, dielectric, optic, mechanical, and thermal properties of the interface. Atomic hetero-coordination makes the interface region completely different from the bulk of the constituent parents [84, 85].

Mixing two or more substances together forms an interface that performs differ-ently from the respective constituent [86, 87]. One can modify the bond strain and charge distribution surrounding the bonded atoms [88, 89] by varying compositions or thermal annealing upon continuous deposition of dissimilar metals [90, 91]. For instances, Ag or Cu atoms can pair with Pd atoms to form alloys with improved heterogeneous catalysts [92] to meet different needs [93]. Cu/Pd is active for CO and alkene oxidation, ethanol decomposition, benzene, toluene, and 1, 3-butadiene hydrogenation. Pd atom is active for CO oxidation while Cu is active for NO dis-sociation. Ag/Pd alloy is a good candidate for reduction or hydrogen reaction and permeation [94]. Both Ag and Cu can grow on Pd in a layer-by-layer fashion at room temperature. Annealing at certain temperatures turns the layered structures into alloys [90, 95]. Reaction with electronegative elements such as F, O, N, C, etc., a metal turns into a semiconductor or into an insulator.

Dispersion of an individual Ir atom into FeO_x enhances greatly the reducibility of the FeO_x, leading to the dispersed single-atom catalysts such as Ir/FeO_x for water gas shift reaction [96]. Likewise, the extremely undercoordinated metallic atoms serve as catalysts for various applications with orders high efficiency than large trunks. The dispersed single-atom catalysts could raise the capability of Co–N–C [97] and Co/N-graphene, Pd/TiO_2 [98] and Pt/ethylenediamine [99] interfaces for hydrogen-generation [100], Pt/Pt_3O_4 for CO oxidation [101], Pt/TiN for electro catalytic oxidation of small organic molecules [99], etc. Figure 1.4 shows the XPS Pd $3d$ spectra and the size resolved energy shift and the ratio of under/full coordinates atomic sites of the cubooctahedral clusters [102]. Atomic undercoordination resolves the core level shift in Au and Pd, irrespective of the substrate of support.

The metal-based catalysts suffer from high cost, low selectivity, poor durability, susceptibility to gas poisoning and detrimental environmental impact. To overcome these limitations, a new class of catalyst based on earth-abundant carbon materials was discovered as an efficient, low-cost, metal-free alternative to platinum for oxygen

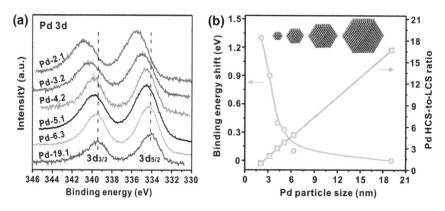

Fig. 1.4 Pd particle size resolved Pd/Al_2O_3 **a** XPS spectra and **b** the Pd $3d_{5/2}$ binding energy shift with reference to the 19.1-nm sized Pd sample, and the estimated high/low-coordinated site ratio. Reprinted with permission [102]

reduction in fuel cells [103–105]. Carbon-based metal-free catalysts with heteroco-ordinated doped atoms have been demonstrated to be effective for an increasing number of catalytic processes [103]. The combination of the ultralow coordination and hetero-coordination results in the unexpectedly strong quantum entrapment of the core electrons and polarization of the dangling bond electrons, which stems the catalytic ability of the metal-free substance [2].

As the wall materials in a fusion device, Be/W alloy protects efficiently nuclear radiation in the International Thermonuclear Experimental Reactor (ITER) [106–110]. Due to the electronic structure difference between W and Be, a strong influence of alloying on both the core and the valence electrons and the interfacial energy may happen but the verification and quantification of this expectation remains unrealistic. Understanding the nature of the bond and the behavior of the energetic electrons at the alloying interface is crucial to designing such functional materials for nuclear radiation protection and catalysis application, particularly.

Therefore, experimental distillation of the atomistic, local, dynamic, and quantitative information and theoretical clarification of the physical origin about the bonding and electronic dynamics at the irregularly coordinated atomic site are of paramount importance to advancing the subjects of Hetero-coordination Chemistry and Undercoordination Physics.

1.1.2 Challenges Faced by Existing Probing Technologies

Formation of bonds between hetero-coordinated atoms and relaxation of bonds between undercoordinated atoms and the associated electron entrapment and polarization mediate the performance of a substance at these irregularly-coordinated atomic sites [1]. However, resolving such atomistic, dynamic, local, and quantitative information is beyond the scope of available techniques though the invention of STM/S has advanced surface science enormously. STM/S maps electrons of the outermost layer of atoms within the energy window of a few eVs crossing E_F [111]. For instances, atomic vacancies at a graphite(0001) skin [42] and at the zigzag edge of a graphene exhibit the same topological and the same Fermi resonant signatures [112]. The resonant states, named Dirac Fermions, serve as carriers for topological insulators exhibiting the anomalous quantum Hall effect [33, 110]. The Dirac Fermions are almost massless with nonzero spin and infinitely large group velocity.

STM also probed the "check-board" like protrusions from Cu(100)–O^- surface [113]. The "check-board" protrusions evolve subsequently into the "dumb-bell" like protrusions when the O^- adsorbate transits into the O^{2-} [114]. The latter exhibits the characteristic antibonding density of states (DOS) featured at +2.0 eV above E_F and the nonbonding states at −2.1 eV below [115]. The former corresponds to the polarization of the Cu^p dipoles by the O^- and the latter to the Cu^p dipoles by the electron lone pair of the O^{2-} [45].

STM/S probed the volume expansion and electron polarization for Au atoms at ends of the monatomic chain [52], for Ag atoms added on Ag(111) skin [116], and for Cu atoms added on Cu(111) skin [117]. However, using STM/S alone, one can hardly elucidate quantitatively how the local chemical bond relaxes and how to derive the STM/S signatures.

In contrast, a photoelectron spectroscopy (PES, such as UPS and XPS) collects the statistic and volumetric information of electron binding energy in the valence band and below within a nanometer-thick skin [118–123]. The penetration depth of the electron beam varies with its incident energy. If the beam energy is in the range of 50–100 eV, the depth is 5–10 Å; if it is 500–880 eV, the depth is 10–20 Å; if the beam energy reaches 3–6 keV, the penetration depth is 30–60 Å [124]. One can hardly discriminate contribution of the monolayer skin from that of its bulk mixture. The effect of atomic undercoordination and bond nature alteration in chemisorption is dominant at most the outermost four atomic layers [3, 4, 45, 46, 125].

Experimental conditions create artifacts in measurements. Varying the incident beam energy or the polar angle between the incident light or the emission electron beam and the surface normal during PES experiment also modulates the spectral peak energy and intensity. Increasing the incident beam energy or decreasing the polar angle (close to the surface normal) could raise the intensity of the high-energy component (close to E_F) [126–130]. For instance, at the same polar angle, increasing slightly the beam energy from 370 to 380 eV increases the intensity of the Rh $3d_{5/2}$ peak at 306.42 eV (bulk) more than that of the 307.18 eV (skin) component [131]. Likewise, at the same beam energy, decreasing the polar angle from 35° to 20° reduces the intensity of the Rh(111) skin component at 307.18 eV [132]. Both beam energy increase and polar angle reduction collect more information from the bulk than from the skin. Experimental artifacts also include the charging effect and the ionization, initial-final state relaxation that present throughout the course of measurements.

XPS profiles of Nb(100) [126], Tb(0001) [128], Al(100) [133, 134], Ta(100) [130], and Be skins [135–138] follow the same polar angle and beam energy dependency. At higher beam energies or at smaller emission angles, the incident beam excites more electrons from deeper layers compared with the otherwise [139]. Thus, one can readily discriminate the contribution of the skin from that of the bulk by varying the beam energy or the emission angle though the derived information was qualitative.

Strikingly, the $2p$ energy of a Si-diode shifts significantly during its operation under both forward and reverse bias [140, 141] or under high frequency charging and discharging the specimen [142, 143]. Photon illumination also enhances the Si $2p$ energy shift [144]. This approach traces potential variation with the concentration and location of dopants.

Since the advent of the third generation synchrotron light sources providing soft X-rays up to 2 keV, XPS becomes a powerful tool for studying surface chemical and physical properties at an unprecedented precision level [145]. The high resolution allows for identifying various surface species, small molecules, and even the vibrational fine structures [146]. The near-ambient pressure XPS mimic the ambient environment for examining the chemical effect on the core band shift of a substance such as gold [147]. The liquid microjets PES and ultrafast liquid-jet UPS provide information on the electron binding energy for skins of liquid water and sized droplet [148, 149].

However, what one can measure using a PES are the convoluted peaks of the core bands contributed by intra-atomic trapping, interatomic binding, crystal orientation, defect vacancy, surface relaxation, nanosolid formation, surface passivation and adsorption, with the mixture of bulk information. An XPS spectral peak often contains multiple components of energy shift with uncertainty of their separation, direction, component integral and width, and the reference point from which these shifts proceed.

Based on the conventional exercises of spectral decomposition analysis, for instance, one can hardly discriminate the contribution of adatoms from that of the skin or the bulk to the PES spectrum with little information regarding the relaxation dynamics of the local bonds and the energetic electrons pertaining to adatoms [132, 150]. Combining the most advanced laser cooling and XPS technologies, one can only measure the energy separation between two levels of the slowly moving gaseous atoms of lower bulk meting point [151]. Therefore, discriminating contributions of different sublayers or oriented crystals from those of the bulk, and separating the contribution of the interatomic interaction from that of the intra-atomic interaction to the spectrum is still a challenge. Determination of the individual energy level of an isolated atom and its shift with coordination and chemical environment remains as a "bottle net" in study [2].

The valence DOS profile provides direct information about charge transportation during reaction from one constituent to the other (ionization), polarization and entrapment of valence charge, which evolves the interatomic potentials from the initial form due to potential screening and splitting effect [45]. Understanding the evolution of the valence DOS needs proper models to relate the valence DOS change to the dynamics of bond formation and relaxation. Analyzing the valence DOS shift is much more complicated than analyzing the core-band shift, as the latter simply fingerprints the binding energy change of a particular energy level of a specific element by interatomic interaction [45].

Quantum computation using density function theory (DFT) resolves the local valance DOS and core-level shift (CLS) with optimization of crystal geometries and interatomic distances [152, 153]. However, the accuracy and reliability of the computational derivatives are often algorithm and boundary-condition sensitive [154]. Employing the ideally periodic or the free boundary condition deviates the calculation outcome from the true situations to a certain extent. Structural distortion happens at boundaries associated with local strain, charge densification, polarization, and quantum entrapment. These happenings in turn modify the local potentials that shift the electronic binding energy accordingly [9].

1.1.3   Known Mechanisms for Binding Energy Shift

Understanding the effect of irregular coordination on the CLS and valence band evolution has been an issue of long debating from various hypothetic mechanisms through the formulation and quantification of the CLS remains unrealistic in most cases:

(1) The "initial-final state" relaxation, core-hole screening, and atomic covalency effect has been a long dominant mechanism for the CLS of both solid skins [127, 128, 155, 156] and nanoparticles [157]. The initial-final state notion suggests that the irradiation of the incident energetic beam ionizes the initially neutral surface atom into the finally ionized one. The ionized atom becomes a 'Z + 1 impurity' sitting on the neutral substrate of Z metal atoms. The difference in cohesive energy between the Z atom and the Z + 1 impurity amounts to the CLS by BE = E_{z+1}(final) − E_z(initial). The CLS may be negative, positive, or their mixture, depending on the Z value that is tunable in computation. However, correlation between the effective Z and the coordination and chemical environment remains yet unknown.

(2) Another opinion is the skin interlayer contraction that creates the low-energy CLS component [126, 130, 139]. A 12% contraction of the first layer spacing deepens the Nb(100) $3d_{3/2}$ level by 0.50 eV [126] and a (10 ± 3)% contraction of the Ta(100) first layer spacing deepens the $4f_{5/2(7/2)}$ level by 0.75 eV [130]. Interlayer charge densification might enhance the incident light resonance, which results in the positive CLS, according to this mechanism.

(3) Electronic configuration variation (the distribution of electrons among $s, p,$ and d orbits) mechanism [158–161] suggests that the chemical shifts and the structural relaxations contribute to the atom-bulk CLS. The narrowing and shifting of the surface DOS is responsible for the CLS [156, 162].

(4) A metal-to-nonmetal transition mechanism explains that this transition shifts positively the CLS for nanocrystals [161, 163, 164]. However, the metal-nonmetal transition takes place only in the size range of 1–2 nm diameter consisting of 300 ± 100 atoms [165, 166] but the CLS extends continually when a solid grows from an isolated atom into the bulk [167].

(5) Mechanisms of charge transfer between dissimilar metals of an alloy [160], size-enhanced ionicity of copper and oxygen [168], and dipole formation between the substrate and the particles [169] explain the hetero-coordination-induced CLS. The density and the momentum of the dipoles might increase with reduction of particle size. The formation of surface oxygen vacancy provides possible mechanism for the O-coverage dependent CLS of Au on the MgO(100) and on the TiO_2(110) support [170].

(6) Surface electron structure and the increase of the effective size of surface O anions [171] are suggested to be responsible for the 0.75 eV positive (deeper) shift of the Mg 2p energy and the constant O 1s level for the MgO(100) skin.

(7) The latest development [2, 8, 139, 172, 173] suggests that atomic irregular coordination perturbs the Hamiltonian that determines intrinsically the CLS and

the amount of the CLS is proportional to the bond energy or the crystal potential at equilibrium. Bond contraction and bond energy gain deepens the CLS for the undercoordinated systems like atomic clusters, defects, solid skins. Bond nature alteration determines the CLS for the hetero-coordinated systems such as alloys, interfaces, dopants, embedded nanocrystals, and impurities. Polarization of electrons splits and screens the local potentials, which shifts CLS negatively. Therefore, CLS could be positive, negative, or their mixture.

(8) However, rules for the irregular coordination mediated valence DOS stay unclear. The valence charge densification, localization, polarization, and transformation upon alloy and compound formation mediate the performance of the substance such as the band gap, catalytic ability, electroaffinity, work function, etc. The valance DOS evolution and the CLS are strongly correlated. Therefore, a holistic exploration of both of them with quantitative derivative is necessary [2].

1.2 Motivation and Objectives

Resolving the coordination-resolved, dynamic, local, and quantitative information regarding the performance of bonds and electrons of the irregularly-coordinated atoms remains as a "dead corner" though the interplay of STM/S, PES, and DFT has advanced this subject area tremendously. Bond formation and relaxation and the associated energetics, localization, entrapment, and polarization of electrons mediate the electronic binding energy and the performance of a substance accordingly [1]. Therefore, one urgently needs to identify what "seeds" the STM/S signatures and how the length and strength of the nearby bonds relax, which forms the subject of the presented Electron Spectrometrics for the irregularly coordinated atoms, particularly.

This part aims to feature the development and applications of the coordination-resolved bond and electron spectrometrics with focus on the following:

(1) To combine the BOLS-NEP-TB notion [153], the enabled ZPS [8] and APECS strategies towards comprehensive and quantitative information on the local bonding and electronic dynamics.
(2) To clarify the rules and factors controlling the CLS and valence DOS evolution.
(3) To correlate the STM/S, PES, and AES attributes of a substance.
(4) To formulate the CLS dependence on local bond relaxation, quantum entrapment, and polarization.
(5) To quantify the local bond length, bond energy, BE density, atomic cohesive energy, energy level of an isolated atom and its CN-resolved shift pertaining to the irregularly coordinated atoms.

Consistent understandings gained insofar is very promising. It is clear that a perturbation to the Hamiltonian by bond contraction, bond nature alteration, and electron polarization, dictates intrinsically and uniquely the energy shift of electrons in the core band—entrapment or polarization. The extent of entrapment is proportional to the local bond energy at equilibrium. Polarization of the nonbonding states by the densely locally entrapped bonding electrons or by the nonbonding electrons can screen and split the crystal potential, which in turn offsets the entrapped states in the core band negatively.

Most strikingly, the entrapment dominance entitles the Pt adatoms and the Cu/Pd alloy acceptor-type catalysts while polarization dominance makes the Rh adatoms and the Ag/Pd alloy donor-type catalysts. Isolation and polarization of the dangling σ bond electrons by the entrapped bonding electrons create the Dirac-Fermi polaritons surrounding graphite atomic vacancies and graphene zigzag edges.

1.3 Scope

The part 1 starts with Chapter 1 overviewing the significance of chemical bonds and energetic electrons in discriminating the performance, for catalytic ability and Dirac-Fermion generation instances, of a substance at the irregularly- (under- and hetero-) coordinated atomic site from that of the ideally full coordinated atoms in the bulk or from that of an isolated atom. Chapter 1 also summarizes advantages and limitations of existing electron spectroscopic techniques, and available mechanisms for the electron binding energy shift.

Chapter 2 describes notions of the bond order-length-strength (BOLS) correlation [2, 3], the nonbonding electron polarization (NEP) [9], and the local bond average (LBA) approach [4], as well as correlation between the valence and the core electrons. Incorporation of the original BOLS-NEP-LBA approach into the tight-binding (TB) theory [174] has enabled the correlation, clarification, formulation, and quantification of the binding energies of the core and the valence electrons pertaining to the irregularly-coordinated atoms. The variation of bond length and bond energy, and the core electron entrapment and the subjective valence electron polarization with the coordination and chemical environment originate intrinsically the energy shift of electrons in various bands [45, 139], while the extrinsic effects of charging and the initial-final state relaxation could be treated as background to be corrected.

Chapter 3 presents the atomistic zone-resolved photoelectron spectrometrics (ZPS) [8]. Complementing the scanning tunneling microscopy/spectroscopy (STM/S), Auger electron spectroscopy (AES), and photoelectron emission spectroscopy (PES including UPS using UV light and XPS using x-ray source, or synchrotron radiation of various wave lengths), the ZPS overcomes their limitations to gain the specific information. By differencing two spectra of the same specimen collected under different conditions, the ZPS purifies the effect of conditioning on bond relaxation and binding energy shift within the monolayer skin and at sites surrounding point defects. Besides, the combination of AES and XPS results in the Auger photoelectron coincidence spectroscopy (APECS) that probes simultaneously the energy shifts of two bands of a surface—one lower and one upper. The energy shift of the Auger parameter equals twice the shift of the upper energy band rather than the sum of both, as conventionally thought. An extension of the Wagner plot enables the APECS to give information of inter-orbital screening and interatomic charge transporting during chemical conditioning of the specimen. The NEXAS follows the same principle of the APECS for the pre-edge energy shift.

Chapters 4–6 present case studies of adatoms, defects, solid skins, chain ends, terrace edges, atomic clusters, adsorbates, and nanocrystals with derivatives of local bond length and energy, binding energy density, and atomic cohesive energy associated with under-coordinated atoms. ZPS purification clarifies that quantum entrapment dominance of all electrons makes some undercoordinated metals like Pt adatoms acceptor-like catalysts and the polarization dominance of the valence electrons makes metals like Rh adatoms donor-like catalysts.

Chapter 7 deals with carbon allotropes including graphene nanoribbons, nanotubes, graphite, and diamond with purification of energy states for graphene edge, graphite point defect and graphite monolayer skin with determination of the coordination-resolved bond length and bond energy. This purification clarifies why Dirac-Fermi polaritons creation prefers sites surrounding point defects and graphene zigzag edges and why the arm chaired edges are semiconductor like. Only one neighbor short of defect atoms turns the valence electrons from quantum entrapment dominance at the flat surface to the polarization dominance, showing the resonant Dirac-Fermion signatures in the STS measurements.

Chapter 8 deals with hetero-junction interfaces and shows that quantum entrapment dominance makes Cu/Pd acceptor-like catalyst while the valence polarization dominance makes Ag/Pd and Zn/Pd acceptor-like catalyst. The stronger interface polarization and the high interface energy density enabled Be/W alloy capable of protecting nuclear thermal radiation effectively. Examination of the Si, Ge, C, Sn, Cu interface binding energies, revealed that entrapment dominance increases the interface mechanical strength and polarization does it contrastingly.

Chapter 9 features the electronic structures of C, N, O chemisorption induced valence DOS of bonding states, nonbonding lone pairs, ionic holes, and antibonding dipoles. The derived DOS features are detectable using STS (around E_F), IPES (inverse PES, $E > E_F$), and PES ($E < E_F$). New bond formation also modifies the crystal potential and hence results in further core-level entrapment.

Chapter 10 discusses the coupling effect of under- and hetero-coordination on the band gap, electroaffinity, work function, photocatalytic ability of defected TiO_2 and ZnO nanocrystals. Defects improve the photocatalytic ability of TiO_2 by band gap and work function reduction and by carrier life elongation and electroaffinity elevation. Size reduction turns the ZnO_2 crystals from the entrapment dominance to the polarization dominance at a critical size of 8 nm across. The electroaffinity refers to separation between the vacuum level and the conduction band bottom edge, which describes ability of a specimen to keep an electron caught from other specimen. The work function is the separation between the vacuum level and the top edge of the conduction band, which features the ability of an electron escspe from the specimen.

Chapter 11 shows the STM/S, XPS, XAS, SFG, DPS, ultrafast UPS, and ultrafast FTIR observation and quantum theory calculation spectral evidence for the bond−electron−phonon correlation in the confined and the hydrating water that is in the supersolid state. The supersolidity is characterized by the shorter and stiffer H−O bond and longer and softer O:H nonbond, O1s energy entrapment, hydrated electron polarization and the longer lifetime of photoelectrons and phonons. The supersolid phase is less dense, elastoviscous, mechanically and thermally more stable with the hydrophobic and frictionless surface. The O:H−O bond cooperative relaxation disperses outwardly the quasisolid phase boundary to raise the melting point and meanwhile lower the freezing temperature of the quasisolid phase.

Chapter 12 concludes on advantages and attainments, limitations and precautions, prospects and perspectives of the present electron emission exercise. Experiments need to be done under ultra-high vacuum condition and applies only conductors and semiconductors. Extension of this set of strategies to *in situ* monitoring skin and molecular dynamics and to phonon and photon relaxation dynamics could contribute significantly to engineering the coordination bonds and valence electrons. Here we use the term of skin more often than surface, as the former is more appealing with involvement of the thickness than the latter. This set of experimental, numerical and theoretical strategies has provoked the present subject area of coordination and multi-field resolved spectrometrics of bond-electron-phonon-phonon relaxation dynamics. For readers' convenience, each section starts with key points synopsizing discussions.

References

1. C.Q. Sun, Relaxation of the chemical bond. Springer Ser. Chem. Phys. **108,** 807 (2014)
2. X.J. Liu, M.L. Bo, X. Zhang, L. Li, Y.G. Nie, H. Tian, Y. Sun, S. Xu, Y. Wang, W. Zheng, C.Q. Sun, Coordination-resolved electron spectrometrics. Chem. Rev. **115**(14), 6746–6810 (2015)
3. C.Q. Sun, Size dependence of nanostructures: impact of bond order deficiency. Prog. Solid State Chem. **35**(1), 1–159 (2007)
4. C.Q. Sun, Thermo-mechanical behavior of low-dimensional systems: the local bond average approach. Prog. Mater Sci. **54**(2), 179–307 (2009)
5. W.T. Zheng, C.Q. Sun, Underneath the fascinations of carbon nanotubes and graphene nanoribbons. Energy Environ. Sci. **4**(3), 627–655 (2011)
6. X. Zhang, Y. Huang, Z. Ma, L. Niu, C.Q. Sun, From ice superlubricity to quantum friction: electronic repulsivity and phononic elasticity. Friction **3**(4), 294–319 (2015)
7. C.Q. Sun, Y. Sun, Y.G. Ni, X. Zhang, J.S. Pan, X.H. Wang, J. Zhou, L.T. Li, W.T. Zheng, S.S. Yu, L.K. Pan, Z. Sun, Coulomb repulsion at the nanometer-sized contact: a force driving superhydrophobicity, superfluidity, superlubricity, and supersolidity. J. Phys. Chem. C **113**(46), 20009–20019 (2009)
8. Z. Ma, Z. Zhou, Y. Huang, Y. Zhou, C. Sun, Mesoscopic superelasticity, superplasticity, and superrigidity. Sci. China Phys. Mech. Astron. **55**(6), 963–979 (2012)
9. C.Q. Sun, Dominance of broken bonds and nonbonding electrons at the nanoscale. Nanoscale **2**(10), 1930–1961 (2010)
10. C.Q. Sun, Y.Q. Fu, B.B. Yan, J.H. Hsieh, S.P. Lau, X.W. Sun, B.K. Tay, Improving diamond-metal adhesion with graded TiCN interlayers. J. Appl. Phys. **91**(4), 2051–2054 (2002)
11. N. Koch, A. Gerlach, S. Duhm, H. Glowatzki, G. Heimel, A. Vollmer, Y. Sakamoto, T. Suzuki, J. Zegenhagen, J.P. Rabe, F. Schreiber, Adsorption-induced intramolecular dipole: correlating molecular conformation and interface electronic structure. J. Am. Chem. Soc. **130**(23), 7300–7304 (2008)

12. J.J. Blackstock, C.L. Donley, W.F. Stickle, D.A.A. Ohlberg, J.J. Yang, D.R. Stewart, R.S. Williams, Oxide and carbide formation at titanium/organic monolayer interfaces. J. Am. Chem. Soc. **130**(12), 4041–4047 (2008)

13. T. He, H.J. Ding, N. Peor, M. Lu, D.A. Corley, B. Chen, Y. Ofir, Y.L. Gao, S. Yitzchaik, J.M. Tour, Silicon/molecule interfacial electronic modifications. J. Am. Chem. Soc. **130**(5), 1699–1710 (2008)

14. C.G. Long, J.D. Gilbertson, G. Vijayaraghavan, K.J. Stevenson, C.J. Pursell, B.D. Chandler, Kinetic evaluation of highly active supported gold catalysts prepared from monolayer-protected clusters: an experimental Michaelis-Menten approach for determining the oxygen binding constant during CO oxidation catalysis. J. Am. Chem. Soc. **130**(31), 10103–10115 (2008)

15. S.W. Lee, S. Chen, J. Suntivich, K. Sasaki, R.R. Adzic, Y. Shao-Horn, Role of surface steps of Pt nanoparticles on the electrochemical activity for oxygen reduction. J. Phys. Chem. Lett. **1**(9), 1316–1320 (2010)

16. G.A. Hope, D.P. Schweinsberg, P.M. Fredericks, Application of FT-Raman spectroscopy to the study of the benzotriazole inhibition of acid copper corrosion. Spectrochim. Acta A **50**(11), 2019–2026 (1994)

17. T. Kravchuk, L. Vattuone, L. Burkholder, W.T. Tysoe, M. Rocca, Ethylene decomposition at undercoordinated sites on Cu(410). J. Am. Chem. Soc. **130**(38), 12552–12553 (2008)

18. A.S. Foster, T. Trevethan, A.L. Shluger, Structure and diffusion of intrinsic defects, adsorbed hydrogen, and water molecules at the surface of alkali-earth fluorides calculated using density functional theory. Phys. Rev. B **80**(11), 115421 (2009)

19. O. Seitz, A. Vilan, H. Cohen, C. Chan, J. Hwang, A. Kahn, D. Cahen, Effect of doping on electronic transport through molecular monolayer junctions. J. Am. Chem. Soc. **129**(24), 7494–7495 (2007)

20. Q.S. Wei, K. Tajima, Y.J. Tong, S. Ye, K. Hashimoto, Surface-segregated monolayers: a new type of ordered monolayer for surface modification of organic semiconductors. J. Am. Chem. Soc. **131**(48), 17597–17604 (2009)

21. J. Lee, J. Lee, T. Tanaka, H. Mori, In situ atomic-scale observation of melting point suppression in nanometer-sized gold particles. Nanotechnology **20**(47), 475706 (2009)

22. L.-M. Wang, Y. Tian, R. Liu, W. Wang, A "universal" criterion for metallic glass formation. Appl. Phys. Lett. **100**(26), 261913 (2012)

23. L.M. Wang, Y. Zhao, M.D. Sun, R.P. Liu, Y.J. Tian, Dielectric relaxation dynamics in glass-forming mixtures of propanediol isomers. Phys. Rev. E **82**(6), 062502 (2010)

24. L.K. Pan, C.Q. Sun, T.P. Chen, S. Li, C.M. Li, B.K. Tay, Dielectric suppression of nanosolid silicon. Nanotechnology **15**(12), 1802–1806 (2004)

25. S. Veprek, M.G.J. Veprek-Heijman, The formation and role of interfaces in superhard nc-MenN/a-Si$_3$N$_4$ nanocomposites. Surf. Coat. Technol. **201**(13), 6064–6070 (2007)

26. C.R. Chen, Y.W. Mai, Comparison of cohesive zone model and linear elastic fracture mechanics for a mode I crack near a compliant/stiff interface. Eng. Fract. Mech. **77**(17), 3408–3417 (2010)

27. C. Lu, Y.W. Mai, P.L. Tam, Y.G. Shen, Nanoindentation-induced elastic-plastic transition and size effect in alpha-Al$_2$O$_3$(0001). Philos. Mag. Lett. **87**(6), 409–415 (2007)

28. Z. Zhao, B. Xu, X.-F. Zhou, L.-M. Wang, B. Wen, J. He, Z. Liu, H.-T. Wang, Y. Tian, Novel superhard carbon: C-centered orthorhombic C$_8$. Phys. Rev. Lett. **107**(21), 215502 (2011)

29. K. Bian, W. Bassett, Z. Wang, T. Hanrath, The strongest particle: size-dependent elastic strength and debye temperature of PbS nanocrystals. J. Phys. Chem. Lett. **5**(21), 3688–3693 (2014)

30. Y.B. Zhang, T.T. Tang, C. Girit, Z. Hao, M.C. Martin, A. Zettl, M.F. Crommie, Y.R. Shen, F. Wang, Direct observation of a widely tunable bandgap in bilayer graphene. Nature **459**(7248), 820–823 (2009)

31. Y.G. Nie, J.S. Pan, W.T. Zheng, J. Zhou, C.Q. Sun, Atomic scale purification of Re surface kink states with and without oxygen chemisorption. J. Phys. Chem. C **115**(15), 7450–7455 (2011)

32. Y. Wang, Y.G. Nie, L.K. Pan, Z. Sun, C.Q. Sun, Potential barrier generation at the BeW interface blocking thermonuclear radiation. Appl. Surf. Sci. **257**(8), 3603–3606 (2011)
33. D. Hsieh, D. Qian, L. Wray, Y. Xia, Y.S. Hor, R.J. Cava, M.Z. Hasan, A topological Dirac insulator in a quantum spin Hall phase. Nature **452**(7190), 970–974 (2008)
34. S. Li, O. Prabhakar, T.T. Tan, C.Q. Sun, X.L. Wang, S. Soltanian, J. Horvat, S.X. Dou, Intrinsic nanostructural domains: possible origin of weaklinkless superconductivity in the quenched reaction product of Mg and amorphous B. Appl. Phys. Lett. **81**(5), 874–876 (2002)
35. S. Li, T. White, K. Laursen, T.T. Tan, C.Q. Sun, Z.L. Dong, Y. Li, S.H. Zho, J. Horvat, S.X. Dou, Intense vortex pinning enhanced by semicrystalline defect traps in self-aligned nanostructured MgB2. Appl. Phys. Lett. **83**(2), 314–316 (2003)
36. D.A. Ricci, T. Miller, T.C. Chiang, Controlling the thermal stability of thin films by interfacial engineering. Phys. Rev. Lett. **95**(26), 266101 (2005)
37. Q. Jiang, H.M. Lu, Size dependent interface energy and its applications. Surf. Sci. Rep. **63**(10), 427–464 (2008)
38. M. Paneru, C. Priest, R. Sedev, J. Ralston, Static and dynamic electrowetting of an ionic liquid in a solid/liquid/liquid system. J. Am. Chem. Soc. **132**(24), 8301–8308 (2010)
39. A. Hodgson, S. Haq, Water adsorption and the wetting of metal surfaces. Surf. Sci. Rep. **64**(9), 381–451 (2009)
40. X. Zhang, Y. Huang, Z. Ma, Y. Zhou, W. Zheng, J. Zhou, C.Q. Sun, A common supersolid skin covering both water and ice. Phys. Chem. Chem. Phys. **16**(42), 22987–22994 (2014)
41. X. Zhang, Y. Huang, Z. Ma, Y. Zhou, J. Zhou, W. Zheng, Q. Jiang, C.Q. Sun, Hydrogen-bond memory and water-skin supersolidity resolving the Mpemba paradox. Phys. Chem. Chem. Phys. **16**(42), 22995–23002 (2014)
42. M.M. Ugeda, I. Brihuega, F. Guinea, J.M. Gómez-Rodríguez, Missing atom as a source of carbon magnetism. Phys. Rev. Lett. **104**, 096804 (2010)
43. Y. Niimi, T. Matsui, H. Kambara, H. Fukuyama, STM/STS measurements of two-dimensional electronic states trapped around surface defects in magnetic fields. Physica E **34**(1–2), 100–103 (2006)
44. A.J. Cox, J.G. Louderback, L.A. Bloomfield, Experimental-observation of magnetism in Rhodium clusters. Phys. Rev. Lett. **71**(6), 923–926 (1993)
45. C.Q. Sun, Oxidation electronics: bond-band-barrier correlation and its applications. Prog. Mater Sci. **48**(6), 521–685 (2003)
46. W.T. Zheng, C.Q. Sun, Electronic process of nitriding: mechanism and applications. Prog. Solid State Chem. **34**(1), 1–20 (2006)
47. C.Q. Sun, The sp hybrid bonding of C, N and O to the fcc(001) surface of nickel and rhodium. Surf. Rev. Lett. **7**(3), 347–363 (2000)
48. Z. He, J. Zhou, X. Lu, B. Corry, Ice-like water structure in carbon nanotube (8,8) induces cationic hydration enhancement. J. Phys. Chem. C **117**(21), 11412–11420 (2013)
49. K. Nakada, M. Fujita, G. Dresselhaus, M.S. Dresselhaus, Edge state in graphene ribbons: nanometer size effect and edge shape dependence. Phys. Rev. B **54**(24), 17954–17961 (1996)
50. C.Q. Sun, S.Y. Fu, Y.G. Nie, Dominance of broken bonds and unpaired nonbonding pi-electrons in the band gap expansion and edge states generation in graphene nanoribbons. J. Phys. Chem. C **112**(48), 18927–18934 (2008)
51. V.S. Stepanyuk, A.N. Klavsyuk, L. Niebergall, P. Bruno, End electronic states in Cu chains on Cu(111): ab initio calculations. Phys. Rev. B **72**(15), 153407 (2005)
52. J.N. Crain, D.T. Pierce, End states in one-dimensional atom chains. Science **307**(5710), 703–706 (2005)
53. T. Fauster, C. Reuss, I.L. Shumay, M. Weinelt, F. Theilmann, A. Goldmann, Influence of surface morphology on surface states for Cu on Cu(111). Phys. Rev. B **61**(23), 16168–16173 (2000)
54. T. Eguchi, A. Kamoshida, M. Ono, M. Hamada, R. Shoda, T. Nishio, A. Harasawa, T. Okuda, T. Kinoshita, Y. Hasegawa, Surface states of a Pd monolayer formed on a Au(111) surface studied by angle-resolved photoemission spectroscopy. Phys. Rev. B **74**(7), 073406 (2006)

55. S. Link, M.A. El-Sayed, Shape and size dependence of radiative, non-radiative and photothermal properties of gold nanocrystals. Int. Rev. Phys. Chem. **19**(3), 409–453 (2000)
56. J. Jiang, X. Sun, X. Chen, B. Wang, Z. Chen, Y. Hu, Y. Guo, L. Zhang, Y. Ma, L. Gao, F. Zheng, L. Jin, M. Chen, Z. Ma, Y. Zhou, N.P. Padture, K. Beach, H. Terrones, Y. Shi, D. Gall, T.-M. Lu, E. Wertz, J. Feng, J. Shi, Carrier lifetime enhancement in halide perovskite via remote epitaxy. Nat. Commun. **10**(1), 4145 (2019)
57. J. Xie, H. Zhang, S. Li, R. Wang, X. Sun, M. Zhou, J. Zhou, X.W. Lou, Y. Xie, Defect-rich MoS_2 ultrathin nanosheets with additional active edge sites for enhanced electrocatalytic hydrogen evolution. Adv. Mater. **25**(40), 5807–5813 (2013)
58. L. Pan, S. Xu, X. Liu, W. Qin, Z. Sun, W. Zheng, C.Q. Sun, Skin dominance of the dielectric electronic-phononic-photonic attribute of nanoscaled silicon. Surf. Sci. Rep. **68**(3–4), 418–445 (2013)
59. M. Eddaoudi, J. Kim, N. Rosi, D. Vodak, J. Wachter, M. O'Keeffe, O.M. Yaghi, Systematic design of pore size and functionality in isoreticular MOFs and their application in methane storage. Science **295**(5554), 469–472 (2002)
60. N. Stock, S. Biswas, *Synthesis of metal-organic frameworks (MOFs): routes to various MOF topologies, morphologies, and composites*. Chem. Rev. **112**(2), 933–969 (2011)
61. F.X.L. i Xamena, A. Abad, A. Corma, H. Garcia, MOFs as catalysts: activity, reusability and shape-selectivity of a Pd-containing MOF. J. Catal. **250**(2), 294–298 (2007)
62. M.S. Chen, D.W. Goodman, The structure of catalytically active gold on titania. Science **306**(5694), 252–255 (2004)
63. N. Lopez, T. Janssens, B. Clausen, Y. Xu, M. Mavrikakis, T. Bligaard, J.K. Nørskov, On the origin of the catalytic activity of gold nanoparticles for low-temperature CO oxidation. J. Catal. **223**(1), 232–235 (2004)
64. X. Zhang, C.Q. Sun, H. Hirao, Guanine binding to gold nanoparticles through nonbonding interactions. PCCP **15**(44), 19284–19292 (2013)
65. S. Dahl, A. Logadottir, R.C. Egeberg, J.H. Larsen, I. Chorkendorff, E. Tornqvist, J.K. Norskov, Role of steps in N_2 activation on Ru(0001). Phys. Rev. Lett. **83**(9), 1814–1817 (1999)
66. J. Woisetschlager, K. Gatterer, E.C. Fuchs, Experiments in a floating water bridge. Exp. Fluids **48**(1), 121–131 (2010)
67. W.P. Tong, N.R. Tao, Z.B. Wang, J. Lu, K. Lu, Nitriding iron at lower temperatures. Science **299**(5607), 686–688 (2003)
68. M. Asscher, G.A. Somorjai, The remarkable surface structure sensitivity of the ammonia synthesis over rhenium single crystals. Surf. Sci. Lett. **143**(1), L389–L392 (1984)
69. M.J. Gladys, I. Ermanoski, G. Jackson, J.S. Quinton, J.E. Rowe, T.E. Madey, A high resolution photoemission study of surface core-level shifts in clean and oxygen-covered Ir(2 1 0) surfaces. J. Electron Spectrosc. Relat. Phenom. **135**(2–3), 105–112 (2004)
70. H. Wang, A.S.Y. Chan, W. Chen, P. Kaghazchi, T. Jacob, T.E. Madey, Facet stability in oxygen-induced nanofaceting of Re(12$\bar{3}$1). ACS Nano **1**(5), 449–455 (2007)
71. A. Kokalj, N. Bonini, C. Sbraccia, S. de Gironcoli, S. Baroni, Engineering the reactivity of metal catalysts: a model study of methane dehydrogenation on Rh(111). J. Am. Chem. Soc. **126**(51), 16732–16733 (2004)
72. G. Fratesi, S. de Gironcoli, Analysis of methane-to-methanol conversion on clean and defective Rh surfaces. J. Chem. Phys. **125**(4), 044701 (2006)
73. S. Abbet, A. Sanchez, U. Heiz, W.D. Schneider, A.M. Ferrari, G. Pacchioni, N. Rosch, Acetylene cyclotrimerization on supported size-selected Pd_n clusters (1 <= n <= 30): one atom is enough! J. Am. Chem. Soc. **122**(14), 3453–3457 (2000)
74. S. Abbet, U. Heiz, H. Hakkinen, U. Landman, CO oxidation on a single Pd atom supported on magnesia. Phys. Rev. Lett. **86**(26), 5950–5953 (2001)
75. C.J. Zhang, P. Hu, The possibility of single C–H bond activation in CH_4 on a MoO_3-supported Pt catalyst: a density functional theory study. J. Chem. Phys. **116**(10), 4281–4285 (2002)
76. E. Roduner, Size matters: why nanomaterials are different. Chem. Soc. Rev. **35**(7), 583–592 (2006)

77. P. Jakob, M. Gsell, D. Menzel, Interactions of adsorbates with locally strained substrate lattices. J. Chem. Phys. **114**(22), 10075–10085 (2001)
78. M. Gsell, P. Jakob, D. Menzel, Effect of substrate strain on adsorption. Science **280**(5364), 717–720 (1998)
79. J. Wintterlin, T. Zambelli, J. Trost, J. Greeley, M. Mavrikakis, Atomic-scale evidence for an enhanced catalytic reactivity of stretched surfaces. Angew. Chem. Int. Ed. **42**(25), 2850–2853 (2003)
80. B. Richter, H. Kuhlenbeck, H.J. Freund, P.S. Bagus, Cluster core-level binding-energy shifts: the role of lattice strain. Phys. Rev. Lett. **93**(2), 026805 (2004)
81. Y. Chen, Z. Huang, X. Gu, Z. Ma, J. Chen, X. Tang, Top-down synthesis strategies: maximum noble-metal atom efficiency in catalytic materials. Chin. J. Catal. **38**(9), 1588–1596 (2017)
82. T. Andersson, C. Zhang, O. Björneholm, M.H. Mikkelä, K. Jänkälä, D. Anin, S. Urpelainen, M. Huttula, M. Tchaplyguine, Electronic structure transformation in small bare Au clusters as seen by x-ray photoelectron spectroscopy. J. Phys. B: At. Mol. Opt. Phys. **50**(1), 015102 (2016)
83. B. Hammer, J.K. Norskov, Why gold is the noblest of all the metals. Nature **376**(6537), 238–240 (1995)
84. J.A. Rodriguez, D.W. Goodman, The nature of the metal metal bond in bimetallic surfaces. Science **257**(5072), 897–903 (1992)
85. P. Kamakoti, B.D. Morreale, M.V. Ciocco, B.H. Howard, R.P. Killmeyer, A.V. Cugini, D.S. Sholl, Prediction of hydrogen flux through sulfur-tolerant binary alloy membranes. Science **307**(5709), 569–573 (2005)
86. E.B. Fox, S. Velu, M.H. Engelhard, Y.H. Chin, J.T. Miller, J. Kropf, C.S. Song, Characterization of CeO_2-supported Cu–Pd bimetallic catalyst for the oxygen-assisted water-gas shift reaction. J. Catal. **260**(2), 358–370 (2008)
87. L.H. Bloxham, S. Haq, Y. Yugnet, J.C. Bertolini, R. Raval, trans-1,2-dichloroethene on Cu50Pd50(110) alloy surface: dynamical changes in the adsorption, reaction, and surface segregation. J. Catal. **227**(1), 33–43 (2004)
88. D.H. Zhang, W. Shi, Dark current and infrared absorption of p-doped InGaAs/AlGaAs strained quantum wells. Appl. Phys. Lett. **73**(8), 1095–1097 (1998)
89. Y.X. Dang, W.J. Fan, S.T. Ng, S.F. Yoon, D.H. Zhang, Study of interdiffusion in GaIn-NAs/GaAs quantum well structure emitting at 1.3 mu m by eight-band k center dot p method. J. Appl. Phys. **97**(10), 103718 (2005)
90. G. Liu, T.P. St Clair, D.W. Goodman, An XPS study of the interaction of ultrathin Cu films with Pd(111). J. Phys. Chem. B **103**(40), 8578–8582 (1999)
91. M. Khanuja, B.R. Mehta, S.M. Shivaprasad, Geometric and electronic changes during interface alloy formation in Cu/Pd bimetal layers. Thin Solid Films **516**, 5435–5439 (2008)
92. M.A. Newton, The oxidative dehydrogenation of methanol at the CuPd[85: 15]{110} p(2x1) and Cu{110} surfaces: effects of alloying on reactivity and reaction pathways. J. Catal. **182**(2), 357–366 (1999)
93. A.M. Venezia, L.F. Liotta, G. Deganello, Z. Schay, L. Guczi, Characterization of pumice-supported Ag–Pd and Cu–Pd bimetallic catalysts by X-ray photoelectron spectroscopy and X-ray diffraction. J. Catal. **182**(2), 449–455 (1999)
94. H. Amandusson, L.G. Ekedahl, H. Dannetun, Hydrogen permeation through surface modified Pd and PdAg membranes. J. Membr. Sci. **193**(1), 35–47 (2001)
95. C.L. Lee, Y.C. Huang, L.C. Kuo, High catalytic potential of Ag/Pd nanoparticles from self-regulated reduction method on electroless Ni deposition. Electrochem. Commun. **8**(6), 1021–1026 (2006)
96. J. Lin, A. Wang, B. Qiao, X. Liu, X. Yang, X. Wang, J. Liang, J. Li, J. Liu, T. Zhang, Remarkable performance of Ir1/FeOx single-atom catalyst in water gas shift reaction. J. Am. Chem. Soc. **135**(41), 15314–15317 (2013)
97. W. Liu, L. Zhang, W. Yan, X. Liu, X. Yang, S. Miao, W. Wang, A. Wang, T. Zhang, Single-atom dispersed Co–N–C catalyst: structure identification and performance for hydrogenative coupling of nitroarenes. Chem. Sci. **7**(9), 5758–5764 (2016)

98. P. Liu, Y. Zhao, R. Qin, S. Mo, G. Chen, L. Gu, D.M. Chevrier, P. Zhang, Q. Guo, D. Zang, Photochemical route for synthesizing atomically dispersed palladium catalysts. Science **352**(6287), 797–800 (2016)

99. G. Chen, C. Xu, X. Huang, J. Ye, L. Gu, G. Li, Z. Tang, B. Wu, H. Yang, Z. Zhao, Interfacial electronic effects control the reaction selectivity of platinum catalysts. Nat. Mater. **15**(5), 564 (2016)

100. H. Fei, J. Dong, M.J. Arellano-Jiménez, G. Ye, N.D. Kim, E.L. Samuel, Z. Peng, Z. Zhu, F. Qin, J. Bao, Atomic cobalt on nitrogen-doped graphene for hydrogen generation. Nat. Commun. **6**, 8668 (2015)

101. R. Bliem, J.E. van der Hoeven, J. Hulva, J. Pavelec, O. Gamba, P.E. de Jongh, M. Schmid, P. Blaha, U. Diebold, G.S. Parkinson, Dual role of CO in the stability of subnano Pt clusters at the Fe_3O_4 (001) surface. Proc. Natl. Acad. Sci. **113**(32), 8921–8926 (2016)

102. H. Wang, X.-K. Gu, X. Zheng, H. Pan, J. Zhu, S. Chen, L. Cao, W.-X. Li, J. Lu, Disentangling the size-dependent geometric and electronic effects of palladium nanocatalysts beyond selectivity. Sci. Adv. **5**(1), eaat6413 (2019)

103. X. Liu, L. Dai, Carbon-based metal-free catalysts. Nat. Rev. Mater. **1**, 16064 (2016)

104. S.S. Yu, W.T. Zheng, Effect of N/B doping on the electronic and field emission properties for carbon nanotubes, carbon nanocones, and graphene nanoribbons. Nanoscale **2**(7), 1069–1082 (2010)

105. S. Yu, W. Zheng, C. Wang, Q. Jiang, Nitrogen/boron doping position dependence of the electronic properties of a triangular graphene. ACS Nano **4**(12), 7619–7629 (2010)

106. R.R. Parker, ITER in-vessel system design and performance. Nucl. Fusion 473–484 (2000)

107. R. Doerner, M. Baldwin, J. Hanna, C. Linsmeier, D. Nishijima, R. Pugno, J. Roth, K. Schmid, A. Wiltner, Interaction of beryllium containing plasma with ITER materials. Phys. Scr. **2007**(T128), 115 (2007)

108. J. Garai, A. Laugier, The temperature dependence of the isothermal bulk modulus at 1 bar pressure. J. Appl. Phys. **101**(2), 2424535 (2007)

109. M.P. Halsall, P. Harmer, P.J. Parbrook, S.J. Henley, Raman scattering and absorption study of the high-pressure wurtzite to rocksalt phase transition of GaN. Phys. Rev. B **69**(23), 235207 (2004)

110. Y. Wang, Y. Huang, Y. Song, X.Y. Zhang, Y.F. Ma, J.J. Liang, Y.S. Chen, Room-temperature ferromagnetism of graphene. Nano Lett. **9**(1), 220–224 (2009)

111. G.F. Reiter, A. Deb, Y. Sakurai, M. Itou, V.G. Krishnan, S.J. Paddison, Anomalous ground state of the electrons in nanoconfined water. Phys. Rev. Lett. **111**(3), 036803 (2013)

112. T. Enoki, Y. Kobayashi, K.I. Fukui, Electronic structures of graphene edges and nanographene. Int. Rev. Phys. Chem. **26**(4), 609–645 (2007)

113. T. Fujita, Y. Okawa, Y. Matsumoto, K.-I. Tanaka, Phase boundaries of nanometer scale c (2× 2)-O domains on the Cu (100) surface. Phys. Rev. B **54**(3), 2167 (1996)

114. F. Jensen, F. Besenbacher, E. Laegsgaard, I. Stensgaard, Dynamics of oxygen-induced reconstruction on Cu(100) studied by scanning tunneling microscopy. Phys. Rev. B **42**(14), 9206–9209 (1990)

115. F.M. Chua, Y. Kuk, P.J. Silverman, Oxygen chemisorption on Cu(110): An atomic view by scanning tunneling microscopy. Phys. Rev. Lett. **63**(4), 386–389 (1989)

116. A. Sperl, J. Kroger, R. Berndt, A. Franke, E. Pehlke, Evolution of unoccupied resonance during the synthesis of a silver dimer on Ag(111). New J. Phys. **11**(6), 063020 (2009)

117. S. Folsch, P. Hyldgaard, R. Koch, K.H. Ploog, Quantum confinement in monatomic Cu chains on Cu(111). Phys. Rev. Lett. **92**(5), 056803 (2004)

118. T.T. Fister, D.D. Fong, J.A. Eastman, H. Iddir, P. Zapol, P.H. Fuoss, M. Balasubramanian, R.A. Gordon, K.R. Balasubramaniam, P.A. Salvador, Total-reflection inelastic X-ray scattering from a 10-nm thick $La_{0.6}Sr_{0.4}CoO_3$ thin film. Phys. Rev. Lett. **106**(3), 037401 (2011)

119. S. Hebboul. *X-raying the Skin* (2011). Available from http://physics.aps.org/synopsis-for/10.1103/PhysRevLett.106.037401?referer=rss

120. G. Speranza, L. Minati, The surface and bulk core line's in crystalline and disordered polycrystalline graphite. Surf. Sci. **600**(19), 4438–4444 (2006)

121. D.Q. Yang, E. Sacher, Carbon 1s X-ray photoemission line shape analysis of highly oriented pyrolytic graphite: the influence of structural damage on peak asymmetry. Langmuir **22**(3), 860–862 (2006)

122. I. Aruna, B.R. Mehta, L.K. Malhotra, S.M. Shivaprasad, Size dependence of core and valence binding energies in Pd nanoparticles: interplay of quantum confinement and coordination reduction. J. Appl. Phys. **104**(6), 064308 (2008)

123. S.A. Chambers, Elastic-scattering and interface of backscattered primary, Auger and X-ray photoelectrons at high kinetic-energy—principles and applications. Surf. Sci. Rep. **16**(6), 261–331 (1992)

124. D.-L. Feng, Photoemission spectroscopy: deep into the bulk. Nat. Mater. **10**(10), 729–730 (2011)

125. W.J. Huang, R. Sun, J. Tao, L.D. Menard, R.G. Nuzzo, J.M. Zuo, Coordination-dependent surface atomic contraction in nanocrystals revealed by coherent diffraction. Nat. Mater. **7**(4), 308–313 (2008)

126. B.S. Fang, W.S. Lo, T.S. Chien, T.C. Leung, C.Y. Lue, C.T. Chan, K.M. Ho, Surface band structures on Nb(001). Phys. Rev. B **50**(15), 11093–11101 (1994)

127. M. Alden, H.L. Skriver, B. Johansson, Ab-initio surface core-level shifts and surface segregation energies. Phys. Rev. Lett. **71**(15), 2449–2452 (1993)

128. E. Navas, K. Starke, C. Laubschat, E. Weschke, G. Kaindl, Surface core-level shift of 4f states for Tb(0001). Phys. Rev. B **48**(19), 14753 (1993)

129. T. Balasubramanian, J.N. Andersen, L. Wallden, Surface-bulk core-level splitting in graphite. Phys. Rev. B **64**, 205420 (2001)

130. R.A. Bartynski, D. Heskett, K. Garrison, G. Watson, D.M. Zehner, W.N. Mei, S.Y. Tong, X. Pan, The 1st interlayer spacing of Ta(100) determined by photoelectron diffraction. J. Vac. Sci. Technol. A **7**(3), 1931–1936 (1989)

131. J.N. Andersen, D. Hennig, E. Lundgren, M. Methfessel, R. Nyholm, M. Scheffler, Surface core-level shifts of some 4d-metal single-crystal surfaces: experiments and ab initio calculations. Phys. Rev. B **50**(23), 17525–17533 (1994)

132. A. Baraldi, L. Bianchettin, E. Vesselli, S. de Gironcoli, S. Lizzit, L. Petaccia, G. Zampieri, G. Comelli, R. Rosei, Highly under-coordinated atoms at Rh surfaces: interplay of strain and coordination effects on core level shift. New J. Phys. **9**, 143 (2007)

133. R. Nyholm, J.N. Andersen, J.F. Vanacker, M. Qvarford, Surface core-level shifts of the Al(100) and Al(111) surfaces. Phys. Rev. B **44**(19), 10987–10990 (1991)

134. N. Benito, R.E. Galindo, J. Rubio-Zuazo, G.R. Castro, C. Palacio, High- and low-energy x-ray photoelectron techniques for compositional depth profiles: destructive versus non-destructive methods. J. Phys. D Appl. Phys. **46**(6), 065310 (2013)

135. L.I. Johansson, H.I.P. Johansson, J.N. Andersen, E. Lundgren, R. Nyholm, Surface-shifted core levels on Be(0001). Phys. Rev. Lett. **71**(15), 2453–2456 (1993)

136. L.I. Johansson, H.I.P. Johansson, E. Lundgren, J.N. Andersen, R. Nyholm, Surface core-level shift on Be(11-20). Surf. Sci. **321**(3), L219–L224 (1994)

137. L.I. Johansson, H.I.P. Johansson, Unusual behavior of surface shifted core levels on Be (0001) and Be(10-10). Nucl. Instru. Methods Phys. Res. B **97**, 430–435 (1995)

138. L.I. Johansson, P.A. Glans, T. Balasubramanian, Fourth-layer surface core-level shift on Be(0001). Phys. Rev. B **58**(7), 3621–3624 (1998)

139. L.K. Pan, Y.K. Ee, C.Q. Sun, G.Q. Yu, Q.Y. Zhang, B.K. Tay, Band-gap expansion, core-level shift, and dielectric suppression of porous silicon passivated by plasma fluorination. J. Vac. Sci. Technol. B **22**(2), 583–587 (2004)

140. S. Suzer, XPS investigation of a Si-diode in operation. Anal. Methods **4**(11), 3527–3530 (2012)

141. S. Suzer, H. Sezen, A. Dana, Two-dimensional X-ray photoelectron spectroscopy for composite surface analysis. Anal. Chem. **80**(10), 3931–3936 (2008)

142. S. Suzer, H. Sezen, G. Ertas, A. Dana, XPS measurements for probing dynamics of charging. J. Electron Spectrosc. Relat. Phenom. **176**(1–3), 52–57 (2010)

143. S. Suzer, E. Abelev, S.L. Bernasek, Impedance-type measurements using XPS. Appl. Surf. Sci. **256**(5), 1296–1298 (2009)

144. H. Sezen, S. Suzer, Communication: enhancement of dopant dependent x-ray photoelectron spectroscopy peak shifts of Si by surface photovoltage. J. Chem. Phys. **135**(14), 141102 (2011)

145. F. Lin, Y. Liu, X. Yu, L. Cheng, A. Singer, O.G. Shpyrko, H.L. Xin, N. Tamura, C. Tian, T.-C. Weng, X.-Q. Yang, Y.S. Meng, D. Nordlund, W. Yang, M.M. Doeff, Synchrotron X-ray analytical techniques for studying materials electrochemistry in rechargeable batteries. Chem. Rev. **117**(21), 13123–13186 (2017)

146. C. Papp, H.-P. Steinrück, In situ high-resolution X-ray photoelectron spectroscopy—fundamental insights in surface reactions. Surf. Sci. Rep. **68**(3–4), 446–487 (2013)

147. A.Y. Klyushin, T.C.R. Rocha, M. Havecker, A. Knop-Gericke, R. Schlogl, A near ambient pressure XPS study of Au oxidation. Phys. Chem. Chem. Phys. **16**, 7881–7886 (2014)

148. K.R. Wilson, B.S. Rude, T. Catalano, R.D. Schaller, J.G. Tobin, D.T. Co, R.J. Saykally, X-ray spectroscopy of liquid water microjets. J. Phys. Chem. B **105**(17), 3346–3349 (2001)

149. K.R. Siefermann, Y. Liu, E. Lugovoy, O. Link, M. Faubel, U. Buck, B. Winter, B. Abel, Binding energies, lifetimes and implications of bulk and interface solvated electrons in water. Nat. Chem. **2**, 274–279 (2010)

150. L. Bianchettin, A. Baraldi, S. de Gironcoli, E. Vesselli, S. Lizzit, L. Petaccia, G. Comelli, R. Rosei, Core level shifts of undercoordinated Pt atoms. J. Chem. Phys. **128**(11), 114706 (2008)

151. W.D. Phillips, Laser cooling and trapping of neutral atoms. Rev. Mod. Phys. **70**(3), 721–741 (1998)

152. X. Zhang, J.L. Kuo, M.X. Gu, X.F. Fan, P. Bai, Q.G. Song, C.Q. Sun, Local structure relaxation, quantum trap depression, and valence charge polarization induced by the shorter-and-stronger bonds between under-coordinated atoms in gold nanostructures. Nanoscale **2**(3), 412–417 (2010)

153. X. Zhang, Y.G. Nie, W.T. Zheng, J.L. Kuo, C.Q. Sun, Discriminative generation and hydrogen modulation of the Dirac-Fermi polarons at graphene edges and atomic vacancies. Carbon **49**(11), 3615–3621 (2011)

154. C.Q. Sun, X. Zhang, J. Zhou, Y. Huang, Y. Zhou, W. Zheng, Density, elasticity, and stability anomalies of water molecules with fewer than four neighbors. J. Phys. Chem. Lett. **4**, 2565–2570 (2013)

155. P.S. Bagus, E.S. Ilton, C.J. Nelin, The interpretation of XPS spectra: insights into materials properties. Surf. Sci. Rep. **68**(2), 273–304 (2013)

156. B. Johansson, N. Martensson, Core-level binding-energy shifts for the metallic elements. Phys. Rev. B **21**(10), 4427–4457 (1980)

157. G.K. Wertheim, H.J. Guggenhe, D.N. Buchanan, Size effect in ionic charge relaxation following auger effect. J. Chem. Phys .**51**(5), 1931–1934 (1969)

158. A.R. Williams, N.D. Lang, Core-level binding-energy shifts in metals. Phys. Rev. Lett. **40**(14), 954–957 (1978)

159. E. Pehlke, M. Scheffler, Evidence for site-sensitive screening of core holes at the Si and Ge (001) surface. Phys. Rev. Lett. **71**(14), 2338–2341 (1993)

160. Y. Wang, X. Zhang, Y.G. Nie, C.Q. Sun, Under-coordinated atoms induced local strain, quantum trap depression and valence charge polarization at W stepped surfaces. Phys. B-Condens. Matter **407**(1), 49–53 (2012)

161. V. Vijayakrishnan, A. Chainani, D.D. Sarma, C.N.R. Rao, Metal-insulator transitions in metal-clusters—a high-energy spectroscopy study of Pd and Ag clusters. J. Phys. Chem. **96**(22), 8679–8682 (1992)

162. P.H. Citrin, G.K. Wertheim, Photoemission from surface-atom core levels, surface densities of states, and metal-atom clusters: a unified picture. Phys. Rev. B **27**(6), 3176–3200 (1983)

163. M.G. Mason, *In Cluster Models for Surface and Bulk Phenomena*, ed. G. Pacchioni (Plenum, New York, 1992)

164. C.N.R. Rao, G.U. Kulkarni, P.J. Thomas, P.P. Edwards, Size-dependent chemistry: properties of nanocrystals. Chem. Eur. J **8**(1), 29–35 (2002)

165. H.N. Aiyer, V. Vijayakrishnan, G.N. Subbanna, C.N.R. Rao, Investigations of Pd clusters by the combined use of HREM, STM, high-energy spectroscopies and tunneling conductance measurements. Surf. Sci. **313**(3), 392–398 (1994)

166. H.-G. Boyen, T. Herzog, G. Kästle, F. Weigl, P. Ziemann, J. Spatz, M. Möller, R. Wahrenberg, M. Garnier, P. Oelhafen, X-ray photoelectron spectroscopy study on gold nanoparticles supported on diamond. Phys. Rev. B **65**(7), 075412 (2002)

167. D.Q. Yang, E. Sacher, Platinum nanoparticle interaction with chemically modified highly oriented pyrolytic graphite surfaces. Chem. Mater. **18**(7), 1811–1816 (2006)

168. K. Borgohain, J.B. Singh, M.V.R. Rao, T. Shripathi, S. Mahamuni, Quantum size effects in CuO nanoparticles. Phys. Rev. B **61**(16), 11093–11096 (2000)

169. D. Schmeisser, O. Bohme, A. Yfantis, T. Heller, D.R. Batchelor, I. Lundstrom, A.L. Spetz, Dipole moment of nanoparticles at interfaces. Phys. Rev. Lett. **83**(2), 380–383 (1999)

170. Z. Yang, R. Wu, Origin of positive core-level shifts in Au clusters on oxides. Phys. Rev. B **67**(8), 081403 (2003)

171. C.J. Nelin, F. Uhl, V. Staemmler, P.S. Bagus, Y. Fujimori, M. Sterrer, H. Kuhlenbeck, H.-J. Freund, Surface core-level binding energy shifts for MgO (100). Phys. Chem. Chem. Phys. **16**(40), 21953–21956 (2014)

172. C.Q. Sun, Y. Nie, J. Pan, X. Zhang, S.Z. Ma, Y. Wang, W. Zheng, Zone-selective photoelectronic measurements of the local bonding and electronic dynamics associated with the monolayer skin and point defects of graphite. RSC Adv. **2**(6), 2377–2383 (2012)

173. Y. Nie, Y. Wang, X. Zhang, J. Pan, W. Zheng, C.Q. Sun, Catalytic nature of under- and heterocoordinated atoms resolved using zone-selective photoelectron spectroscopy (ZPS). Vacuum **100**, 87–91 (2014)

174. M.A. Omar, *Elementary Solid State Physics: Principles and Applications* (Addison-Wesley, New York, 1993)

Chapter 2
Theory: Bond-Electron-Energy Correlation

Abstract Electron binding energy shift directly features the change of bond energy with coordination environments and chemical conditions, from which one can evaluate the local and quantitative information on the local bond length, bond energy, core charge entrapment and valence electron polarization. Bonds and electrons associated with undercoordinated adams, point defects, skins, and nanostructures follows the BOLS-NEP notion but bonds associated with the hetero-coordinated and the tetrahedrally-coordinated impurities and interfaces may subject to bond nature alteration and the local electrons may subject to entrapment or polarization.

Highlights

- One can shift elelctron binding energy from that of an isolated atom by perturbing Hamiltonian.
- The core-level shifts (CLS) with the bond energy transiting from one equilibrium to the other.
- Atomic irregular-CN shifts the CLS positively by bond contraction, core electron entrapment.
- Nonbonding electron polarization screens the local potential and offsets the CLS contrastingly.

2.1 Atomic Coordination Classification

Table 2.1 classifies substances according to their atomic coordination environments. Atomic undercoordination means an atom with fewer neighbors than it is in the bulk fcc structural standard of $z_b = 12$. For an isolated atom, $z = 0$, which happens only at 0 K ideally [1] or in the gaseous phase. Undercoordination ($0 < z < 12$) is ubiquitous to adatoms, defects, terrace edges, grain boundaries, skins, of solid species and nanostructures of various shapes and sizes. Monatomic chains and monolayer atomic sheets are ideal cases of one- and two-dimensional undercoordination systems, which possess tremendously revolutionary properties [2].

© The Editor(s) (if applicable) and The Author(s), under exclusive license
to Springer Nature Singapore Pte Ltd. 2020
C. Q. Sun, *Electron and Phonon Spectrometrics*,
https://doi.org/10.1007/978-981-15-3176-7_2

Table 2.1 Classification, origin, annotation, and the size dependency and size-induced emerging properties of a substance at the nanoscales, edges, skins and interfaces associated with bond relaxation and potential trap or barrier formation

Tetrahedral CN (C, N, O, F)	Isolated atom	Atomic chain, sheet, tube, wire, dot, skin, nanocrystal, ...	Ideal bulk	Interface/impurity
Interaction	0	1	1	1
Atomic CN	0	<12 (under)	12 (full)	Hetero
Notions		• Interaction between undercoordinated atoms differs in nature nanostructures from an isolated atom or the bulk in performance		
		• Bond between undercoordinated atoms becomes shorter and stronger		
		• Defects, skins, adatoms, nanostructures of various shapes share the same attribute of undercoordination		
		• Skin atoms/bonds dictate the size dependency of nanocrystals in various properties		
		• Interface potential barrier/trap formation results in local entrapment and/or polarization		
Electronic behavior		Atomic irregular coordination causes local bond relaxation or bond nature alteration, which results in local quantum entrapment of core electrons and polarization of nonbonding electrons		
		The energetic behaviors of bonds and electrons mediate the BE shift and the functionalities of a substance		

Hetero-coordination means that an atom has different kinds of neighbors such as alloys, compounds, dopants, impurities, and interfaces. The A–B type exchange interaction will come into play in addition to the A–A and B–B type potentials for an AB alloy. If one specimen, like amorphous high-entropy states, has n constituents, there will be $n!/(2!(n-2)!)$ types of interatomic interactions. New bond formation with bond nature alteration changes locally the Hamiltonian with an addition of such exchange interaction terms.

Tetrahedral coordination refers to sp^3-orbital hybridization of C, N, O, and F upon interaction with atoms of less electronegative element. This configuration creates four valence states of bonding, nonbonding lone pair, electron-hole pair, and antibonding dipole, which turns a metal into a semiconductor or an insulator and expands the band gap of a semiconductor [3, 4]. For the molecular crystals like water, one has to consider the inter- and intramolecular interactions and their coupling effect such as the O:H–O (HB) bond that serves as a coupled oscillator pair [5]. The ':' represents the electron lone pair of oxygen. For solutions and energetic explosives, the repulsive H↔H anti-HB and the O:⇔:O super-HB are involved [6–8].

2.2 Core Band Energy Dispersion

The following single-body Hamiltonian and eigen wavefunction describe an electron moving in the νth orbit of an atom in the ideal bulk [9],

$$H = H_0 + H'$$

$$with \quad \begin{cases} H_0 = -\frac{\hbar^2 \nabla^2}{2m} + V_{atom}(r) & \text{(Total energy for an electron of an isolated atom)} \\ H' = V_{cryst}(r) & \text{(Inter - atomic interaction)} \end{cases}$$

$$|\nu, i\rangle \cong u(r) \exp(ikr) \qquad \text{(Bloch wave - function)} \qquad (2.1)$$

H_0 is the Hamiltonian for an isolated atom, which sums the kinetic energy and intra-atomic potential energy experienced by the specific electron. The interatomic potential $V_{cryst}(r)$ sums all interactions with neighboring atoms and electrons. There are might be multiple constituents for the $V_{cryst}(r)$ but the spectroscopy collect them inclusively in a convoluted form without needing any decomposition. The $V_{atom}(r)$ $= V_{atom}(r + R) < 0$, the $V_{cryst}(r) = V_{cryst}(r + R) < 0$, and the Bloch wavefunctions are periodic in real space, where R is the lattice constant. Because of the localization nature of the core electrons, the eigen wavefunction $|\nu, i\rangle$ meets the following criterion, where i and j denote atomic positions:

$$\langle \nu, j | \nu, i \rangle = \delta_{ij} = \begin{cases} 1 & (i = j) \\ 0 & (i \neq j) \end{cases}$$

The coupling of the potentials experienced by an electron represented by the *and its* Bloch wavefunction determines the energy shift. The energy of an electron in an ideal bulk disperses in the following manner (with atomic CN or $z_b = 12$ for an fcc-structured bulk standard):

$$E_\nu(z_b) = E_\nu(0) + (\alpha_\nu + z_b \beta_\nu) + 2z_b \beta_\nu \Phi_\nu(k, R)$$

$$with \quad \begin{cases} E_\nu(0) = -\langle \nu, i | H_0 | \nu, i \rangle & \text{(Atomic core level)} \\ \alpha_\nu = -\langle \nu, i | H' | \nu, i \rangle \propto E_b & \text{(Exchange integral)} \\ \beta_\nu = -\langle \nu, i | H' | \nu, j \rangle \propto E_b & \text{(Overlap integral)} \end{cases} \qquad (2.2)$$

The $V_{atom}(r)$ defines the energy level of an isolated atom $E_\nu(0)$, from which the core band shifts. As intrinsic constants, the $E_\nu(0)$ reduces its value with the quantum number ν from 10^3 to 10^0 eV until the vacuum level $E_0 = 0$ as the ν increases, or as one moves from the innerest orbit outwardly of an atom.

The CLS fingerprints the variation of interatomic interaction that changes with chemical and coordination environment. The involvement of the $V_{cryst}(r)$ upon bulk or liquid formation deepens the $E_\nu(0)$ by an amount of $\Delta E_\nu(z_b) = E_\nu(z_b) - E_\nu(0) = \alpha_\nu + z_b \beta_\nu$, and meanwhile, turns the CL into a band of $E_{\nu W} = 2z_b \beta_\nu \Phi_\nu(k, R)$ width. Both the exchange integral α_ν and the overlap integral β_ν are proportional to the cohesive energy per bond at equilibrium, E_b, or the zeroth approximation of the crystal potential in a Taylor series. The term $\Phi_v(k, R) \cong \sin^2(kR/2) \le 1$, for the fcc structure instance, is a distribution function.

The exchange integral dominates the CLS. Typically,

$$\Delta E_v(z_b) = \alpha_v + z_b\beta_v = \alpha_v\left(1 + z_b\beta_v/\alpha_v\right) \propto E_b\left(1 + E_{vW}/2E_b\right)$$
$$\approx 3.0(1 + 0.2/6.0) = 3.0(1 + 3\%).$$

A 3% contribution from the overlap integral to the CLS is negligibly small. For the deeper bands, this ratio is even smaller. Thus, the bulk CLS depends mainly on the bond energy E_b in the first order approximation. Any relaxation of the interatomic bond changes directly the E_{vW} and the CLS accordingly.

Atomic ionization by X-ray radiation or by charging accumulation on the sample changes the crystal potential throughout the course of measurements, which serves as a removable background in calibration. In fact, charging effect exists only for thick insulating samples in measurement because of the non-conductive character of the specimen. The charging effect for conductors or thinner insulators becomes negligible [10]. The charging effect can be minimized by grounding the specimen during experiment [2, 11]. The electronic multi-body interaction also serves as an average background as it exists throughout the specimen at quest. The final-initial state relaxation effect exists throughout all the measurements so this effect can also be averaged as background in the date processing.

The band width $E_{vW} = 2z_b\beta_v\Phi_v(k, R) \propto z_bE_b$ is proportional to the atomic cohesive energy. As illustrated in Fig. 2.1a, if one moves from the valence band downward, the $\Delta E_v(z_b)$ shift will turn from 10^0 to 10^{-1} eV and the E_{vW} will approach

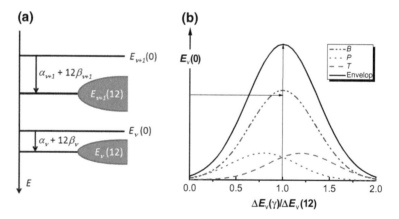

Fig. 2.1 **a** Evolution of the vth atomic energy level $E_v(0)$ to the band $E_v(z_b = 12)$ with a shift of $\Delta E_v(z_b) \propto E_b$ and an expansion $E_{vW}(z_b) \propto z_bE_b$ of the vth band upon bulk formation. Artifacts in measurements and charge polarization may change the amounts of BE shift and the width of the components but not the nature of origin. **b** A typical XPS spectral peak shows components of entrapment (T, or positive shift), polarization (P, or negative shift), and the bulk (B). Reprinted with permission from [46]. Copyright 2006 American Physical Society

lines for the spin-resolved energy levels, such as the $1s$, $2p_{1/2}$, and $2p_{3/2}$ of Cu. Electronic occupation of the core bands often approximates the Gaussian or the Lorentz type functions in decomposing the spectral peaks.

2.3 BOLS-NEP-LBA Notion

2.3.1 Local Hamiltonian Perturbation

Atomic CN is the primary variable that determines the CLS for undercoordinated systems. Any change of the CN will relax the bond in length and energy accordingly, which perturbs the $V_{cryst}(r)$. Possible factors of perturbation include:

(1) Undercoordination induced bond contraction, charge densification, localization, entrapment, and polarization.
(2) Hetero-coordination induced bond nature alteration, bond relaxation, entrapment, or polarization.
(3) Radiation induced ionization associated with the "initial-final states" relaxation during experiment.
(4) External bias caused charge accumulation of the tested specimen.
(5) Mechanical or thermal field induced bond relaxation in length and energy.

The zeroth approximation of the interatomic potential dictates the CLS. Figure 2.2a illustrates bond length and energy (d, E) relaxation along the modulation function $f(x)$ under a certain stimulus x such as the atomic CN, mechanical,

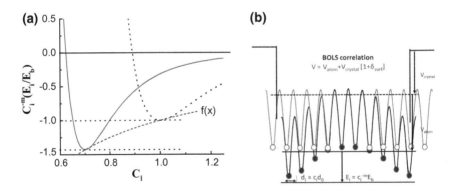

Fig. 2.2 Potentials of a dimer bond relaxing under external stimulus x. **a** Valley of the pairing potential $u(r, E)$ corresponds to bond length and energy (d, E_b) that relax along the modulation function $f(x)$ under stimulus x (x can be pressure, temperature or coordination environment, etc.). **b** Bond order loss shortens and strengthens bonds between undercoordinated atoms at the bonding network terminals such as sites of defects and skins of nanocrystals, which cause local densification and entrapment of the bonding electrons, and polarization of nonbonding electrons. Reprinted with permission from [24]. Copyright 2001 IOP Publishing Ltd.

or thermal excitation. A Taylor series approximates the pairing potential u(r):

$$u(r) = \left.\frac{\partial^n u(r)}{n!\partial r^n}\right|_{r=d} (r-d)^n = E_b + \left.\frac{\partial^2 u(r)}{2\partial r^2}\right|_{r=d} (r-d)^2 + \left.\frac{\partial^3 u(r)}{6\partial r^3}\right|_{r=d} (r-d)^3$$
$$+ 0\left((r-d)^{n\geq 4}\right) \tag{2.3}$$

The zeroth differential (d, E_b) at equilibrium dominates uniquely the BE shift. The second-order differential, or the curvature of the potential $u(r)$ corresponding to the force constant of a dimer undergoing harmonic vibration. The higher-order nonlinear terms contribute to the transport dynamics like lattice expansion and heat conductivity. Bond relaxation refers to the response of the coordinates (d, E_b) at equilibrium to external stimulus such as pressure, temperature, coordination and chemical environment [2]. In place of what said in the text book, atoms vibrate in the potential well and meanwhile the potential well glides along the modulation curve $f(x)$, shown in Fig. 2.2a.

The $V_{cryst}(r)$ experienced by the electron is the sum of $u(r)$ over all nearest neighbors. For the irregularly-coordinated systems, one needs to consider the local perturbation to the crystal potential from $V_{cryst}(r)$ to $V_{cryst}(r)(1 + \Delta_H) \cong E_b(1 + \Delta_H)$ without needing considering perturbation to the wavefunction or the high-order potential terms.

The perturbation shifts the CL positively if $\Delta_H > 0$ or negatively if $\Delta_H < 0$. The former is the quantum entrapment (T) and the latter is the polarization (P). A mixed shift is possible if both the T and P contribute competitively. Perturbation turns Eq. (2.1) into the following form by replacing the z_b with x, representing for the effect of atomic CN(z) variation or polarization P,

$$H' = V_{cryst}(r)(1 + \Delta_H)$$

$$E_v(x) = E_v(0) + \alpha_{vx}(1 + x\beta_{vx}/\alpha_{vx}) + 2x\beta_{vx}\Phi_v(k, R)$$

$$with \begin{cases} E_v(0) = -\langle v, i|H_0|v, i\rangle & \text{(Atomic core level)} \\ \alpha_{vx} = -\langle v, i|H'|v, i\rangle \propto E_b(1 + \Delta_H) \propto E_x & \text{(Exchange integral)} \\ \beta_{vx} = -\langle v, i|H'|v, j\rangle \propto E_b(1 + \Delta_H) \propto E_x & \text{(Overlap integral)} \end{cases}$$
$$\tag{2.4}$$

Perturbation mediates only the integrals without any influencing on the referential $E_v(0)$.

2.3.2 BOLS-NEP Notion for the Undercoordinated Atoms

2.3.2.1 BOLS: Undercoordination Induced Bond Contraction

According to the BOLS correlation premise [2, 12], bond order loss shortens and strengthens bonds between undercoordinated atoms, as illustrated in Fig. 2.2b. Bond contraction densifies the local bonding charge, binding energy, and atomic mass; bond strength gain deepens the potential well and shifts the core levels accordingly. The following formulates the BOLS notion [13],

$$\begin{cases} C_z = d_z/d_b = 2\{1 + \exp[(12 - z)/(8z)]\}^{-1} & \text{(bond length)} \\ C_z^{-m} = E_z/E_b & \text{(bond energy)} \\ C_z^{-(m+\lambda)} = \left(E_z/d_z^\lambda\right)/\left(E_b/d_b^\lambda\right) & \text{(energy density)} \\ z_{ib}C_z^{-m} = zE_z/(z_bE_b) & \text{(atomic cohesive energy)} \end{cases} \quad (2.5)$$

The bond nature index m correlates the bond energy to its length of a specific substance. $\lambda = 1$ defines a monoatomic chain and $\lambda = 2$ the monoatomic sheets or single walled nanotube. $\lambda = 3$ is the general case of three-dimensional solid. The C_z is universal to match the CN-resolved atomic distance of carbon nanotubes, Au nanoparticles, Au, Pt, Ir, Ti, Zr, and Zn atomic chains, as well as skins of Fe, Ni, Ru, Re, W and diamond [12, 14].

The BOLS notion also defines the CN dependence of the reduced bond length d_z, bond energy E_z, BE density E_{den}, and atomic cohesive energy E_{coh} at the under-coordinated atomic site in a dimensionless form. These key quantities determine the behavior of a substance at these under-coordinated atomic sites, as they link the local atomistic bonding identities to macroscopic properties, such as adhesion ability, diffusivity, elasticity, reactivity, strength, wettability, and so on so forth [12].

2.3.2.2 NEP: Strong Localization

Conversely, the densely, locally entrapped bonding and core electrons in turn polarize the nonbonding electrons of the undercoordinated rim atoms [15, 16]. The polarized states (P) in turn screen and split the local potential that offsets the entrapped core bands negatively. Negative CLS may not happen if the specific CL is too deep or the extent of polarization is insufficiently high. Therefore, it is not surprising that some materials show positive and some others show negative or mixed CLS [17].

However, the polarization is subject to the availability of nonbonding electrons [12, 15]. The nonbonding electrons refer to those composed of dipoles induced by the lone pairs of O, N, and F, the unpaired dangling bond of C and Si at defect edges, and the otherwise conduction unpaired electrons of metals at undercoordinated atomic sites. The polarization raises the nonbonding states in energy toward or cross over the E_F [15]. Such CN-resolved local bond strain, electron densification, entrapment, polarization and its effect on potential screening and splitting may explain the "defect

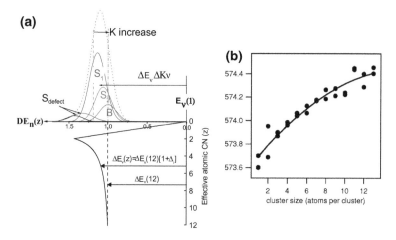

Fig. 2.3 **a** Illustration of the correspondence between the cluster size (K, upper part) to the atomic CN dependence of the CLS (lower part). **b** The number-of-atom dependence of the Cr_n $2p$ energy shifts. Th energy shifts from 537.5 eV for an isolated Cr atom to E_{2p} (n = 13) = 574.4 eV for z = 2 [20], which follows the BOLS prediction in the z ≤ 2 regime of (**a**). Reprinted with permission from [20, 47]. Copyright 2005 and 2009 American Physical Society

induced strong localization and strong correlation" premise of Anderson [18] to the irregularly-coordinated systems including amorphous glasses [19].

Figure 2.3a illustrates the cluster-size or the effective atomic-CN dependency of the CLS. If the solid grows from an isolated atom to a solid of infinitely large, the CL peak shifts along the line in the lower part of (a). Measurements [12, 20, 21] confirmed that when the solid grows from an isolated atom the BE of the vth level deepens from the initial $E_v(0)$ level sharply to a maximum at z = 2, and then restores gradually in a K^{-1} fashion to the bulk $E_v(12)$ [12], as the atomic CN increases [20]. The K is the characteristic size of a nanocrystal. The z = 2 corresponds to the effective CN for an atom in an fcc unit cell or in the monatomic chain. The size resolved $2p$ level shift of Cr_n (n = 2 − 13) clusters (b) does follow the z ≤ 2 prediction [20] and the number-of-layer resolved C $1s$ level shift of graphene follows the z > 2 size trend as well [22].

2.3.2.3 Point Defect and Monolayer Skin

For a point defect or a monolayer solid skin, one only needs to consider the relaxation of one representative bond for all under the same coordination environment. Incorporating Eq. (2.5) into Eq. (2.4) yields:

$$\frac{\Delta E_v(z)}{\Delta E_v(12)} = (1 + \Delta_{Hz}) = \frac{E_z}{E_b} = C_z^{-m}$$
$$\Delta E_v(z \geq 2) = \Delta E_v(12)(1 + \Delta_{Hz}) = [E_v(12) - E_v(0)]C_z^{-m} \qquad (2.6)$$

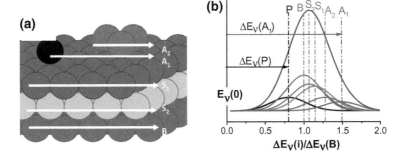

Fig. 2.4 Atomic CN-resolved CLS. **a** Adatoms or defect vacancies, A_i, surface layers S_i on the bulk B crystal, which create the corresponding component in the CLS spectrum **b**. In addition to the main peaks of B, S_2, S_1, there are A_i and P components being the entrapped (T) and the polarized (P) states of the undercoordinated atoms. The energy shift of each component is proportional to the bond energy, which follow the relationship: $\Delta E_v(z_i)/\Delta E_v(12) = E_{zi}/E_{zb} = C_z^{-m}$ ($i = A, S_1, S_2$), if the polarization effect is insignificant; otherwise, the pC_z^{-m} replaces the C_z^{-m} with p being the polarization coefficient

Figure 2.4a Illustrates the layer-counting for a solid skin of three atomic layers with addition of adatoms or vacancy defects. Panel (b) shows the respective CLS component in the XPS spectrum. According to the BOLS-NEP notion, adatoms (A_i), surface skins (S_1, S_2, ...), bulk (B), and electron polarization (P) shift their binding energies each by an amount in the order of: $\Delta E_v(A_i) > \Delta E_v(S_1) > \Delta E_v(S_2) > \Delta E_v(B) > \Delta E_v(P)$ with respect to the atomic $E_v(0)$ reference. The component for the least-coordinated adatom ($z = 2$) shifts most.

2.3.2.4 BOLS-LBA for Atomic Cluster and Nanocrystal

For crystals of different sizes, one has to consider the weighted sum over the outer-most three atomic layers based on the LBA approach and the core-shell configuration premise [14]. The Fourier transformation principle inspires the LBA, which indicates that, for a given specimen, no matter whether it is a crystal, non-crystal, with or without defects, the nature and the total number of bonds remain unchanged unless phase transition occurs. However, the length and strength of all the involved bonds will respond to the applied stimulus in the same manner simultaneously. Therefore, one can focus on the length and energy response of the representative bond to the external stimulus and its effect on the CLS for the entire solid.

However, the hydrogen bond (O:H–O) in water and ice is an exception [5]. Because of the asymmetrical, short-range interactions and the O–O Coulomb repulsion coupling, the O:H nonbond and the H–O covalent bond segment relax oppositely in length and energy. The stronger H–O bond (4.0 eV level) dictates the CL as the weaker O:H nonbond (0.1 eV level) contributes only 3% or less to the crystal binding energy [23].

Generally, the experimentally observed size (K), shape (τ), and bond nature (m) dependence of the $\Delta E_v(\tau, K, m)$, of a nanosolid follows a scaling relation based on the core-shell configuration [12],

$$\frac{\Delta E_v(K) - \Delta E_v(12)}{\Delta E_v(12)} = \begin{cases} bK^{-1} & \text{(Experiment)} \\ \Delta_H & \text{(Theory)} \end{cases}$$

$$with \begin{cases} \Delta_H = \sum_{i \leq 3} \gamma_i \left(\Delta E_{zi} / E_{zb} \right) = \sum_{i \leq 3} \gamma_i \left(C_{zi}^{-m} - 1 \right) \text{ (Skin resolved perturbation)} \\ \gamma_i = N_i / N = V_i / V = \tau K^{-1} C_i \leq 1 \qquad \text{(Fraction of skin atoms)} \end{cases}$$

$$(2.7)$$

As the dimensionless form of size, the K is the number of atoms lined along the radius of a spherical dot $(\tau = 3)$, or a cylindrical rod $(\tau = 2)$, or cross the thickness of a thin plate $(\tau = 1)$ where τ is the shape factor. N_i is the number of atoms and V_i the volume of the ith atomic layer, respectively. $E_v(K)$ is the peak energy of the vth band for a K-sized solid. E_{zi} is the energy of a bond in the ith layer between z-coordinated atoms. The weighting factor, γ_i, represents the fraction of undercoordinated atoms in the ith shell of a K-sized and τ-shaped nanosolid. Subscript i counts from the outermost layer inward up to three as no bond order loss happens at $i > 3$.

Generally, the size dependent CN for a spherical dot follows empirically [12],

$$\begin{cases} z_1 = 4\left(1 - 0.75K^{-1}\right) \\ z_2 = z_1 + 2 \\ z_3 = z_2 + 4 \end{cases} \qquad (2.8)$$

$K > 0$ corresponds to a solid and $K < 0$ to a cavity. $K = 0$ to a flat skin and $K = \infty$ to an atom inside the ideal bulk. The z_1 changes with the curvature in the order: $1 = z_{dimer} < z_{cluster} < z_{nanocrystal} < z_{flat-skin} < z_{cavity} < z_{bluk} = 12$. At $K \leq 0.75$, the solid degenerates into an isolated atom. For a spherical dot at the lower end of the size limit, $K = 1.5$ ($Kd = 0.43$ nm for an Au spherical dot example, or an fcc unit cell), $z_1 = 2$, which is equivalent to an atomic chain, the edge of a monolayer graphene, and to the primary fcc unit cell having 13 atoms. Bonds between atoms of the same z values perform identically. This LBA expression covers all sizes and shapes varying from a dimer, to a monatomic chain, a monolayer atomic sheet, a hollow cavity, a flat skin and bulk solid. The BOLS-NEP notion applies to all the undercoordinated systems without discrimination of the stricture phase or bond nature.

2.3.2.5 BOLS-LBA Versus Quantum Confinement

Figure 2.2b illustrates the BOLS-NEP notion for the sized matter [24] compared with the theory of Quantum Confinement (QC) that was firstly proposed by chemists in 1982 for the light emission and band gap modulation by crystal size reduction. Electron-hole pairs, or excitons, moving in the confinement box in the potential above the open circles in the diagram, which modulates the band gap E_g of a sized semiconductor (K) in the empirical form of,

$$E_g(K) - E_g(\infty) = A/K^2 + B/K + C$$

where $E_g(\infty)$ is the bulk value and A, B, C are adjustable parameters to reproduce experimental results. The first term in the right hand side describes the kinetic energy of the excitons and the second the Coulomb attraction and the third term the background correction. The excitons move inside the box with electron-hole separation of the potential box size K.

There are two limitations of the QC theory. One is the localization of the covalent electrons of a semiconductors such as diamond and silicon. The excited electrons recombine with holes almost spontaneously to emit light whose wave length depending on the band gap size that is intrinsically determined by the Hamiltonian in the nearly-free electron approximation [9]. The other one is the global size dependence of a crystal at the nanometer scale for all detectable properties such as the core level shift, dielectric constant, elastic modulus, melting point [12]. The global variation of properties with solid size is beyond the scope of QC description. It would be responsible to seek for the common mechanism of the size dependency rather simply refer to the QC scheme.

In contrast, the BOLS-NEP notion considers the bond relaxation from one equilibrium to another by the degree of freedom of atomic coordination number and its consequences on potential well [12]. The valley of the potential corresponds to the bond length and energy, whose relaxation modulates the potential function in the Hamiltonian and atomic cohesive energy. The bond order loss shortens and strengthens the remaining bonds between undercoordinated atoms at the bonding network terminals such as sites of defects and skins of nanocrystals, which cause local densification and entrapment of the bonding electrons, and polarization of nonbonding electrons. The relaxation of th bond and the associated energetics, localization, densification and polarization govern the performance of a sized crystals. Therefore, the BOLS-NEP notion is much more revealing, pertinent and comprehensive than the hypothetic QC approach.

2.3.3 Hetero-coordination: Entrapment or Polarization

2.3.3.1 Exchange Interaction

The physical BOLS [12] describes situations of atomic CN deficiency, which is different from the chemical BOLS [25] defined for reaction dynamics in which the bond-order is the number of chemical bonds between a pair of atoms. The chemical BOLS correlates the bond-length d, bond-energy E, and bond-order n for reaction [26, 27]:

$$\begin{cases} d/d_s = 1 - 0.26\ln(n)/d_s \\ E/E_s = \exp[c(d_s - d)] = n^p \end{cases} \tag{2.9}$$

where subscript s stands for the "single bond", c and p are hypothetic coefficients. This BOLS estimates: (i) binding energies released during gas-solid reaction and (ii) activation energies for chemisorption or desorption [28–33].

2.3.3.2 Interface Potential: Entrapment or Polarization

Upon doping or alloying, diffusion of constituent atoms forms a region of graded composition (Fig. 2.5a) [14]. The crystal potential for each constituent in the interface region turns from the $V_{cryst}(r, B)$ to the $V_{cryst}(r, I) = \gamma V_{cryst}(r, B)$. The γ is the ratio of bond energy in the interface region (I) to that in the ideal constituent bulk (B). If $\gamma > 1$, quantum entrapment (T) dominates; otherwise, polarization happens. Hence, the $V_{cryst}(r, I)$ becomes deeper ($\gamma > 1$) or shallower ($\gamma < 1$ for a potential barrier formation) than the respectively $V_{cryst}(r, B)$ of the specific constituent standing alone.

This specification is in accordance with that proposed by Popovic and Satpathy [34] in calculating oxide superlattices. They introduced a wedge-shaped potential well between $SrTiO_3$ and $LaTiO_3$ superlattices to mimic the monolayer sandwiched between them. Electrons in the interface form the airy-function-localized states.

If the atomic CN changes insignificantly in the interface, the bond energy determines the interface CLS, $\Delta E_v(I)$ with respect to the energy shift in the elemental bulk crystal, $\Delta E_v(B)$,

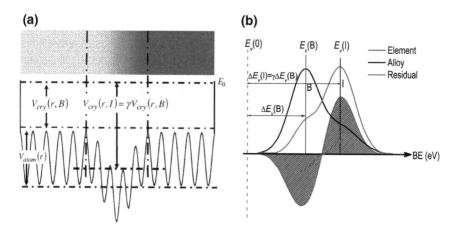

Fig. 2.5 **a** Interface potential variation transits **b** the spectral peak intensities from the $E_v(B)$ dominance to the $E_v(I)$ dominance. If the ratio $\gamma = V_{cry}(r, I)/V_{cry}(r, B) > 1$, interface quantum entrapment dominates, $\Delta E_v(I) > \Delta E_v(B)$; otherwise, interface polarization dominates, $\Delta E_v(I) < \Delta E_v(B)$. The shaded area in **b** is the residual spectrum gained by subtracting the referential spectral peak of the constituent elemental solid from that of the alloy. Reprinted with permission from [48]. Copyright 2010 Elsevier

$$\frac{\Delta E_v(I)}{\Delta E_v(B)} = \gamma = \frac{E_I}{E_b} = 1 + \Delta_{HI} \tag{2.10}$$

Figure 2.5b shows that alloy or compound formation evolves the $E_v(B)$ into the $E_v(I)$ with an intensity inversion of the two XPS components. The peak intensity inversion happens and the total intensity conserves as the total number of electrons in the particular energy level is subject to no loss.

2.3.4 Generalization of the Irregular-Coordination Effect

The following expressions formulate the irregular-coordination induced local entrapment and polarization of electrons in terms of Hamiltonian perturbation,

$$\Delta_H(x_l) = \begin{cases} \Delta_H(z_i) = \frac{E_{zi} - E_b}{E_b} = C_{zi}^{-m} - 1 & \text{(Defect and skin)} \\ \Delta_H(K) = \sum_{i \le 3} \gamma_i \Delta_H(z_i) = \tau K^{-1} \sum_{i \le 3} C_{zi}(C_{zi}^{-m} - 1) & \text{(Nanosolid)} \\ \Delta_H(I) = \frac{E_I - E_b}{E_b} = \gamma - 1 & \text{(Interface)} \\ \Delta_H(P) = [E_v(p) - E_v(0)]/\Delta E_v(12) - 1 & \text{(Polarization)} \end{cases} \tag{2.11}$$

The following correlates energies of component l and l' in an XPS spectrum from a specimen ($l = S_1, S_2, \ldots$),

$$\frac{E_v(x_l) - E_v(0)}{E_v(x_{l'}) - E_v(0)} = \frac{1 + \Delta_{HI}}{1 + \Delta_{HI'}}, (l' \ne l) \tag{2.12}$$

This formulation yields immediately [17],

$$\begin{cases} E_v(0) = [E_v(x_l)(1 + \Delta_{HI'}) - E_v(x_{l'})(1 + \Delta_{HI})]/(\Delta_{HI'} - \Delta_{HI}) \\ \Delta E_v(12) = E_v(12) - E_v(0) \\ \Delta E_v(x_l) = \Delta E_v(12)(1 + \Delta_{HI}) \end{cases} \tag{2.13}$$

Chemical reaction or coordination variation alters neither the $E_v(0)$ nor the bulk shift $\Delta E_v(12)$. Accuracy of determination of the $E_v(0)$ and the $\Delta E_v(12)$ is subject to calibration of the XPS and to determination in the shape and size of the nanocrystal. Nevertheless, furnished with this formulation, one could elucidate, in principle, the core level positions of an isolated atom $E_v(0)$ and the bulk shift $\Delta E_v(12)$, as well as the local bond length and energy from XPS measurement, as illustrated in subsequent chapters.

2.4   Valence Band and Nonbonding States

2.4.1   Complexity of the Valence DOS

Valence electrons perform quite differently from electrons in the core bands of a specific constituent element. Valence electrons are mixture of all involved elements in the solid, which respond to the chemical environment directly at functioning. The behavior of valence electrons is much more complicated because of the delocalization, polarization, and charge redistribution among constituent elements in reaction. Besides the undercoordination effect featured by BOLS-NEP notion, atomic hetero-coordination due to formation of alloys, compounds, dopants, impurities, interfaces, or glasses also results in the densification, localization, entrapment, and polarization of the valence electrons because of the bond nature alteration. Both under- and hetero-coordination change the valence band substantially and irregularly.

In addition to polarization and localization of the conduction electrons by the densely entrapped core electrons, such as the zigzag-edge of graphene [16] and the Rh adatoms [35], presence of the nonbonding lone pairs and the lone pair-induced dipoles upon reaction with F, O, and N play a role of significance [3, 4, 15].

2.4.2   Tetrahedral-Bonding Mediated Valence DOS

Figure 2.6 illustrates the residual DOS of metals and semiconductors resulting from involvement of N, O, and F [4]. The sp^3-orbit hybridization produces four directional

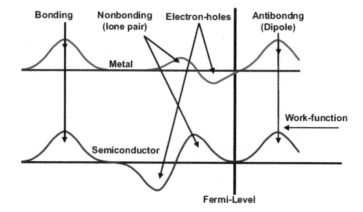

Fig. 2.6 Differential valence DOS of metals (upper) and semiconductors (lower) mediated by N, O, and F addition with four excessive DOS features: bonding pairs ($\ll E_F$), nonbonding lone pairs ($<E_F$), electron holes ($<E_F$), and anti-bonding dipoles ($>E_F$). The DOS features in the vicinity of the E_F are crucial to the performance of a compound. Reprinted with permission from [4]. Copyright 2006 Elsevier

orbits, each capable of being occupied by two electrons, forming a quasi-tetrahedron. These four directional orbits can hence accommodate eight electrons. The central O atom, for example, has six $2s^2 2p^4$ electrons and needs another two to complete its valence shell. Therefore, the O forms two bonds with the nearest neighbors by sharing one electron with each of its two neighbors, while its lone electron pairs occupy its remaining two orbits. Likewise, a nitrogen atom needs three electrons for sharing and generates one lone pair; and an F atom forms a tetrahedron with three lone pairs. The manner of electronic distribution, bond type, bond length and energy surrounding the central F, O, and N atom in the tetrahedron are anisotropic at the atomic scale [2].

Counting from lower to higher energy, the DOS features are the bonding states, electronic holes, electron lone pairs of F^-, O^{2-}, or N^{3-}, and the anti-bonding dipoles of the host. In a semiconductor compound, holes create at the top of the valence band, which further expands the semiconductor's band gap, turning a semiconductor into an insulator, such as Si_3N_4 and SiO_2.

In metallic compounds, holes generate near the Fermi surface, which open a band gap and turns the conductor into an insulator or a semiconductor, such as Al_2O_3, TiO_2, ZnO, and AlN. Nonbonding states located in the band gap form impurity states, while dipoles form antibonding states above the E_F. The production of dipoles shifts the surface potential barrier outwardly with high saturation [36], opposing to the effect of the positively charged ions. These dipoles also screen and split the local potentials, adding excessive features to the core bands. STM probes dipoles as protrusions, while the positive ions are probed as depressions.

2.4.3   Impact of Nonbonding States

The "nonbonding" states also refer to the energetic electrons involved in the anti-bonding dipoles, as well as H-bond like and C–H-bond like. An ionic impurity in a metal also polarizes neighbors to create polarized states [3, 37]. The van der Waals bond having a maximal energy of several tenths of an eV is in this category as it stands for dipole-dipole interaction instead of charge sharing exchange interactions.

Contributing insignificantly to the Hamiltonian or to the atomic cohesive energy, the nonbonding electrons add, however, impurity states near E_F, which neither follow the regular dispersion relations nor occupy the allowed states of the valence band and below. They are located at energies within the energy window of an STM/S. In additional to the weak interactions with energies of ~50 meV, as detected using Raman and electron energy loss spectroscopy (EELS) [3], these lone pairs, however, polarize the neighboring atoms, instead, causing their neighbors changing to dipoles.

The impact of the nonbonding lone pairs and the antibonding dipoles is ubiquitously abundant and profound. For example, the presence of the antibonding dipoles lowers drastically the work function by more than 1 eV [38], which aids greatly the electron emission for imaging and display [39, 40]. Nitrogenating of diamond and carbon nanotubes, oxidation and fluorination of metals, etc., have the same effect of polarization and work function reduction. Presence of these localized nonbonding states is responsible for the size derivacy of defects and nanostructures such as the Dirac-Fermions creation, dilute magnetism, catalytic enhancement, super-hydrophobicity, etc. Such nonbonding and antibonding states play significant roles in the topologic insulators, thermoelectric materials, and high-T_C superconductors [15]. The nonbonding electrons, the O:H interaction and its cooperativity with H-O intra-molecular bond dictates the mysteries of water and ice [5, 23, 41].

2.5 Numerical Strategies: Formulation and Quantification

2.5.1 Irregular-Coordination Effect

With the BOLS-NEP notion and its formulation, one can decompose a skin XPS spectrum accordingly. Parameters of concern are the number and order of components and their intensities, energies, and their full width at half maximum (FWHM) if one describes the peak using, but not limited to, Gaussian distribution function. Ideally, the following provide guidelines:

$$\begin{cases} I = \sum_i I_{zi} \exp\left\{ -\left[\frac{E - E_v(z_i)}{E_v W(z_i)} \right]^2 \right\} & \text{(Spectral intensity)} \\ \frac{E_v(z_i) - E_v(0)}{E_v(12) - E_v(0)} = \frac{E_{zi}}{E_b} = C_{zi}^{-m} & \text{(Component peak energy)} \\ \frac{E_{vwi}(z_i)}{E_{vwb}(12)} = \frac{z_i E_{zi}}{z_b E_b} = z_{ib} C_{zi}^{-m} & \text{(Component width)} \end{cases} \quad (2.14)$$

The overall intensity is a superposition of that for all components. The spectral components follow the positive CLS according to the TB approximation unless polarization becomes dominance. The energy level of an isolated atom $E_v(0)$ is the unique reference from which the CLS proceeds. The bulk component width $E_{vwb}(12)$ is to be optimized in spectral decomposition. One can determine the number of peaks and their ordering positions, widths, and intensities in each spectrum with the known $E_v(0)$, $E_{vwb}(12)$ of the specific spectrum. The $E_v(0)$ and $E_{vwb}(12)$ are available by decomposing a set of as many as possible spectra collected from different registries of the same substance, as exemplified shortly.

Equation (2.14) provides only guidelines for decomposition but the actual spectral intensity and shape are subject to polarization and artifacts in measurements. Furthermore, if the bond nature index m value is sufficiently large, the width of the skin component may be broader than that of the bulk. Therefore, fine-tuning all components collectively in spectral decomposition is necessary.

Another issue is that the effective CN for the same registered sublayer of the same crystal geometry conserves regardless of chemical composition of the material. For example, the CN of the first sublayer of the fcc(100) skin holds 4.0 for all fcc-structured specimen such as gold, copper, and Rhodium [12, 42, 43].

Proper calibration of the equipment can remove the artifacts like ionization-induced core-hole relaxation and radiation-induced surface charging. Upon background correction and peak-area normalization, one can decompose a skin XPS spectrum into the B and the S_i components in a positive CLS order. The best fit to the spectral peak can estimate the number of components and the width of each for a given spectrum. One can take the effective $z_1 = 4$ for the outermost layer of the fcc(100) flat skin as the standard reference. A fine-tuning of the component energies of all involved sublayers can optimize their z values collectively. The optimized set of z values are only geometry dependent. Repeated fine-tuning optimizes the z value, intensity, energy, and width of each component. The least deviation σ justifies the accuracy and precision of the spectral analysis derivatives.

An incorporation of the BOLS-LBA-NEP notion into the TB approach and the PES measurements not only clarifies the physical origin of the CLS but also enables quantification of several parameters. These parameters include the energy level of an isolated atom $E_v(0)$ and its bulk shift $\Delta E_v(12)$, the CN dependent CLS $\Delta E_v(z)$ and component width $E_{vW}(z)$, local bond length d_z, bond energy E_z, energy density E_{den}, atomic cohesive energy E_{coh}, in addition to the charge distribution—entrapment or polarization. Polarization happens only to atoms with an effective CN smaller than that of an atom in the flat skin.

2.5.2 Local Energy Density and Atomic Cohesive Energy

Traditionally, the surface free energy (γ_s) is defined as the energy required for cutting a given crystal into two halves, or the energy costs in making a unit area of a surface [44]. The interface energy (γ_I) is what required to form the unit area of an interface. The unit of the γ_s and γ_I is sometimes in eV/nm^2 (density perunit area) and some other times in eV/atom (cohesive enenrgy). Magnitude of the former is often higher than that of the latter for the same substance. The γ_s, the γ_I, and their functionalities arise from nothing more than interaction between the irregularly-coordinated atoms.

In fact, instead of the energy cost for forming a surface or an interface, the energy-gain per unit volume E_{den} or the cohesive energy remnant per atom E_{coh} in the skin or in the interface region dictates the performance of atoms and electrons in these irregularly-coordinated atomic sites. To effectively describe phenomena and processes at the skin and interface, the following concepts are necessary to complement the conventional terms of surface and interface free energy. Table 2.2 summarizes the formulation, physical origin, and the functionality of E_{den} and E_{coh} in the irregularly-coordinated atomic sites based on the BOLS notion [45]. The concept of skin of a certain thickness is much more meaningful than the surface or interface in two-dimension [2].

Table 2.2 Definition, formulation, origin, and functionality of the E_D and the E_C in the irregularly-coordinated atomic sites [45]

Definition	Skin	Interface ($A_x B_{1-x}$ alloy)
Energy density E_{den} (eV/nm^3)	$E_{DS} = \int_0^{d_3} (E_{zi}/d_{zi}^3)dy \Big/ \int_0^{d_3} dy$	$E_{DI} = N_{cell} z_I E_I / V_{cell}(d_I)$ $with \begin{cases} d_I = xd_A + (1-x)d_B; \ z_I \cong z_b \\ E_I = xE_A + (1-x)E_B + x(1-x)\sqrt{E_A E_B} \end{cases}$
Physical origin	Energy gain per unit area of ($d_1 + d_2 + d_3$) thick skin due to BOLS	Energy gain due to bond nature alteration and exchange interaction
Functionality	Surface stress; elasticity; surface optics; dielectrics; electron and photon transport dynamics; work function, etc.	Interface mechanics, joining, tunneling junction, etc.
Atomic cohesive energy E_{coh} (eV/atom)	$E_{CS} = \int_0^3 d(z_i E_{zi})/3$	$E_{CI} = z_I E_I$
Physical origin	Energy remnant per discrete atom upon surface/interface formation	
Functionality	Thermal stability, wettability, diffusivity, reactivity, self-assembly, reconstruction	

2.6 Summary

Formation of the bond between hetero-coordinated atoms and relaxation of the bond between undercoordinated atoms shift intrinsically the electronic BE in the core band and in the valence band of a substance. The process of charge entrapment, localization, and polarization modify the valence and the core band consistently. The BOLS-NEP-TB describes the CLS adequately while the artifact of "initial-final" state relaxation serves as background. One could determine the local bond length d_z, bond energy E_z, energy density E_{den}, and atomic cohesive energy E_{coh} associated with irregularly-coordinated atoms from spectral analysis based on the framework of the presented BOLS-NEP-LBA-TB notion.

References

1. C.Q. Sun, Y. Wang, B. Tay, S. Li, H. Huang, Y. Zhang, Correlation between the melting point of a nanosolid and the cohesive energy of a surface atom. J. Phys. Chem. B **106**(41), 10701–10705 (2002)
2. C.Q. Sun, *Relaxation of the chemical bond* Springer Ser. Chem. Phys. **108**, 807 (2014)
3. C.Q. Sun, Oxidation electronics: bond-band-barrier correlation and its applications. Prog. Mater Sci. **48**(6), 521–685 (2003)
4. W.T. Zheng, C.Q. Sun, Electronic process of nitriding: mechanism and applications. Prog. Solid State Chem. **34**(1), 1–20 (2006)
5. Y.L. Huang, X. Zhang, Z.S. Ma, Y.C. Zhou, W.T. Zheng, J. Zhou, C.Q. Sun, Hydrogen-bond relaxation dynamics: resolving mysteries of water ice. Coord. Chem. Rev. **285**, 109–165 (2015)

6. L. Zhang, C. Yao, Y. Yu, S.-L. Jiang, C.Q. Sun, J. Chen, Stabilization of the dual-aromatic cyclo-$N_5{}_-$ anion by acidic entrapment. J. Phys. Chem. Lett. **10**, 2378–2385 (2019)

7. C.Q. Sun, Unprecedented O:⇔: O compression and H↔H fragilization in Lewis solutions (perspective). Phys. Chem. Chem. Phys. **21**, 2234–2250 (2019)

8. C.Q. Sun, *Solvation dynamics: a notion of charge injection.* Springer Ser. Chem. Phys. **121**, 316 (2019)

9. M.A. Omar, *Elementary Solid State Physics: Principles and Applications* (Addison-Wesley, New York, 1993)

10. R. Egerton, P. Li, M. Malac, Radiation damage in the TEM and SEM. Micron **35**(6), 399–409 (2004)

11. C.-Y. Lin, H.W. Shiu, L.Y. Chang, C.-H. Chen, C.-S. Chang, F.S.-S. Chien, Core-level shift of graphene with number of layers studied by microphotoelectron spectroscopy and electrostatic force microscopy. J. Phys. Chem. C **118**(43), 24898–24904 (2014)

12. C.Q. Sun, Size dependence of nanostructures: Impact of bond order deficiency. Prog. Solid State Chem. **35**(1), 1–159 (2007)

13. X.J. Liu, M.L. Bo, X. Zhang, L. Li, Y.G. Nie, H. Tian, Y. Sun, S. Xu, Y. Wang, W. Zheng, C.Q. Sun, Coordination-resolved electron spectrometrics. Chem. Rev. **115**(14), 6746–6810 (2015)

14. C.Q. Sun, Thermo-mechanical behavior of low-dimensional systems: the local bond average approach. Prog. Mater Sci. **54**(2), 179–307 (2009)

15. C.Q. Sun, Dominance of broken bonds and nonbonding electrons at the nanoscale. Nanoscale **2**(10), 1930–1961 (2010)

16. W.T. Zheng, C.Q. Sun, Underneath the fascinations of carbon nanotubes and graphene nanoribbons. Energy Environ. Sci. **4**(3), 627–655 (2011)

17. C.Q. Sun, Surface and nanosolid core-level shift: impact of atomic coordination-number imperfection. Phys. Rev. B **69**(4), 045105 (2004)

18. E. Abrahams, P.W. Anderson, D.C. Licciardello, T.V. Ramakrishnan, scaling theory of localization: absence of quantum diffusion in two dimensions. Phys. Rev. Lett. **42**(10), 673–676 (1979)

19. R.A. Street, *Hydrogenated Amorphous Silicon* (Cambridge University Press, 1991)

20. M. Reif, L. Glaser, M. Martins, W. Wurth, Size-dependent properties of small deposited chromium clusters by x-ray absorption spectroscopy. Phys. Rev. B **72**(15), 155405 (2005)

21. C.Q. Sun, C.M. Li, S. Li, B.K. Tay, Breaking limit of atomic distance in an impurity-free monatomic chain. Phys. Rev. B **69**(24), 245402 (2004)

22. C.Q. Sun, Y. Sun, Y.G. Nie, Y. Wang, J.S. Pan, G. Ouyang, L.K. Pan, Z. Sun, Coordination-resolved C–C bond length and the C 1s binding energy of carbon allotropes and the effective atomic coordination of the few-layer graphene. J. Phys. Chem. C **113**(37), 16464–16467 (2009)

23. C.Q. Sun, Y. Sun, The attribute of water: single notion, multiple myths. Springer Ser. Chem. Phys. **113**, 494 (2016)

24. C.Q. Sun, T.P. Chen, B.K. Tay, S. Li, H. Huang, Y.B. Zhang, L.K. Pan, S.P. Lau, X.W. Sun, An extended 'quantum confinement' theory: surface-coordination imperfection modifies the entire band structure of a nanosolid. J. Phys. D Appl. Phys. **34**(24), 3470–3479 (2001)

25. L. Pan, S. Xu, X. Liu, W. Qin, Z. Sun, W. Zheng, C.Q. Sun, Skin dominance of the dielectric electronic-phononic-photonic attribute of nanoscaled silicon. Surf. Sci. Rep. **68**(3–4), 418–445 (2013)

26. L. Pauling, Atomic radii and interatomic distances in metals. J. Am. Chem. Soc. **69**(3), 542–553 (1947)

27. H.S. Johnston, C. Parr, Activation energies from bond energies: 1. Hydrogen transfer reactions. J. Am. Chem. Soc. **85**(17), 2544–2551 (1963)

28. W.H. Weinberg, The Bond-Energy Bond-Order (BEBO) model of chemisorption. J. Vac. Sci. Technol. **10**(1), 89–94 (1973)

29. E. Shustorovich, Chemisorption phenomena: analytic modeling based perturbation theory and bond-order conservation. Surf. Sci. Rep. **6**, 1–63 (1986)

30. H. Gross, C.T. Campbell, D.A. King, Metal-carbon bond energies for adsorbed hydrocarbons from calorimetric data. Surf. Sci. **572**(2–3), 179–190 (2004)

31. W.H. Weinberg, R.P. Merrill, Crystal-field surface orbital—bond-energy bond-order (CFSO-BEBO) model for chemisorption—application to hydrogen adsorption on a Platinum (111) surface. Surf. Sci. **33**(3), 493–515 (1972)

32. C.T. Campbell, D.E. Starr, Metal adsorption and adhesion energies on MgO(100). J. Am. Chem. Soc. **124**(31), 9212–9218 (2002)

33. V.A. Bondzie, S.C. Parker, C.T. Campbell, The kinetics of CO oxidation by adsorbed oxygen on well-defined gold particles on TiO2(110). Catal. Lett. **63**(3–4), 143–151 (1999)

34. Z.S. Popovic, S. Satpathy, Wedge-shaped potential and Airy-function electron localization in oxide superlattices. Phys. Rev. Lett. **94**(17), 176805 (2005)

35. C.Q. Sun, Y. Wang, Y.G. Nie, Y. Sun, J.S. Pan, L.K. Pan, Z. Sun, Adatoms-induced local bond contraction, quantum trap depression, and charge polarization at Pt and Rh surfaces. J. Phys. Chem. C **113**(52), 21889–21894 (2009)

36. C.Q. Sun, O-Cu(001): I. Binding the signatures of LEED, STM and PES in a bond-forming way. Surface Rev. Lett. **8**(3–4), 367–402 (2001)

37. B.C. Stipe, M.A. Rezaei, W. Ho, Single-molecule vibrational spectroscopy and microscopy. Science **280**(5370), 1732–1735 (1998)

38. W.T. Zheng, C.Q. Sun, B.K. Tay, Modulating the work function of carbon by N or O addition and nanotip fabrication. Solid State Commun. **128**(9–10), 381–384 (2003)

39. J.J. Li, W.T. Zheng, C.Z. Gu, Z.S. Jin, Y.N. Zhao, X.X. Mei, Z.X. Mu, C. Dong, C.Q. Sun, Field emission enhancement of amorphous carbon films by nitrogen-implantation. Carbon **42**(11), 2309–2314 (2004)

40. W.T. Zheng, J.J. Li, X. Wang, X.T. Li, Z.S. Jin, B.K. Tay, C.Q. Sun, Electron emission of carbon nitride films and mechanism for the nitrogen-lowered threshold in cold cathode. J. Appl. Phys. **94**(4), 2741–2745 (2003)

41. X. Zhang, Y. Huang, Z. Ma, Y. Zhou, W. Zheng, J. Zhou, C.Q. Sun, A common supersolid skin covering both water and ice. Phys. Chem. Chem. Phys. **16**(42), 22987–22994 (2014)

42. B.S. Fang, W.S. Lo, T.S. Chien, T.C. Leung, C.Y. Lue, C.T. Chan, K.M. Ho, Surface band structures on Nb(001). Phys. Rev. B **50**(15), 11093–11101 (1994)

43. R.A. Bartynski, D. Heskett, K. Garrison, G. Watson, D.M. Zehner, W.N. Mei, S.Y. Tong, X. Pan, The 1st interlayer spacing of Ta(100) determined by photoelectron diffraction. J. Vac. Sci. Technol. A **7**(3), 1931–1936 (1989)

44. R. Dingreville, J. Qu, M. Cherkaoui, Surface free energy and its effect on the elastic behavior of nano-sized particles, wires and films. J. Mech. Phys. Solids **53**(8), 1827–1854 (2005)

45. M.W. Zhao, R.Q. Zhang, Y.Y. Xia, C. Song, S.T. Lee, Faceted silicon nanotubes: Structure, energetic, and passivation effects. J. Phys. Chem. C **111**(3), 1234–1238 (2007)

46. C.Q. Sun, Y. Shi, C.M. Li, S. Li, T.C.A. Yeung, Size-induced undercooling and overheating in phase transitions in bare and embedded clusters. Phys. Rev. B **73**(7), 075408 (2006)

47. Y. Sun, Y. Wang, J.S. Pan, L.L. Wang, C.Q. Sun, Elucidating the 4f binding energy of an isolated Pt atom and its bulk shift from the measured surface- and size-induced Pt 4f core level shift. J. Phys. Chem. C **113**(33), 14696–14701 (2009)

48. Y. Nie, Y. Wang, Y. Sun, J.S. Pan, B.R. Mehta, M. Khanuja, S.M. Shivaprasad, C.Q. Sun, CuPd interface charge and energy quantum entrapment: a tight-binding and XPS investigation. Appl. Surf. Sci. **257**(3), 727–730 (2010)

Chapter 3
Probing Methods: STM/S, PES, APECS, XAS, ZPS

Abstract A set of analytical strategies has enabled atomistic, local, dynamic, and quantitative information on the bonding and electronic energetics induced by atomic under- and hetero-coordination. With the aids of the ZPS, one can purify the energy states with high precision without needing decomposition of the spectral peaks. APECS and NEXAS probe simultaneously the shifts of a core and the valence energy bands with provision of the screening and recharging information. Quantitative information includes the bond length, bond energy, core level shift, core charge entrapment and valence electron polarization, atomic cohesive energy, and binding energy density. Such a collection of information is fundamentally crucial to designing and synthesizing functional materials.

Highlights

- STM/S maps electrons from a superficial skin in the energy window surrounding the E_F.
- PES scopes BE information of the core and the valence electrons from a mixture of bulk and skin.
- APECS detects two-band BE shifts with information on potential screening and charge transition.
- ZPS offers CN-resolved information on bond relaxation, valence and core band, BE shift, etc.

3.1 Energy Band Structure and Electronic Dynamics

Figure 3.1 illustrates the energy bands of a solid. When a solid forms with N atoms, the discrete energy levels expand into bands that contain each N sublevels because of the degeneration by involvement of crystal potentials. $E_0 = 0$ is the vacuum level and E_F the Fermi energy. The separation between the E_0 and the E_F is the work function ϕ, which determines the easiness of the substance ejecting an electron. The separation

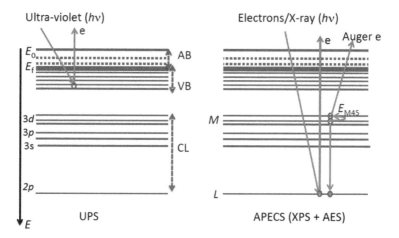

Fig. 3.1 Schematic illustration of the energy bands of an ideal solid and principles for the PES and the APECS (XPS and AES (*LMM* for instance; $L = 2p$, $M = 3d$). From top, there are the vacuum level $E_0 = 0$, unoccupied antibonding band (AB, broken lines), Fermi level E_F, valence or conduction band (VB or CB), and the core bands (3d, 3s, 2p,...). The incident beams of the PES or AES carry $h\nu$ energy with h being the Planck's constant and ν the frequency of light. The rule of energy conservation allows the PES and the AES to determine the BE and the DOS evolution in the respective bands

between the E_0 and the bottom edge of the conduction band (CB) is the electroaffinity, which determines the ability of a specimen in holding electrons captured during reaction. Conversely, the electronegativity is the intrinsic attribute of an element, which determines the easiness of electron transferring between elements of different negativity. The electronegativity difference between two elements determines the nature of the bond between them. Quantum entrapment enlarges the electroaffinity and polarization reduces the work function. Electroaffinity and work function are key indicators for the catalytic behavior and the toxicity of a substance, particularly at the nanometer scale or for atoms with even fewer CN than those at the flat skin [1].

From E_F downward, there is the CB for a conductor or the valence band (VB) for a semiconductor and then the core bands. The center of a core band shifts down by an amount that is proportional to the bond energy at equilibrium [2]. The width of the band becomes narrower and narrower and the energy shift of the band center becomes smaller and smaller, as moving from the VB downward. Electrons occupy the lowest energy level and then gradually up until the VB, which follows the rule of Pauli's repulsion. Electrons in the CB of a conductor are more delocalized than those are in the core band. Electrons in the VB of a semiconductor are more localized because of the covalent bond nature such as Silicon and Germanium. The energy shifts of the VB and below fingerprints the change of bond energy with the coordination and chemical environment.

At sites nearby defects with even less-coordinated atoms, localization and polarization of the dangling bond electrons occurs to semiconductors such as graphene edges, and conduction electrons of metallic adatoms or terrace edges, which produce antibonding (AB) dipole states occupying the upper edge of the VB or above the E_F [3]. For a solid in an amorphous phase, impurity states present with energies closing to the E_F and tails of both the VB and the VB [4, 5].

3.2 STM/S: Nonbonding and Anti-bonding States

Figure 3.2 illustrates the principle of an STM/S that maps a surface by probing the tunneling current, I, crossing the vacuum gap between the tip and the surface of conductors or semiconductors under bias V. STS records the dI/dV − V or the d(lnI)/d(lnV) − V profile at a fixed atomic site. The energy window of an STS is only a few eVs crossing E_F. The STS features the local DOS on an atomic scale in the lateral and subatomic in vertical direction [6]. If the sample is negatively biased, charge will flow from the specimen to the tip, probing the occupied states of the sample; otherwise, charge flows inversely, probing the unoccupied energy states above the E_F. Therefore, STS features below E_F (reference 0) represent the occupied DOS of the surface while features above the E_F represent the allowed, yet unoccupied, DOS of the surface atoms.

Figure 3.2b shows the typical STS spectra of a clean (spectrum A) and an O–Cu(110) surface (spectra B and C) with tip at different locations along the 'O^{2-}:Cu^P:O^{2-}:' chain (inset) with Cu^P being protruding dipoles [6]. Upon O^{2-}

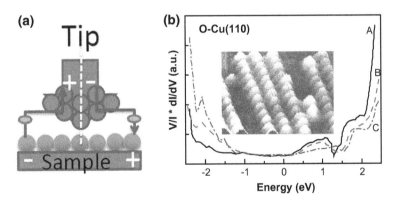

Fig. 3.2 Illustration of **a** the STM/S principle and **b** typical profiles collected from the clean (A) and the Cu(110)-(2 × 1)-O^{2-} surface at the bridge (B), on-top (C) site along the 'O^{2-}:Cu^P:O^{2-}:' chain (inset shows the STM image). The differential conductance dI/dV or VdI/IdV versus bias V is proportional to the local DOS mapped by an STS. Under sample positive bias, current flows from the tip to the surface, the STS peak represents the unoccupied DOS above E_F; otherwise, the STS peak represents the occupied DOS below E_F. Reprinted with permission from [6, 7]. Copyright 2003 Elsevier. Copyright 1989 American Physical Society

chemisorption, the initially empty DOS peaking at 0.5~1.8 eV above E_F is partially occupied by the Cu^P electrons, which weakens the intensity of the unoccupied states. Meanwhile, the lone pair of O^{2-} creates additional DOS features at -2.0 eV. The presence of the lone pair and the dipole states is the same to that detected using angular-resolved PES and using the de-excitation spectroscopy of metastable atoms of the same surface [8].

3.3 PES and AES: Valence- and Core-Band Shift

3.3.1 General Description

Angular-resolved PES means the PES data collection by changing the azimuth angle between the incident beam and often the <10> directional reference in the real lattice space or the ΓX direction in the Brillouin zone, as shown in Fig. 3.3. The angular variation only resolves the anisotropy of the crystal structure and the valence electrons distributed in real space, particularly for the reconstructed surfaces. This technique has formed the state-of-art means for investigating the electronic structure of high-T_C superconductors and topological insulators [9, 10].

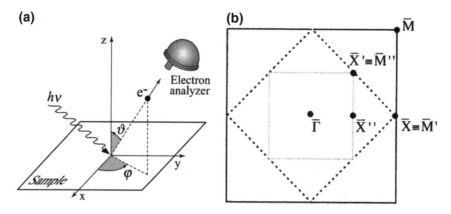

Fig. 3.3 Angle-resolved photoemission spectroscopy (ARPES): **a** the geometry of an ARPES experiment in which the emission direction of the photoelectron is specified by the polar (θ) and the azimuthal (ϕ) angles and **b** the Brillouin zones of a two dimensional square reciprocal lattice, which has the same symmetry of the real lattice [11]

3.3.2 PES and AES

Using light sources of different energies, PES measures energy states of different bands. With energy less than 50 eV He-II radiation at 40.8 eV, UPS monitors the change of the valence band consisting electrons of all constituents due to charge transportation, entrapment, polarization, and electron-hole production. The inverse IPES (He-I radiation at 21.2 eV) probes the occupation of the unoccupied DOS about E_F by antibonding dipoles, which functionalize the same as STS.

X-ray Photoelectron Spectroscopy (XPS), also known as ESCA, is a surface analysis technique that uses an X-ray beam to eject electrons from a surface. The kinetic energies of the ejected electrons (photoelectrons) are then analyzed to produce a spectrum, in which photoelectron peaks coming from different chemical elements can be identified by their characteristic energies. The typical sample areas that can be analyzed are between 1 mm^2 and 1 cm^2, which makes XPS a useful technique for studying a wide variety of samples. Figure 3.4 shows the full XPS scan survey of an oxided-InAs sample for composition determination.

However, the XPS measures the elemental CLS as a mixture of the skin and the bulk components. The XPS monitors the consequence of crystal field change due to bond relaxation and bond nature alteration upon reaction on the CLS that is proportional to the bond energy at equilibrium. Compared with the UPS for the valence band and above, the analysis of XPS data is relatively simpler [13].

In the PES process, the incident radiation of $h\nu$ energy excites an electron in the specific νth level. The excited electrons will overcome the bound (work function ϕ and the BE of the respective CL (L level $E_{B(L,VB)}$, for example) and escape from the specimen to become free with kinetic energy $E_{K(L,VB)}$.

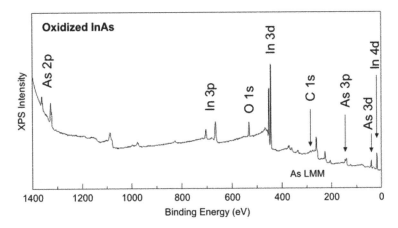

Fig. 3.4 XPS survey spectrum of an oxidized InAs surface. The XPS profile showed characteristic peaks for In, As, C and O presenting on the surface. A series of As LMM Auger peaks appears between 200 and 400 eV [12]

The Auger electron spectroscopy (AES) [14, 15] uses energetic electron source in replace of X-ray in the XPS. The AES processes include the following:

(1) The process of ionization is the same to that in the PES. The incident electron beams ionize atoms of the skin by kicking out an electron in the deeper energy level. The ejected electron flies away from the skin and leaves a hole behind the lower $L(2p)$ level, for a Cu instance.
(2) The ionization of the core electron relaxes all energy levels because of the screen weakening of both the intra- and the inter-atomic potentials.
(3) An electron then transits from the relaxed $M(3d_{5/2})$ upper level to filling the hole in the initial L level. This transition releases energy that equals the separation between the two levels, $\Delta E_R = E_{BL} - E_{BM}$.
(4) The ΔE_R kicks another electron off the further relaxed and degenerated $M(3d_{3/2})$ level to form the Auger electron that overcomes the E_{BM} and then escapes from the solid with kinetic energy E_{KM}.

The electronic energies in the PES and the AES processes conserve,

$$\begin{cases} h\nu = E_{B(L,VB)} + \phi + E_{K(L,VB)} & (PES) \\ E_{BL} - E_{BM} = E_{M45} + E_{BM} + \phi + E_{KM} & (AES) \end{cases} \quad (1)$$

The E_{M45} is the energy separation between the spin-resolved $3d_{3/2}$ and $3d_{5/2}$ lines of Cu, for instance.

3.4 APECS/XAS: Dual Energy-Level Shifts

3.4.1 AES: Auger Parameter

Differentiating Eq. (1) for the L and the M levels yields immediately,

$$\begin{cases} \Delta E_{BL} = -\Delta E_{KL} \\ 2\Delta E_{BM} = \Delta E_{BL} - \Delta E_{KM} \end{cases}$$
$$\Delta\alpha' = -(\Delta E_{KL} + \Delta E_{KM}) = 2\Delta E_{BM} \quad (2)$$

The ionization and relaxation changes the $\Delta\phi \cong 0$ and the $\Delta E_{M45} = \Delta(E_{3d5/2} - E_{3d3/2}) \cong 0$ insignificantly for the same specimen. Therefore, this expression correlates the energy shifts of the APECS involved L and M levels, for the first time.

Figure 3.5 illustrates the mechanism of APECS as a combination of AES and XPS. Summing the kinetic energy shifts of the XPS and the AES electrons ($\Delta E_{KL} + \Delta E_{KM}$), conventionally called Auger parameter $\Delta\alpha'$, equals twice the shift of the M (3d) level shift in magnitude. Therefore, the APECS resolves simultaneously both the L and the M levels without any approximation or assumption.

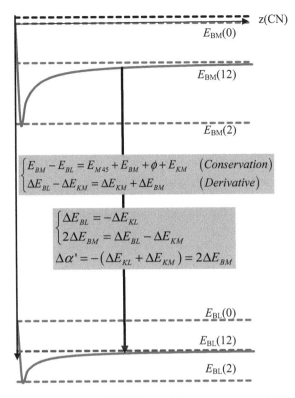

Fig. 3.5 Energy conservation in the AES-PES coincidence spectroscopy (APECS) processes for the CN-resolved energy shifts of the $L_3M_{45}M_{45}$ lines for a crystal. The energy levels, $E_{BM}(z)$ and $E_{BL}(z)$ shift simultaneously with the variation of atomic CN to a maxim at $z = 2$ and revert to the bulk values. The $E_{BM}(z)$ and $E_{BL}(z)$ and their separation, represented by the solid lines, are measurable while the atomic $E_{BM}(0)$ and $E_{BL}(0)$ and their bulk shift (dotted lines) can be obtained with the present analytical method. The work function ϕ and the sublevel separation E_{45} change insignificantly in the APECS processes. Inset show the energy conservation relationships. Reprinted with permission from [21]. Copyright 2006 Elsevier

Traditionally, two terms describe the energy shifts of the APECS-involved levels. One is the Auger parameter and the other is the Wagner plot [16, 17]:

$$\begin{cases} \Delta\alpha' = |\Delta E_{BL}| + |\Delta E_{KM}| \cong 2\Delta R^{ex} & (Auger - parameter) \\ E_{KM,x} - E_{KM,1} = \beta(E_{BL,x} - E_{BL,1}) & (Wagner - plot) \end{cases} \tag{3}$$

As justified above, it is improper to assume that the energy shifts of the L and the M level contribute equally to the $\Delta\alpha'$. The $\Delta\alpha'$ is greater than the ΔE_{BL} or the ΔE_{KM} alone, and therefore, the $\Delta\alpha'$ is more sensitive to the chemical or the coordination stimulus. The R^{ex} is a hypothetic parameter that corresponds to extra energies of atomic relaxation or polarization, coming from neighbors of the core-ionized atom.

Actually, Eq. (2) proves that the $\Delta\alpha'$ is exactly twice of the ΔE_{BM} resulting from the relaxation of the chemical bond by the irregular-coordination effect:

$$\Delta\alpha' = |\Delta E_{BL}| + |\Delta E_{KM}| = -(\Delta E_{KL} + \Delta E_{KM}) = 2\Delta E_{BM}, \qquad (4)$$

Therefore, one can measure the energy shift of the upper M level (or the valence band) in the form of [16]:

$$\Delta E_{BM} = -(\Delta E_{KL} + \Delta E_{KM})/2$$

The slope β of the Wagner plot correlates the shift of the Auger kinetic energy ΔE_{KM} and the photoelectron kinetic energy ΔE_{KL} (or the ΔE_{BL} in magnitude) [18, 19]. Subscript x represents the concentration of the element in the compound.

3.4.2 APECS: Dual-Level Shifts

With two independent measurements of the Auger parameters, for example, $\alpha'_1 = E_B(1s) + E_K(KLL)$ and $\alpha'_2 = E_B(2p) + E_K(KLL)$, of compounds containing the same constituent with different x concentrations, one could estimate the separation between two levels, $E_B(1s) - E_B(2p)$, rather than either of them alone [16, 20, 21].

3.4.3 Extended Wagner Plots: Screening and Recharging

Although the absolute energy shift varies from level to level, the relative shift of each level across from their reference should be identical if no chemical process is involved. One may extend the Wagner plot to the atomic CN (particle size) domain to correlate the relative shifts of the APECS involved levels. This extension discriminates the effect of crystal-field screening from that of the valence recharging (initial-final state and charge transport in reaction), in the following manner:

$$\begin{cases} \kappa_{ML} = \left\{ \frac{E_{BM}(z) - E_{BM}(0)}{E_{BM}(12) - E_{BM}(0)} \right\} \Big/ \left\{ \frac{E_{BL}(z) - E_{BL}(0)}{E_{BL}(12) - E_{BL}(0)} \right\} = 1 \ (Extended - Wagner - plot) \\ \eta_{ML} = \frac{E_{BM}(12) - E_{BM}(0)}{E_{BL}(12) - E_{BL}(0)} > 1 \qquad\qquad (Screening - coefficient) \end{cases}$$

$$(5)$$

The slope κ_{ML} correlates the relative shift of the two levels, which represents valence recharging effect when the coordination or chemical environment changes. Ideally, $\kappa_{ML} = 1$, if no chemical reaction occurs. Therefore, κ_{ML} can be indicative of reaction and its value relates directly to the interatomic bond. The screening coefficient, η_{ML}, represents the relative BE between these two levels in the bulk of an elemental solid. Therefore, the presently extended Wagner plot is more convenient and revealing than its conventional form because it employs the dimension of crystal size or the effective atomic CN with a focus on the relative energy shift [13].

3.4.4   *XAS/XES: Dual-Level Resultant Shift*

Figure 3.6 Compares the electronic process of XAS/XES with processes of XPS. The XPS requires source of fixed energy while the XAS/XES needs sources of flexible energy that can be reached in the syncroelectronics [22]. XAS probes the density of unoccupied states, XES probes that of the occupied states. It is highly attractive in time domain studies as it senses the changes in electronic and spin structure that are induced by photoexcitation of the valence electrons. It is particularly attractive in the case of spin state changes. XES also provides information about the nature of chemical bonds and their distances around a given atom.

The pre-edge and rising edge shift of the XAS/XES results from the combined shifts of the core band and the valence band under perturbation to the crystal potential, see Fig. 3.5, for instance, as

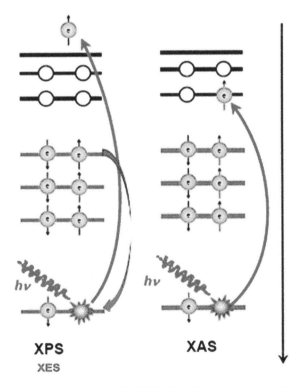

Fig. 3.6 Comparison of the processes for the XPS and the X-ray absorption/emission spectroscopy (XAS/XES). Reproduced with permission from [22]

$$\Delta E_{edge} = \Delta E_{core} - \Delta E_{VB} = \begin{cases} < 0 \ (polarization) \\ > 0 \ (entrapment) \end{cases}.$$

Because of the screening effect, $|\Delta E_{core}| < |\Delta E_{VB}|$. Quantum entrapment dominance will shift both energy level downwardly, and otherwise, polarization dominance shifts both energy level upwardly. The ΔE_{edge} shifts in a direction being opposite to that of the XPS by the same origin-entrapment or polarization. The XAS and XES are often used in studying water and solutions under various stimuli [23, 24].

3.5   ZPS: Atomic CN-Resolved Bond Relaxation

3.5.1   Experimental and Analytical Procedures

One can imagine what will happen to the outcome by differentiating two spectra collected under any of the following conditions from:

(1) the same defect-free surface at different emission angles;
(2) the same surface after and before conditioning such as defect creation, atom addition, or chemisorption under the same probing conditions; or
(3) specimens containing the same constituent but different concentrations.

Upon the standard processes of background correction and spectral peak area normalization, the ZPS in (i) distills the spectral features due to the monolayer skin by filtering out the bulk information, as the XPS collects more information from the surface at larger emission angles [25, 26]. Likewise, the ZPS in (ii) purifies merely the spectral features due to conditioning. The ZPS in (iii) resolves the alloying effect on the energy shifts of the respective levels of a specific element. The ZPS also filters out all artifacts such as the surface charging and the "initial-final states" relaxation that exist throughout the course of measurements. This ZPS strategy can monitor the surface and interface processes such as crystal growth, defect generation, chemical reaction, alloy formation, etc., both statically and dynamically with high sensitivity and accuracy without needing any approximation or assumption or the conventionally tedious processes of spectral peak decomposition.

The integrated intensity of a spectral peak collected at a larger emission angle or from a rougher skin is weaker than that of the otherwise because of the scattering effect. The mean free path of the ejected electrons is generally longer than the penetration depth of the incident beams [27]. Conversely, the peak area integral of the specific peak is proportional to the total number of electrons emitted from the specimen under the same probing conditions. To make all spectra quantitatively comparable, area normalization of the same peak collected under different conditions can minimize the influence of the scattering and artificial effects. Having all the spectra been normalized, one can subtract the referential spectrum from the ones collected from the same specimen upon conditioning.

3.5.2 Quantitative Information

The ZPS distils the DOS gain as components presenting above the lateral axis and features the DOS loss as valleys below the lateral axis. This process removes the commonly shared spectral area that provides little useful information. Ideally, the resultant ZPS components conserve as the spectral areas above and areas below the lateral axis are identical. Any improper background correction or spectral normalization may asymmetrize the spectral gain and loss compared with the ideal situation. With these criteria, one can readily gain quantitative information of local bond length, bond energy, charge entrapment and polarization, etc., under various working conditions.

Conventionally, one needs to correct the spectral background using the standard Tougaard method [28–30] by employing Gaussian-, Lorentz-, or Doniach-Sunjic-type functions before decomposing the XPS profiles. However, ZPS saves such tedious procedures of background correction, component specification, and peak energy fine-tuning. ZPS gives directly the skin or conditioned component as emerging peaks and the bulk component as valleys.

3.6 Summary

Complementing the STM/S and PES, the ZPS resolves directly the local bond relaxation and the associated CLS and valence charge evolution without involvement of trial-error optimization in analyzing the coordination effect. Extracted information includes the local binding energy density, atomic cohesive energy, charge quantum entrapment and polarization, energy level of an isolated atom. Reformulation of the APECS derive the energy shift of two energy levels simultaneously with provision of the coefficients of charge sharing and potential field screening. It is particularly rectified that the shift of Auger parameter equals twice of the shift of the upper energy level instead of the sum of both the deeper and the upper levels. The involvement of the atomic CN effect results in the quantitative information on the screening and charge transporting information of the specimen. The XAS pre-edge shift results from the resultant of the dual level shift of the core band and the valence band. The XAS pre-edge shifts by a perturbation always in an opposite direction of the single XPS core level. The combination of the STS, XAS, ZPS, and APECS empowers the currently available spectroscopic techniques to resolve the atomistic, local, dynamic and quantitative information on bond relaxation and the associated energetics of bonding and nonbonding electrons pertaining to irregularly-coordinated atoms. Interplay of these spectrometric techniques would be more revealing than using any of them alone.

References

1. C.Q. Sun, *Relaxation of the Chemical Bond* (Springer Ser. Chem. Phys.), vol 108 (Springer, Heidelberg, 2014), 807pp
2. C.Q. Sun, Surface and nanosolid core-level shift: Impact of atomic coordination-number imperfection. Phys. Rev. B **69**(4), 045105 (2004)
3. E. Abrahams, P.W. Anderson, D.C. Licciardello, T.V. Ramakrishnan, Scaling theory of localization: absence of quantum diffusion in two dimensions. Phys. Rev. Lett. **42**(10), 673–676 (1979)
4. R.A. Street, *Hydrogenated Amorphous Silicon* (Cambridge University Press, 1991)
5. C.Q. Sun, Dominance of broken bonds and nonbonding electrons at the nanoscale. Nanoscale **2**(10), 1930–1961 (2010)
6. F.M. Chua, Y. Kuk, P.J. Silverman, Oxygen chemisorption on Cu(110): an atomic view by scanning tunneling microscopy. Phys. Rev. Lett. **63**(4), 386–389 (1989)
7. C.Q. Sun, Oxidation electronics: bond-band-barrier correlation and its applications. Prog. Mater Sci. **48**(6), 521–685 (2003)
8. W. Jacob, V. Dose, A. Goldmann, Atomic adsorption of oxygen on Cu (111) and Cu (110). Appl. Phys. A **41**(2), 145–150 (1986)
9. D.-L. Feng, Photoemission spectroscopy: deep into the bulk. Nat. Mater. **10**(10), 729–730 (2011)
10. A. Damascelli, Z. Hussain, Z.-X. Shen, Angle-resolved photoemission studies of the cuprate superconductors. Rev. Mod. Phys. **75**(2), 473 (2003)
11. M.A. Omar, *Elementary Solid State Physics: Principles and Applications* (Addison-Wesley, New York, 1993)
12. D. Petrovykh, J. Sullivan, L. Whitman, *Quantification of Discrete Oxide and Sulfur Layers on Sulfur-Passivated InAs by XPS*. DTIC Document (2005)
13. X.J. Liu, M.L. Bo, X. Zhang, L. Li, Y.G. Nie, H. Tian, Y. Sun, S. Xu, Y. Wang, W. Zheng, C.Q. Sun, Coordination-resolved electron spectrometrics. Chem. Rev. **115**(14), 6746–6810 (2015)
14. W. Qin, Y. Wang, Y.L. Huang, Z.F. Zhou, C. Yang, C.Q. Sun, Bond order resolved 3d(5/2) and valence band chemical shifts of Ag surfaces and nanoclusters. J. Phys. Chem. A **116**(30), 7892–7897 (2012)
15. C.Q. Sun, L.K. Pan, H.L. Bai, Z.Q. Li, P. Wu, E.Y. Jiang, Effects of surface passivation and interfacial reaction on the size-dependent 2p-level shift of supported copper nanosolids. Acta Mater. **51**(15), 4631–4636 (2003)
16. G. Moretti, Auger parameter and Wagner plot in the characterization of chemical states by X-ray photoelectron spectroscopy: a review. J. Electron Spectrosc. Relat. Phenom. **95**(2–3), 95–144 (1998)
17. G. Moretti, The Wagner plot and the Auger parameter as tools to separate initial-and-final-state contributions in X-ray photoemission spectroscopy. Surf. Sci. **618**, 3–11 (2013)
18. M. Satta, G. Moretti, Auger parameters and Wagner plots. J. Electron Spectrosc. Relat. Phenom. **178**, 123–127 (2010)
19. C.D. Wagner (eds.), *Practical Surface Analysis by Auger and X-Ray Photoelectron Spectroscopy*, 2nd edn, ed. by D. Briggs, M.P. Seah (Wiley, Chichester, 1990)
20. K. McEleney, C.M. Crudden, J.H. Horton, X-ray photoelectron spectroscopy and the Auger parameter as tools for characterization of Silica-supported Pd catalysts for the Suzuki-Miyaura reaction. J. Phys. Chem. C **113**(5), 1901–1907 (2009)
21. C.Q. Sun, L.K. Pan, T.P. Chen, X.W. Sun, S. Li, C.M. Li, Distinguishing the effect of crystal-field screening from the effect of valence recharging on the $2P_{3/2}$ and $3d_{5/2}$ level energies of nanostructured copper. Appl. Surf. Sci. **252**(6), 2101–2107 (2006)
22. M. Chergui, Emerging photon technologies for chemical dynamics. Faraday Discuss. **171**, 11–40 (2014)
23. M. Nagasaka, H. Yuzawa, N. Kosugi, Interaction between water and alkali metal ions and its temperature dependence revealed by oxygen K-edge X-ray absorption spectroscopy. J. Phys. Chem. B **121**(48), 10957–10964 (2017)

24. M. Nagasaka, H. Yuzawa, N. Kosugi, Development and application of in situ/operando soft X-ray transmission cells to aqueous solutions and catalytic and electrochemical reactions. J. Electron Spectrosc. Relat. Phenom. **200**, 293–310 (2015)

25. T. Balasubramanian, J.N. Andersen, L. Wallden, Surface-bulk core-level splitting in graphite. Phys. Rev. B **64**, 205420 (2001)

26. C.J. Nelin, F. Uhl, V. Staemmler, P.S. Bagus, Y. Fujimori, M. Sterrer, H. Kuhlenbeck, H.-J. Freund, Surface core-level binding energy shifts for MgO (100). Phys. Chem. Chem. Phys. **16**(40), 21953–21956 (2014)

27. D. Wallin, I. Shorubalko, H. Xu, A. Cappy, Nonlinear electrical properties of three-terminal junctions. Appl. Phys. Lett. **89**(9), 092124 (2006)

28. S. Hajati, S. Coultas, C. Blomfield, S. Tougaard, XPS imaging of depth profiles and amount of substance based on Tougaard's algorithm. Surf. Sci. **600**(15), 3015–3021 (2006)

29. M.P. Seah, I.S. Gilmore, S.J. Spencer, Background subtraction -II. General behaviour of REELS and the Tougaard universal cross section in the removal of backgrounds in AES and XPS. Surf. Sci. **461**(1–3), 1–15 (2000)

30. X.B. Zhou, J.L. Erskine, Surface core-level shifts at vicinal tungsten surfaces. Phys. Rev. B **79**(15), 155422 (2009)

Chapter 4
Solid and Liquid Skins

Abstract Decomposition of the XPS profiles into components of sublayers derives information on the local bond length, bond energy, atomic cohesive energy, binding energy density, and the energy levels $E_\nu(0)$ of an isolated atom and its shift with the coordination environment. The $E_\nu(0)$ and $E_\nu(12)$ remain constant and the atomic CN varies only with the layer order and surface registry, regardless of the skin chemical constituent. Atomic undercoordination induced bond contraction drives relaxation and reconstruction of the surface of a crystal.

Highlights

- BOLS-TB-XPS not only quantifies the $E_\nu(0)$ and $E_\nu(12)$ but also the CN-resolved d_z, E_z, E_{den}, and E_{coh}.
- Atomic CN varies only with the sublayer order and the surface registry regardless of chemical composition.
- Atomic undercoordination shortens the skin bonds and densely entraps core electrons, deepens the CL.
- The skin of a substance is generally denser, stiffer, yet chemically and thermally more active than the bulk.

4.1 XPS Derivatives

It has long been controversial regarding the skin CLS that may be in a positive, a negative, or a mixed order. Rules are lacking for such order assignment despite intuitive or calculation derivative using the "final-initial" state relaxation scheme. Nb(100) [1, 2], graphite [3], Tb(0001) [4], Ta(100) [5], Ta(110) [6], Mg(10$\bar{1}$0) [7], and Ga(0001) [8] skins follow the positive CLS order (E_F, B, ..., S_2, S_1) with E_F as the reference point. Be(0001) [9], Be(10$\bar{1}$0) [7, 10, 11], Ru(10$\bar{1}$0) [12], Mo(110) [13], Al(100) [14], W(110) [15], W(320) [16], and Pd(110), (100) and (111) [17] skins follow the negative CLS order (E_F, S_1, S_2, ..., B). However, Si(111) [18],

C. Q. Sun, *Electron and Phonon Spectrometrics*,
https://doi.org/10.1007/978-981-15-3176-7_4

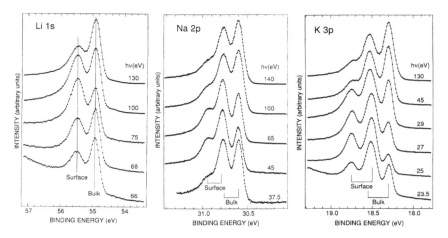

Fig. 4.1 Beam energy dependence of the relative intensity of the surface component to the bulk of Li, Na, and K crystals. Higher incident beam energy collects more bulk information and raises the intensity of the culk component. Reprinted with permission from [25]. Copyright 1992 of American Physical Society

Si(113) [19], Ge(100) [20], Ge(111) [21], Ru(0001) [22], and Be(10$\bar{1}$0) [23] skins are assigned in the mixed order (S_2, S_1, ..., B, E_F).

The XPS peak energies and intensities vary with the crystal orientation or with the atomic density of the skin. On one hand, a high-density skin blocks not only the incident beam from penetrating into deeper layers but also electrons from being ejected from the deeper sublayers and hence weakens the peak intensity of the bulk component; on the other, densely-packed skin atoms shift the CL lesser than the loosely-packed one does because of the CN-resolved bond energy relaxation [24]. Higher skin density means higher atomic CN, which weakens the skin chemical bond, according to the BOLS notion. The skin atomic densities of the fcc(110), (001), and (111) skins are in the order of $1/\sqrt{2}:1:2/\sqrt{3}$. The relative componental intensity for Li, Na, and K surfaces shown in Fig. 4.1 demonstrate that the incident beam energy collects indeed more information from the bulk interior and the intensity of the bulk increases with incident beam energy [25]. This observation provides a direct criterion for assigning the surface and bulk components.

Therefore, the S_1 component of the (111) skin with slightly higher CN value will shift lesser from the $E_v(0)$ than the S_1 of the (110) skin and lesser than the S_1 for the (100) skin that serves as the reference. The S_1 peak intensity for the (111) skin is relatively higher than the same S_1 component of the loosely packed (110) skin, as observed by Anderson et al. [17]. Figure 4.2 shows that two components present at 334.35 (B) and ~334.92 eV (S_1) for the Pd(110), (100) and (111) skins under 390 eV X-ray excitation. The intensity of the S_1 component decreases when the skin turns from (111) to (110) orientation. On the other hand, the (111) skin shifts the SCL indeed lesser than that of the (100) and the (110) does in general. This trend agrees with derivatives of quantum computations and theoretical predictions based on the

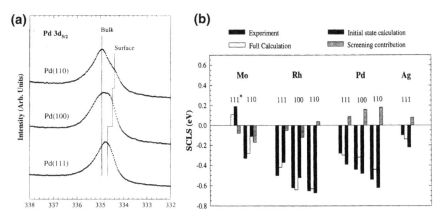

Fig. 4.2 Surface registry-resolved **a** Pd $3d_{5/2}$ spectra from Pd(111), Pd(100) and Pd(110) surfaces measured at a photon energy of 390 eV. **b** Comparison of the measured and calculated $3d$ CLS for the Mo, Rh, Pd and Ag surfaces on the base of the "initial-final" state relaxation scheme. One may note that the resultant peak shifts deeper if moves from (111) to (100) and (110) surface because of their effective atomic CN difference, so the deeper component addresses the surface, instead. Reprinted with permission from [17], American Physical Society 1994

bond counting scheme [26]. This trend holds true if the mean-free-path of the ejected electrons, in the 40 nm or around, is greater than the penetration depth of nanometer scale [27, 28].

The BOLS-TB formulation has fortunately enabled the resolution of the $E_v(0)$, $E_v(12)$, and the z-resolved $E_v(z)$, and the d_z, E_z, E_{den}, E_{coh} from the XPS measurements. Results clarify that all skins follow the positive CLS order without any exception, and that the effective CN for each sublayer depends only on the geometric orientation and the sublayer order, irrespective of the chemical composition of the material [24, 29, 30]. The $E_v(0)$ and $E_v(12)$ derived from the skin XPS analysis provides the reference to the subsequent quantification of bonding identities for defects, nanostructures, and interfaces.

4.2 BOLS-TB Formulation

Considering a sublayer of a solid skin, the z-resolved CLS,

$$\begin{cases} E_v(0) = \left[C_{z'}^m E_v(z') - C_z^m E_v(z) \right] / \left(C_{z'}^m - C_z^m \right) \\ E_v(z) = E_v(0) + \Delta E_v(12) C_z^{-m} \end{cases} \tag{1}$$

If the polarization and entrapment couple pronouncedly, the term pC_z^{-m} will replace the C_z^{-m} in the $E_v(z)$ expression. Polarization will offset the originally entrapped states T back by p fold, which may result in the mixed or the negative shift.

In calculations, one needs to determine the $E_v(0)$ first. If l (> 2) sublayers are involved in a set of XPS spectra collected from skins of a specific substance of different registries, the $E_v(0)$ and the $\Delta E_v(12)$ take the mean value of $N = C(l, 2) = l! / [(l - 2)! 2!]$ possible combinations with the standard deviation σ,

$$
\begin{cases}
E_v(x_l) & = < E_v(0) > \pm \sigma + \Delta E_v(12)(1 + \Delta_{Hl}) \\
< E_v(0) > & = \sum_N E_{vl}(0)/N \\
\sigma & = \sqrt{\sum_{C(l,2)} [E_{vl}(0) - < E_v(0) >]^2 / N(N + 1)}
\end{cases}
\tag{2}
$$

For instance, one may collect a set of XPS spectra from each of the fcc(100), (110), and (111) skins of the same substance. There is a total of $l = 1 + 3 \times 3 = 10$ sublayers (S_1, S_2, S_3 components in each skin and a common component of B). The B component conserves in energy and exists in all spectra. There is a combination of $N = C(10, 2) = 45$ possible $E_v(0)$ solutions for averaging. Higher N number means higher <$E_v(0)$> accuracy and reliability.

One needs to know the bond nature index m before analysis. A fitting to the size-dependent melting point $T_m(K)$ of the same substance with known shape (τ), for instance, gives the m value for this specimen [31].

$$
\frac{T_m(K) - T_m(\infty)}{T_m(\infty)} = \tau K^{-1} \sum_{i \leq 3} C_i \left(z_{ib} C_i^{-m} - 1 \right)
$$

where $T_m(K)$ and K are detectable and $T_m(\infty)$ is the known bulk reference. The relative atomic CN, z_{ib} is the z_i normalized by the bulk standard 12.

With the derived m value, one can optimize the $E_v(0)$ towards least σ deviation ($< 10^{-3}$ in general) using the following relation by tuning all $E_v(z)$ values collectively,

$$
\frac{E_v(z) - E_v(0)}{E_v(z') - E_v(0)} = \frac{C_z^{-m}}{C_{z'}^{-m}}
\tag{3}
$$

The average <$E_v(0)$> with the least σ approaches true situation. Therefore, one can resolve these skin XPS spectra simultaneously with quantification of the $E_v(0)$ and the geometric-orientation and sublayer-order dependent z value and the CLS, $\Delta E_v(2 \leq z \leq 12)$.

4.3 Registry and Sublayer-Order Resolution

4.3.1 Fcc-Structured Al, Ag, Au, Ir, Rh, and Pd

Data sourced from the literature has been corrected by removing background and noises. Including the B component that counts only once, there are a total of $l = 7$ components for these fcc(100), (110), and (111) skins, as each peak contains the S_1 and S_2 components [32, 33]. There will be $C(7, 2) = 21$ different $E_v(0)$ values for averaging. The decomposition refines the effective CN of atoms in each sublayer. Figure 4.3, 4.4, 4.5 and 4.6 show the decomposed spectra for Rh, Pd, Al, Ag, Au, and Ir skins of different orientations. Table 4.1 summarizes the decomposition outcome. Derivatives include the effective CN, bond strain ε_z, bond energy E_z, energy density E_{den}, and atomic cohesive energy E_{coh}, relative to their bulk values. These elemental quantities determine the performance of the skin. For instance, E_{den} determines the mechanical strength and elasticity and the E_{coh} determine the diffusivity and thermal stability such as the critical temperature for phase transition [31].

Figure 4.3 displays that the Pd $3d_{5/2}$ spectra exhibit one symmetric convoluted peak but the Rh $3d_{5/2}$ spectra display two majors. The asymmetry at the deeper edge indicates the possible presence of defects or adatoms that induce quantum entrapment of binding energy. Electronic configurations of Pd($5s^04d^{10}$), Rh($5s^14d^8$), and Ir($6s^25d^7$) may affect their spectral patterns as they are in the same geometry and similar atomic cross-section area for scattering [24]. Atoms at the sharp edges may retain partly their discreted orbital nature in isolation.

Results indicate indeed that the amount of the CLS for the (110) skin is greater than that for the (001) and the (111) skins because of its lower atomic CN of the (110). The CLS of the (111) skin shifts least because of its highest effective CN among them [17]. Bonds in the outermost layer are shortest and strongest than those in the subsequent sublayers, which agrees with what discovered by Matsui et al. [35] from Ni surface. They resolved that the Ni $2p$ levels of the outermost three atomic layers shift positively to deeper BE, with the outermost layer shifting the most.

It is clear now that atomic undercoordination creates the local strain and stress (gradient of bond energy) that reconstructs laterally and relaxes inward vertically the skin, which is intrinsically unavoidable. The skin of the fcc-structured metals consisting of at most three atomic layers or two interatomic spacings performs differently from the bulk. Therefore, the concept of skin is more meaningful and practical than the concept of surface without thickness being involved [31].

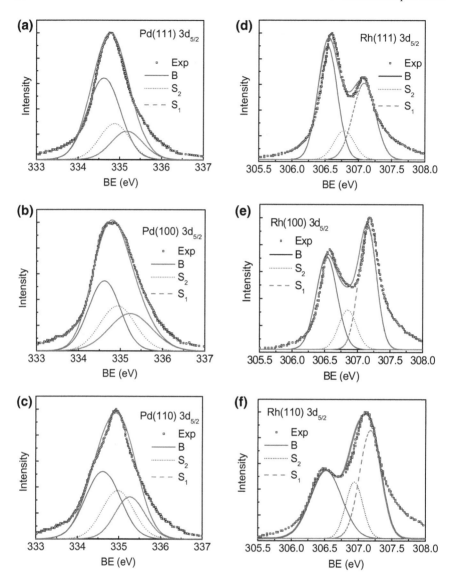

Fig. 4.3 Decomposition of the $3d_{5/2}$ spectra for the fcc(111), (100), and (110) skins of (**a–c**) Pd and (**d–f**) Rh [17, 34] using S_1, S_2, and B components. Table 4.1 features the derived information of strain and bond energy. The asymmetrical tails at the deeper edge arise from the entrapment of vacancy defects or adatoms. Reprinted with permission from [24]. Reproduced by permission of the PCCP Owner Societies

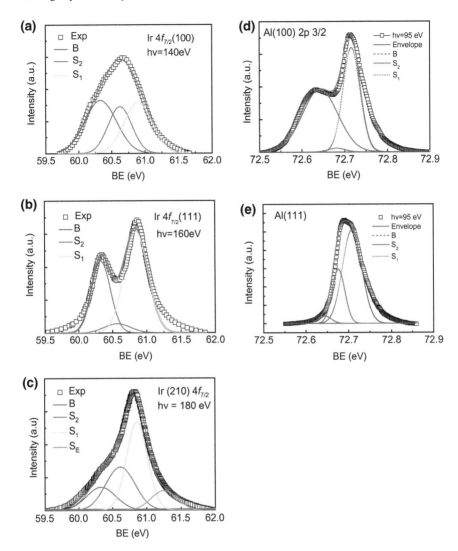

Fig. 4.4 Decomposition of (**a–c**) the XPS $4f_{7/2}$ spectra for Ir(111), Ir(001), and Ir(210) skins [39–41] and (**d–e**) the XPS $2p_{3/2}$ spectra for Al(001) and Al(111) skins [42] with Gaussian components representing the bulk B, S_2 and S_1 sublayers and defect D from higher to lower BE. Reprinted with permission from [43, 44]. Copyright 2012 American Chemical Society. Copyright 2014 Elsevier

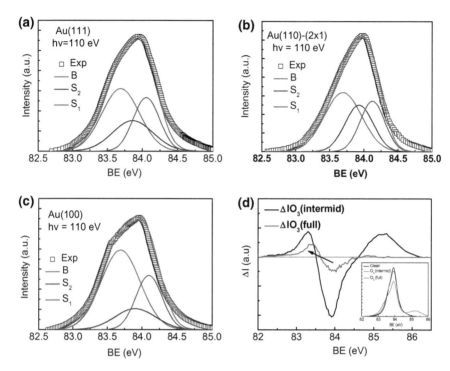

Fig. 4.5 Decomposition of the XPS (**a–c**) $4f_{7/2}$ spectra for Au(111), Au(110), and Au(100) skins [36], **d** $4f_{7/2}$ ZPS of gold foil after O_3 treatment at 100 °C at different stages [37]. Oxidation polarizes the skin (<83.5 eV) and an over dosage oxygen creates H-bond surface network that annihilates partially the skin dipoles [38]. Reprinted with permission from [45]. Copyright 2015 Wiley-VCH

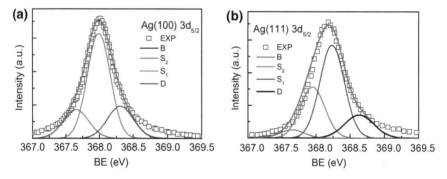

Fig. 4.6 Decomposition of the $3d_{5/2}$ spectra of **a** Ag(001) and **b** Ag(111) skins [17, 46, 47] with the bulk B, S_2, and S_1 sublayers and defect D component. Reprinted with permission from [44]. Copyright 2012 American Chemical Society

Table 4.1 Layer-order and crystal-orientation resolved atomic CN(z), optimal $E_\nu(z)$, atomic $E_\nu(0)$ and its bulk shift $\Delta E_\nu(12)$. Sublayers S_i of the same registry share their common z value regardless of the chemical composition

		z	Pd $3d_{5/2}$	Rh $3d_{5/2}$	Ir $4f_{7/2}$	Al $2p_{3/2}$	Au $4f_{7/2}$	Ag $3d_{5/2}$	$-\varepsilon_z$ (%)	δE_z (%)	δE_{den} (%)	$-\delta E_{coh}$ (%)
m			1 [17]	1 [34]	1 [39–41, 43]	1 [42]	1 [36]	1 [47]				
$E_\nu(0)$		0	330.261	302.163	56.367	72.146	80.726	363.022				
σ			0.003	0.004	0.002	0.003	0.002	0.003				
$E_\nu(12)$	B	12	334.620	306.530	60.332	72.645	83.692	367.650	0	0	0	0
$\Delta E_\nu(12)$	–	–	4.359	4.367	3.965	0.499	2.866	–				
(111)	S_2	6.31	334.88	306.79	60.571	72.675	84.057	367.93	5.63	5.97	26.08	44.28
	S_1	4.26	335.18	307.08	60.84	72.709	83.863	368.24	11.31	12.75	61.60	59.97
	D	3.14	–	–	–	–	–	368.63	17.45	21.15	115.39	68.30
(001)	S_2	5.73	334.94	306.85	60.624	72.682	84.099	367.99	6.83	7.33	32.70	48.75
	S_1	4.00	335.24	307.15	60.898	72.716	83.902	368.31	12.44	14.20	70.09	61.93
(110)	S_2	5.40	334.98	306.89	–	–	84.122	–	7.62	8.25	37.33	51.29
	S_1	3.87	335.28	307.18	–	–	83.929	–	13.05	15.02	74.99	62.91
(210)	S_3	5.83	–	–	60.613	–	–	–	6.60	7.07	31.43	47.98
	S_2	4.16	–	–	60.861	–	–	–	11.72	13.28	64.68	60.73
	S_1	2.97	–	–	61.251	–	–	–	18.78	23.12	129.77	69.53

D denotes defect. With the optimal z and known m value, one can readily derive the bond strain $\varepsilon_z = C_z$, -1, and the atomic cohesive energy $\delta E_{coh} = z_{ib}C_z^{-m} - 1$ in the respective sublayer accordingly. One can obtain binding energy density $\delta E_{den} = C_z^{-(m+\tau)} - 1$, relative increases of bond energy $\delta E_z = C_z^{-m} - 1$, these quantities in the subsequent analysis. The energy unit is in eV

Figures 4.4 and 4.5 show the BOLS-TB decomposed $4f_{7/2}$ spectra for the Ir, Al, and Au(001), (111), and (110) skins [36]. Table 4.1 summarizes the optimal $4f_{7/2}(0, z, 12)$ component and the bulk shift $\Delta E_{4f5/2}(12)$ for Au skins under the common z values for the fcc structures. One can readily derive the local strain, bond length, bond energy, E_{coh} and E_{den} according to practice discussed in Sect. 5. Figure 4.5d shows the ZPS $4f_{7/2}$ profiles of gold foil after O_3 oxidation at different stages and at 100 °C [37]. Clearly, oxidation polarizes the skin but over dosage of oxygen creates H-bond network at the surface, which annihilates the skin dipoles and prevents the skin from further oxidation [38]. Figure 4.6 shows the BOLS-TB decomposed $3d$ spectra for the Ag(100) and Ag(111) skins using the same optimized CN set for the fcc geometries.

4.3.2 Bcc-Structured W, Mo and Ta

The apparent atomic CN in the bcc bulk is 8 instead of 12 for the fcc standard. However, one can normalize the CN by applying $z = 12 \times CN_{bcc}/8$. The documented best fit using the B, S_2, and S_1 components provides reference for fine-tuning in the present XPS spectral analysis [13, 48–50]. Figures 4.7 and 4.8 show the decomposed XPS spectra for the bcc (001), (110), and (111) skins of W $4f_{7/2}$, Ta $4f_{7/2}$, and Mo $3d_{5/2}$ with derivatives given in Table 4.2.

4.3.3 Diamond-Structured Si and Ge

Figure 4.9 shows the sublayer-resolved XPS Si $2p_{3/2}$ spectra [57, 58] and Ge $3d_{5/2}$ spectra [21, 59] for their (100) and (111) skins. Table 4.3 lists information derived from the best fit. Fine-tuning results in $z_1 = 5.08$ instead of 4.0 for the (100) skin because the diamond structure is an interlock of two fcc unit cells. The layer-resolved Si $2p$ [57, 58] and Ge $3d$ [21, 59] spectra for the (100) and (111) skins show consistently that atomic undercoordination results in the local quantum entrapment.

4.3.4 The hcp-Structured Be, Re, and Ru

Figure 4.11a–c shows the decomposed 1s spectra for the Be(10$\bar{1}$0), (0001), and (11$\bar{2}$0) skins [9, 11, 63, 64]. The XPS spectrum of Be(0001) surface contains four components. An S_1 addition to the deeper end of the Be(10$\bar{1}$0) spectrum represents the undercoordinated Be atom in the Be(10$\bar{1}$0) kink edge. Including the B component that was counted only once, there are a total of n = 12 components for the Be skins and 55 possible $E_v(0)$ values for averaging.

According to the decomposition criteria, the common B component must exist and keep constant in all skins of the same substance regardless of geometrical orientation.

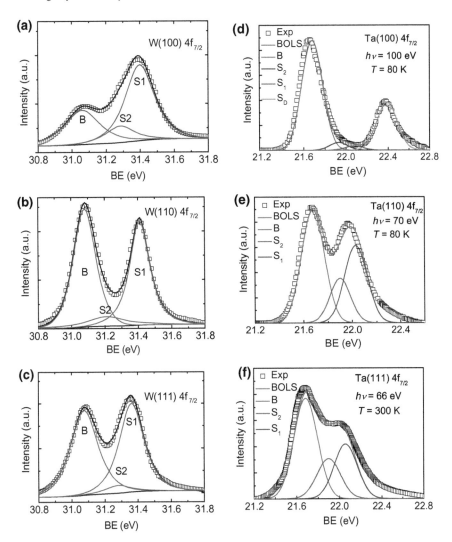

Fig. 4.7 Decomposition of the $4f_{7/2}$ spectra of (**a–c**) W(100), W(110), and W(111) [48, 49] and (**d–f**) Ta (001) [55], Ta(110) [6], and Ta(111) [52] using B, S_2 and S_1 components. The baselines correspond to the spectral background correction. Deconvolution also confirms the positive SCLS and leads to the quantitative information, as featured in Table 4.2. Reprinted with permission from [6, 52]. Copyright 1982 and 1984 American Physical Society. Reprinted with permission from [54, 56]. Copyright 2012 Elsevier. Reproduced by permission from the PCCP Owner Societies

However, the B component is invisible to the Be($11\bar{2}0$) spectrum. This invisibility indicates that the higher atomic density of the Be($11\bar{2}0$) skin prevents the incident beam from penetrating into the bulk or prevents electrons from ejecting off the bulk. Deeper sublayer electrons are not energetic enough to escape from the crystal. This observation indicates that all the Be skins contain four atomic layers with possible

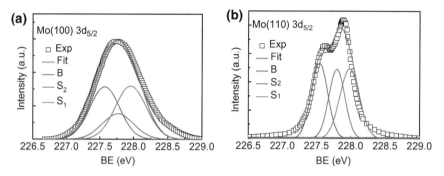

Fig. 4.8 Decomposition of the $3d_{5/2}$ spectra of **a** Mo(100) and **b** (110) [13, 50] skins using B, S_2 and S_1 components with derivatives shown in Table 4.2

Table 4.2 Sublayer-resolved atomic CN(z), $E_v(z)$, $E_v(0)$, bulk shift $\Delta E_v(12)$, strain ε_r, relative bond energy δE_z, relative cohesive energy δE_{coh}, and relative energy density δE_{den}, for the bcc-structured W [48, 49, 51], Mo [13, 50] and Ta [6, 52]

	i	z	W $4f_{7/2}$	Mo $3d_{5/2}$	Ta $4f_{7/2}$	$-\varepsilon_z$ (%)	δE_z	$-\delta E_{coh}$	δE_{den}
m			1 [53]	1 [54]	1 [6, 52]				
$E_v(0)$	–	0	28.889	224.868	19.368				
$E_v(12)$	B	12	31.083	227.567	21.650	0	0	0	0
$\Delta E_v(12)$		–	2.194	2.699	2.282				
σ			0.002	0.002	0.002				
(100)	S_2	5.16	31.283	227.813	21.855	8.27	9.01	53.13	41.21
	S_1	3.98	31.398	227.957	21.977	12.53	14.32	62.08	70.81
(110)	S_2	5.83	31.240	227.761	21.811	6.60	7.07	47.98	31.43
	S_1	3.95	31.402	227.962	21.981	12.67	14.51	62.31	71.92
(111)	S_2	5.27	31.275	–	21.847	7.96	8.65	52.28	39.37
	S_1	4.19	31.370	–	21.949	11.60	13.12	60.50	63.73

addition of defects or adatoms, rather than two in the fcc and bcc structures, because of the smaller Be atomic cross-section and the high packing density. The lacking of bulk information in the Be($11\bar{2}0$) spectrum also indicates the penetration depth of the 135 eV X-ray beam limiting only to the outermost three atomic layers – less than 1 nm thick.

Figure 4.10 shows the decomposition of the $3d_{5/2}$ spectra for the Ru(0001) and Ru($10\bar{1}0$) [22, 65] skins with derived information featured in Table 4.4. Figure 4.11d–f show the decomposed 4f spectra for the Re(0001) and Re($12\bar{3}1$) skins. With respect to the data source, these skins are decomposed into 3 and 4 components, respectively, because of the Re($12\bar{3}1$) kinks. There is a total of n = 8 components and 28 possible $E_v(0)$ values for average. The optimized z value of 2.836 for the Re($12\bar{3}1$) outermost

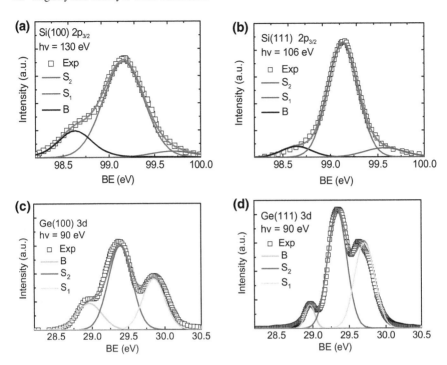

Fig. 4.9 Sublayer-resolved XPS spectra for the (100) and (111) skins of (**a, b**) Si $2p_{3/2}$ [57, 58] and (**c, d**) Ge $3d_{5/2}$ [21, 59]. Three Gaussian components represent the B, S_2, and S_1 states. Table 4.2 summarizes the respectively optimized z, $E_v(z)$, $E_v(0)$, and $\Delta E_v(12)$. Reprinted with permission from [62]. Copyright 2014 American Institute of Physics

Table 4.3 Sublayer-resolved atomic CN(z), $E_v(z)$, $E_v(0)$, bulk shift $\Delta E_v(12)$, strain ε_r, relative bond energy δE_z, cohesive energy δE_C, and energy density δE_D, for the diamond-structured Si [57, 58] and Ge [21, 59] skins

	i	z	Si $2p_{3/2}$	Ge $3d_{5/2}$	$-\varepsilon_z$ (%)	δE_z	$-\delta E_{coh}$	δE_{den}
m			4.88 [60]	5.47 [61]				
$E_v(0)$	–	0	96.089	27.579				
$E_v(12)$	B	12	98.550	28.960	0	0	0	0
$\Delta E_v(12)$		–	2.461	1.381				
σ			0.003	0.002				
(100)	S_2	6.76	99.224	29.391	4.84	31.18	26.10	52.24
	S_1	5.08	99.884	29.823	8.49	62.50	31.21	112.08
(111)	S_2	7.08	99.143	29.339	4.34	27.47	24.79	45.62
	S_1	5.39	99.719	29.713	7.65	54.54	30.58	96.22

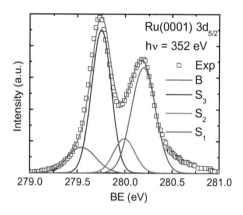

Fig. 4.10 Decomposition of the $3d_{5/2}$ spectra for the Ru (0001) and Ru($10\bar{1}0$) skins [22, 65] with derived information featured in Table 4.4. Reprinted with permission from [30]. Copyright 2009 American Chemical Society

Table 4.4 Sublayer-resolved atomic CN(z), $E_v(z)$, $E_v(0)$, bulk shift $\Delta E_v(12)$, strain ε_r, relative bond energy δE_z, cohesive energy δE_C, and energy density δE_D, for the hcp-structured Be, Re, and Ru skins

		z	Re 4f 5/2	Ru $3d_{5/2}$	Be 1s	$-\varepsilon_z$ (%)	δE_z	$-\delta E_C$	δE_D
m			1 [29]	1 [30]	1 [68]				
$E_v(0)$/eV		0	40.015	275.883	106.416				
σ			0.003	0.003	0.003				
$E_v(12)$/eV	B	12	42.645	279.544	111.110				
$\Delta E_v(12)$/eV			2.629	4.661	3.694				
(0001)	S_3	6.50	42.794	279.749	111.370	5.28	5.58	42.81	24.25
	S_2	4.39	42.965	279.992	111.680	10.79	12.10	58.99	57.90
	S_1	3.50	43.110	280.193	111.945	15.06	17.73	65.66	92.14
($10\bar{1}0$)	S_4	6.97	–	279.719	111.330	4.51	4.72	39.18	20.26
	S_3	4.80	–	279.921	111.590	9.35	10.31	55.88	48.08
	S_2	3.82	–	280.105	111.830	13.30	15.35	63.28	77.01
	S_1	3.11	–	280.329	112.122	17.68	21.47	68.52	117.74
($11\bar{2}0$)	S_4	6.22	–	–	111.400	5.80	6.16	44.97	27.01
	S_3	4.53	–	–	111.650	10.27	11.45	57.93	54.26
	S_2	3.71	–	–	111.870	13.88	16.11	64.10	81.76
	S_1	2.98	–	–	112.190	18.70	22.99	69.46	128.84
($12\bar{3}1$)	S_4	6.78	42.779	–	–	4.81	5.05	40.65	21.79
	S_3	4.88	42.910	–	–	9.09	10.00	55.27	46.43
	S_2	3.55	42.100	–	–	14.77	17.33	65.29	89.49
	S_1	2.84	43.305	–	–	19.89	24.83	70.46	142.80

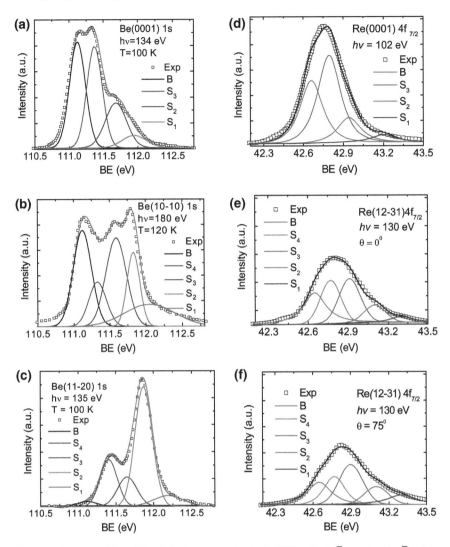

Fig. 4.11 Decomposition of (**a–c**) the 1 *s* spectra for the Be(0001), Be(10$\bar{1}$0), and Be(11$\bar{2}$0) skins [9, 11, 63, 64] and **d** the $4f_{5/2}$ spectra for the Re(0001) and Re(12$\bar{3}$1) skins [66, 67] collected at **e** 0° and **f** 75° polar angles. The relative change of the S_3/B intensity with emission angle evidences the positive SCLS. Table 4.4 features the derived information. Reprinted with permission from [29, 68]. Copyright 2011 American Chemical Society. Reproduced by permission of the PCCP Owner Societies

layer is lower than the value of 3.50 for the Re(0001) top layer. The low-energy tails in Fig. 4.11 correspond to the quantum entrapment of the even lower coordinated defects or edge atoms.

Table 4.5 Bond nature (m) dependence of the relative E_{den} (eV/nm^3) and E_{coh} (eV/atom) of a skin with respect to their bulk values. Subscript 1 and 2 refers to the outermost first and the second skin sublayer. Surface free energy, E_{Ds}/E_{Db}, approximates $1 - E_{CS}/E_{Cb}$ according to traditional definition of the surface energy per unit area required to cutting a bulk into two halves [72]

m	E_D(Bulk) (eV/nm^3)	E_D(Skin) (eV/nm^3)	E_{Ds}/E_{Db}	E_C(Bulk) (eV/atom)	E_{coh} (Skin) (eV/atom)	E_{Cs}/E_{Cb}
1 (Metal, Cu)	155.04	198.60	1.468	4.39	2.00	0.455
2.56 (Diamond)	1307.12	2262.63	1.713	7.37	3.86	0.524
4.88 (Si)	164.94	357.09	2.165	4.63	3.00	0.649

4.4 Local Binding Energy Density and Atomic Cohesive Energy

As the standard reference, the flat fcc(100) skin having $z_1 = 4.00$, $z_2 = 5.73$, and $z_{i \geq 3} = 12$, the bond contracts from the bulk value to $C_1 = 0.88$, $C_2 = 0.92$, and $C_{i \geq 3} = 1$, accordingly. For metals such as Au, Ag, and Cu, $m = 1$; for carbon, $m = 2.56$ [69]; for Si, $m = 4.88$ [70]. For other alloys and compounds, the m value may vary. With the given m values and the known bond energy for Cu (4.39 eV/atom), Diamond (7.37 eV/atom) and Si (4.63 eV/atom) [71], one can easily calculate the skin geometrical-orientation and sublayer-order resolved energy density E_{den} (in eV/nm^3 unit) and atomic cohesive energy E_{coh} (in eV/atom unit), as shown in Table 4.5.

Results indicate that the skin E_{DS} is always higher and the skin E_{CS} is lower than their respective bulk values and they are in different units. Figure 4.12 shows the consistency between the BOLS predicted (solid curves) and the XPS derived z-dependent bond strain, $\Delta E_v(12)$, E_{coh}, and E_{den} of solid skins. These derivatives empower the XPS in revealing such local quantitative information that is critical to devising materials.

4.5 Summary

The BOLS-TB-XPS strategy has enabled unification of the crystal-orientation and sublayer-order dependency of the CLS for skins of the fcc, bcc, hcp, and diamond structured solids with derivative of quantitative information about the skin bond length, bond energy, BE density, atomic cohesive energy. This strategy derives the energy levels of an isolated atom and their shift due to bulk formation and atomic undercoordination. The orientation- and layer-order resolved effective CN conserves for the same crystal geometry regardless of the chemical composition.

The SCLS is always positive without any exception. Negative and mixed shifts may be possible due to the splitting and screening of the crystal potentials by skin

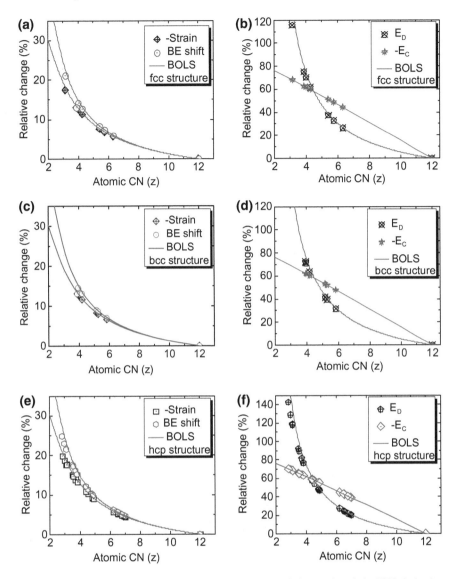

Fig. 4.12 Consistency between the BOLS-TB predicted (solid curves) and the XPS derived z-dependent bond strain, $\Delta E_v(12)$, atomic cohesive energy E_{coh}, and energy density E_{den} of (**a, b**) fcc-, (**c, d**) bcc-, and (**e, f**) hcp-structured solid skins. Reprinted with permission from [73] Copyright 2015 American Chemical Society

polarization but the weak polarization is not detectable for the flat surface. Shorter and stronger bond between undercoordinated atoms not only raises the local energy density but also deepens the local potentials, which perturbs the Hamiltonian that determines uniquely the CLS. However, as artifacts of measurements, the "initial-final states" relaxation and the charging effect present throughout the course of measurements. One can correct these artifacts through background correction.

References

1. B.S. Fang, W.S. Lo, T.S. Chien, T.C. Leung, C.Y. Lue, C.T. Chan, K.M. Ho, Surface band structures on Nb(001). Phys. Rev. B **50**(15), 11093–11101 (1994)
2. M. Alden, H.L. Skriver, B. Johansson, Ab-initio surface core-level shifts and surface segregation energies. Phys. Rev. Lett. **71**(15), 2449–2452 (1993)
3. T. Balasubramanian, J.N. Andersen, L. Wallden, Surface-bulk core-level splitting in graphite. Phys. Rev. B **64**, 205420 (2001)
4. E. Navas, K. Starke, C. Laubschat, E. Weschke, G. Kaindl, Surface core-level shift of 4f states for Tb(0001). Phys. Rev. B **48**(19), 14753 (1993)
5. R.A. Bartynski, D. Heskett, K. Garrison, G. Watson, D.M. Zehner, W.N. Mei, S.Y. Tong, X. Pan, The 1st interlayer spacing of Ta(100) determined by photoelectron diffraction. J. Vac. Sci. Technol., A **7**(3), 1931–1936 (1989)
6. D.M. Riffe, G.K. Wertheim, Ta(110) surface and subsurface core-level shifts and 4f7/2 line shapes. Phys. Rev. B **47**(11), 6672–6679 (1993)
7. J.H. Cho, K.S. Kim, S.H. Lee, M.H. Kang, Z.Y. Zhang, Origin of contrasting surface core-level shifts at the Be(10-10) and Mg(10-10) surfaces. Phys. Rev. B **61**(15), 9975–9978 (2000)
8. A.V. Fedorov, E. Arenholz, K. Starke, E. Navas, L. Baumgarten, C. Laubschat, G. Kaindl, Surface shifts of 4f electron-addition and electron-removal states on Gd(0001). Phys. Rev. Lett. **73**(4), 601–604 (1994)
9. L.I. Johansson, H.I.P. Johansson, J.N. Andersen, E. Lundgren, R. Nyholm, Surface-shifted core levels on Be(0001). Phys. Rev. Lett. **71**(15), 2453–2456 (1993)
10. S. Lizzit, K. Pohl, A. Baraldi, G. Comelli, V. Fritzsche, E.W. Plummer, R. Stumpf, P. Hofmann, Physics of the Be(10-10) surface core level spectrum. Phys. Rev. Lett. **81**(15), 3271–3274 (1998)
11. H.I.P. Johansson, L.I. Johansson, E. Lundgren, J.N. Andersen, R. Nyholm, Core-level shifts on Be(10-10). Phys. Rev. B **49**(24), 17460–17463 (1994)
12. A. Baraldi, S. Lizzit, G. Comelli, A. Goldoni, P. Hofmann, G. Paolucci, Core-level subsurface shifted component in a 4d transition metal: Ru(10(1)over-bar-0). Phys. Rev. B **61**(7), 4534–4537 (2000)
13. E. Lundgren, U. Johansson, R. Nyholm, J.N. Andersen, Surface core-level shift of the Mo(110) surface. Phys. Rev. B **48**(8), 5525–5529 (1993)
14. R. Nyholm, J.N. Andersen, J.F. Vanacker, M. Qvarford, Surface core-level shifts of the Al(100) and Al(111) surfaces. Phys. Rev. B **44**(19), 10987–10990 (1991)
15. D.M. Riffe, B. Kim, J.L. Erskine, Surface core-level shifts and atomic coordination at a stepped W(110) surface. Phys. Rev. B **50**(19), 14481–14488 (1994)
16. J.H. Cho, D.H. Oh, L. Kleinman, Core-level shifts of low coordination atoms at the W(320) stepped surface. Phys. Rev. B **64**(11), 115404 (2001)
17. J.N. Andersen, D. Hennig, E. Lundgren, M. Methfessel, R. Nyholm, M. Scheffler, Surface core-level shifts of some 4d-metal single-crystal surfaces: Experiments and ab initio calculations. Phys. Rev. B **50**(23), 17525–17533 (1994)
18. G.A. Slack, S.F. Bartram, Thermal expansion of some diamondlike crystals. J. Appl. Phys. **46**(1), 89–98 (1975)

19. T.R. Hart, R.L. Aggarwal, B. Lax, Temperature dependence of Raman scattering in silicon. Phys. Rev. B **1**(2), 638–642 (1970)
20. M.S. Liu, L.A. Bursill, S. Prawer, R. Beserman, Temperature dependence of the first-order Raman phonon lime of diamond. Phys. Rev. B **61**(5), 3391–3395 (2000)
21. M. Kuzmin, M.J.P. Punkkinen, P. Laukkanen, J.J.K. Lång, J. Dahl, V. Tuominen, M. Tuominen, R.E. Perälä, T. Balasubramanian, J. Adell, B. Johansson, L. Vitos, K. Kokko, I.J. Väyrynen, Surface core-level shifts on Ge(111)c(2×8): experiment and theory. Phys. Rev. B **83**(24), 245319 (2011)
22. S. Lizzit, A. Baraldi, A. Groso, K. Reuter, M.V. Ganduglia-Pirovano, C. Stampl, M. Scheffler, M. Stichler, C. Keller, W. Wurth, D. Menzel, Surface core-level shifts of clean and oxygen-covered Ru(0001). Phys. Rev. B **63**(20), 205419 (2001)
23. J.B. Cui, K. Amtmann, J. Ristein, L. Ley, Noncontact temperature measurements of diamond by Raman scattering spectroscopy. J. Appl. Phys. **83**(12), 7929–7933 (1998)
24. Y. Wang, Y.G. Nie, J.S. Pan, L.K. Pan, Z. Sun, L.L. Wang, C.Q. Sun, Orientation-resolved 3d(5/2) binding energy shift of Rh and Pd surfaces: anisotropy of the skin-depth lattice strain and quantum trapping. Phys. Chem. Chem. Phys. **12**(9), 2177–2182 (2010)
25. G. Wertheim, D.M. Riffe, N. Smith, P. Citrin, Electron mean free paths in the alkali metals. Phys. Rev. B **46**(4), 1955 (1992)
26. L.K. Pan, Y.K. Ee, C.Q. Sun, G.Q. Yu, Q.Y. Zhang, B.K. Tay, Band-gap expansion, core-level shift, and dielectric suppression of porous silicon passivated by plasma fluorination. J. Vac. Sci. Technol., B **22**(2), 583–587 (2004)
27. C.J. Powell, A. Jablonski, Progress in quantitative surface analysis by X-ray photoelectron spectroscopy: current status and perspectives. J. Electron Spectrosc. Relat. Phenom. **178–179**, 331–346 (2010)
28. J. Chawla, X. Zhang, D. Gall, Effective electron mean free path in TiN (001). J. Appl. Phys. **113**(6), 063704 (2013)
29. Y.G. Nie, J.S. Pan, W.T. Zheng, J. Zhou, C.Q. Sun, Atomic scale purification of Re surface kink states with and without oxygen chemisorption. J. Phys. Chem. C **115**(15), 7450–7455 (2011)
30. Y. Wang, Y.G. Nie, L.L. Wang, C.Q. Sun, Atomic-layer- and crystal-orientation-resolved 3d$_{5/2}$ binding energy shift of Ru(0001) and Ru(1010) surfaces. J. Phys. Chem. C **114**(2), 1226–1230 (2010)
31. C.Q. Sun, *Relaxation of the Chemical Bond* (Springer Ser. Chem. Phys.), vol 108 (Springer, Heidelberg, 2014) 807 pp
32. W.J. Huang, R. Sun, J. Tao, L.D. Menard, R.G. Nuzzo, J.M. Zuo, Coordination-dependent surface atomic contraction in nanocrystals revealed by coherent diffraction. Nat. Mater. **7**(4), 308–313 (2008)
33. W.H. Qi, B.Y. Huang, M.P. Wang, Bond-length and -energy variation of small Gold nanoparticles. J. Comput. Theor. Nanosci. **6**(3), 635–639 (2009)
34. A. Baraldi, L. Bianchettin, E. Vesselli, S. de Gironcoli, S. Lizzit, L. Petaccia, G. Zampieri, G. Comelli, R. Rosei, Highly under-coordinated atoms at Rh surfaces: interplay of strain and coordination effects on core level shift. New J. Phys. **9**, 143 (2007)
35. F. Matsui, T. Matsushita, Y. Kato, M. Hashimoto, K. Inaji, F.Z. Guo, H. Daimon, Atomic-layer resolved magnetic and electronic structure analysis of Ni thin film on a Cu(001) surface by diffraction spectroscopy. Phys. Rev. Lett. **100**(20), 207201 (2008)
36. P. Heimann, J.F. van der Veen, D.E. Eastman, Structure-dependent surface core level shifts for the Au(111), (100), and (110) surfaces. Solid State Commun. **38**(7), 595–598 (1981)
37. A.Y. Klyushin, T.C.R. Rocha, M. Havecker, A. Knop-Gericke, R. Schlogl, A near ambient pressure XPS study of Au oxidation. Phys. Chem. Chem. Phys. **16**, 7881–7886 (2014)
38. C.Q. Sun, Oxidation electronics: bond-band-barrier correlation and its applications. Prog. Mater Sci. **48**(6), 521–685 (2003)
39. M. Bianchi, D. Cassese, A. Cavallin, R. Comin, F. Orlando, L. Postregna, E. Golfetto, S. Lizzit, A. Baraldi, Surface core level shifts of clean and oxygen covered Ir(111). New J. Phys. **11**, 063002 (2009)

40. N. Barrett, C. Guillot, B. Villette, G. Treglia, B. Legrand, Inversion of the core level shift between surface and subsurface atoms of the iridium $(100)(1 \times 1)$ and $(100)(5 \times 1)$ surfaces. Surf. Sci. **251**, 717–721 (1991)
41. M.J. Gladys, I. Ermanoski, G. Jackson, J.S. Quinton, J.E. Rowe, T.E. Madey, A high resolution photoemission study of surface core-level shifts in clean and oxygen-covered Ir(2 1 0) surfaces. J. Electron Spectrosc. Relat. Phenom. **135**(2–3), 105–112 (2004)
42. M. Borg, M. Birgersson, M. Smedh, A. Mikkelsen, D.L. Adams, R. Nyholm, C.-O. Almbladh, J.N. Andersen, Experimental and theoretical surface core-level shifts of aluminum (100) and (111). Phys. Rev. B **69**(23), 235418 (2004)
43. M. Bo, Y. Wang, Y. Huang, X. Yang, Y. Yang, C. Li, C.Q. Sun, Coordination-resolved local bond relaxation, electron binding-energy shift, and Debye temperature of Ir solid skins. Appl. Surf. Sci. **320**, 509–513 (2014)
44. W. Qin, Y. Wang, Y.L. Huang, Z.F. Zhou, C. Yang, C.Q. Sun, Bond order resolved 3d(5/2) and valence band chemical shifts of Ag surfaces and nanoclusters. J. Phys. Chem. A **116**(30), 7892–7897 (2012)
45. W. Yu, M. Bo, Y. Huang, Y. Wang, C. Li, C.Q. Sun, Coordination-resolved bond and electron spectrometrics of Au atomic clusters, solid skins, and oxidized foils. Chem. Phys. Chem. **16**, 2159–2164 (2015)
46. M. Rocca, L. Savio, L. Vattuone, U. Burghaus, V. Palomba, N. Novelli, F.B. de Mongeot, U. Valbusa, R. Gunnella, G. Comelli, A. Baraldi, S. Lizzit, G. Paolucci, Phase transition of dissociatively adsorbed oxygen on Ag(001). Phys. Rev. B **61**(1), 213–227 (2000)
47. Y. Zhu, Q. Qin, F. Xu, F. Fan, Y. Ding, T. Zhang, B.J. Wiley, Z.L. Wang, Size effects on elasticity, yielding, and fracture of silver nanowires: In situ experiments. Phys. Rev. B **85**(4), 045443 (2012)
48. J. Jupille, K.G. Purcell, D.A. King, W(100) clean surface phase transition studied by core-level-shift spectroscopy: order-order or order-disorder transition. Phys. Rev. B **39**(10), 6871–6879 (1989)
49. K.G. Purcell, J. Jupille, G.P. Derby, D.A. King, Identification of underlayer components in the surface core-level spectra of W(111). Phys. Rev. B **36**(2), 1288–1291 (1987)
50. E. Minni, F. Werfel, Oxygen interaction with Mo (100) studied by XPS, AES and EELS. Surf. Interface Anal. **12**(7), 385–390 (1988)
51. X.B. Zhou, J.L. Erskine, Surface core-level shifts at vicinal tungsten surfaces. Phys. Rev. B **79**(15), 155422 (2009)
52. J. Van der Veen, F. Himpsel, D. Eastman, Chemisorption-induced 4 f-core-electron binding-energy shifts for surface atoms of W (111), W (100), and Ta (111). Phys. Rev. B **25**(12), 7388 (1982)
53. Y.G. Nie, X. Zhang, S.Z. Ma, Y. Wang, J.S. Pan, C.Q. Sun, XPS revelation of tungsten edges as a potential donor-type catalyst. Phys. Chem. Chem. Phys. **13**(27), 12640–12645 (2011)
54. W. Zhou, M. Bo, Y. Wang, Y. Huang, C. Li, C.Q. Sun, Local bond-electron-energy relaxation of Mo atomic clusters and solid skins RSC. Advances **5**, 29663–29668 (2015)
55. D.M. Riffe, W. Hale, B. Kim, J. Erskine, Conduction-electron screening in the bulk and at low-index surfaces of Ta metal. Phys. Rev. B **51**(16), 11012 (1995)
56. Y. Wang, X. Zhang, Y.G. Nie, C.Q. Sun, Under-coordinated atoms induced local strain, quantum trap depression and valence charge polarization at W stepped surfaces. Phys. B-Condens. Matter **407**(1), 49–53 (2012)
57. F. Himpsel, P. Heimann, T. Chiang, D. Eastman, Geometry-dependent Si(2p) surface core-level excitations for Si(111) and Si(100) surfaces. Phys. Rev. Lett. **45**(13), 1112–1115 (1980)
58. F. Himpsel, G. Hollinger, R. Pollak, Determination of the Fermi-level pinning position at Si(111) surfaces. Phys. Rev. B **28**(12), 7014–7018 (1983)
59. R. Niikura, K. Nakatsuji, F. Komori, Local atomic and electronic structure of Au-adsorbed Ge (001) surfaces: scanning tunneling microscopy and x-ray photoemission spectroscopy. Phys. Rev. B **83**(3), 035311 (2011)
60. L. Pan, S. Xu, X. Liu, W. Qin, Z. Sun, W. Zheng, C.Q. Sun, Skin dominance of the dielectric electronic-phononic-photonic attribute of nanoscaled silicon. Surf. Sci. Rep. **68**(3–4), 418–445 (2013)

61. L. Wu, M. Bo, Y. Guo, Y. Wang, C. Li, Y. Huang, C.Q. Sun, Skin bond electron relaxation dynamics of germanium manipulated by interactions with H_2, O_2, H_2O, H_2O_2, HF, and Au. ChemPhysChem **17**(2), 310–316 (2016)

62. M. Bo, Y. Wang, Y. Huang, X. Zhang, T. Zhang, C. Li, C.Q. Sun, Coordination-resolved local bond contraction and electron binding-energy entrapment of Si atomic clusters and solid skins. J. Appl. Phys. **115**(14), 4871399 (2014)

63. L.I. Johansson, P.A. Glans, T. Balasubramanian, Fourth-layer surface core-level shift on Be(0001). Phys. Rev. B **58**(7), 3621–3624 (1998)

64. L.I. Johansson, H.I.P. Johansson, Unusual behavior of surface shifted core levels on Be (0001) and Be(10-10). Nucl. Instrum. Methods Phys. Res. B **97**, 430–435 (1995)

65. A. Baraldi, S. Lizzit, G. Comelli, G. Paolucci, Oxygen adsorption and ordering on Ru(10(1)over-bar-0). Phys. Rev. B **63**(11), 115410 (2001)

66. N. Martensson, H.B. Saalfeld, H. Kuhlenbeck, M. Neumann, Structural dependence of the 5d-metal surface energies as deduced from surface core-level shift measurements. Phys. Rev. B **39**(12), 8181–8186 (1989)

67. A.S.Y. Chan, G.K. Wertheim, H. Wang, M.D. Ulrich, J.E. Rowe, T.E. Madey, Surface atom core-level shifts of clean and oxygen-covered Re(1231). Phys. Rev. B **72**(3), 035442 (2005)

68. Y. Wang, Y.G. Nie, J.S. Pan, L. Pan, Z. Sun, C.Q. Sun, Layer and orientation resolved bond relaxation and quantum entrapment of charge and energy at Be surfaces. Phys. Chem. Chem. Phys. **12**(39), 12753–12759 (2010)

69. W.T. Zheng, C.Q. Sun, Underneath the fascinations of carbon nanotubes and graphene nanoribbons. Energy Environ. Sci. **4**(3), 627–655 (2011)

70. C.Q. Sun, Size dependence of nanostructures: impact of bond order deficiency. Prog. Solid State Chem. **35**(1), 1–159 (2007)

71. C. Kittel, *Intruduction to Solid State Physics*, 8th edn. (Willey, New York, 2005)

72. M. Zhao, W.T. Zheng, J.C. Li, Z. Wen, M.X. Gu, C.Q. Sun, Atomistic origin, temperature dependence, and responsibilities of surface energetics: an extended broken-bond rule. Phys. Rev. B **75**(8), 085427 (2007)

73. X.J. Liu, M.L. Bo, X. Zhang, L. Li, Y.G. Nie, H. Tian, Y. Sun, S. Xu, Y. Wang, W. Zheng, C.Q. Sun, Coordination-resolved electron spectrometrics. Chem. Rev. **115**(14), 6746–6810 (2015)

Chapter 5
Adatoms, Defects, and Kink Edges

Abstract Atoms with even fewer neighbors perform both atomic like and bulk like associated with shorter and stronger interatomic bonds. The bond contraction raises the local charge and energy density and the bond strength gain deepens the local potential well and entraps the core electrons. The locally and densely entrapped core electrons in turn polarize the valence electrons. The subjective valence electron polarization occurs to those atoms with unpaired lone electrons in the s orbitals such as Rh, Au, Ag, Cu and the unpaired $4f^{14}5d^46s^2$ ($5d^56s^1$ seems to be stable) electrons of the W adatoms and Mo($4d^55s^1$) as well. However, the Co($3d^74s^2$) with fully-occupied s electrons and the Re($5d^56s^2$) with semi-occupied d electrons exhibit entrapment dominance. The undercoordination resolved valence electron entrapment or polarization laid foundations for the extraordinary catalytic ability of the excessively undercoordinated atoms and the dispersed single atom.

Highlights

- Atoms with even fewer neighbors than that at the flat skin enhance the BOLS-NEP effect.
- Bond between a metallic adatom and its substrate is 18% shorter and 21% stronger.
- Shared by Re and Co edge atoms, quantum entrapment makes Pt adatom an acceptor-type catalyst.
- Shared by Au, Ag, Cu, Mo, and W adatoms, polarization entitles Rh adatom a donor-type catalyst.

5.1 Observations

5.1.1 XPS Detection

Interaction between undercoordinated atoms distinguishes nanocrystals from their bulk parents in terms of size dependency of known properties and size derivacy of new properties. Size dependence means that the quantities such as the elastic modulus and the melting point remain no longer constant but change with the shape and size of the

C. Q. Sun, *Electron and Phonon Spectrometrics*,
https://doi.org/10.1007/978-981-15-3176-7_5

crystal. Size derivacy refers properties that the parent bulk never demonstrate. For instance, non-magnetic metals manifest magnetism at the nanoscale [1–3]; phase transition from conductor to insulator occurs at the size of a few nanometers [4]; nanoscale gold meets the demand of local surface plasmonics [5, 6]; the catalytic ability of gold for CO oxidation is greatly enhanced at small sizes [7–9]. Other intriguing properties for applications include RNA-delivery, localized surface plasma resonance for enhancing Raman spectroscopy signals [10, 11], laser applications in medication [12], enhancement of photoluminescence [13], etc. Accompanied with the structure evolution from the fcc bulk to the strained structures such as icosahedral or decahedral [14, 15], substantial bond contraction occurs to the outermost atomic shells [16] that can be described as an elastic sheet of "skin" covering the bulk body [17].

Besides the bond length contraction, potential trap depression, charge and energy density elevation, and electronic configurations occur to nanoparticles. Single-electron tunneling spectroscopy revealed the generation of an energy gap whose width is inversely proportional to the diameter of Au and Pd crystals [4, 18]. STM/S also revealed that the valence DOS of Au monomer and dimer [19], Au-Au chain [19, 20], and Au nanowire [21] moves up in energy, being indication of strong local polarization. Fascinating properties demonstrated by such even undercoordinated atoms are associated with Coulomb blockade [4], size-dependent reaction dynamics [18] and standing wave formation at edges [22].

Atoms with even fewer CNs at sites like terrace edges, point defects, or adatoms modify the CLS significantly and irregularly. Figure 5.1 shows typical XPS spectra for Rh and Pt adatoms. Rh adatoms modify the spectral at energies between 306.6 and 307.1 eV with features belong to neither positive nor negative CLS order [23]. Pt adatoms enhance, however, the intensity of the 71.0 eV peak other than the 70.5 eV peak with positive shift [24]. Terrace edges of W(110) vicinal surfaces [25], Re surfaces [26, 27], Rh(111) vicinal surfaces [28], and Rh(110) missing-row type reconstructed surfaces [29] also modify the XPS profiles irregularly. However, signal from such lower coordinated atoms is weaker compared to that of the skin. Direct spectral decomposition is hardly certain and less reliable.

The thickness dependence of the InSe/Pt in Fig. 5.2 shows the same coordination trend of Pt adatoms. For the Pt $4f_{3/2}$, redshift from 71.6 to 71.1 eV occurs as the thickness increases from 0.05 to 7 ML (left panel). The In $4d$ and Se $3d$ levels also undergo redshift as the thickness increases from 0.03 ML to 2.2 ML. CN reduction induced core-level entrapment holds globally true and it is element independent. However, polarization is quite subjective [31].

5.1.2 STM Observation

Interaction of metallic adatoms on metal oxide supports forms a key area of research in catalysis. In particular, gold and platinum nanoparticles supported on TiO_2 and

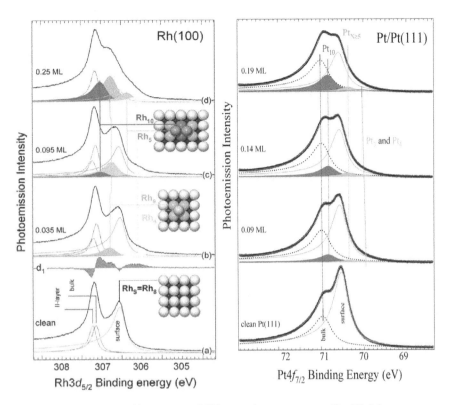

Fig. 5.1 Rh $3d_{5/2}$ and Pt $4f_{7/2}$ spectra of different adatom coverages. The Rh $3d$ spectra were acquired at 380 eV photon energy. Polar emission angles were 35° (spectra a and b) and 20° (spectrum c), with respect to the surface normal. The Pt $4f_{7/2}$ spectra were collected at 125 eV photon energy and at $T = 30$ K and at normal emission. Decomposition assigned the irregular shift caused by the bulk and surfaces. Reprinted with permission from [23, 24]. Copy right of 2008 American Institute of Physics and 2007 Institute of Physics

CeO_2 have received considerable attention as they are an effective low-temperature oxidation catalyst [36]. As a model catalyst, Au adsorption on rutile $TiO_2(110)$ has been extensively studied because this substrate is the most well-characterized metal oxide substrate. When reduced, TiO_2 (110) contains point defects in the form of oxygen vacancies ($O_{b\text{-vacs}}$) as well as Ti interstitial atoms in the bulk.

Figure 5.3 compares the low-temperature STM/TEM images of Au/Pt adatoms on $TiO_2(110)$ surface [32, 34] and of Au on CeO_2 surface [33]. DFT calculations [37, 38] and STM observations show consistently that Au (radius = 1.336–1.439 Å; $5d^{10}6s^1$) adatoms prefer stable sites of atop oxygen vacancies.

At low coverage, individual Au atoms distribute homogenously on the ceria surface and show no preference for binding at step edges. With increasing exposure, characteristic Au aggregation becomes visible, such as upright Au dimers and bilayer and trilayer pyramids. The ultrasmall clusters exhibit pronounced fluxionality; i.e., they easily modify their internal shape and binding position during the scanning

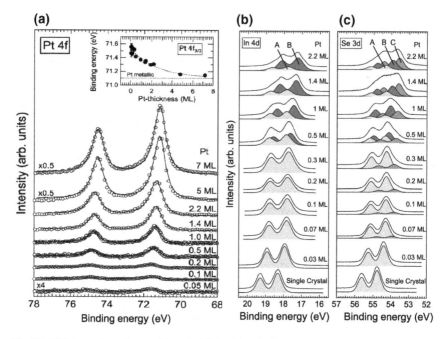

Fig. 5.2 Thickness dependence of the **a** Pt 4*f*, **b** In 4*d*, and **c** Se 3*d* XPS spectra of InSe/Pt interfaces formed by Pt deposition. C1 and C2 indicate spin-resolved core level shift. All the Pt 4*f*, In 4*d* and Se 3*d* core levels undergo redshift when the Pt adlayer thickness increases. Reprinted with copy right permission from [30]

process. This observation suggests the presence of various iso-energetic Au isomers on the surface and points to a relatively weak metal–oxide interaction. At higher Au exposures, tall 3D particles develop on the ceria surface. Conductance spectroscopy on these deposits reveals a set of unoccupied states localized in the energy region of the Au 6*p* levels. Neither the topographic nor the spectroscopic data indicate a charging of Au species on the defect-poor CeO$_2$(111) surface, suggesting that Au mainly binds in the neutral charge state [33]. However, scanning TEM revealed that Pt (radius = 1.290–1.385 Å; 5*d*^{9}6*s*^{1} or 5*d*^{10}6*s*^{0}) adatoms prefer sites of bridging subsurface oxygen vacancies.

Both the real and Laplace filtered TEM images [35] in Fig. 5.3d show clearly the Ni polarization as bright spot at graphene edges. This TEM observation and force field MD and DFT calculations captured the catalytic action of individual Ni atoms at the edges of a growing graphene flake at the millisecond time scale, which unveil the mechanism governing the activity of a single-atom catalyst at work—Ni adatoms at graphene edge lowers substantially the corresponding reaction barriers.

In contrast, XPS measurements revealed that Pt atoms equally incorporate into two trigonal-prismatic intralayer positions existing within the InSe layer, although, at low Pt coverage, Pt atoms prefer one of these sites, where they have a lower interaction with Se atoms. At initial stages of Pt diffusion, isolated Pt atoms act as a surface

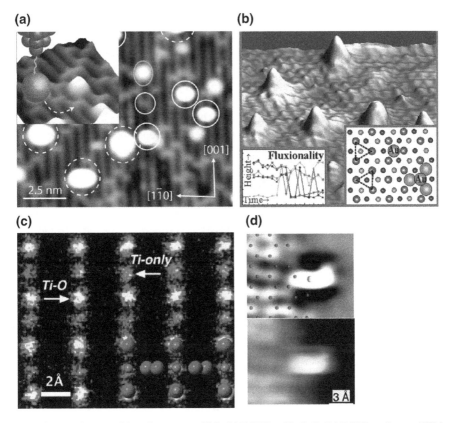

Fig. 5.3 STM images of Au adatoms on **a** TiO$_2$(110) [32] and **b** CeO$_2$(111) [33] surfaces. **c** TEM image of Pt [34] adatoms on TiO$_2$(110) surfaces. Au prefers sites of atop oxygen vacancy with high protrusion but Pt prefers the subsurface positions contrastingly. **d** Laplace filtered and the real TEM image of Ni grown on graphene edge [35]. Reprinted with permission from [32–35]

acceptor which turns the interface into intrinsic. Beyond a certain submonolayer Pt coverage, Pt-InSe reaction gives rise to localized states within the InSe band gap coming from the InSe valence band maximum. The Pt $4f$, In $4d$, and the Se $3d$ core levels show consistent negative shift when the thickness of the Pt adlayer increases [30].

5.2 ZPS of Pt and Rh Adatoms: Catalytic Nature

Application of the BOLS-ZPS strategy shown in this section resolves readily the mechanism of single-atom catalysis and the acceptor- and donor-like catalyst. The ZPS distills directly the $E_v(12)$ as a spectral valley, and the entrapment and polarization as the DOS gain for the even undercoordinated adatoms, defects, kinks, chemisorbed, and reconstructed skins. The ZPS resolves a tiny change of atomic CN and a trace of adsorption.

In order to examine the spectral features due to homogeneous epitaxy coverage of less than one monolayer (ML), one can appeal to the ZPS to purify bond-and-electron information due to adatoms without needing pre-specification of any spectral components. The ZPS as shown in Fig. 5.4 purifies the adatom effect with a DOS gain and the bulk/skin effect as a DOS loss. The ZPS shows clearly that Pt adatoms induce only entrapment (T at 71.00 eV) at the deeper edge of the core band. The valley (70.49 eV) corresponds to the bulk component. The effective CN of the Pt adatoms is estimated 3.15, which is lower than the CN of 4.0 for an atom at the flat skin. The bond between the Pt adatoms and the Pt substrate is 17.5% shorter and the bond is 21% stronger compared with those in the bulk, according to the BOLS formulation.

However, the XPS profiles for Rh adatoms are much more complicated. In addition to the entrapped states at energies corresponding to $z = 4$–6, the polarized P state centered at 306.20 eV presents rendering the upward shift of the originally entrapped

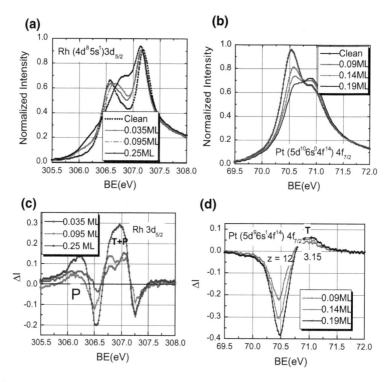

Fig. 5.4 Adatom coverage (ML) resolved **a** Rh $3d_{5/2}$ [23] and **b** Pt $4f_{7/2}$ [24] XPS profiles and the corresponding. **c, d** ZPS profiles [40]. The ZPS valley confirms the bulk component at 70.49 for Pt and at 306.53 eV for Rh. The ZPS reveals unambiguously the atomic undercoordination-induced **c** coexistence and coupling of entrapment (T) and polarization (P) for Rh $3d_{5/2}$ and **d** the entrapment dominance of Pt adatoms with effective $z = 3.15$ Reprinted with permission from [40]. Copyright 2009 American Chemical Society

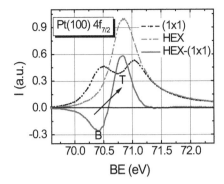

Fig. 5.5 ZPS for the hexagonally-reconstructed Pt(100) edges and the smooth Pt(100) − (1 × 1) surfaces [39] showing the bulk valley (B) at 70.45 eV and the edge entrapment (T) at 70.80 eV. Reprinted with permission from [40]. Copyright 2009 American Chemical Society

state that is supposed to be at z ~ 3. The P states are located above the bulk valley at 306.55 eV. Most strikingly, the ZPS (see Sect. 3.5) affirms directly that the Pt bulk component centered at 70.49 eV and that for the Rh bulk at 306.53 eV, which is consistent to the derivative from the skin XPS analysis.

The valley at 307.25 eV for Rh arises from the screening and splitting of the crystal potential by adatom dipoles, which offset the entrapped states upwardly from effective CN = 3.15 to 4–6. The conduction electrons of Rh adatoms are fully polarized, which screens the crystal potential and hence moves the core DOS up to z = 4–6 and creates the P + T states shifting up to pC_z^{-m}. The absence of the P states in the Pt ($5d^{10}6s^04f^{14}$ instead of the unlike $5d^96s^1$) $4f_{7/2}$ spectra may indicate that the empty 6s and the fully occupied $4f^{14}$ states are hardly polarizable.

The residual $4f_{7/2}$ states of the hexagonally-reconstructed Pt(100) surface with respect to that of the Pt(100) − (1 × 1) surface [39], as shown in Fig. 5.5, manifest the same entrapment dominance of adatoms. Due to the Pt–Pt distance contraction, the top layer of the reconstructed skin accommodates about 25% more edge atoms than the (100) − (1 × 1) layer. This fact further supports the BOLS-TB-ZPS derivatives regarding the structural relaxation and quantum entrapment by the shorter and stronger bonds between undercoordinated Pt edge atoms.

The difference in the ZPS derivatives between the Pt and the Rh adatoms confirms the BOLS-NEP notion that the otherwise conductive half-filled s-electron Rh($4d^85s^1$) can be polarized and locked as adatom dipoles, making less contribution to the conductivity. These locally polarized electrons are responsible for the dilute magnetism of the small clusters as well compared with their nonmagnetic parent bulk [41–43]. This observation evidences that the polarized unpaired electrons of the adatom dipoles are responsible for the magnetism of nanocrystals [43]. However, hydrogenation annihilates the unpaired dipoles and cluster size inflation lowers the fraction of the skin dipoles of the clusters [43, 44].

It is also clear now why the Pt and Rh adatoms perform differently in the catalytic reaction from the electronic structure point of view. Entrapment dominance entitles the undercoordinated Pt adatoms to serve as an acceptor-type catalyst that is beneficial to oxidation but the Rh adatoms as a donor-type catalyst for reduction. During the reaction, Pt adatoms tend to capturing electrons from the reactant while the Rh adatoms tend to donating. Strikingly, the undercoordination-induced entrapment dominance of Pt and the polarization dominance may explain why Au and Pt adatoms prefer different oxygen vacancy sites on the $TiO_2(110)$ surface (see Fig. 5.3). Along with this guideline, it is possible to devise and search for new catalysts at different needs using the ZPS spectrometrics.

5.3 ZPS of Rh, W, and Re Kink Edges

5.3.1 Atomic Arrangement at Edges

Figure 5.6a illustrates the vicinal fcc(111) surfaces of (151513) and (553) kinks with an effective coverage 1/L(L is the edge separation) of 0.07 and 0.26 ML, respectively. The θ is the angle between the vicinal and the ideal (111) surface. There are adatom (A), edge atom (E), surface (1, 2), and bulk (B) atoms. Their effective CNs are in the order of: $z_A < z_E < z_1 < z_2 < z_B = 12$. Figure 5.6b shows the reconstructed fcc(110) − (1 × 2) and the (1 × 1) + (1 × 2) surfaces. Every other raw of atoms is missing in

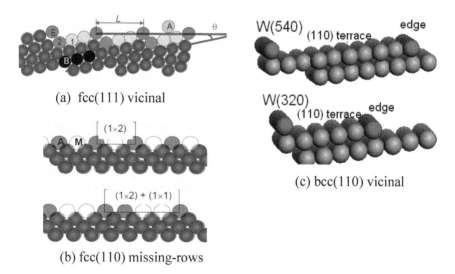

(a) fcc(111) vicinal

(b) fcc(110) missing-rows

(c) bcc(110) vicinal

Fig. 5.6 Schematic illustration of **a** the fcc(111) vicinal (151513) and (553) skins with 0.07 and 0.26 ML edge-atom coverage (1/L). Denoted are adatoms (A), edge (E), surface (1, 2), and bulk (B) atoms with effective CNs in the order of: $z_A < z_E < z_1 < z_2 < z_B = 12$. The θ is the angle between the vicinal and the ideal (111) surface; **b** the (1 × 2) and (1 × 1) + (1 × 2) missing-row type reconstructed fcc(110) skins have the same 0.5 ML coverage but slightly different atomic CNs. M is the vacancy. **c** The bcc(110) vicinal (320) surface with edge density of 0.28 ML and (540) skin with edge density of 0.16 ML

the (1×2) while every other pairing-raw of atoms is missing in the $(1 \times 1) + (1 \times 2)$ surface. A is the atom near the missing-row vacancy M.

Although the surface coverage remains 0.5 ML, the effective CN of the E atom near the (1×2) single-missing-row vacancy is slightly lower than that of the other. Panel (c) is the bcc(110) vicinal surfaces with indicated red edge atoms. In contrast to the smooth hcp(0001) surface, the $(12\bar{3}1)$ surface is much rougher with higher fraction of even undercoordinated kink atoms. These even undercoordinated atoms will enhance the BOLS-NEP effect on entrapment and polarization, giving rise unexpected properties such as the Dirac-Fermion carrier of the topological insulators. ZPS can resolve the effect of tiny CN difference on the CLS.

5.3.2 Rh(110) and (111) Vicinal Edges

Figure 5.7 shows the normalized XPS profiles and the ZPS spectra for the Rh(111) and the (110) vicinal surfaces. The ZPS reveals consistently the expected T and P features in addition to the bulk valley (B at 306.53 eV). For a surface with edge atoms, the peak intensity of the bulk component attenuates because of the polarization offsetting. Therefore, the ZPS reveals only one valley of B. The undercoordinated Rh atoms generally add two DOS features. One is the polarization above the B component and the other is due to the coupling of T and P at energies corresponding to $z = 4$–6, rendering the upward shift of originally trapped states at $z \sim 3$ by pC_z^{-m}.

Despite the general features of both the quantum entrapment and the polarization of the core electrons, the (151513) (0.07 ML) in Fig. 5.7a and the $(1 \times 2) + (1 \times 1)]$ (0.5 ML) in (b) showing weak polarization. The polarization for the former is too weak to influence the core band; the atomic CN for the latter is not low enough

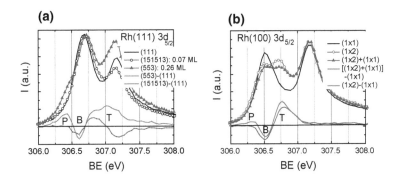

Fig. 5.7 Comparison of the XPS/ZPS for **a** the Rh(111) vicinal surfaces [28] and **b** Rh(110) missing-row type reconstructed surfaces [29]. The valley at 306.53 eV corresponds to the bulk component. Being able to resolve the tiny CN difference, ZPS spectra revealed the generality of polarization screening on the core band. Reprinted with permission from [45]. Reproduced by permission of the PCCP Owner Societies

to induce the polarization though the overall entrapment does happen in both cases. Although the (1 × 2) and the (1 × 2) + (1 × 1), in (b), share the same 0.5 ML coverage, the doubly-edged adatoms show stronger polarization screening effect than the singly-edged atoms (Fig. 5.7b). Therefore, the ZPS is insofar most sensitive to a tiny change of atomic CN.

5.3.3 W(110) Vicinal Edges

From the original 4f spectra in Fig. 5.8a for the W(110), (320), and (540) skins collected under the same probing conditions [25], one can hardly discriminate one spectral feature from another. The ZPS however makes a great difference, as shown in Fig. 5.8b, proceeded by subtracting the spectrum for the un-edged W(110) spectrum from that of the edged W(540) and (320) skins. Unexpectedly, two extra components centered at 30.95 and 31.31 eV and two valleys centered at 31.08 and 31.45 eV appear. The emergence of these spectral features indicates that the electronic structure for the edge atoms is indeed different from those at the bulk interior or at the flat surface. These spectral features are the same to that uncovered from undercoordinated Rh adatoms—polarization dominance results in potential screening and splitting. The entrapped core electrons are subject to the polarization, which shift up the entrapped component.

Derivatives of the CN-resolved SCLS in Sect. 5.4.4 clarify the origin of the ZPS features in Fig. 5.8. The ZPS identifies the bulk valley at 31.083 eV directly without needing any hypothetic assignment, which rectifies the bulk component that has long been mistaken at 31.45 eV [47]. The T component below the B results from the edge quantum entrapment. ZPS confirms that the locally densely entrapped T states

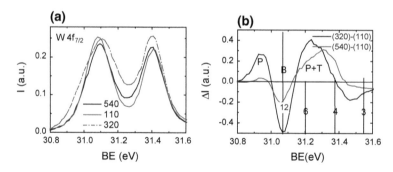

Fig. 5.8 Peak area normalized $4f_{7/2}$ spectra of the **a** W(110), (540), and (320) skins [25]. **b** ZPS resolves unambiguously the edge states with the P and the P + T excessive states and an additional valley at the bottom edge [46], which is the same to that of Rh edges and adatoms—polarization dominance. Reprinted with permission from [46]. Reproduced by permission of the PCCP Owner Societies

polarize the valence electrons of the edge atoms, which in turn screens and splits the crystal potential and offsets the entrapped T states in the core band upwardly.

The polarization results in the P component at 30.95 eV. One can estimate the polarization coefficient with the known energies of the P and the B components, $p = [E_{4f7/2}(p) - E_{4f7/2}(0)]/[E_{4f7/2}(12) - E_{4f7/2}(0)] = (30.945-28.889)/2.194 = 93.7\%$, which means that the polarization weakens 6.3% the crystal potential experienced by the $4f$ electrons in the bulk.

The otherwise T component turns to be T + P with an additional valley at the bottom edge of the core band because of the coupling of T and P, which follows the same mechanism for the Rh adatoms and terrace edges. The screening effect also applies to the trapped states, and therefore, this valley presents, and the T component becomes P + T. If the C_3^{-1} is replaced with $pC_3^{-1} = C_{3.75}^{-1}$, which means that the original edge states located at $z = 3$ shift up to energy being equivalent to $z = 3.75$. The edge bond is strengthened by $[E_{4f}(3.75) - E_{4f}(0)]/[E_{4f}(12) - E_{4f}(0)] = (31.310-28.889)/2.194 = 1.103$, or 10.3%, because of the joint T + P effect.

The ZPS of W edges share the same attributes to those of Rh edges or adatoms [40], which has been identified as donor-type catalyst. From the electronic configuration viewpoint, the undercoordinated W edges perform the same to Rh adatoms as donor-type catalyst.

DFT calculations on the flat W(110) and the edged W(320) and (540) skins focusing on the local valence DOS evolution [48] verified the BOLS-NEP expectation on the skin bond contraction and valence charge polarization [48, 50]. DFT calculations also confirmed the size-reduction induced $4s$ core band entrapment and the $(4d^3 5s^1)$ valence polarization of Mo_n clusters. Results in Fig. 5.9 and Table 5.1 confirmed this expectation.

5.3.4 Re(0001) and (12$\bar{3}$1) Kink Edges

Figure 5.10a shows the ZPS of the original Re (12$\bar{3}$1) $4f_{5/2}$ and $4f_{7/2}$ bands[48,127,227–228,258–261] gained by subtracting the spectrum collected at 0° from that collected at 75° emission angle after background correction and spectral area normalization [26, 27, 51–56]. The obvious valleys correspond to the bulk (B) and peaks to the entrapped kink edges (T). The outermost two atomic layers with the effective z values of 3.6 (S_2) and 2.8 (S_1) create the T components, as indicated in Fig. 5.10b. The small feature at the upper edge of the $4f_{7/2}$ band arises from the $5p_{3/2}$ band overlap of the kinks, instead of polarization.

Figure 5.10b shows the ZPS of the specific $4f_{5/2}$ level. Two valleys centered at 42.645 and 42.910 eV correspond to the B and S_3 components, which indicate that the B valley contains the S_4 (centered at 42.778 eV) and the S_3 components. Therefore, the ZPS reveals information of the outermost two atomic layers of the low-index (12$\bar{3}$1) skin. Except for the valleys, a broad peak at the bottom edge of the $4f_{5/2}$ band represents the gain of electronic energy when the S_2 and S_1 kinks are involved. The

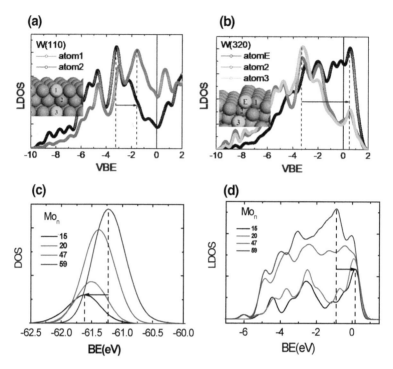

Fig. 5.9 DFT derived coordination-resolved W($5d^56^1$) valence DOS for **a** W(110) and **b** W(320) skins. Polarization of the edge atom (atom E) becomes more significant than that of atoms in the flat surface (atom 1) and those in the subsequent sublayers. Size-resolved **c** 4s core band entrapment and **d** $4d^55s^1$ valence band polarization of Mo$_n$ clusters. Reprinted with permission from [48, 50]. Copyright 2012 Elsevier. Copyright 2015 The Royal Society of Chemistry

Table 5.1 DFT derived W skin bond contraction [48], which follows the trend of BOLS prediction though DFT usually underestimates the extent of bond strain [49]

	Bond position	Strain (%)
W(110)	1–2	−3.28
	2–3	−0.36
W(320)	E–1	−5.84
	E–2	−5.47
	2–3	−0.07
W(540)	E–1	−4.74
	E–2	−5.84
	2–3	−0.29

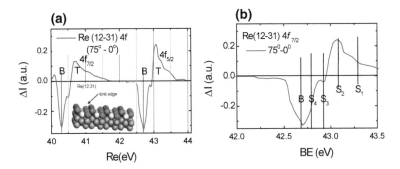

Fig. 5.10 ZPS of **a** the kinked Re $(12\bar{3}1)$ $4f$ bands [27]. The B valleys correspond to bulk (B) components and the T peaks to the kink-induced quantum entrapment. **b** The ZPS of the $4f_{5/2}$ band with the valley centered at 42.645 eV includes contribution from the B, the S_4 and the S_3 sublayers. The spectral gains at the bottom edge correspond to the quantum entrapment dominated by the outermost two layers of kink atoms. Reprinted with permission from [57]. Copyright 2011 American Chemical Society

ZPS could therefore separate the kinks from their bulk mixture without needing any hypothetic assignment of the B or the skin component.

From the spectral bandwidth of the B valley, one can find that the overlap integral β is indeed negligible compared with the exchange integral α in the TB approximation. The width of a core band is $2z\beta$ and the shift of the band from that of an isolated atom is $\alpha + z\beta$. Based on the width of the B valley for both the $4f_{7/2}$ and the $4f_{5/2}$ bands of ~0.25 eV and the bulk shift of $\alpha + z\beta = 2.629$ eV $(\beta/\alpha \approx 0.01/2.63 < 0.4\%)$ only, in the present case.

5.3.5 O-Re $(12\bar{3}1)$ Kink Edge and Chemisorbed States

Figure 5.11a, b decompose the $4f_{7/2}$ band for the Re $(12\bar{3}1)$ surface with and without oxygen chemisorption [27]. A direct decomposition of the spectrum for oxygen-adsorbed Re $(12\bar{3}1)$ surface using the coordination scheme is no longer valid because of oxide bond formation. The O-Re bond formation will increase the local crystal potential in the adsorbed region, which deepens the surface states, as shown in Fig. 5.11b [52, 56]. The O-induced Re $4f_{7/2}$ shift is in line with what detected from oxygen-chemisorbed surfaces [58], in which the O $2p$ states shift positively by an amount of ~0.5 eV upon oxide formation.

Figure 5.11c compares the ZPS for the Re $(12\bar{3}1)$ surface with and without oxygen adsorption. A BO valley centered at 40.40 eV is different from the B (40.30 eV) and the S_4 in the referential ZPS of the clean Re surface. This difference indicates that the O-Re bond is much stronger and that the synchrotron beam of 90 eV collects less information from the bulk or the S_4 region. The extra states extend to energy that is even lower than the energy states of the clean surface. The ZPS has thus

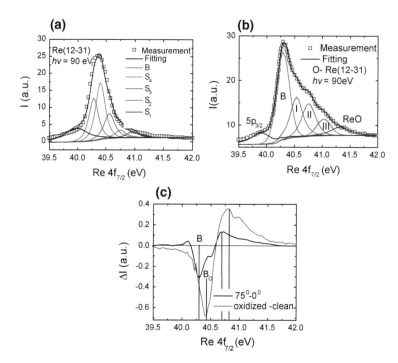

Fig. 5.11 Decomposition of the $4f_{7/2}$ spectra collected from **a** the clean and **b** the O-adsorbed Re $(12\bar{3}1)$ skins [27]. **c** The ZPS for the O-Re $(12\bar{3}1)$ (red line) and for the Re $(12\bar{3}1)$ $4f_{7/2}$ kink states (dark line) resolve the DOS gain because of oxygen chemisorption that deepens the bulk component slightly from 40.3 to 40.4 eV. The T peak corresponds to the outermost two atomic layers, which indicates that O prefers the central position of the tetrahedron [58]. Reprinted with permission from [57]. Copyright 2011 American Chemical Society

enabled discrimination of the chemisorbed surface states from its mixture with the bulk and thus allowed one to estimate the depth of two atomic layers involved in oxygen adsorption. Oxygen adsorbate tends to locate between the first and the second sublayers preferring the central position of the tetrahedron by interacting with two metal atoms through bonding and another two through nonbonding lone pairs of electrons [58].

5.4 Summary

Table 5.2 summarizes BOLS-NEP-ZPS derived information in comparison with that derived from traditional practice induced by the even-undercoordinated atoms. Most importantly, the polarization at sites of even lower atomic CN than a flat surface is beyond expectation.

Table 5.2 The BOLS-NEP-ZPS analysis of the electron BE shift of the even undercoordinated atoms leads to quantitative information about the energy level of an isolated atom and the associated entrapment and polarization

Energy levels	Site	z	C_z	$d_z(\text{Å})$	E_z/eV	$E_\nu(z)/eV$
Rh(100) $3d_{3/2}$ [23]	Atom	0	–			302.16
	Bulk	12.0	1.00	2.68	0.48	306.53
	(100) S_1	4.0	0.88	2.36	0.54	307.15
	Adatom	3.0	0.82	2.20	0.58	307.51
	Polaron					306.20
W(110) $4f_{7/2}$ [25]	Atom	0				31.083
	Bulk	12	1.00	2.79	1.11	33.256
	(110)S_1(p)	6	0.94	2.61	1.19	33.401
	Kink(p)	4	0.88	2.44	1.29	33.565
	Polaron					28.910
Pt (111) $4f_{7/2}$ [24]	Atom	0	–			68.10
	Bulk	12.00	1.00	2.77	0.4867	70.49
	S_1	4.25	0.89	2.47	0.5468	70.91
	Adatom	3.15	0.83	2.20	0.5935	71.18
Re ($12\bar{3}1$) $4f_{5/2}$ [27]	Atom	0				40.014
	Bulk	12	1.00	2.75	0.67	42.643
	S_1	3.6	0.86	2.35	0.78	43.088
	Kink	2.8	0.80	2.19	0.84	43.310
Re ($12\bar{3}1$) $4f_{7/2}$ [27]	S_1	3.6	0.86	2.35	0.78	40.30
	O-($12\bar{3}1$) S_1	O sits between the outermost two atomic layers				40.65
Energy levels	Site	z	C_z	$d_z(\text{Å})$	E_z/eV	$E_\nu(z)/eV$
Rh(100) $3d_{3/2}$ [23]	Atom	0	–			302.16
	Bulk	12.0	1.00	2.68	0.48	306.53
	(100) S_1	4.0	0.88	2.36	0.54	307.15
	Adatom	3.0	0.82	2.20	0.58	307.51
	Polaron					306.20
W(110) $4f_{7/2}$ [25]	Atom	0				31.083
	Bulk	12	1.00	2.79	1.11	33.256
	(110)S_1(p)	6	0.94	2.61	1.19	33.401
	Kink(p)	4	0.88	2.44	1.29	33.565
	Polaron					28.910
Pt (111) $4f_{7/2}$ [24]	Atom	0	–			68.10
	Bulk	12.00	1.00	2.77	0.4867	70.49
	S_1	4.25	0.89	2.47	0.5468	70.91

(continued)

Table 5.2 (continued)

Energy levels	Site	z	C_z	$d_z(Å)$	E_z/eV	$E_v(z)$/eV
	Adatom	3.15	0.83	2.20	0.5935	71.18
Re $(12\bar{3}1)$ $4f_{5/2}$ [27]	Atom	0				40.014
	Bulk	12	1.00	2.75	0.67	42.643
	S_1	3.6	0.86	2.35	0.78	43.088
	Kink	2.8	0.80	2.19	0.84	43.310
Re $(12\bar{3}1)$ $4f_{7/2}$ [27]	S_1	3.6	0.86	2.35	0.78	40.30
	O-$(12\bar{3}1)$ S_1	O sits between the outermost two atomic layers				40.65

With the derived z, d_z, and E_z, one is able to find E_{den} and E_{coh} (the skin and adatom components and their effective CNs are subject to offset by polarization pC_z^{-m})

The following summarizes the physical understandings of the even undercoordinated atoms $(z < 4)$:

(1) Adatoms, terrace edges, and kinks share the same attributes of global quantum entrapment and subjective polarization, of which atoms of the bulk or at a flat skin do not demonstrate.

(2) Bonds between such undercoordinated atoms and the substrate are even shorter and stronger than those are in the flat skin. The lower the atomic CN, the shorter the bond length and the deeper the potential well will be, and hence, the higher extent of polarization occurs to the even lower coordinated atoms.

(3) The polarization dominance makes Rh adatoms or terrace edges to be donor-type catalyst. The same trend of W edges may make W an alternative for catalytic reduction applications.

(4) The entrapment dominance makes Pt adatoms and edges acceptor-type catalysis. Re could be an alternative for acceptor-like catalysis because of the same DOS entrapment feature.

(5) Oxygen chemisorption deepens the Re $(12\bar{3}1)$ SCL due to the stronger O-Re bond formation in the outermost two atomic layers.

(6) Being sensitive to the small change of atomic CN and adsorption, ZPS gives direct information of the B valley without needing any assumptions as usually do in spectral decomposition.

References

1. J.S. Garitaonandia, M. Insausti, E. Goikolea, M. Suzuki, J.D. Cashion, N. Kawamura, H. Ohsawa, I. Gil de Muro, K. Suzuki, F. Plazaola, Chemically induced permanent magnetism in Au, Ag, and Cu nanoparticles: localization of the magnetism by element selective techniques. Nano Lett. **8**(2), 661–667 (2008)

2. Y. Yamamoto, T. Miura, M. Suzuki, N. Kawamura, H. Miyagawa, T. Nakamura, K. Kobayashi, T. Teranishi, H. Hori, Direct observation of ferromagnetic spin polarization in gold nanoparticles. Phys. Rev. Lett. **93**(11), 116801 (2004)

3. R. Magyar, V. Mujica, M. Marquez, C. Gonzalez, Density-functional study of magnetism in bare Au nanoclusters: evidence of permanent size-dependent spin polarization without geometry relaxation. Phys. Rev. B **75**(14), 144421 (2007)

4. B. Wang, X.D. Xiao, X.X. Huang, P. Sheng, J.G. Hou, Single-electron tunneling study of two-dimensional gold clusters. Appl. Phys. Lett. **77**(8), 1179–1181 (2000)

5. O.A. Yeshchenko, I.M. Dmitruk, A.A. Alexeenko, M.Y. Losytskyy, A.V. Kotko, A.O. Pinchuk, Size-dependent surface-plasmon-enhanced photoluminescence from silver nanoparticles embedded in silica. Phys. Rev. B **79**(23), 235438 (2009)

6. J. Sancho-Parramon, Surface plasmon resonance broadening of metallic particles in the quasi-static approximation: a numerical study of size confinement and interparticle interaction effects. Nanotechnology **20**(23), 235706 (2009)

7. M. Turner, V.B. Golovko, O.P. Vaughan, P. Abdulkin, A. Berenguer-Murcia, M.S. Tikhov, B.F. Johnson, R.M. Lambert, Selective oxidation with dioxygen by gold nanoparticle catalysts derived from 55-atom clusters. Nature **454**(7207), 981–983 (2008)

8. M. Valden, X. Lai, D.W. Goodman, Onset of catalytic activity of gold clusters on titania with the appearance of nonmetallic properties. Science **281**(5383), 1647–1650 (1998)

9. A. Wittstock, B.R. Neumann, A. Schaefer, K. Dumbuya, C. Kübel, M.M. Biener, V. Zielasek, H.-P. Steinrück, J.M. Gottfried, J.R. Biener, Nanoporous Au: an unsupported pure gold catalyst? J. Phys. Chem. C, **113**(14), 5593–5600 (2009)

10. Y.B. Zheng, L. Jensen, W. Yan, T.R. Walker, B.K. Juluri, L. Jensen, T.J. Huang, Chemically tuning the localized surface plasmon resonances of gold nanostructure arrays. J. Phys. Chem. C **113**(17), 7019–7024 (2009)

11. A. Elbakry, A. Zaky, R. Liebl, R. Rachel, A. Goepferich, M. Breunig, Layer-by-layer assembled gold nanoparticles for siRNA delivery. Nano Lett. **9**(5), 2059–2064 (2009)

12. V. Pustovalov, V. Babenko, Optical properties of gold nanoparticles at laser radiation wavelengths for laser applications in nanotechnology and medicine. Laser Phys. Lett. **1**(10), 516–520 (2004)

13. M.X. Gu, C.Q. Sun, C.M. Tan, S.Z. Wang, Local bond average for the size and temperature dependence of elastic and vibronic properties of nanostructures. Int. J. Nanotechnol. **6**(7–8), 640–652 (2009)

14. K.P. McKenna, Gold nanoparticles under gas pressure. Phys. Chem. Chem. Phys. **11**(21), 4145–4151 (2009)

15. P.D. Jadzinsky, G. Calero, C.J. Ackerson, D.A. Bushnell, R.D. Kornberg, Structure of a thiol monolayer-protected gold nanoparticle at 1.1 angstrom resolution. Science, **318**(5849), 430–433 (2007)

16. W.H. Qi, B.Y. Huang, M.P. Wang, Bond-length and -energy variation of small gold nanoparticles. J. Comput. Theor. Nanosci. **6**(3), 635–639 (2009)

17. C.Q. Sun, Thermo-mechanical behavior of low-dimensional systems: the local bond average approach. Prog. Mater Sci. **54**(2), 179–307 (2009)

18. B. Wang, K.D. Wang, W. Lu, J.L. Yang, J.G. Hou, Size-dependent tunneling differential conductance spectra of crystalline Pd nanoparticles. Phys. Rev. B **70**(20), 205411 (2004)

19. J.N. Crain, D.T. Pierce, End states in one-dimensional atom chains. Science **307**(5710), 703–706 (2005)

20. N. Nilius, T.M. Wallis, W. Ho, Development of one-dimensional band structure in artificial gold chains. Science **297**(5588), 1853–1856 (2002)

21. K. Schouteden, E. Lijnen, D.A. Muzychenko, A. Ceulemans, L.F. Chibotaru, P. Lievens, C.V. Haesendonck, A study of the electronic properties of Au nanowires and Au nanoislands on Au(111) surfaces. Nanotechnology **20**(39), 395401 (2009)

22. B.G. Briner, P. Hofmann, M. Doering, H.P. Rust, E.W. Plummer, A.M. Bradshaw, Charge-density oscillations on Be(10-10): screening in a non-free two-dimensional electron gas. Phys. Rev. B **58**(20), 13931–13943 (1998)

23. A. Baraldi, L. Bianchettin, E. Vesselli, S. de Gironcoli, S. Lizzit, L. Petaccia, G. Zampieri, G. Comelli, R. Rosei, Highly under-coordinated atoms at Rh surfaces: interplay of strain and coordination effects on core level shift. New J. Phys. **9**, 143 (2007)

24. L. Bianchettin, A. Baraldi, S. de Gironcoli, E. Vesselli, S. Lizzit, L. Petaccia, G. Comelli, R. Rosei, Core level shifts of undercoordinated Pt atoms. J. Chem. Phys. **128**(11), 114706 (2008)

25. X.B. Zhou, J.L. Erskine, Surface core-level shifts at vicinal tungsten surfaces. Phys. Rev. B **79**(15), 155422 (2009)

26. N. Martensson, H.B. Saalfeld, H. Kuhlenbeck, M. Neumann, Structural dependence of the 5d-metal surface energies as deduced from surface core-level shift measurements. Phys. Rev. B **39**(12), 8181–8186 (1989)

27. A.S.Y. Chan, G.K. Wertheim, H. Wang, M.D. Ulrich, J.E. Rowe, T.E. Madey, Surface atom core-level shifts of clean and oxygen-covered Re(1231). Phys. Rev. B **72**(3), 035442 (2005)

28. J. Gustafson, M. Borg, A. Mikkelsen, S. Gorovikov, E. Lundgren, J.N. Andersen, Identification of step atoms by high resolution core level spectroscopy. Phys. Rev. Lett. **91**(5), 056102 (2003)

29. A. Baraldi, S. Lizzit, F. Bondino, G. Comelli, R. Rosei, C. Sbraccia, N. Bonini, S. Baroni, A. Mikkelsen, J.N. Andersen, Thermal stability of the Rh(110) missing-row reconstruction: Combination of real-time core-level spectroscopy and ab initio modeling. Phys. Rev. B **72**(7), 075417 (2005)

30. J. Sánchez-Royo, J. Pellicer-Porres, A. Segura, S. Gilliland, J. Avila, M. Asensio, O. Safonova, M. Izquierdo, A. Chevy, Buildup and structure of the In Se/Pt interface studied by angle-resolved photoemission and x-ray absorption spectroscopy. Phys. Rev. B **73**(15), 155308 (2006)

31. X.J. Liu, M.L. Bo, X. Zhang, L. Li, Y.G. Nie, H. Tian, Y. Sun, S. Xu, Y. Wang, W. Zheng, C.Q. Sun, Coordination-resolved electron spectrometrics. Chem. Rev. **115**(14), 6746–6810 (2015)

32. A. Mellor, D. Humphrey, C.M. Yim, C.L. Pang, H. Idriss, G. Thornton, Direct visualization of Au Atoms Bound to TiO2 (110) O-Vacancies. J. Phys. Chem. C **121**(44), 24721–24725 (2017)

33. Y. Pan, Y. Cui, C. Stiehler, N. Nilius, H.-J. Freund, Gold Adsorption on CeO2 Thin Films Grown on Ru(0001). J. Phys. Chem. C **117**(42), 21879–21885 (2013)

34. T.-Y. Chang, Y. Tanaka, R. Ishikawa, K. Toyoura, K. Matsunaga, Y. Ikuhara, N. Shibata, Direct imaging of pt single atoms adsorbed on TiO2 (110) surfaces. Nano Lett. **14**(1), 134–138 (2013)

35. L.L. Patera, F. Bianchini, C. Africh, C. Dri, G. Soldano, M.M. Mariscal, M. Peressi, G. Comelli, Real-time imaging of adatom-promoted graphene growth on nickel. Science **359**(6381), 1243–1246 (2018)

36. M. Haruta, Size-and support-dependency in the catalysis of gold. Catal. Today **36**(1), 153–166 (1997)

37. D. Matthey, J. Wang, S. Wendt, J. Matthiesen, R. Schaub, E. Lægsgaard, B. Hammer, F. Besenbacher, Enhanced bonding of gold nanoparticles on oxidized TiO$_2$ (110). Science **315**(5819), 1692–1696 (2007)

38. S. Chrétien, H. Metiu, Density functional study of the interaction between small Au clusters, Au n (n = 1–7) and the rutile TiO$_2$ surface. II. Adsorption on a partially reduced surface. J. Chem. Phys. **127**(24), 244708 (2007)

39. A. Baraldi, E. Vesselli, L. Bianchettin, G. Comelli, S. Lizzit, L. Petaccia, S. de Gironcoli, A. Locatelli, T.O. Mentes, L. Aballe, J. Weissenrieder, J.N. Andersen, The (1x1)-> hexagonal structural transition on Pt(100) studied by high-energy resolution core level photoemission. J. Chem. Phys. **127**(16), 164702 (2007)

40. C.Q. Sun, Y. Wang, Y.G. Nie, Y. Sun, J.S. Pan, L.K. Pan, Z. Sun, Adatoms-induced local bond contraction, quantum trap depression, and charge polarization at Pt and Rh surfaces. J. Phys. Chem. C **113**(52), 21889–21894 (2009)

41. C.Q. Sun, Size dependence of nanostructures: Impact of bond order deficiency. Prog. Solid State Chem. **35**(1), 1–159 (2007)

42. A.J. Cox, J.G. Louderback, S.E. Apsel, L.A. Bloomfield, Magnetic in 4d-transition metal-clusters. Phys. Rev. B **49**(17), 12295–12298 (1994)

43. E. Roduner, Size matters: why nanomaterials are different. Chem. Soc. Rev. **35**(7), 583–592 (2006)

44. C.Q. Sun, Dominance of broken bonds and nonbonding electrons at the nanoscale. Nanoscale **2**(10), 1930–1961 (2010)

45. W. Zheng, J. Zhou, C.Q. Sun, Purified rhodium edge states: undercoordination-induced quantum entrapment and polarization. Phys. Chem. Chem. Phys. **12**(39), 12494–12498 (2010)

46. Y.G. Nie, X. Zhang, S.Z. Ma, Y. Wang, J.S. Pan, C.Q. Sun, XPS revelation of tungsten edges as a potential donor-type catalyst. Phys. Chem. Chem. Phys. **13**(27), 12640–12645 (2011)

47. D.M. Riffe, B. Kim, J.L. Erskine, Surface core-level shifts and atomic coordination at a stepped W(110) surface. Phys. Rev. B **50**(19), 14481–14488 (1994)

48. Y. Wang, X. Zhang, Y.G. Nie, C.Q. Sun, Under-coordinated atoms induced local strain, quantum trap depression and valence charge polarization at W stepped surfaces. Phys. B-Condensed Matter **407**(1), 49–53 (2012)

49. X. Zhang, Y.G. Nie, W.T. Zheng, J.L. Kuo, C.Q. Sun, Discriminative generation and hydrogen modulation of the Dirac-Fermi polarons at graphene edges and atomic vacancies. Carbon **49**(11), 3615–3621 (2011)

50. W. Zhou, M. Bo, Y. Wang, Y. Huang, C. Li, C.Q. Sun, Local bond-electron-energy relaxation of Mo atomic clusters and solid skins RSC. Advances **5**, 29663–29668 (2015)

51. M. Asscher, J. Carrazza, M. Khan, K. Lewis, G. Somorjai, The ammonia synthesis over rhenium single-crystal catalysts: kinetics, structure sensitivity, and effect of potassium and oxygen. J. Catal. **98**(2), 277–287 (1986)

52. A.S.Y. Chan, W. Chen, H. Wang, J.E. Rowe, T.E. Madey, Methanol reactions over oxygen-modified re surfaces: influence of surface structure and oxidation. J. Phys. Chem. B **108**(38), 14643–14651 (2004)

53. R. Ducros, J. Fusy, Core level binding energy shifts of rhenium surface atoms for a clean and oxygenated surface. J. Electron Spectrosc. Relat. Phenom. **42**(4), 305–312 (1987)

54. B. Johansson, N. Martensson, Core-level binding-energy shifts for the metallic elements. Phys. Rev. B **21**(10), 4427–4457 (1980)

55. D. Spanjaard, C. Guillot, M.-C. Desjonquères, G. Tréglia, J. Lecante, Surface core level spectroscopy of transition metals: a new tool for the determination of their surface structure. Surf. Sci. Rep. **5**(1–2), 1–85 (1985)

56. H. Wang, A.S.Y. Chan, W. Chen, P. Kaghazchi, T. Jacob, T.E. Madey, Facet stability in oxygen-induced nanofaceting of Re(12$\bar{3}$1). ACS Nano **1**(5), 449–455 (2007)

57. Y.G. Nie, J.S. Pan, W.T. Zheng, J. Zhou, C.Q. Sun, Atomic scale purification of Re surface kink states with and without oxygen chemisorption. J. Phys. Chem. C **115**(15), 7450–7455 (2011)

58. C.Q. Sun, Oxidation electronics: bond-band-barrier correlation and its applications. Prog. Mater Sci. **48**(6), 521–685 (2003)

Chapter 6
Atomic Chains, Clusters, and Nanocrystals

Abstract Like adatoms, monoatomic chain ends, and atomic clusters with even less-coordinated atoms demonstrate extraordinary properties due to dominance of stronger quantum entrapment and polarization. Consistency between quantum calculations and XPS/STS observations resolves the origin of the unusual performance of such even undercoordinated atoms. A combination of the XPS and AES, called APECS, refines the energy shifts of both the core band and the valence band with derived information of the screening effect and charge transport during reaction.

Highlights

- Au, Ag and Cu adatoms exhibit polarization dominance but Co and Si nanocrystals entrapment.
- ZPS resolves the global entrapment and subjective polarization of even undercoordinated atoms.
- APECS resolves Ag, Cu, and Ni particle-substrate interaction and two-band cooperative shift.
- Curvature enhanced entrapment and polarization entitles nanocrystals with novel properties.

6.1 Observations

When the atomic CN increases from zero to the fcc bulk standard 12, the CL shifts from the $E_v(0)$ to a maximum at $z = 2$ and then recovers in a K^{-1} fashion to the bulk value of $E_v(12)$. The amount of the CLS depends not only on the specific $E_v(0)$ value but also on the shape-and-size of the crystal. Au, Ag, Ni, Cu, Pd, Si, C, and their compounds do follow this size CLS trend [1–12]. Calculations revealed that bonds in Ag, Cu, Ni, and Fe atomic chains contracted by 12.5–18.5% with 0.5–2.0 eV gain of bond energy [13, 14]. The $2p$ level of Cu_{18} and Ni_{18} clusters undergo positive shifts from their bulk values by 0.7–0.8 eV while the average strain of the clusters increases

© The Editor(s) (if applicable) and The Author(s), under exclusive license
to Springer Nature Singapore Pte Ltd. 2020
C. Q. Sun, *Electron and Phonon Spectrometrics*,
https://doi.org/10.1007/978-981-15-3176-7_6

from 0 to −6% [15]. These observations evidence the shortening and strengthening of bonds between undercoordinated atoms.

The S $2s$ and the S $2p$ band of ZnS and CdS nanosolids exhibit each three components [16, 17]. These components correspond to the outermost capping layer, the surface layer, and the core of the nanosolid. The capping and the surface layer are each 0.2–0.3 nm thick. Particle size reduction enhances the intensities of the capping and the surface components rendering the intensity of the core component, which follows the size-dependence of the surface-to-volume ratio of a nanosolid. These observations evidence that cluster size reduction enhances globally the CLS for nanostructures, regardless of composition or structure phase of the nanocrystals [5].

An incorporation of the BOLS-NEP scheme to STM/S, PES, APECS, and ZPS measurements as well as DFT calculations leads to consistent understanding of the performance of atomic clusters and nanocrystals from the perspective of local bond contraction, quantum entrapment, and nonbonding (valence) electron polarization.

6.2   BOLS-TB Formulation

The size dependent CLS of a nanocluster follows the core-shell configuration [18, 19]:

$$\frac{E_v(K) - E_v(12)}{E_v(12) - E_v(0)} = \begin{cases} \Delta_H(\tau, m, K) \ (BOLS) \\ B_v/K \qquad (Measurement) \end{cases}$$

$$\begin{cases} \Delta_H(\tau, m, K) = \sum_{i \leq 3} \gamma_i (C_i^{-m} - 1) \ (perturbation) \\ \gamma_i = \tau C_i K^{-1} \qquad\qquad (surface\text{-}to\text{-}volume) \end{cases} \tag{1}$$

The perturbation Δ_H is bond nature m, cluster size K, and geometrical shape τ dependent. The B_v is the slope of the linearization of the measured size dependent CLS for nanocrystals. The perturbation counts the weighted contribution of the outermost three atomic layers. The $C_i^{-m} - 1$ term is exact what used for defects and solid skins with different z values. Therefore, nanostructure is an extension of the point defects and solid skins of varied curvatures and core-shell configuration.

The following illustrates how to resolve the cluster size dependent CLS when the τ, m, $E_v(12)$, and $E_v(0)$ are yet to be known. Firstly, one can obtain the $E_v(12)$ by linearizing the measurements $E_v(K) = b + B'/K$. The intercept at the vertical axis is the $E_v(12)$. Equaling the BOLS prediction to the measurement yields, $\Delta_H = B_v/K$ and $B_v = \tau \sum_{i \leq 3} C_i (C_i^{-m} - 1)$. One can determine the m, τ, and $E_v(0)$ from measurements using this relationship with the known $C_i(z)$ given in Eq. (1) and the known z_i in Eq. (1). The accuracy of quantities estimated from analyzing the XPS data for nanostructures is often one order lower with respect to that derived from skin XPS analysis as the former is subject to the accuracy and uniformity of particle sizes. The following shows typical examples.

6.3 Gold

6.3.1 STM/S-DFT: End and Edge Polarization

STM/S profiles in Fig. 6.1a, b revealed that the chain-end Au atoms on Si(553) substrate are topographically higher than those in the chain interior associated with a new DOS feature at -0.5 eV below E_F [20]. This observation indicates the presence of edge atom polarization by the densely entrapped electrons of bond between the edge atom and its nearest neighbors. The polarization increases the atomic volume and raises the energy of the local valence states. The dI/dV profiles for wires of different thicknesses in Fig. 6.1c, d further evidence the size trends of polarization. The wire of narrower (red) cross-section exhibits stronger polarization [21].

Using electron cohesive diffraction, Huang et al. [22] uncovered that the Au–Au bond contracts only in the outermost two atomic layers of a gold nanosolid in a radial way, which is further supported by molecular dynamics (MD) simulations [23]. The Au–Au bond of a 3.5 nm size Au crystal contracting by 7% (Fig. 6.1b) [22] is below BOLS expectation. The equilibrium Au–Au bond in the monatomic chain contracts by 30% from the bulk value of 0.29 to 0.21 nm at 4 K, according to the thermally-resolved strain limits of the chain [24]. The Au–Au bond contracts by 12% in the fcc(100) skin. The CN-resolved bond contraction is insensitive to the types of substrate support [25] or to the structural phase or particular elements or the nature of the bond [26, 27].

DFT calculations of Au clusters with 13–147 atoms have confirmed the BOLS-NEP predications (Fig. 6.1e, f) [10]:

(1) The Au–Au bond contracts by up to 30%;
(2) The valence charge transfers from the inner to the outer atomic shells of the crystal;
(3) The valence charge transits from lower to higher binding energies and the extent of polarization is more significant for smaller clusters than for larger ones;
(4) Such polarization should be responsible for the enhanced catalytic ability of Au adatoms.

Further DFT calculations for the binding mechanism between gold nanoparticles and DNA bases confirmed that [28] negative charges transfer from the inner volume to the skin of Au nanoparticle as a result of the local quantum entrapment and the valence states shift up toward the Fermi level due to polarization. Thereby Au dipoles participate more actively in the binding to guanine. These effects are more prominent in a smaller nanoparticle.

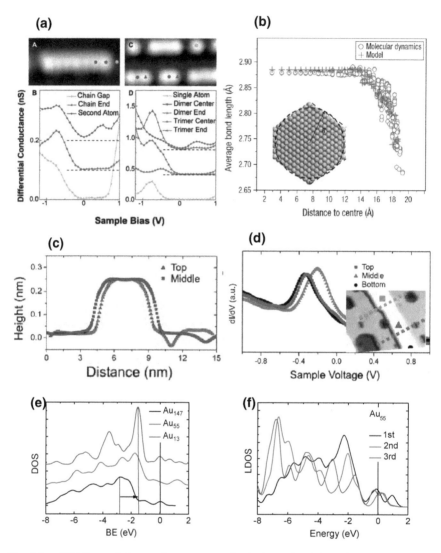

Fig. 6.1 a STM/S morphology and energy states discriminates the chain-end Au atoms from that in the interior of the monotomic chain [20]. **b** Coherent electron diffraction and MD derived Au skin bond contraction in a 3.5 nm crystal (inset) [22]. STM/S (**c**) line scans (inset in d) of an Au nanowire of different thickness and **d** the respective dI/dV spectrum. Results show the width effect on the polarization of the valence DOS [21]. DFT derived polarization of the **e** size-resolved valence DOS and **f** shell-resolved LDOS of an Au75 cluster (inset denotes the layer number). Reprinted with permission from [10, 20–22]. Copyright 2005 American Association for the Advancement of Science. Copyright 2008 Nature Publishing Group. Copyright 2009 IOP Publishing Ltd. Copyright 2010 The Royal Society of Chemistry

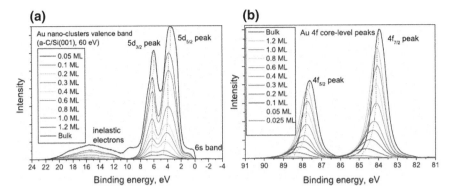

Fig. 6.2 Cluster size induced positive BE shift of the spin-degenerated **a** $5d_{3/2,5/2}$ valence band (60 eV beam energy) and **b** the $4f_{7/2,5/2}$ core band (140 eV energy) of Au clusters grown on amorphous carbon. Reprinted with permission from [29]. Copyright 2011 American Physical Society

6.3.2 PES: The 4f and the 5d Bands

Visikovskiy et al. [29] measured the size dependence of the $5d$ valence band and the $4f$ core band of Au nanoclusters deposited on amorphous carbon. They examined d-band width $W(d)$, d-band center position $E(d)$, and the apparent $5d_{3/2, 5/2}$ spin-orbit splitting $E(SO)$, as a function of the number of Au atoms per cluster, Au_n, and an average atomic CN in the size range of $11 < n < 1600$. Figure 6.2 shows that both the $5d$ valence band and the $4f$ core band shift positively when the cluster size is reduced without apparent polarization.

Figure 6.3 shows the size-induced positive shift of the $4f$ band for the thiol-capped Au [30] and Au clusters deposited on octane dithiol [12], TiO_2 [31] and Pt [32] substrate, which agrees with those obtained from Au deposited on CNTs [1, 33], HOPG [34, 35], SiO_2 [36–39], GaN [40, 41], and Re [32] substrates. However, the $4f$ of Au on Pt [32] shifts oppositely and the $4f$ of Au on NiO(100) remains unchanged [42] because of the involvement of interface alloying effect.

6.3.3 BOLS-TB Quantification

With the know size dependent CN: $z_1 = 4(1 - 0.75/K)$, $z_2 = z_1 + 2$, and $m = 1$ [43], BOLS formulation of the measured $\Delta E_{4f}(K)$ of Au on Octane gives $E_{4f}(0) = 81.50$ eV and $\Delta E_{4f}(\infty) = 2.86$ eV, agreeing with that derived from the skin XPS analysis (Sect. 6.4.3). Formulation, see Fig. 6.3a, revealed that Au growth on TiO_2 and on Pt(100) substrates proceeds in a layer-by-layer fashion [31, 32]. The size-induced deepening of the inner muffin-tin potential of gold NPs in Fig. 6.3b [44] follows the same size trend.

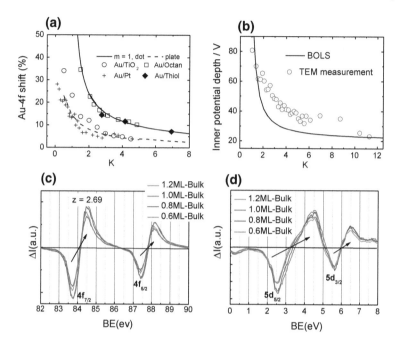

Fig. 6.3 BOLS (solid line) theory reproduction of the measured (scattered) size dependence of **a** the $\Delta E_{4f}(K)/\Delta E_{4f}(\infty) - 1$ relative shift [45] and **b** the muffin-tin inner potential of gold clusters [10, 44]. The thiol-capped Au [30] and Au on Octane dithiol [12] show three-dimensional features while Au on TiO_2 [31] and Pt [32] show two-dimensional pattern of formation.**c,d** ZPS distilled BE shift of the spin-degenerated $5d$ and $4f$ band of Au grown on amorphous carbon (raw data in Fig. 6.2) [29]. Reprinted with permission from [5, 45]. Copyright 2004 American Chemical Society and Copyright 2004 American Physical Society

The ZPS profiles of the $5d$ and $4f$ spectra, in Fig. 6.3c, d, of selected coverages show that both the $5d$ and the $4f$ bands undergo quantum entrapment but by different amounts. As expected, the upper $5d$ band shifts more than the inner $4f$ band does. The ZPS resolves the relative shift of the spin-degenerated $4f$ band as $[E_{4f5/2}(K) - E_{4f7/2}(K)]/[E_{4f5/2}(\infty) - E_{4f7/2}(\infty)] \approx (88.2 - 84.5)/(87.4 - 83.7) = 1$ and the $5d$ band as $(6.5 - 4.4)/(5.6 - 2.5) = 2.1/3.1 \approx 2/3$, respectively. The valence $5d$ band is more sensitive to the chemical and coordination environment than the inner $4f$ band that is subject to the screen shielding of the 5d electrons. Table 6.1 summarizes information gained from duplicating the $4f$ size tends of Au deposited on various substrates.

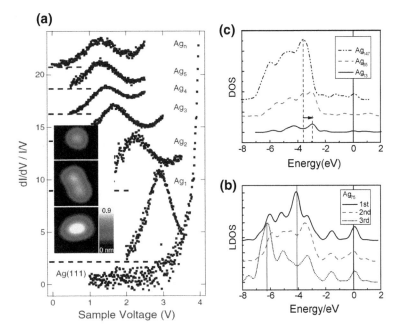

Fig. 6.4 **a** STS spectra of Ag clusters deposited on Ag(111) surface with inset showing the STM images of Ag_1, Ag_2, and a quasi-dimer (middle) [46, 47]. DFT derived shell-resolved,**b** Ag_{75} LDOS and **c** size-resolved DOS polarization for Ag_n clusters [11]. Reprinted with permission from Copyright 2008 American Physical Society. Copyright 2009 IOP Publishing Ltd. Copyrght of the PCCP Owner Societies

Table 6.1 The $E_{4f}(0)$ of an isolated Au atom and its bulk shift $\Delta E_{4f}(12)$ obtained from decoding the size-dependent $E_{4f}(K)$ of Au nanosolids [5] in comparison to that derived from skin XPS analysis (Sect. 6.4.3)

	Au/Octane	Au/TiO$_2$	Au/Pt	Au skins
m [5]	1			
τ	3	1	1	1
$E_{4f}(0)$	81.504	81.506	81.504	80.726
$E_{4f}(12)$	84.370			83.692
$\Delta E_{4f}(12)$	2.866	2.864	2.866	2.866

The charging effect of nanocrystals offset the $E_{4f}(0)$ but not the $\Delta E_{4f}(12)$ [45]

6.4 Silver

6.4.1 STM/S-DFT: Adatom Polarization

STM/S profiles in Fig. 6.4a show the polarization (bias positive) of Ag atoms added onto Ag(111) surface [46, 47]. The extent of polarization increases with the reduction of cluster size. The shell-resolved LDOS and the size-resolved DOS in Fig. 6.4b, c follow the same trend of BOLS-NEP prediction of polarization. The outermost shell of a cluster and the smallest dot polarize the valence electrons most [11]. The size-enhanced polarization explains why Ag participles could enhance the Raman signal or plasmonics with respect to their bulk parent [48] and why the size and shape influence its biocide property [49] from the perspective of core electron local quantum entrapment and valence charge polarization.

6.4.2 APECS: 3d and 5s Band Cooperative Shift

Figure 6.5 shows the Ag $3d$ core level shift as a function of thickness. The STM image shows the patterns of the Ag clusters on CeO_2 surface [50]. The APECS profiles, i.e., (a) XPS core level shift, (b) valence band, and (c) Auger parameter, in Fig. 6.6, for Ag particles deposited on CeO_2 surface [50–53] are the same to that of Ag clusters on Al_2O_3 surface [54–57]. Results show consistently that the $\Delta E_{3d5/2}$ increases and the Auger parameter ($E_{3d} + E_K$) drops in magnitude when the crystal size is reduced. Ag particle size reduction polarizes the valence charge slightly. The $3d$ states of Ag particles on HOPG [58] and on TiO_2 [59] follow the same size trend of core level

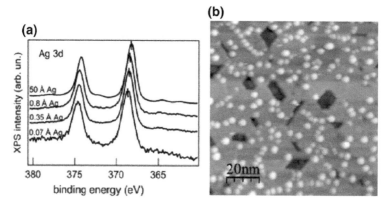

Fig. 6.5 a XPS spectra and **b** STM image of Ag clusters deposited on CeO_2 surface. Reprinted with permission from [50]. Copyright 2011 American Chemical Society

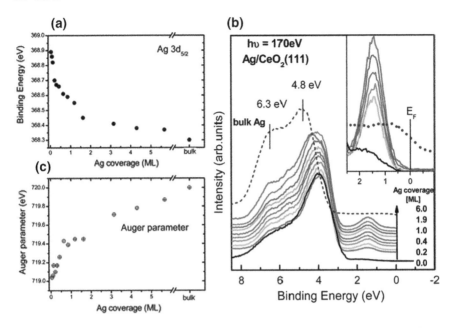

Fig. 6.6 Size dependence of **a** the $3d_{5/2}$ BE entrapment, **b** valence band polarization, and **c** Auger parameter reduction of Ag particles (coverage) on $CeO_2(111)$ surface at 300 K. Reprinted with permission from [51]. Copyright 2011 American Chemical Society

quantum entrapment and valence band polarization. Conversely, the CL shifts more at lower measuring temperature [58].

6.4.3 BOLS-TB Formulation and Derivatives

To formulate the size trends of the APECS involved energies for Ag particles, we need to obtain the dimensionality τ and the bond nature index m of Ag clusters on different substrates. Because there is no charge transfer between the CeO_2 substrate and the Ag nanoclusters [51], $m = 1$ holds for Ag/CeO_2 clusters [26]. Linearization of the measured size trend $E_{3d5/2} - K$ turns out the slope $B_v = 5.567$ and the intersection $E_{3d}(\infty) = 368.25$ eV. The slope gives the dimensionality $\tau = 1.45$, see Eq. (1) based on the known $z_i(K)$ relationship, $z_1 = 4(1 - 0.75/K)$. Because Ag particles have the same shape when grown on CeO_2 and Al_2O_3 substrates [54], the value τ holds for both [26]. However, the bond nature of Ag derived from the experiment is different because of charge transfer between Ag and Al_2O_3 substrate [57]. Similarly, linearization of the measurement results in the B_v and $E_{3d}(\infty)$ gives $m = 3.82$ for Ag/Al_2O_3 clusters. The higher m value indicates that the Ag bonds strongly to the Al_2O_3 substrate, as revealed by DFT calculations [57].

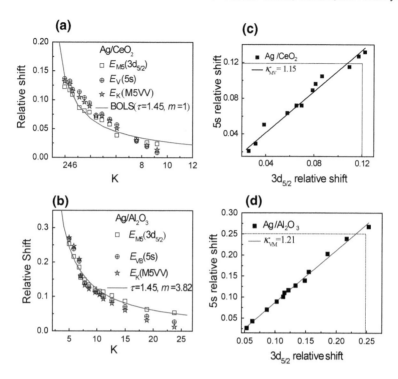

Fig. 6.7 BOLS reproduction (solid line) of the size dependence of the E_{3d}, E_{5s}, and E_k (scattered data) for Ag particles deposited on **a** CeO$_2$ [51] and **b** Al$_2$O$_3$ [54] substrates. The extended Wagner plots (**c, d**) correlate the relative shift of the $[\Delta E_{5s}(K)/\Delta E_{5s}(12) - 1]/[\Delta E_{3d}(K)/\Delta E_{3d}(12) - 1]$ = κ_{VM} (M = 3d, V = 5s) for Ag clusters. Reprinted with permission from [18]. Copyright 2012 American Chemical Society

With the derived m and τ values, BOLS formulation resolves the size dependence of the $3d_{5/2}$, the valence $5s$ band, and the Auger E_K of Ag particles. Figure 6.7a, b show the BOLS theoretical reproduction of the measurements and Table 6.2 features the derivatives. Figure 6.7c, d shows the extended Wagner plots of Ag nanostructures on CeO$_2$ and Al$_2$O$_3$ substrates. The $E_{3d}(0)$ and $E_{3d}(12)$ derived from the XPS of clean Ag skin (see Table 6.2) and from the APECS are consistent, which evidence that the BOLS governing the energy shift as probed using both XPS and APECS.

Parameters m, κ_{VM}, η_{VM} characterize the interface reaction or skin chemical conditioning. A higher m value means that stronger chemical bond forms between the cluster and the substrate. The κ_{VM} describes the effect of valence recharging during processing. The κ_{VM} in Table 6.2 is almost independent of the shape and size of the particle but it is sensitive to the chemical treatment. If charge transfers from one constituent to another, the κ_{VM} will increase, otherwise it remains unity. The η_{VM} $\cong 1.21/1.15$ means that the strength of crystal binding to the VB($5s$) level is η_{VM} times stronger than that to the deeper M5($3d$) level. Polarization happens only to the antibonding vacant states or to the upper edge of the valence band in this case. The

Table 6.2 BOLS-APECS derived $E_{3d}(z)$, $E_{5s}(z)$, and $E_k(z)$ for Ag clusters [18] and BOLS-XPS resolved CN(z), $E_{3d5/2}(0)$, and $\Delta E_{3d5/2}(12)$ for Ag(100) and (111) skins [60, 61]

(APECS)	Ag/Al$_2$O$_3$	Ag/CeO$_2$	Ag skin
m [18]	3.82	1	1
τ	1.45 (dome)		1
$E_{3d}(12)$	367.59	368.25	367.650
$E_{3d}(0)$	363.02	363.02	363.022
$\Delta E_{3d}(12)$	4.57	5.23	4.628
$E_{5s}(12)$	7.56	8.32	
$E_{5s}(0)$	0.36	0.36	
$\Delta E_{5s}(12)$	7.20	7.96	
$E_K(12)$	352.45	351.61	
$E_K(0)$	362.30	362.30	
$\Delta E_K(12)$	−9.85	−10.69	
η_{VM}	1.55	1.52	
κ_{VM}	1.21	1.15	

$K = \infty$ is equivalent to $z = 12$ for the ideal bulk. The screening coefficient η_{VM} shows the relative BE of the upper $5s$ valence levels (V) to the deeper $3d$ (M). The recharging coefficient κ_{VM} describes the chemical process. One can also derive the ε_z, E_z, E_{coh} and E_{den} for the skins and the nanoparticles with the known relationship. Reprinted with permission from [18]. Copyright 2012 American Chemical Society

$$\begin{cases} \Delta E_v(K) = \Delta E_v(\infty) \times (1 + \Delta_H) \\ \Delta_H(\tau, m, K) = \tau K^{-1} \sum_{i \le 3} C_i (C_i^{-m} - 1) \end{cases}$$

polarization of Ag is not strong enough to split and screen the local potential as the valence electrons of Rh and W specimens do.

6.5 Copper

6.5.1 STM/S-PES-DFT: Entrapment and Polarization

Figure 6.8 shows the low-temperature STM/S profiles for Cu atoms added on Cu(111) surface [62] and the size-resolved DOS of Cu$_n$ clusters and the shell-resolved LDOS of a Cu$_{75}$ cluster. Both STM/S and DFT confirm consistently that the locally densely entrapped bonding electrons polarize the valence and the conduction states. PES measurements, in Fig. 6.9, further confirm (a) the valance DOS polarization [63] and (b) the $2p$ band entrapment [64], which is the same to that of Au and Ag clusters.

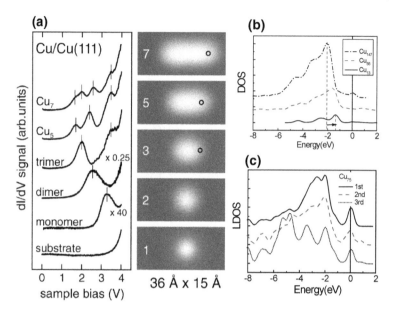

Fig. 6.8 STM/S positive-bias profiles of **a** Cu adatoms on Cu(111) surface, **b** Cu$_n$-cluster size-resolved valence DOS and, **c** Au$_{75}$ shell-resolved LDOS with respect to E$_F$ = 0 reference. Results show consistently the undercoordination induced polarization. Reprinted with permission from [11, 62]. Copyright 2004 American Physical Society. Reproduced by permission of the PCCP Owner Societies

6.5.2 APECS: Interface 2p and 3d Energy Shift

Yang and Sacher [65–67] systematically studied the 2p shift of Cu nanoparticles deposited on HOPG and Cyclotene 3022 (CYCL, a polymer of microelectronic industry interest) under various conditions using APECS. They also examined the effect of Ar$^+$ and N$^+$ bombardment on the 2p energy shift of Cu nanocrystals. Figure 6.10a, b show that both the 2p$_{3/2}$ and the Auger parameter shift following the K^{-1} scaling relation Eq. (1): $\Delta E_\nu(K) = K^{-1}\left[\tau E_\nu(\infty) \sum_{i\leq 3} C_i\left(C_i^{-m} - 1\right)\right] = B_\nu K^{-1} \pm \sigma(0.01 \sim 0.02)$, where B$_\nu$ is the slope and σ the standard deviation of linearization,

Wu et al. [68] examined the 2p shift for Cu films deposited on Al$_2$O$_3$ substrate at 80 and 300 K substrate temperatures. They found that both the 2p$_{3/2}$ and the Auger parameter shift not only with the film thickness but also with the substrate temperature, see Fig. 6.10c, d. Heating weakens interatomic interaction, which reduces the 2p$_{3/2}$ shift but increases the Auger energy E$_{KM}$.

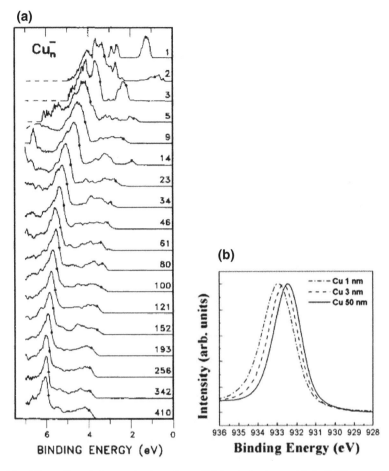

Fig. 6.9 PES profiles of size-resolved **a** valance DOS polarization and **b** $2p$ DOS entrapment of Cu clusters. Reprinted with permission from [63, 64]. Copyright 1990 American Physical Society. Copyright 2006 Elsevier

6.5.3   BOLS-TB Formulation and Derivatives

The BOLS-TB formulation extracts bonding and electronic information from these nice sets of data. The linearization of the size dependent APECS peaks of Cu/HOPG gives the slope B_v and the interception that corresponds to the $E_{2p}(\infty) = 932.7$ eV [69]. The slope corresponds to $m = 1$ for the Cu spherical dot ($\tau = 3$). Fitting to the $E_L(K)$, the $\alpha'(K)$, the $E_M(K)$, and the $E_K(K)$ curves, one can obtain the responsive $E_v(0)$ and $E_v(\infty)$ values for the Cu/HOPG clusters. Figure 6.11 presents the fitting to the size dependent APECS lines.

Table 6.3 presents information derived from the formulation of observations. The atomic $2p$ level energy $E_{2p}(0) = 931.0$ eV and its bulk shift $\Delta E_{2p}(\infty) = 1.70$ eV.

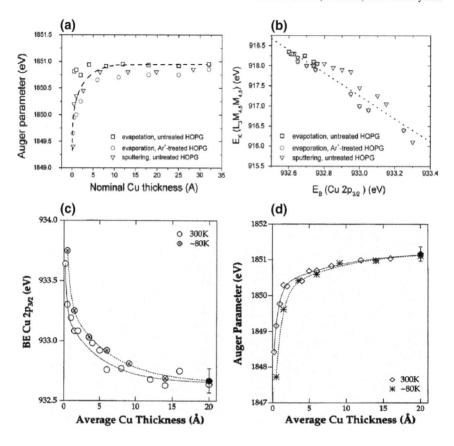

Fig. 6.10 Size dependence of the **a** Auger parameter α′(K) and the **b** Wagner plot for Cu/HOPG clusters. Size and temperature induced shift of **c** the $2p_{3/2}$ and **d** the α′(K) for Cu on Al_2O_3 substrate. Atomic undercoordination deepensthe Cu $2p$ level, reduces the Auger parameter. Reprinted with percussion from [65, 68]. Copyright 2002 Elsevier. Copyright 1996 American Institute of Physics

Likewise, the $E_{3d}(0) = 5.11$ and the $\Delta E_{3d}(\infty) = 2.12$ eV. The $E_{3d}(0)$ corresponds to E_F and $E_{3d}(\infty) = E_{3d}(0) + \Delta E_{3d}(\infty) = 7.23$ eV is the upper edge of the Cu 3d band [71–73].

Further formulating the size-induced $\Delta E_{2p}(K)$ for Cu/CYCL clusters by taking $\Delta E_{2p}(\infty) = 1.70$ eV reference turns out m = 1.82, which indicates that the Cu/CYCL interface interaction is stronger than that of Cu/HOPG. Repeating the iteration using the same values of $\Delta E_{2p}(\infty) = 1.70$ and $\Delta E_{3d}(\infty) = 2.12$ eV to the APECS lines of other chemically treated Cu samples results in m values that vary with the processing conditions, as shown in Fig. 6.12.

As compared in Fig. 6.12 and Table 6.3, BOLS-TB duplication of the measured APECS lines of the as grown and the chemically conditioned Cu nanocrystals on different substrates revealed the following:

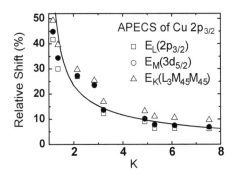

Fig. 6.11 BOLS reproduction (solid line) of the measured (scattered) size dependenc of the $E_L(K)$, $E_M(K)$, and $E_K(K)$ of Cu nanostructures deposited on HOPG [65] results in the atomic $E_L(0)$, $E_M(0)$, and $E_K(0)$ and their bulk shifts. Reproduction also results in the screening $\eta_{ML} = 1.25$ and the valence recharging $\kappa_{ML} = 1.05$ coefficients, as featured in Table 6.3. Reprinted with permission from [70]. Copyright 2006 Elsevier

Table 6.3 The $M(3d_{5/2})$ and the $L(2p_{3/2})$ energies of an isolated Cu atom and their bulk shifts derived from BOLS formulation of the size dependent APECS ($E_K(L_3M_{45}M_{45})$) lines

Substrate	HOPG	CYCL	CYCL	CYCL	Al$_2$O$_3$	Al$_2$O$_3$
Conditions	Ar$^+$	–	Ar$^+$	N$^+$	80 K	300 K
m	1	1.30	1.82	1.96	1.27	1.94
τ	3					1
B_v/E_v (∞)	2.08	2.78	4.03	4.39	1.01	1.61
$E_{M5}(\infty)$ (eV)	7.23					
$E_{M5}(0)$	5.11					
$\Delta E_{M5}(\infty)$	2.12					
$E_L(\infty)$	932.70 [70]					
$E_L(0)$	931.00					
$\Delta E_L(\infty)$	1.70					
η_{ML}	1.25					
κ	1.05	1.42	1.15	2.05	1.04	1.23

Size dependent energy shift for both levels:

$$E_v(z) = < E_v(0) > \pm\sigma + \Delta E_v(12)(1 + \Delta_H)$$

$$= \begin{cases} 931.00 \pm 0.01 + 1.70(1 + \Delta_H) & (2p) \\ 5.11 \pm 0.02 + 2.12(1 + \Delta_H) & (5s) \end{cases} \text{(eV)}$$

Linearization of the XPS data [68] using $\Delta E_{2p}(K) = b/K \pm \sigma$ (0.01 ~ 0.02) for the chemically treated Cu/HOPG, Cu/CYCL, and Cu/Al$_2$O$_3$ [65] derives the following quantities [19]

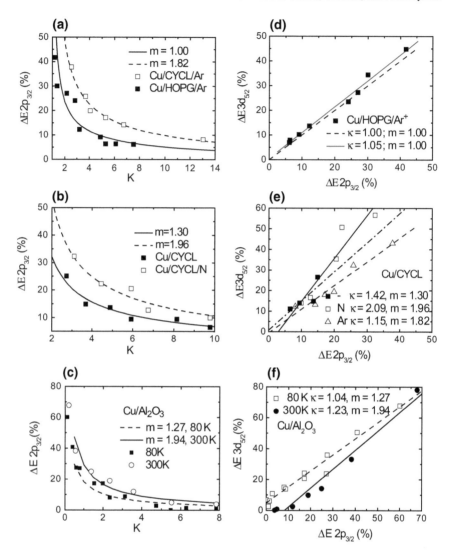

Fig. 6.12 Size dependence of the $2p_{3/2}$ level entrapment with involvement of **a** the Cu/HOPG and Cu/CYCL interface contribution and **b** N^+ plasma passivation [65], and **c** substrate temperature effect on Cu/Al$_2$O$_3$ clusters [68]. Cu prefers a layer-by-layer growth mode on Al$_2$O$_3$ ($\tau = 1$ and K is the average thickness of Cu film). The extended Wagner plots (**d–f**) correlate the energy shifts of the APECS involved M($3d$) and L($2p$) lines and their slopes κ_{ML} as the valence recharging coefficient. Reprinted with permission from [19]. Copyright 2003 Elsevier

(1) The m value increases from unity for Cu/HOPG/Ar$^+$ to 1.30 for Cu/CYCL and to 1.82 for Cu/CYCL/Ar$^+$, which indicates that the Cu is more reactive with CYCL than with HOPG. Cu atoms react hardly with carbon at room temperature [74] but it reacts easily with polymers [75, 76]. Although the Ar$^+$ does not react with Cu, Ar$^+$ bombardment promotes the Cu/CYCL reaction by heating and impacting.

(2) The m value increases from 1.3 for Cu/CYCL to 1.96 for Cu/CYCL/N$^+$. N$^+$-bombardment alters the surface bond from metallic to ionic that enhances the crystal potential [77].

(3) The layer-by-layer growth fashion ($\tau = 1$) of Cu films on Al$_2$O$_3$ could fit the results better though dispute remains on the growth mode of metal on oxide surfaces ([78] and references therein). The m increase from 1.27 at 80 K to 1.94 at 300 K for Cu/Al$_2$O$_3$ indicates that Cu bonds to oxygen atom more easily at room temperature than at lower temperatures [65].

(4) The $\eta_{ML} \cong 1.25$ means that the strength of crystal binding to the M($3d$) level is 25% stronger than that to the deeper L($2p$) level.

(5) The κ_{ML} in Fig. 6.12d–f is independent of the shape and size of the particle but it is sensitive to the chemical treatment. If charge transfers from one constituent to another, the κ_{ML} will change, otherwise it is unity, as compared in Table 6.3 for the conditioned Cu crystals.

6.6 Nickel

6.6.1 NEXAFS-XPS: Shell-Resolved Entrapment

Figure 6.13 shows the near edge X-ray absorption fine structure (NEXAFS) spectra, which revealed the sublayer-resolved Ni 2p shift [79]. The $2p$ shifts positively and the outermost layer shifts most. Low-energy electron diffraction (LEED) measurement also revealed that the first Ni(110) layer spacing contracts by $9.8 \pm 1.8\%$ with respect to the bulk lattice constant [80]. These observations evidence the BOLS prediction on the skin bond contraction and the associated local quantum entrapment.

6.6.2 APECS: 2p and 3d Band Cooperative Shift

Figure 6.14 displays the XPS $2p_{3/2}$ spectra and (b) the TEM images of Ni crystals deposited on SiO$_2$ substrates. Results indicate that the $2p$ peak undergoes quantum entrapment due to particle size reduction [81]. The thickness dependent APECS spectra in Fig. 6.15 demonstrate that the (a) $E_{2p1/2}$, (b) $E_{2p3/2}$, (c) E_k(LMM), and (d) $E_{3d5/2}$ shift positively but the E_K negatively when the Ni/TiO$_2$ film becomes thinner [82]. The broad $3d_{5/2}$ band contains the $4s$ band contribution. NEXAFS,

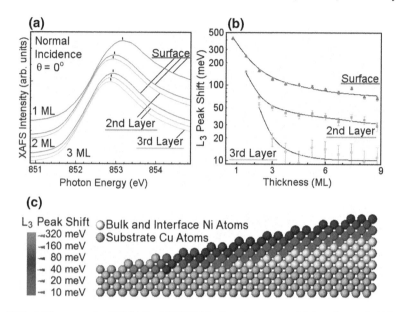

Fig. 6.13 Sublayer-resolved Ni(110) **a** NEXAFS $L_3(2p_{3/2})$ spectra and **b** L_3-edge peak shift with respect to the bulk component. Atomic undercoordination causes the core level entrapment. Panel **c** illustrates the respective Ni(110) sublayers. Reprinted with permission from [79]. Copyright 2008 American Physical Society

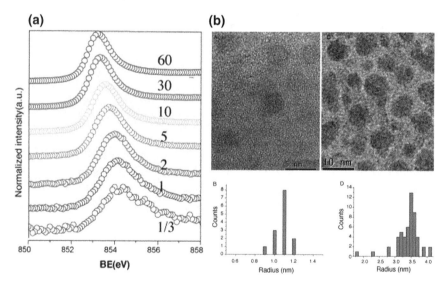

Fig. 6.14 **a** Deposition time (min) dependent Ni $2p_{3/2}$ CLS and **b** typical TEM images and the histogram for Ni/SiO$_2$ growing 2 and 30 min, showing the cluster size reduction induced positive core level shift. Reprinted with permission from [81]. Copyright 2010 Elsevier

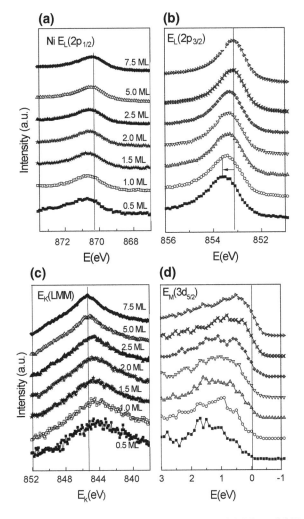

Fig. 6.15 Thickness dependent APECS **a** $E_{2p1/2}$, **b** $E_{2p3/2}$, **c** E_k(LMM), and **d** $E_{3d5/2}$ spectra for Ni films deposited on TiO$_2$ substrate. Atomic coordination reduction shifts the $2p$ core band and the $3d$ valance band positively and the kinetic energy negatively. Reprinted with permission from [82]. Copyright 2009 American Chemical Society

XPS, and APECS measurements verified consistently the BOLS expectation of Ni skin quantum entrapment (Table 6.4).

6.6.3   BOLS-TB-ZPS Formulation and Derivatives

Subtracting the XPS $2p$ spectrum collected from Ni film deposited for 3 min by
that deposited for 60 min resulted in the ZPS shown in Fig. 6.16a. The valley at
853 eV corresponds to the bulk component and the peak at 854 eV to the size-
induced quantum entrapment. Figure 6.16b shows the BOLS reproduced APECS
lines (scattered data) and (c) the extended Wagner plot with a slope $\kappa_{ML} \sim 1$. The
linear fitting to the $E_v(K)$ curves of the $2p_{1/2}$, $2p_{3/2}$, $E_k(LMM)$, and $3d_{5/2}$ peaks
gives the respective slope and intercept, which allows determination of the $E_v(0)$
and $\Delta E_v(\infty)$ in BOLS convention. (d) BOLS duplication of the size-resolved mean
lattice strain of Ni [13] in comparison to that of Al [89], and Pd [90] films. Table 6.4
summarizes the BOLS-APECS derivatives.

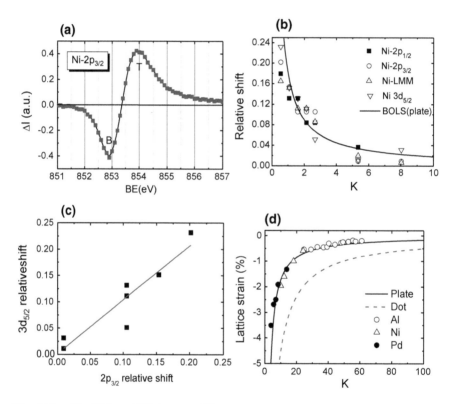

Fig. 6.16 a The Ni $2p_{3/2}$ ZPS (spectral difference between films deposited for 3 min and that
for 60 min on TiO_2) shows the bulk valley (B $\approx$ 853 eV) and the size-induced quantum entrapment (T $\approx$ 854 eV). **b** BOLS reproduction of (**b**) the APECS involved relative energy shifts (Δ_H
= $\Delta E_v(K)/\Delta E_v(\infty)$) with derived information featured in Table 6.4. **c** The unity $\Delta E_M - \Delta E_L$
slope indicates that interaction between the Ni crystals and TiO_2 substrate is negligible. **d** BOLS
reproduction of the size-resolved mean lattice strain of Ni [13] in comparison to that of Al [89], Pt
[91], and Pd [90] films. Reprinted with permission from [82]. Copyright 2009 American Chemical
Society

Table 6.4 BOLS-TB-APECS derived information of E_M, E_L and E_k energy shift for Ni/TiO$_2$ films [82]

Ni (m = 1, τ = 1)	$E_\nu(0)$/eV	$E_\nu(\infty)$/eV	$\Delta E_\nu(\infty)$/eV
$2p_{1/2}$	868.23	870.33	2.10
$2p_{3/2}$	851.10	853.18	2.08
$3d_{5/2}$	0.51	5.49	4.99
E_K(LMM)	853.02	845.45	−7.57

$\eta_{ML} = 4.99/2.08 = 2.40$ indicates that the crystal field experienced by the $3d$ electrons is 2.4 times that of the $2p$ electrons. The $\kappa_{ML} \sim 1$ in Fig. 6.16c shows that the Ni/TiO$_2$ interface effect is negligible

The CN dependent L_1, L_3, and the M_5 shift:

$$E_\nu(z) = \langle E_\nu(0) \rangle + \Delta E_\nu(12)C_z^{-1}$$
$$= \begin{cases} 868.23 + 2.10C_z^{-1} & (2p_{1/2}) \\ 851.10 + 2.08C_z^{-1} & (2p_{3/2}) \text{ eV} \\ 0.51 + 4.99C_z^{-1} & (3d_{5/2}) \end{cases}$$

6.7 Li, Na, K, Rb, and Cs Clusters and Skins

6.7.1 Na 2p and K 3p Entrapment

Figure 6.17 and Table 6.5 feature the decomposition of the $2p$ profiles for Na$_N$ [92] and the $3p$ for K$_N$ [93] atomic clusters. Table 6.6 lists the DFT derived charge transportation and bond strain for atoms located at different positions of I_h-13 and I_h-55 structures. Results indicate that the skin of lower coordinated atoms gain charge from the inner atoms because of the skin quantum entrapment. Bonds between lower coordinated corner or edge atoms contract more than that in the cluster interior [84].

6.7.2 CN Dependent Binding Energy Shift

Figure 6.18 shows the DFT derivatives of atomic-site resolved $2p$ shift for Na$_{13}$ and Na$_{55}$ clusters. Clearly, the fewer the neighbors is, the more the CLS will be. The size trend of the K $3p$ CLS is the same and it is unnecessary to repeat. This observation elaborates sufficiently the size trends for the Na $2p$ and the K $3p$ CLS. Table 6.6 features the local strain, charge gain, and the CLS for atoms at different sites in the Na$_{13}$ and in the Na$_{55}$ clusters. Consistency between measurements and DFT calculations confirms the BOLS Prediction.

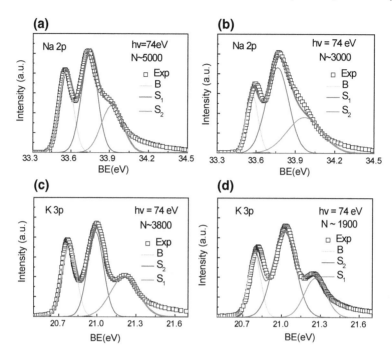

Fig. 6.17 BOLS-TB decomposition of the XPS profiles for **a, b** Na $2p$ and **c, d** K $3p$ atomic clusters with the bulk B, S_2, and S_1. Reprinted with permission from [83, 84]. Copyright 2015 Elsevier. Table 6.5 features the derived information

Table 6.5 The effective atomic CN(z) and the respective component energy of Na$_N$ [92] and K$_N$ [93] clusters (m = 1)

		z	E_{2p}
Na $E_{2p}(0)$/eV		0	31.167
$E_{2p}(12)$/eV	B	12	33.568
$\Delta E_{2p}(12)$/eV			2.401
3000	S_2	5.46	33.762
	S_1	3.63	33.968
5000	S_2	5.71	33.745
	S_1	3.91	33.921
K $E_{3p}(0)$/eV		0	18.034
$E_{3p}(12)$/eV	B	12	20.788
$E_{3p}(12)$/eV			2.754
1900	S_2	5.32	21.020
	S_1	3.58	21.257
3800	S_2	5.74	20.987
	S_1	3.80	21.212

One can obtain the local bond strain ε_z, bond energy, relative E_{coh} and E_{den} in sublayers accordingly [84]. As shown in Fig. 6.19, Li$_N$ follows the same size trend but different slope of core level shift

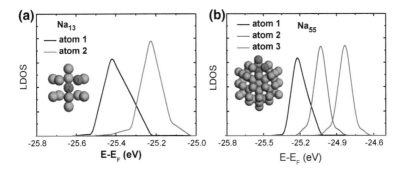

Fig. 6.18 DFT-derived atomic site-resolved BE of the $2p$ level for **a** Na_{13} and **b** Na_{55} clusters ($z_1 < z_2 < z_3$). Atom with the lowest CN shifts positively most. The BE shift for K atoms is in the same trend. Reprinted with permission from [84]. Copyright 2015 Elsevier

Table 6.6 DFT derived local bond strain $C_z - 1$, CLS and charge gain of Na and K atoms at different sites

	$C_z - 1(\%)$ (1–2)	$C_z - 1(\%)$ (2–3)	e (1–2)	e (1)	e (2)	e (3)	CLS (1–2)	CLS (1–3)
Na_{13}	−8.32	–	−0.072	−0.006	0.075	–	0.198	–
Na_{55}	−10.04	−6.65	−0.510	−0.052	−0.011	0.051	0.196	0.393
K_{13}	−7.814	–	−0.324	−0.027	0.323	–	0.324	–
K_{55}	−8.335	−5.188	−1.186	−0.059	−0.009	0.097	0.188	0.338

Negative sign means charge gain otherwise charge loss [84]

6.7.3 Li, Na, K Skins and Size Trends

Figure 6.19 shows the decomposition of the Li $1s$ [86], Na $2p$, [87], K $3p$ [88] energy shifts of the ((110) surfaces and the size-selected free Li_N, Na_N (Exp [92]), and K_N (Exp [94]) clusters based on measurements and DFT calculations. Table 6.7 features the energy shift derived from XPS analysis. Results show consistently that atomic undercoordination deepens the core levels without any discrimination. The solid skin and atomic clusters share the same nature of undercoordination effect. One can obtain the local bond length, bond energy, energy density and atomic cohesive energy by repeating previous iteration.

6.7.4 Rb and Cs Skins and Size Trends

Figure 6.20 shows the BOLS-TB decomposition of the bcc(110) skins of Rb $4p$ and Cs $5p$ and Table 6.8 summarizes the derived information. The incident beam energy

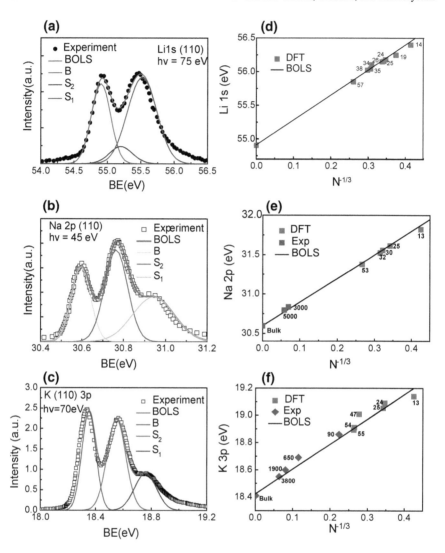

Fig. 6.19 Layer-order-resolved (110) skin XPS of **a** Li 1s [86], **b** Na 2p [87], **c** K 3p [88] and the DFT derived size-selected free **d** Li$_N$, **e** Na$_N$ (exp [92]), and **f** K$_N$ (exp [94]) clusters. Reprinted with permission from [83–85]. Copyright 2015 Elsevier and The Royal Society of Chemistry

resolved ZPS of the inset confirms that lower beam energy reveals more information from the skin the higher energy beams [136].

Figure 6.21 and Table 6.9 features the DFT derived, atomic site-resolved, energy shift of the local DOS for Rb and Cs clusters [85]. Results show consistently that these clusters follow the BOLS Prediction on the lattice strain, bond energy and the core level positions.

Table 6.7 Binding energy information of the bcc(110) skins of Li, Na, and K [83–85]

			Li [86]	Na [87]	K [88]
m = 1	i	z	$E_{1s}(z)$	$E_{2p}(z)$	$E_{3p}(z)$
bcc(110)	Atom	0	50.673	28.194	15.595
	B	12	54.906	30.595	18.354
	S_2	5.83	55.205	30.764	18.551
	S_1	3.95	55.520	30.943	18.757
	$\Delta E_{1s}(12)$	–	4.233	2.401	2.758

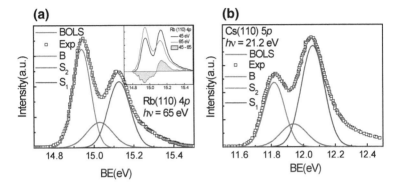

Fig. 6.20 BOLS – TB decomposition of the XPS $4p$ spectrum for **a** Rb(110) skin collected at 65 eV beam energy [86] and the $5p$ spectrum for **b** Cs(110) skin collected at 21.2 eV beam energy [95] with the bulk B, S_2, and S_1 components. The inset (**a**) shows the ZPS (difference between spectra collected using 45 and 65 eV beam energies) distills the bulk valley (B = 14.940 eV) and the monolayer skin component (S_1 = 15.127 eV). Reprinted with permission from [136]

6.8 Diamond Structured Si and Pb

6.8.1 Entrapment of the 2p and the Valence Band of Si

The valence band of the Si_N^+ clusters [96] and Si(210) edge atoms show the quantum entrapment dominance. Figure 6.22a, b plot the size-resolved $2p$ and valence band entrapment of Si_N^+ (N = 5–92) obtained by Vodel et al. [96] using soft X-ray photoionization method. Figure 6.22c is the size dependent $2p$ for porous Si with effective clusters sizes of 1.4–2.1 nm [97]. Results show consistently that molecular undercoordination induces local quantum entrapment, agreeing the skin-XPS derivatives (Sect. 4.5).

The entrapment of both the $2p$ and the valence bands follows the same $N^{-1/3}$ fashion. The cluster size $N = 4\pi K^3/3$ is converted to $K^{-1} = (3N/4\pi)^{-1/3} = 1.61N^{-1/3}$. With respect to the known bulk $E_{2p}(12)$ value at –99.2 eV, a 0.14 eV(= 99.2 – 99.06) and a –3.4 (=99.2 – 102.6) eV offset correction is necessary, which arises from the

Table 6.8 BOLS-ZPS derived effective CN, $\Delta E_v(i)$, ε_z, δE_z, δE_{Coh}, and δE_{Den} in various registries of bcc-structured (110) skins of Rb [86] and Cs [95]

	i	z	Rb(110)		Cs(110)		$-\varepsilon_z$ (%)	δE_z (%)	$-\delta E_{Coh}$ (%)	δE_{Den} (%)
			$E_{4p}(i)$	$\Delta E_{4p}(i)$	$E_{5p}(i)$	$\Delta E_{5p}(i)$				
Atom	—	0	13.654	—	10.284	—	—	—	—	—
Bulk	B	12.00	14.940	1.286	11.830	1.546	0	0	0	0
bcc(110)	S2	5.83	15.029	1.375	11.940	1.656	6.605	7.072	47.981	31.433
	S1	3.95	15.127	1.473	12.053	1.769	12.669	14.507	62.308	71.919

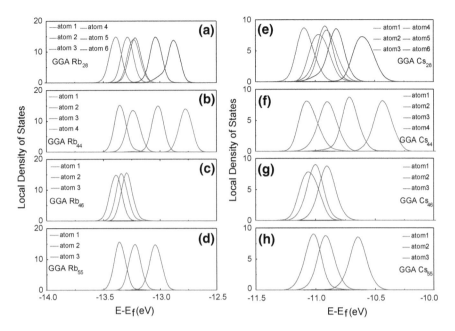

Fig. 6.21 DFT derived atomic site-resolved energy shift of the local DOS for **a, e** C_{I28}, **b, f** O_{h44}, **c, g** C_{3v46}, and **d, h** O_{h55} clusters of Rb and Cs [85]. Atoms counts from the outermost inward with an increase of the effective atomic CN. Table 6.9 list the detailed information on the undercoordination effect on the properties of the Cs and Rb clusters [85]

surface charging during experiment. The missing of one electron in the Si_N^+ cluster weakens the screening on the crystal potential, which enhances the shift accordingly. BOLS theory reproduction of the measurements turns out the $E_{2p}(0)$ of 96.74 eV for an isolated Si atom and its bulk shift $\Delta E_{2p}(12)$ of 2.46 eV.

A fitting to the size trends for porous Si [99] and Si_N^+ clusters [96] results in the K dependent Si $2p$ energy shift of both atomic clusters and nanostructured Si:

$$\begin{cases} E_{2p}(K) = E_{2p}(\infty) + b/K = 99.06 + 9.68/K \quad [p - Si] \\ E_{2p}(K) = 102.60 + 13.8/N^{1/3} = 102.60 + 10.69/K \quad [Si_N^+] \\ E_{vb}(K) = 4.20 + 10.89/K \quad [Si_N^+] \end{cases}$$

Figure 6.22d compares the DFT optimized valence DOS for the (100) and the stepped (210) skins. Bond contraction of the undercoordinated (210) edge bond entraps the valence DOS by about 1.0 eV without polarization. Results show consistently that Si atoms undercoordination in the skin of clusters and the terrace-edge entrapped electrons and deepen both the $2p$ and the valence bands of Si simultaneously. These observations are in consistency with that derived from skin XPS analysis, as shown in Sect. 4.5.

Table 6.9 The average effective CN, core level shift ΔE_z, CN density z_{ib}, bond strain ε_z, relative bond energy δE_z and relative energy density δE_{den} in various registries of Rb and Cs clusters with GGA functions [85]

	Atom position	E_{4p} (i)	z	ΔE_z	z_{ib} (%)	$-\varepsilon_z$ (%)
Rb_{28}	1	13.374	2.197	0.480	18.308	27.186
	2	13.267	2.575	0.373	21.458	22.485
	3	13.214	2.836	0.320	23.633	19.926
	4	13.214	2.836	0.320	23.633	19.926
	5	13.054	4.320	0.160	36	11.066
	6	12.894	12	0	100	0
Rb_{44}	1	13.368	1.949	0.582	16.242	31.160
	2	13.262	2.209	0.476	18.408	27.014
	3	12.998	3.659	0.212	30.492	14.152
	4	12.786	12	0	100	0
Rb_{46}	1	13.403	3.169	0.268	26.408	17.243
	2	13.349	3.638	0.214	30.317	14.268
	3	13.295	4.320	0.160	36	11.066
Rb_{55}	1	13.371	2.836	0.320	23.633	19.926
	2	13.211	4.320	0.160	36	11.066
	3	13.051	12	0	100	0
	Atom position	E_{5p} (i)	z	ΔE_z	z_{ib} (%)	$-\varepsilon_z$ (%)
Cs_{28}	1	11.102	2.527	0.462	21.058	23.010
	2	10.972	3.110	0.332	25.917	17.678
	3	10.918	3.470	0.278	28.917	15.244
	4	10.918	3.470	0.278	28.917	15.244
	5	10.825	4.417	0.185	36.808	10.689
	6	10.640	12	0	100	0
Cs_{44}	1	11.089	2.048	0.646	17.067	29.470
	2	10.905	2.527	0.462	21.058	23.010
	3	10.721	3.470	0.278	28.917	15.244
	4	10.443	12	0	100	0
Cs_{46}	1	11.091	3.470	0.278	28.917	15.244
	2	10.998	4.417	0.185	36.808	10.689
	3	10.906	6.287	0.093	52.397	5.673
Cs_{55}	1	11.013	2.901	0.371	24.175	19.356
	2	10.920	3.470	0.278	28.917	15.244
	3	10.642	12	0	100	0

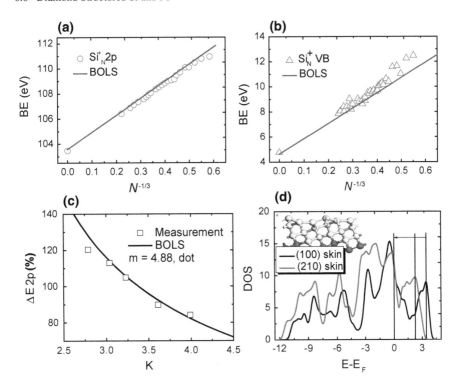

Fig. 6.22 Size trends of **a, b** the 2*p* and the valence band for Si$_N^+$ clusters [96] and **c** the 2*p* band for the porous Si [97]. The p-Si increases its size from K = 2.7 (A, 1.4 nm) to 4.0 (2.1 nm, E) in radius. DFT optimized (**d**) edge-entrapment of valence DOS for Si(210) (inset) with respect to that of the Si(100) surface. DFT revealed distances between atom E and its neighbors denoted 1, 2, and 3 are 0.2577, 0.3572, and 0.3394 nm with respect to the bulk distance of 0.3840, 0.4503, and 0.3840 nm. Reprinted with permission from [98]. Copyright 2014 American Institute of Physics

6.8.2 Pb 5d Binding Energy Shift

Figure 6.23a–d shows the BOLS-TB decomposed Pb $5d_{3/2}$ spectra collected from the (111) skin and Pd film deposited on the Si(100) substrate as well as Pb$_{3000}$ and Pb$_{1000}$ clusters. Figure 6.23e is the BOLS linearization of the measured and the DFT calculated size dependence of the 5*d* CLS with respect to E$_F$ (offset).

Table 6.10 summarizes the optimal z and E$_v$(z) components, and the energy levels E$_v$(0) and their bulk shift ΔE$_v$(12) for Pb and Si skins and clusters. One electron short of the Si$_N^+$ offsets the E$_v$(0) from that derived from the skin by 4.87 eV without changing the ΔE$_v$(12) value of 2.46 eV. This observation exemplifies how to correct the charging effect in calibrating and analyzing the XPS spectra. Reprinted with permission from [103]. Copyright 2014 The Royal Society of Chemistry.

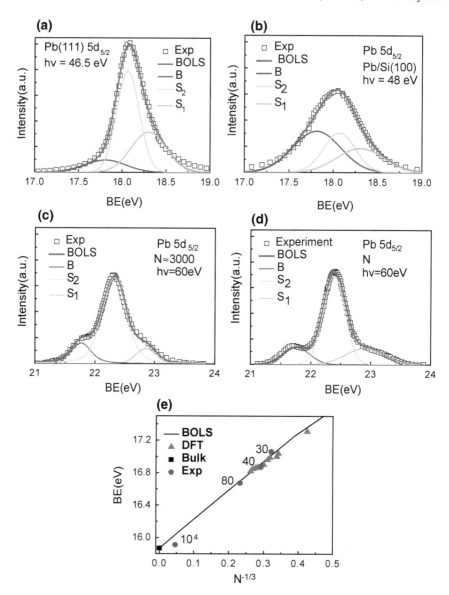

Fig. 6.23 BOLS-TB decomposition of Pb $5d_{5/2}$ spectra for **a** the (111) skin [100], **b** Pb film on Si(100) substrate [101], and **c**, **d** atomic clusters [102]. **e** The linearization of the size-resolved BE shift

Table 6.10 BOLS-TB derived effective z, $E_v(z)$ component, $E_v(0)$, and $E_v(12)$ for Pb skins and Si atomic clusters [96, 100–102]

		z	Pb 5d$_{3/2}$	Si 2p$_{3/2}$ (Si$_N^+$)	Si 2p$_{3/2}$ (Skins)
m			1 [5]	4.88 [98]	
$E_v(0)$/eV		0	14. 334	100.96	96.089
σ			0.005	0.003	0.003
$E_v(12)$/eV	B	12	17. 809	98.550	103.42
$\Delta E_v(12)$/eV	–	–	3.475	2.461	2.460
fcc(111)	S$_2$	6.31	18.016	–	–
	S$_1$	4.26	18.304	–	–
fcc(100)	S$_2$	5.73	18.020	–	–
	S$_1$	4.00	18.254	–	–
N = 3000	S$_2$	3.69	22.325	–	–
	S$_1$	2.45	22.855	–	–
N = 1000	S$_2$	3.47	22.387	–	–
	S$_1$	2.37	22.910	–	–

Derivatives for Si clusters are the same to the Si skin XPS (Sect. 4.5) [103]

6.9 Co, Fe, Pt, Rh, and Pd Nanocrystals

6.9.1 Co Islands: Valence Entrapment

A combination of STM/S measurements and DFT-MD calculations revealed that the mean lattice constant of Co islands contracts by 6% from the bulk value of 0.251–0.236 nm if one moves from the center to the edge of Co islands deposited on copper substrates [104]. Figure 6.24 shows that island size reduction from 22.5 to 4.8 nm entraps the valence DOS by ~0.2 eV [105]. However, the unoccupied states at 0.3 eV remain unchanged, which indicates that atomic undercoordination induces only the entrapment without polarization, which is the same to Pt adatoms [106] and Si terrace edges.

6.9.2 Pd, Fe, Rh, and Pt: Core Level Entrapment

Figure 6.25 shows the BOLS-TB reproduction of the size dependent CLS of Pd, Pt, Fe, and Rh nanocrystals using the same BOLS iteration. Inset (d) show the size-induced lattice strain of Pt and Rh crystals. XRD measurements further confirmed the size dependence of Pt lattice contraction [108]. Table 6.11 summarizes the estimated information form the size trend of XPS. Outcomes show consistently the size induced

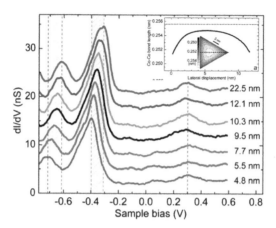

Fig. 6.24 STS of Co islands on Cu(111) surface show the dominance of quantum entrapment without polarization. Inset shows the Co–Co bond length contraction at the island edges. Reprinted with permission from [107]. Copyright 2007 American Physical Society

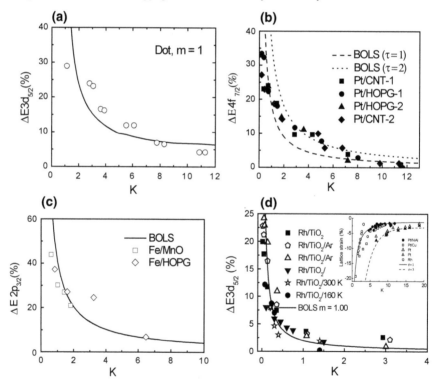

Fig. 6.25 BOLS reproduction of the size dependent energy shift of **a** Pd 3d [119], **b** Pt 4f [120–123], **c** Fe 2p [112, 113] and **d** Rh 3d [114–116] on various substrates with derived information featured in Table 6.11. Inset (**d**) shows BOLS reproduced size dependence of the lattice strains of Pt and Rh nanocrystals. Reprinted with permission from [110, 111, 124]. Copyright 2009 Elsevier and American Chemical Society. Copyright 2003 World Scientific Publishing Co Pte Ltd.

Table 6.11 Estimation of the energy levels of an isolated atom and their bulk shift of Pd [110], Pt [111], Fe [112, 113] and Rh [114–116] nanocrystals deposited on different substrates in comparison with the $\Delta E_v(12)$ (denoted with Refs.) derived from XPS analysis

	Pd 3d	Pt/C 4f	Fe/MgO 2p	Fe/HOPG 2p	Rh/TiO$_2$ 3d
m	1				
τ	3	3	1	1	3
$E_v(12)$/eV	330.34	71.10	706.37	707.02	307.50
$E_v(0)$/eV	334.35	68.10	704.38	704.65	302.16
$\Delta E_v(12)$/eV	3.98	2.99	1.99	2.37	5.34
$\Delta E_v(12)$/eV	4.359 [117]	3.28 [118]	–	–	4.367 [117]

quantum entrapment to the electron BE without any exception albeit the accuracy of derivatives, as further confirmed with DFT calculations for the Rh and Pt clusters [109] and Cu and Ag clusters [11].

6.10 STS of Si and Pd Nanostructures

Figure 6.26 compares the STS conductive spectra for Pd and Si nanowires, which shows the size reduction induced band gap enlargement. As the band gap is proportional to the cohesive energy per bond, the gap will expand when the size of a semiconductor shrinks such as Si nanowires [125]. For metals, the valence band will

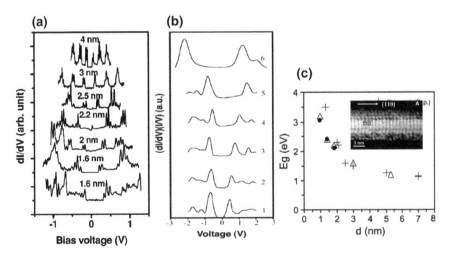

Fig. 6.26 STS conductance of **a** crystalline Pd particles with diameter in ranging of 1.6–4 nm [128] and of **b** Si nanowires with diameter decreases from 7 (curve 1) to 1.3 nm (curve 6) [125]. **c** Band gap expansion from 1.1 to 3.5 eV as the Si nanowire (inset) diameter shrinks from 7 to 1.3 nm

split, generating the artificial band gap, which may explain why a conductor turns to be an insulator or a semiconductor when its size turns to be the nanoscale [126], such as Au [127] and Pd [128] nanostructures. The artificial band gaps for Au and Pd clusters increase with the reducing number of Au and Pd atoms in the clusters.

Without igniting electron-phonon interaction at T = 4 K or electron-hole production or combination, the vehicle for the quantum confinement theory, STS revealed that [125] the E_G of Si nanorods increases from 1.1 to 3.5 eV when the wire diameter is reduced from 7.0 to 1.3 nm. The surface Si–Si bond contracts by ~12% from the bulk value (0.263 nm) to ~0.23 nm. This discovery concurs excitingly with the BOLS expectation: CN-imperfection shortens and strengthens the remaining bonds of the lower-coordinated atoms associated with E_G expansion that is proportional to the single bond energy. Similarly, the size-enlarged E_G of Si nanorods, Si nanodots, Ge nanostructures, and other III–V and II–VI semiconductors at the nanoscale follows closely the BOLS prediction without involving electron-hole interaction, electron-phonon coupling or quantum confinement [26, 129, 130].

Likewise, Vodel et al. [96] measured the energy shifts of the $2p$ core-level and the valence-band of size-selected Si_N^+ (n = 5–70) clusters using soft X-ray photoionization method. They found that, as shown in Fig. 6.27, the binding energies of both the $2p$ and the valence bands shift simultaneously to deeper (away from Fermi energy) in the same $n^{-1/3}$ manner.

Strikingly, the $2p$ energy of a Si-diode shift significantly during its operation under both forward and reverse bias [131, 132] or under high-frequency charging [133, 134]. The Si $2p$ energy shift can also be enhanced by photo illumination due to reduction in surface band bending [135]. This technique traces chemical and location specified surface potential variations as shifts of the peak positions with respect to the magnitude as well as the polarity of the applied voltage bias, which enables one to separate the dopant dependent shifts from those of the chemical ones. Therefore, neither electron-hole pair creation nor electron-phonon coupling comes into play in

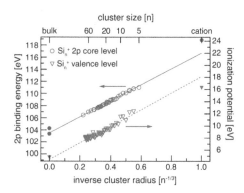

Fig. 6.27 Size-induced energy shift of the $2p$ core-band and the valence-band of Si_N^+ cluster. Reprint with Author permission from [96]

the STM/S experiments at 4 K. Therefore, the Si-E_G expansion and the energy shift of both the $2p$ and the valance band are intrinsically correlated, which further verify the essentiality and universality of the BOLS notion.

6.11   Summary

Consistency in the BOLS-NEP predictions, DFT calculations, STM/S, TEM, XPS, and APECS measurements of CN dependence of the lattice strain, muffin-tin inner potential depression, core level shift, and the chain end and island charge polarization of a number of metallic nanostructures confirmed the BOLS-NEP predictions. Atomic undercoordination is so important that it results in the local strain, quantum entrapment, charge densification and the subjective valence charge polarization, which is responsible for the size dependence of known bulk properties and size derivacy of anomalies that the bulk does never show. These emerging properties include the catalytic enhancement, toxicity, dilute magnetism, and Dirac-Fermi polaritons in topological insulators, as well as conductor-insulator transition, etc. Entrapment enlarges the electroaffinity and the band gap. Polarization lowers the work function. These elemental quantities mediate intrinsically the performance of a substance.

The localized polarization of the nonbonding s-electrons for Au($6s^1$), Ag($5s^1$), Rh($5s^1$) and Cu($4s^1$) adatoms makes the conducting metal to be insulator at the nanoscale. However, Pt($6s^0$) adatoms, reconstructed Pt surface, Co($4s^2$) islands, Co, Si, and Ni nanocrystals show entrapment dominance, instead, because of the lacking of the unpaired s^1 charge, according to the BOLS expectation and the present observations. In contrast, the strong valence polarization suggests the electronic configuration preference of the undercooordinated W($5d^1 6s^1$) and Mo ($4d^5 5s^1$) atoms. The consistence in the Si nanowire bandgap expansion, Pd bandgap generation, and Si cluster $2p$ level and valence band shift evidence the essentiality and universality of the BOLS notion and the associated quantum entrapment and subjective polarization.

References

1. C. Bittencourt, A. Felten, B. Douhard, J. Ghijsen, R.L. Johnson, W. Drube, J.J. Pireaux, Photoemission studies of gold clusters thermally evaporated on multiwall carbon nanotubes. Chem. Phys. **328**(1–3), 385–391 (2006)
2. S.P. Suprun, E.V. Fedosenko, Low-temperature recrystallization of Ge nanolayers on ZnSe. Semiconductors **41**(5), 590–595 (2007)
3. J.G. Tao, J.S. Pan, C.H.A. Huan, Z. Zhang, J.W. Chai, S.J. Wang, Origin of XPS binding energy shifts in Ni clusters and atoms on rutile TiO_2 surfaces. Surf. Sci. **602**(16), 2769–2773 (2008)
4. I. Aruna, B.R. Mehta, L.K. Malhotra, S.M. Shivaprasad, Size dependence of core and valence binding energies in Pd nanoparticles: interplay of quantum confinement and coordination reduction. J. Appl. Phys. **104**(6), 064308 (2008)

5. C.Q. Sun, Surface and nanosolid core-level shift: Impact of atomic coordination-number imperfection. Phys. Rev. B **69**(4), 045105 (2004)
6. B. Balamurugan, T. Maruyama, Size-modified d bands and associated interband absorption of Ag nanoparticies. J. Appl. Phys. **102**(3), 034306 (2007)
7. M. Reif, L. Glaser, M. Martins, W. Wurth, Size-dependent properties of small deposited chromium clusters by X-ray absorption spectroscopy. Phys. Rev. B **72**(15), 155405 (2005)
8. B. Balamurugan, T. Maruyama, Inhomogeneous effect of particle size on core-level and valence-band electrons: size-dependent electronic structure of Cu_3N nanoparticles. Appl. Phys. Lett. **89**, 033112 (2006)
9. S. Kim, M.C. Kim, S.H. Choi, K.J. Kim, H.N. Hwang, and C.C. Hwang, Size dependence of Si 2p core-level shift at Si nanocrystal/SiO_2 interfaces. Appl. Phys. Lett. **91**(10), 103113 (2007)
10. X. Zhang, J.L. Kuo, M.X. Gu, X.F. Fan, P. Bai, Q.G. Song, C.Q. Sun, Local structure relaxation, quantum trap depression, and valence charge polarization induced by the shorter-and-stronger bonds between under-coordinated atoms in gold nanostructures. Nanoscale **2**(3), 412–417 (2010)
11. S. Ahmadi, X. Zhang, Y. Gong, C.H. Chia, C.Q. Sun, Skin-resolved local bond contraction, core electron entrapment, and valence charge polarization of Ag and Cu nanoclusters. Phys. Chem. Chem. Phys. **16**(19), 8940–8948 (2014)
12. T. Ohgi, D. Fujita, Consistent size dependency of core-level binding energy shifts and single-electron tunneling effects in supported gold nanoclusters. Phys. Rev. B **66**(11), 115410 (2002)
13. A. Kara, T.S. Rahman, Vibrational properties of metallic nanocrystals. Phys. Rev. Lett. **81**(7), 1453–1456 (1998)
14. P.J. Feibelman, Relaxation of hcp(0001) surfaces: A chemical view. Phys. Rev. B **53**(20), 13740–13746 (1996)
15. B. Richter, H. Kuhlenbeck, H.J. Freund, P.S. Bagus, Cluster core-level binding-energy shifts: the role of lattice strain. Phys. Rev. Lett. **93**(2), 026805 (2004)
16. J. Nanda, D.D. Sarma, Photoemission spectroscopy of size selected zinc sulfide nanocrystallites. J. Appl. Phys. **90**(5), 2504–2510 (2001)
17. E. Grüneisen, The state of a body, in *Handbook of Physics*, vol. 10, 1–52 (NASA translation RE2-18-59W)
18. W. Qin, Y. Wang, Y.L. Huang, Z.F. Zhou, C. Yang, C.Q. Sun, Bond order resolved 3d(5/2) and valence band chemical shifts of Ag surfaces and nanoclusters. J. Phys. Chem. A **116**(30), 7892–7897 (2012)
19. C.Q. Sun, L.K. Pan, H.L. Bai, Z.Q. Li, P. Wu, E.Y. Jiang, Effects of surface passivation and interfacial reaction on the size-dependent 2p-level shift of supported copper nanosolids. Acta Mater. **51**(15), 4631–4636 (2003)
20. J.N. Crain, D.T. Pierce, End states in one-dimensional atom chains. Science **307**(5710), 703–706 (2005)
21. K. Schouteden, E. Lijnen, D.A. Muzychenko, A. Ceulemans, L.F. Chibotaru, P. Lievens, C.V. Haesendonck, A study of the electronic properties of Au nanowires and Au nanoislands on Au(111) surfaces. Nanotechnology **20**(39), 395401 (2009)
22. W.J. Huang, R. Sun, J. Tao, L.D. Menard, R.G. Nuzzo, J.M. Zuo, Coordination-dependent surface atomic contraction in nanocrystals revealed by coherent diffraction. Nat. Mater. **7**(4), 308–313 (2008)
23. W.H. Qi, B.Y. Huang, M.P. Wang, Bond-Length and -Energy Variation of Small Gold Nanoparticles. J. Comput. Theor. Nanosci. **6**(3), 635–639 (2009)
24. C.Q. Sun, C.M. Li, S. Li, B.K. Tay, Breaking limit of atomic distance in an impurity-free monatomic chain. Phys. Rev. B **69**(24), 245402 (2004)
25. J.T. Miller, A.J. Kropf, Y. Zha, J.R. Regalbuto, L. Delannoy, C. Louis, E. Bus, J.A. van Bokhoven, The effect of gold particle size on Au-Au bond length and reactivity toward oxygen in supported catalysts. J. Catal. **240**(2), 222–234 (2006)
26. C.Q. Sun, Size dependence of nanostructures: Impact of bond order deficiency. Prog. Solid State Chem. **35**(1), 1–159 (2007)

27. C.Q. Sun, Y. Sun, Y.G. Ni, X. Zhang, J.S. Pan, X.H. Wang, J. Zhou, L.T. Li, W.T. Zheng, S.S. Yu, L.K. Pan, Z. Sun, Coulomb repulsion at the nanometer-sized contact: a force driving superhydrophobicity, superfluidity, superlubricity, and supersolidity. J. Phys. Chem. C **113**(46), 20009–20019 (2009)

28. X. Zhang, C.Q. Sun, H. Hirao, Guanine binding to gold nanoparticles through nonbonding interactions. PCCP **15**(44), 19284–19292 (2013)

29. A. Visikovskiy, H. Matsumoto, K. Mitsuhara, T. Nakada, T. Akita, Y. Kido, Electronic d-band properties of gold nanoclusters grown on amorphous carbon. Phys. Rev. B **83**(16), 165428 (2011)

30. P. Zhang, T. Sham, X-ray studies of the structure and electronic behavior of alkanethiolate-capped gold nanoparticles: the interplay of size and surface effects. Phys. Rev. Lett. **90**(24), 245502 (2003)

31. A. Howard, D.N.S. Clark, C.E.J. Mitchell, R.G. Egdell, V.R. Dhanak, Initial and final state effects in photoemission from Au nanoclusters on TiO_2(110). Surf. Sci. **518**(3), 210–224 (2002)

32. M. Salmeron, S. Ferrer, M. Jazzar, G.A. Somorjai, Core-band and valence-band energy-level shifts in small two-dimensional islands of gold deposited on Pt(100)—the effect of step edge, surface, and bulk atoms. Phys. Rev. B **28**(2), 1158–1160 (1983)

33. A. Felten, C. Bittencourt, J.J. Pireaux, Gold clusters on oxygen plasma functionalized carbon nanotubes: XPS and TEM studies. Nanotechnology **17**(8), 1954–1959 (2006)

34. A. Tanaka, Y. Takeda, T. Nagasawa, K. Takahashi, Chemical states of dodecanethiolate-passivated Au nanoparticles: synchrotron-radiation photoelectron spectroscopy. Solid State Commun. **126**(4), 191–196 (2003)

35. D.C. Lim, R. Dietsche, M. Bubek, T. Ketterer, G. Gantefor, Y.D. Kim, Chemistry of mass-selected Au clusters deposited on sputter-damaged HOPG surfaces: the unique properties of Au-8 clusters. Chem. Phys. Lett. **439**(4–6), 364–368 (2007)

36. H.G. Boyen, A. Ethirajan, G. Kastle, F. Weigl, P. Ziemann, G. Schmid, M.G. Garnier, M. Buttner, P. Oelhafen, Alloy formation of supported gold nanoparticles at their transition from clusters to solids: does size matter? Phys. Rev. Lett. **94**(1), 016804 (2005)

37. S.B. DiCenzo, S.D. Berry, E.H. Hartford, Photoelectron spectroscopy of single-size Au clusters collected on a substrate. Phys. Rev. B **38**(12), 8465 (1988)

38. H. Yasuda, H. Mori, Spontaneous alloying of zinc atoms into gold clusters and formation of compound clusters. Phys. Rev. Lett. **69**(26), 3747–3750 (1992)

39. D.C. Lim, I. Lopez-Salido, R. Dietsche, M. Bubek, Y.D. Kim, Electronic and chemical properties of supported Au nanoparticles. Chem. Phys. **330**(3), 441–448 (2006)

40. C.W. Zou, B. Sun, G.D. Wang, W.H. Zhang, P.S. Xu, H.B. Pan, F.Q. Xu, Initial interface study of Au deposition on GaN(0001). Physica B-Condens. Matter **370**(1–4), 287–293 (2005)

41. A. Barinov, L. Casalis, L. Gregoratti, M. Kiskinova, Au/GaN interface: initial stages of formation and temperature-induced effects. Phys. Rev. B **63**(8), 085308 (2001)

42. T. Okazawa, M. Fujiwara, T. Nishimura, T. Akita, M. Kohyama, Y. Kido, Growth mode and electronic structure of Au nano-clusters on NiO(001) and TiO2(110). Surf. Sci. **600**(6), 1331–1338 (2006)

43. P. Buffat, J.P. Borel, Size effect on melting temperature of gold particles. Phys. Rev. A **13**(6), 2287–2298 (1976)

44. P. Donnadieu, S. Lazar, G.A. Botton, I. Pignot-Paintrand, M. Reynolds, S. Perez, Seeing structures and measuring properties with transmission electron microscopy images: a simple combination to study size effects in nanoparticle systems. Appl. Phys. Lett. **94**(26), 263116 (2009)

45. C.Q. Sun, H.L. Bai, S. Li, B.K. Tay, C. Li, T.P. Chen, E.Y. Jiang, Length, strength, extensibility, and thermal stability of a Au-Au bond in the gold monatomic chain. J. Phys. Chem. B **108**(7), 2162–2167 (2004)

46. A. Sperl, J. Kroger, N. Neel, H. Jensen, R. Berndt, A. Franke, E. Pehlke, Unoccupied states of individual silver clusters and chains on Ag(111). Phys. Rev. B (Condens. Matter Mater. Phys.) **77**(8), 085422–085427 (2008)

47. A. Sperl, J. Kroger, R. Berndt, A. Franke, E. Pehlke, Evolution of unoccupied resonance during the synthesis of a silver dimer on Ag(111). New J. Phys. **11**(6), 063020 (2009)
48. D. Roy, Z. Barber, T. Clyne, Ag nanoparticle induced surface enhanced Raman spectroscopy of chemical vapor deposition diamond thin films prepared by hot filament chemical vapor deposition. J. Appl. Phys. **91**(9), 6085–6088 (2002)
49. S. Pal, Y.K. Tak, J.M. Song, Does the antibacterial activity of silver nanoparticles depend on the shape of the nanoparticle? A study of the gram-negative bacterium *Escherichia coli*. Appl. Environ. Microbiol. **73**(6), 1712–1720 (2007)
50. P. Luches, F. Pagliuca, S. Valeri, F. Illas, G. Preda, G. Pacchioni, Nature of Ag islands and nanoparticles on the $CeO_2(111)$ surface. J. Phys. Chem. C **116**, 1122–1132 (2011)
51. D.D. Kong, G.D. Wang, Y.H. Pan, S.W. Hu, J.B. Hou, H.B. Pan, C.T. Campbell, J.F. Zhu, Growth, structure, and stability of Ag on CeO(2)(111): synchrotron radiation photoemission studies. J. Phys. Chem. C **115**(14), 6715–6725 (2011)
52. J.A. Farmer, J.H. Baricuatro, C.T. Campbell, Ag adsorption on reduced $CeO_2(111)$ Thin Films†. J. Phys. Chem. C **114**(40), 17166–17172 (2010)
53. M.A.M. Branda, N.C. Hernández, J.F. Sanz, F. Illas, Density functional theory study of the interaction of Cu, Ag, and Au atoms with the regular CeO_2 (111) surface. J. Phys. Chem. C **114**(4), 1934–1941 (2010)
54. K. Luo, X. Lai, C.W. Yi, K.A. Davis, K.K. Gath, D.W. Goodman, The growth of silver on an ordered alumina surface. J. Phys. Chem. B **109**(9), 4064–4068 (2005)
55. G. Moretti, Auger parameter and Wagner plot in the characterization of chemical states by X-ray photoelectron spectroscopy: a review. J. Electron Spectrosc. Relat. Phenom. **95**(2–3), 95–144 (1998)
56. M. Ohno, Many-electron effects in the Auger-photoelectron coincidence spectroscopy spectra of the late 3d-transition metals. J. Electron Spectrosc. Relat. Phenom. **136**(3), 229–234 (2004)
57. N.C. Hernández, J. Graciani, A. Márquez, J.F. Sanz, Cu, Ag and Au atoms deposited on the α-Al2O3(0001) surface: a comparative density functional study. Surf. Sci. **575**(1–2), 189–196 (2005)
58. I. Lopez-Salido, D.C. Lim, Y.D. Kim, Ag nanoparticles on highly ordered pyrolytic graphite (HOPG) surfaces studied using STM and XPS. Surf. Sci. **588**(1–3), 6–18 (2005)
59. K. Luo, T.P. St Clair, X. Lai, D.W. Goodman, Silver growth on TiO2(110)(1 × 1) and (1 × 2). J. Phys. Chem. B **104**(14), 3050–3057 (2000)
60. M. Rocca, L. Savio, L. Vattuone, U. Burghaus, V. Palomba, N. Novelli, F.B. de Mongeot, U. Valbusa, R. Gunnella, G. Comelli, A. Baraldi, S. Lizzit, G. Paolucci, Phase transition of dissociatively adsorbed oxygen on Ag(001). Phys. Rev. B **61**(1), 213–227 (2000)
61. J.N. Andersen, D. Hennig, E. Lundgren, M. Methfessel, R. Nyholm, M. Scheffler, Surface core-level shifts of some 4d-metal single-crystal surfaces: experiments and ab initio calculations. Phys. Rev. B **50**(23), 17525–17533 (1994)
62. S. Folsch, P. Hyldgaard, R. Koch, K.H. Ploog, Quantum confinement in monatomic Cu chains on Cu(111). Phys. Rev. Lett. **92**(5), 056803 (2004)
63. O. Cheshnovsky, K.J. Taylor, J. Conceicao, R.E. Smalley, Ultraviolet photoelectron-spectra of mass-selected copper clusters—evolution of the 3d band. Phys. Rev. Lett. **64**(15), 1785–1788 (1990)
64. D.-W. Shin, C. Dong, M. Mattesini, A. Augustsson, S. Mao, C. Chang, C. Persson, R. Ahuja, J. Nordgren, S.X. Wang, Size dependence of the electronic structure of copper nanoclusters in SiC matrix. Chem. Phys. Lett. **422**(4), 543–546 (2006)
65. D.Q. Yang, E. Sacher, Initial- and final-state effects on metal cluster/substrate interactions, as determined by XPS: copper clusters on Dow Cyclotene and highly oriented pyrolytic graphite. Appl. Surf. Sci. **195**(1–4), 187–195 (2002)
66. D.Q. Yang, L. Martinu, E. Sacher, A. Sadough-Vanini, M. Grp Couches, Nitrogen plasma treatment of the dow Cyclotene 3022 surface and its reaction with evaporated copper. Appl. Surf. Sci. **177**(1–2), 85–95 (2001)
67. D.Q. Yang, E. Sacher, Argon ion treatment of the dow cyclotene 3022 surface and its effect on the adhesion of evaporated copper. Appl. Surf. Sci. **173**(1–2), 30–39 (2001)

68. Y.T. Wu, E. Garfunkel, T.E. Madey, Initial stages of Cu growth on ordered Al_2O_3 ultrathin films. J. Vacuum Sci. Technol. A **14**(3), 1662–1667 (1996)

69. J. Bearden, A. Burr, Reevaluation of X-ray atomic energy levels. Rev. Mod. Phys. **39**(1), 125 (1967)

70. C.Q. Sun, L.K. Pan, T.P. Chen, X.W. Sun, S. Li, C.M. Li, Distinguishing the effect of crystal-field screening from the effect of valence recharging on the $2P_{3/2}$ and $3d_{5/2}$ level energies of nanostructured copper. Appl. Surf. Sci. **252**(6), 2101–2107 (2006)

71. C.Q. Sun, in *Relaxation of the Chemical Bond*. The Springer Series in Chemical Physics, vol. 108 (Springer, Heidelberg, 2014), 807pp

72. R. DiDio, D. Zehner, E. Plummer, An angle-resolved UPS study of the oxygen-induced reconstruction of Cu (110). J. Vaccum Sci. Technol. A **2**(2), 852–855 (1984)

73. C.Q. Sun, Oxidation electronics: bond-band-barrier correlation and its applications. Prog. Mater Sci. **48**(6), 521–685 (2003)

74. W.F. Egelhoff Jr., G.G. Tibbetts, Growth of copper, nickel, and palladium films on graphite and amorphous carbon. Phys. Rev. B **19**(10), 5028 (1979)

75. J.M. Burkstrand, Substrate effects on the electronic structure of metal overlayers—an XPS study of polymer-metal interfaces. Phys. Rev. B **20**(12), 4853 (1979)

76. M. Chtaib, J. Ghijsen, J. Pireaux, R. Caudano, R. Johnson, E. Orti, J. Bredas, Photoemission study of the copper/poly (ethylene terephthalate) interface. Phys. Rev. B **44**(19), 10815 (1991)

77. C.Q. Sun, A model of bonding and band-forming for oxides and nitrides. Appl. Phys. Lett. **72**(14), 1706–1708 (1998)

78. K. Borgohain, J.B. Singh, M.V.R. Rao, T. Shripathi, S. Mahamuni, Quantum size effects in CuO nanoparticles. Phys. Rev. B **61**(16), 11093–11096 (2000)

79. F. Matsui, T. Matsushita, Y. Kato, M. Hashimoto, K. Inaji, F.Z. Guo, H. Daimon, Atomic-layer resolved magnetic and electronic structure analysis of Ni thin film on a Cu(001) surface by diffraction spectroscopy. Phys. Rev. Lett. **100**(20), 207201 (2008)

80. M.L. Xu, S.Y. Tong, The structure of overlayer adsorption on Ni(001) by high-resolution electron-energy loss spectroscopy. J. Vacuum Sci. Technol. A **4**(3), 1302–1303 (1986)

81. Y.G. Nie, J.S. Pan, Z. Zhang, J.W. Chai, S.J. Wang, C.S. Yang, D. Li, C.Q. Sun, Size dependent 2p(3/2) binding-energy shift of Ni nanoclusters on SiO_2 support: skin-depth local strain and quantum trapping. Appl. Surf. Sci. **256**(14), 4667–4671 (2010)

82. Y. Sun, J.S. Pan, J.G. Tao, Y.G. Nie, C.H.A. Huan, Z. Zhang, J.W. Chai, D. Li, S.J. Wang, C.Q. Sun, Size dependence of the 2p(3/2) and 3d(5/2) binding energy shift of Ni nanostructures: skin-depth charge and energy trapping. J. Phys. Chem. C **113**(25), 10939–10946 (2009)

83. T. Zhang, M. Bo, Y. Guo, Y. Huang, H. Chen, C. Li, C.Q. Sun, Coordination-resolved atomistic local bonding and 3p electronic energetics of K(110) skin and atomic clusters. Appl. Surf. Sci. **325**, 33–38 (2015)

84. M. Bo, Y. Wang, Y. Huang, Y. Liu, C. Li, C.Q. Sun, Atomistic spectrometrics of local bond-electron-energy pertaining to Na and K clusters. Appl. Surf. Sci. **325**, 33–38 (2015)

85. M. Bo, Y. Guo, Y. Huang, Y. Liu, Y. Wang, C. Li, C.Q. Sun, Coordination-resolved bonding and electronic dynamics of Na atomic clusters and solid skins. RSC Adv. **5**(44), 35274–35281 (2015)

86. G. Wertheim, D.M. Riffe, N. Smith, P. Citrin, Electron mean free paths in the alkali metals. Phys. Rev. B **46**(4), 1955 (1992)

87. D.M. Riffe, G. Wertheim, P. Citrin, Enhanced vibrational broadening of core-level photoemission from the surface of Na (110). Phys. Rev. Lett. **67**(1), 116–119 (1991)

88. G. Wertheim, D.M. Riffe, Evidence for crystal-field splitting in surface-atom photoemission from potassium. Phys. Rev. B **52**(20), 14906 (1995)

89. J. Woltersdorf, A. Nepijko, E. Pippel, Dependence of lattice parameters of small particles on the size of the nuclei. Surf. Sci. **106**(1), 64–69 (1981)

90. R. Lamber, S. Wetjen, N.I. Jaeger, Size dependence of the lattice parameter of small palladium particles. Phys. Rev. B **51**(16), 10968 (1995)

91. M. Zhao, X. Zhou, Q. Jiang, Comparison of different models for melting point change of metallic nanocrystals. J. Mater. Res. **16**(11), 3304–3308 (2001)

92. S. Peredkov, G. Öhrwall, J. Schulz, M. Lundwall, T. Rander, A. Lindblad, H. Bergersen, A. Rosso, W. Pokapanich, N. Mårtensson, S. Svensson, S. Sorensen, O. Björneholm, M. Tchaplyguine, Free nanoscale sodium clusters studied by core-level photoelectron spectroscopy. Phys. Rev. B **75**(23), 235407 (2007)

93. A. Rosso, G. Öhrwall, I.L. Bradeanu, S. Svensson, O. Björneholm, M. Tchaplyguine, Photoelectron spectroscopy study of free potassium clusters: core-level lines and plasmon satellites. Phys. Rev. A: At. Mol. Opt. Phys. **77**, 043202 (2008)

94. M.-H. Mikkelä, M. Tchaplyguine, K. Jänkälä, T. Andersson, C. Zhang, O. Björneholm, M. Huttula, Size-dependent study of Rb and K clusters using core and valence level photoelectron spectroscopy. Eur. Phys. J. D—Atomic Mol. Opt. Plasma Phys. **64**(2), 347–352 (2011)

95. G. Wertheim, D. Buchanan, Conduction-electron screening and surface properties of Cs metal. Phys. Rev. B **43**(17), 13815 (1991)

96. M. Vogel, C. Kasigkeit, K. Hirsch, A. Langenberg, J. Rittmann, V. Zamudio-Bayer, A. Kulesza, R. Mitrić, T. Möller, B. v. Issendorff, J. Lau, 2p core-level binding energies of size-selected free silicon clusters: Chemical shifts and cluster structure. Phys. Rev. B **85**(19), 195454 (2012)

97. L.K. Pan, Y.K. Ee, C.Q. Sun, G.Q. Yu, Q.Y. Zhang, B.K. Tay, Band-gap expansion, core-level shift, and dielectric suppression of porous silicon passivated by plasma fluorination. J. Vac. Sci. Technol. B **22**(2), 583–587 (2004)

98. M. Bo, Y. Wang, Y. Huang, X. Yang, Y. Yang, C. Li, C.Q. Sun, Coordination-resolved local bond relaxation, electron binding-energy shift, and Debye temperature of Ir solid skins. Appl. Surf. Sci. **320**, 509–513 (2014)

99. L. Pan, S. Xu, X. Liu, W. Qin, Z. Sun, W. Zheng, C.Q. Sun, Skin dominance of the dielectric electronic-phononic-photonic attribute of nanoscaled silicon. Surf. Sci. Rep. **68**(3–4), 418–445 (2013)

100. J. Dalmas, H. Oughaddou, G. Le Lay, B. Aufray, G. Tréglia, C. Girardeaux, J. Bernardini, J. Fujii, G. Panaccione, Photoelectron spectroscopy study of Pb/Ag (111) in the submonolayer range. Surf. Sci. **600**(6), 1227–1230 (2006)

101. G. Le Lay, K. Hricovini, J. Bonnet, Ultraviolet photoemission study of the initial adsorption of Pb on Si (100) 2 × 1. Phys. Rev. B **39**(6), 3927 (1989)

102. S. Peredkov, S. Sorensen, A. Rosso, G. Öhrwall, M. Lundwall, T. Rander, A. Lindblad, H. Bergersen, W. Pokapanich, S. Svensson, O. Björneholm, N. Mårtensson, M. Tchaplyguine, Size determination of free metal clusters by core-level photoemission from different initial charge states. Phys. Rev. B **76**(8), 081402 (2007)

103. M. Bo, Y. Wang, Y. Huang, W. Zhou, C. Li, C.Q. Sun, Coordination-resolved local bond relaxation and electron binding-energy shift of Pb solid skins and atomic clusters. J. Mater. Chem. C **2**(30), 6090–6096 (2014)

104. O. Mironets, H.L. Meyerheim, C. Tusche, V.S. Stepanyuk, E. Soyka, H. Hong, P. Zschack, N. Jeutter, R. Felici, J. Kirschner, Bond length contraction in cobalt nanoislands on Cu(001) analyzed by surface X-ray diffraction. Phys. Rev. B **79**, 035406 (2009)

105. O. Mironets, H.L. Meyerheim, C. Tusche, V.S. Stepanyuk, E. Soyka, P. Zschack, H. Hong, N. Jeutter, R. Felici, J. Kirschner, Direct evidence for mesoscopic relaxations in cobalt nanoislands on Cu(001). Phys. Rev. Lett. **100**, 096103 (2008)

106. L. Bianchettin, A. Baraldi, S. de Gironcoli, E. Vesselli, S. Lizzit, L. Petaccia, G. Comelli, R. Rosei, Core level shifts of undercoordinated Pt atoms. J. Chem. Phys. **128**(11), 114706 (2008)

107. M.V. Rastei, B. Heinrich, L. Limot, P.A. Ignatiev, V.S. Stepanyuk, P. Bruno, J.P. Bucher, Size-dependent surface states of strained cobalt nanoislands on Cu(111). Phys. Rev. Lett. **99**(24), 246102–246104 (2007)

108. I. Leontyev, A. Kuriganova, N. Leontyev, L. Hennet, A. Rakhmatullin, N. Smirnova, V. Dmitriev, Size dependence of the lattice parameters of carbon supported platinum nanoparticles: X-ray diffraction analysis and theoretical considerations. RSC Adv. **4**(68), 35959–35965 (2014)

109. S. Ahmadi, X. Zhang, Y. Gong, C.Q. Sun, Atomic under-coordination fascinated catalytic and magnetic behavior of Pt and Rh nanoclusters. Phys. Chem. Chem. Phys. **16**(38), 20537–20547 (2014)

110. C.Q. Sun, Atomic-coordination-imperfection-enhanced Pd-3d(5/2) crystal binding energy. Surf. Rev. Lett. **10**(6), 1009–1013 (2003)
111. Y. Sun, Y. Wang, J.S. Pan, L.L. Wang, C.Q. Sun, Elucidating the 4f binding energy of an isolated Pt atom and its bulk shift from the measured surface- and size-induced Pt 4f core level shift. J. Phys. Chem. C **113**(33), 14696–14701 (2009)
112. D.-Q. Yang, E. Sacher, Characterization and oxidation of Fe nanoparticles deposited onto highly oriented pyrolytic graphite, using X-ray photoelectron spectroscopy. J. Phys. Chem. C **113**(16), 6418–6425 (2009)
113. V. Di Castro, S. Ciampi, XPS study of the growth and reactivity of FeMnO thin films. Surf. Sci. **331**, 294–299 (1995)
114. A. Berkó, I. Ulrych, K. Prince, Encapsulation of Rh nanoparticles supported on TiO_2 (110)-(1 × 1) surface: XPS and STM studies. J. Phys. Chem. B **102**(18), 3379–3386 (1998)
115. H.R. Sadeghi, V.E. Henrich, Rh on TiO_2: model catalyst studies of the strong metal-support interaction. Appl. Surf. Sci. **19**(1), 330–340 (1984)
116. L. Óvári, J. Kiss, Growth of Rh nanoclusters on TiO_2(110): XPS and LEIS studies. Appl. Surf. Sci. **252**(24), 8624–8629 (2006)
117. Y. Wang, Y.G. Nie, J.S. Pan, L.K. Pan, Z. Sun, L.L. Wang, C.Q. Sun, Orientation-resolved 3d(5/2) binding energy shift of Rh and Pd surfaces: anisotropy of the skin-depth lattice strain and quantum trapping. Phys. Chem. Chem. Phys. **12**(9), 2177–2182 (2010)
118. C.Q. Sun, Y. Wang, Y.G. Nie, Y. Sun, J.S. Pan, L.K. Pan, Z. Sun, Adatoms-induced local bond contraction, quantum trap depression, and charge polarization at Pt and Rh surfaces. J. Phys. Chem. C **113**(52), 21889–21894 (2009)
119. H.N. Aiyer, V. Vijayakrishnan, G.N. Subbanna, C.N.R. Rao, Investigations of Pd clusters by the combined use of HREM, STM, high-energy spectroscopies and tunneling conductance measurements. Surf. Sci. **313**(3), 392–398 (1994)
120. P. Marcus, C. Hinnen, XPS study of the early stages of deposition of Ni, Cu and Pt on HOPG. Surf. Sci. **392**(1–3), 134–142 (1997)
121. D.Q. Yang, E. Sacher, Platinum nanoparticle interaction with chemically modified highly oriented pyrolytic graphite surfaces. Chem. Mater. **18**(7), 1811–1816 (2006)
122. D.Q. Yang, E. Sacher, Strongly enhanced interaction between evaporated Pt nanoparticles and functionalized multiwalled carbon nanotubes via plasma surface modifications: effects of physical and chemical defects. J. Phys. Chem. C **112**(11), 4075–4082 (2008)
123. C. Bittencourt, M. Hecq, A. Felten, J.J. Pireaux, J. Ghijsen, M.P. Felicissimo, P. Rudolf, W. Drube, X. Ke, G. Van Tendeloo, Platinum-carbon nanotube interaction. Chem. Phys. Lett. **462**(4–6), 260–264 (2008)
124. Y. Wang, L.L. Wang, C.Q. Sun, The $2p_{3/2}$ binding energy shift of Fe surface and Fe nanoparticles. Chem. Phys. Lett. **480**(4–6), 243–246 (2009)
125. D.D.D. Ma, C.S. Lee, F.C.K. Au, S.Y. Tong, S.T. Lee, Small-diameter silicon nanowire surfaces. Science **299**(5614), 1874–1877 (2003)
126. E. Roduner, Size matters: why nanomaterials are different. Chem. Soc. Rev. **35**(7), 583–592 (2006)
127. B. Wang, X.D. Xiao, X.X. Huang, P. Sheng, J.G. Hou, Single-electron tunneling study of two-dimensional gold clusters. Appl. Phys. Lett. **77**(8), 1179–1181 (2000)
128. B. Wang, K.D. Wang, W. Lu, J.L. Yang, J.G. Hou, Size-dependent tunneling differential conductance spectra of crystalline Pd nanoparticles. Phys. Rev. B **70**(20), 205411 (2004)
129. L.K. Pan, C.Q. Sun, Coordination imperfection enhanced electron-phonon interaction. J. Appl. Phys. **95**(7), 3819–3821 (2004)
130. L.K. Pan, Z. Sun, C.Q. Sun, Coordination imperfection enhanced electron-phonon interaction and band-gap expansion in Si and Ge nanocrystals. Scripta Mater. **60**(12), 1105–1108 (2009)
131. S. Suzer, XPS investigation of a Si-diode in operation. Anal. Methods **4**(11), 3527–3530 (2012)
132. S. Suzer, H. Sezen, A. Dana, Two-dimensional X-ray photoelectron spectroscopy for composite surface analysis. Anal. Chem. **80**(10), 3931–3936 (2008)

133. S. Suzer, H. Sezen, G. Ertas, A. Dana, XPS measurements for probing dynamics of charging. J. Electron Spectrosc. Relat. Phenom. **176**(1–3), 52–57 (2010) .

134. S. Suzer, E. Abelev, S.L. Bernasek, Impedance-type measurements using XPS. Appl. Surf. Sci. **256**(5), 1296–1298 (2009)

135. H. Sezen, S. Suzer, Communication: enhancement of dopant dependent X-ray photoelectron spectroscopy peak shifts of Si by surface photovoltage. J. Chem. Phys. **135**(14), 141102 (2011)

136. Y. Guo, M. Bo, Y. Wang, Y. Liu, Y. Huang, C. Q. Sun, Atomistic bond relaxation, energy entrapment, and electron polarization of the Rb and Cs clusters (N $\leq$ 58). Phys. Chem. Chem. Phys. **17**(45), 30389–30397 (2015)

Chapter 7
Carbon Allotropes

Abstract A combination of STM/S and ZPS confirms the BOLS-NEP prediction that the C–C bond contraction, core electron entrapment, and valence electron subjective polarization occur when the atomic CN is reduced. Valence charge polarization becomes dominance only at sites surrounding point defects or along the zigzag edges of $\sqrt{3}$d atomic distance. Triple or quasi-triple bond formation prevents charge polarization at the graphite monolayer skin and the armcharied or reconstructed zigzag edges of graphene. As a carrier of topological insulators, Dirac-Fermi polariton forms from the isolation and polarization of the unpaired, spin-resolved, dangling σ bond electron at the excessively undercoordinated atomic sites.

Highlights

- Two-coordinated C–C bonds contract by 30% associated with 150% energy gain and polarization.
- Three-coordinated C–C bonds shorten by 19% with 68% energy gain and entrapment dominance.
- Dirac-Fermion forms along the zizag-edge by dangling-bond charge isolation and polarization.
- Shorter C–C length inhibits Dirac-Fermion formation at the armchaired or reconstructed z-edge.

7.1 Introduction

7.1.1 Wonders of CNTs and GNRs

Since the discovery of carbon nanotubes (CNTs) in the earlier 1990s [1], there has been ever-increasing interest in the new forms of carbon because of not only the novel structures and properties that the bulk graphite or diamond do not demonstrate but also the potentially important applications in scientific and engineering thrusts such

C. Q. Sun, *Electron and Phonon Spectrometrics*,
https://doi.org/10.1007/978-981-15-3176-7_7

as atomic-force microscope tips [2], cathode field emitters [3, 4], electronic circuit devices [5, 6], hydrogen storage [7–10], chemical sensors [11, 12], energy storage and management [10, 13–15], and phonon and electronic transportation devices [16–19].

Unrolling a single-walled CNT (SWCNT) generates a graphene nanoribbon (GNR) [20, 21] with high fraction of undercoordinated carbon atoms located at the open edges. The undercoordinated edge atoms and the abnormal performances of electrons surrounding the edges have inspired even more increasing interest because of the edge-associated intriguing phenomena. The edge associated anomalies can be seen from neither the SWCNTs nor the infinitely large graphene sheets (LGSs) [22–30].

Graphene is a wonder material with many superlatives to its name. It is the thinnest known material in the universe and the strongest ever measured. Its charge carriers or Dirac fermions (or call them Dirac-Fermi polaritons as will be justified in later section) exhibiting giant intrinsic mobility can travel for micrometers without scattering at room temperature. Graphene can sustain current densities six orders of magnitude higher than that of copper, shows record thermal and electric conductivity, is impermeable to gases, and reconciles such conflicting qualities as brittleness and ductility.

Serving as vehicles for the quantum spin Hall-effect in topological insulators [27, 31–37], Dirac-Fermi polaritons [38, 39] exhibit unique electrical supercurrent properties [40] on account of its reduced dimensionality and "relativistic" band structure [41]. When contacted with two superconducting electrodes, graphene can support Cooper pair transport, resulting in the well-known Josephson effect [42]. STM/S measurements [43–45] have uncovered the Dirac-Fermi polaritons as high protrusions in image and as sharp resonant peak at E_F in spectrum from sites surrounding atomic vacancies, the edges of monolayer graphite terrace and graphene nanoribbons [46–49]. These polaritons demonstrate anomalies including the extremely low effective mass [50], extremely high group velocity, and a net ½ spin [51, 52], following the Dirac equation, and a nearly linear dispersion (Dirac cone) with energies crossing Fermi energy [27, 31–34, 53–60]. Electron transport in graphene allows the investigation of relativistic quantum phenomena in a bench top experiment. These phenomena and the strip-width-induced band gap expansion demonstrated by the AGNR can never be observed in the SWCNT, graphene or graphite crystal [23, 61].

7.1.2 Challenges and Objectives

Overwhelming experimental efforts have been exerted primarily in the CNTs and GNRs growth, characterization, and functioning for practical applications. Considerable theoretical efforts have been made on the performance of the Dirac fermions that follow the relativistic Dirac equations and the energetic and structural optimization. However, physical insight into the origin behind the fascinations and their interdependence of the CNTs and the GNRs remain challenging. Opening questions may be exampled as the following:

1. GNRs and CNTs are mechanically stronger yet chemically and thermally less stable. CNTs exhibit extremely high strength yet relatively lower chemical and thermal stability compared to their bulk counterparts. Compared with the bulk value of 1.05 TPa, the elastic modulus of the SWCNT was measured to vary from 0.5 to 5.5 TPa depending on the presumption of the wall thickness of the CNT [62–69]. The Young's modulus of the multi-walled CNTs (MWCNTs) drops with the inverse wall thickness and it is less sensitive to the outermost radius of the MWCNTs if the wall thickness remains unchanged [70, 71]. Atoms in the open edge of a SWCNT coalesce at 1593 K [72] and a ~280% extensibility of the CNT occurs at ~2000 K [73]. Under the flash of an ordinary camera, the SWCNT burns under the ambient conditions [74]. Generally, for bulk materials, the elastic modulus is always proportional to their melting points. The mechanism behind the paradox of elastic enhancement and T_m suppression of the CNTs is still a puzzle.

2. The wall thickness, the C–C bond length and energy, and the role of atomic under-coordination remain challenge. The wall thickness and the Young's modulus of the C–C bond in the SWCNTs are correlated, which leads to the uncertainty in both quantities. Although atoms that surround defects or are located at the tip ends or at the surface are expected to play some unusual, yet unclear, roles in dominating the mechanical and thermal properties of CNTs and GNRs. A consistent insight into the mechanism behind the fascinations from the perspective of atomic under-coordination is necessary.

3. Mechanisms for the metallic and magnetic ZGNR and the semiconductive AGNR remain unclear. Compared with the LGS or CNTs, the ZGNRs possess strongly-localized edge states [75] with magnetic and metallic nature, whereas the AGNRs have larger band gap (E_G) with semiconductive nature. The E_G is roughly proportional to the inverse width of the GNRs [76–78]. However, discrepancy remains between the theory and the experimental derived E_G of GNR. Measurements [77] and theoretical calculations [51, 78, 79] showed less consistent in the band gap opening of GNR. Mechanisms remain yet unclear regarding the generation of the localized edge states and the expansion trends of the E_G in spite of the possible mechanisms such as doping [80], defects forming [51, 81, 82], symmetry breaking [83], substrate interaction [84], edge distortion [76], strain [85], quantum confinement [48], and the staggered sublattice potentials modulation [86].

4. Mechanism for the edge selective generation and hydrogen annihilation of the Dirac fermions remains opening. The Dirac-Fermi polaritons [44, 45] generate at sites surrounding atomic vacancies [43, 87], the edges of monolayer graphite terrace and the ZGNRs [46–49], other than edges of the AGNR or the rec-AGNR. It remains unclear why the edge and site discriminate the generation of Dirac fermions.

5. Origin of and correlation between the positive C 1s core level shift of the GNR edge, GNR interior and the associated work function reduction with the number of GNR layers are ambiguous. Three XPS C 1s components have been resolved from graphene flakes produce [88], corresponding, respectively, from

lower (larger value) to higher binding energies, to the contributions from the GNR edge, the mono-layer GNR or the surface of the triple-layered graphene, and the bulk graphite in multi-layered graphene. The C 1s spectrum of the multilayered graphene is dominated by the surface and the bulk components while the spectra for the mono- and the triple-layers are dominated by the surface and edge components. It has been found [89] from the epitaxial few-layer graphene that the work function decreases from 4.6 to 4.3 eV and that the C 1s core level shifts positively from 284.42 to 284.83 eV simultaneously when the number of graphene layers is decreased from ten to one, which is consistent with the reported thickness dependence of the Dirac point energy. The same thickness trend has also been observed from the C_{60} [90]. Unfortunately, few theoretical models are available to account for the origin and interdependence of the coordination-resolved C 1s binding energy shift and the associated work function reduction.

6. When the number-of-layer, strain, temperature, and pressure change, the vibration frequencies of the GNR and carbon alltorpes shift abnormally. Formulation of the lattice dynamics is necessary.

7. The common origin of these anomalies and their interdependence need to be established. The difference between the graphite and the GNRs or the CNTs is nothing more than atomic undercoordination that could be the point of starting. From the observation of the atomic dynamics of carbon at the edge of a hole in a suspended, single atomic layer of graphene, Girit et al. [49] found the rearrangement of bonds and the electron beam-induced ejection of carbon atoms as the vacancy hole grows in their high-resolution *in situ* TEM studies. They observed the edge reconstruction and the stability of the "zigzag" edge configuration, revealing the complex behavior of atoms preferentially occurring at the boundary. Therefore, atomic undercoordination and its consequences on the bond length, bond energy, and the associated electronic dynamics should be the origin for the anomalies and their interdependence.

In order to harness the GNRs and CNTs, one has to get these concerns be understood. It should be clear what the advantages are and what the limitations would be, and how to make use of the advantages and to overcome the limitations in practical applications. In fact, properties of a substance are determined by the process and consequences of bond and nonbond formation, dissociation, relaxation and vibration, and the associated energetics and dynamics of charge repopulation, polarization, densification, and localization—the theme of bonding and electronic dynamics [91]. From this perspective, this section addresses the above challenging issues with a focus on the fundamentals behind the fascinations and their interdependence. It is demonstrated that the atomic-undercoordination-induced local bond contraction and quantum entrapment, the polarization of the unpaired dangling σ-bond sp^2 electrons by the entrapped core and bond charges at the atomic vacancy and the ZGNR edges, and the formation of the pseudo-π-bond between the nearest dangling σ-bond electrons along the AGNR and the rec-ZGNR edges result in the fascinations. Theoretical reproduction of the experimentally observed elastic modulus enhancement, melting point depression, C 1s core-level shift, band gap expansion, edge and defect Dirac-Fermi polarons generation and the associated magnetism consistently confirmed that the shorter and stronger bonds between undercoordinated carbon atoms modulate locally the atomic cohesive energy and the Hamiltonian which alter the detectable bulk properties. The polarization of the unpaired sp^2 electrons by the densely, deeply, and locally entrapped core and bonding electrons generates the massless, magnetic and mobile Dirac-Fermi polarons at sites surrounding defects and ZGNR edges. The pseudo-π-bond formation at edges discriminates the AGNR and the rec-ZGNR from the AGNR in the electronic and magnetic anomalies.

7.2 Experimental Observations

7.2.1 STM/S-DFT: GNR Edge and Defect Polarization

Bond-order variation and the versatility of the sp-orbital hybridization enabled carbon allotropes a group of amazing materials varying from diamond, graphite, fullerene C_{60}, nanotube (CNT), nanobud (CNB), graphene, and graphene nanoribbons (GNRs) with different topological edges. Graphite is an electronic conductor and opaque but diamond is an insulator yet transparent to light of almost all wavelengths; the former shares nonbonding unpaired (or π-bond) electrons due to sp^2–orbital hybridization compared with the latter of an ideal sp^3-hybridization. GNR performs quite differently from CNT or from an infinitely large sheet of graphene because of the involvement of the two-coordinated edge atoms [20, 37, 49, 92–97].

Graphite point defects and GNRs with different types of edges demonstrate many fascinating properties that neither the large graphene sheet nor the bulk graphite displays. One of such properties is the edge-selective generation of the Dirac Fermions (DFs) with unexpectedly low effective mass, extremely high mobility [38], non-zero spin [98–101], demonstrating the spin quantum Hall effect [48, 102]. The DFs perform abnormally in many aspects, which is beyond the description of Schrödinger equation, but they follow Dirac equation of motion with a nearly linear dispersion crossing E_F [35, 55]. Because of the polarization and localization [93], it would be comprehensive to name DFs as Dirac-Fermi polaritons (DFPs) that are associated with atoms at the zigzag GNR edges or surrounding point defects with $\sqrt{3}d$ lattice spacing [92]. Performing differently from those unpaired nonbonding electrons in the GNR interior, the DFPs determine the catalytic, electric, magnetic, optic, and transport properties of the edged graphenes [48, 103, 104]. The zigzag-edged GNR performs metallic like while the armchair-edged GNR semiconductor like.

STM/S probed the graphitic DFs as bright protrusions with resonant peak at E_F from sites surrounding atomic vacancies, shown in Fig. 7.1a [43–45, 87], from edges of monolayer graphite terraces, and from graphene nanoribbons, see Fig. 7.1b [47–49]. Resonant current flows between the STM tip and the GNR edge under zero bias.

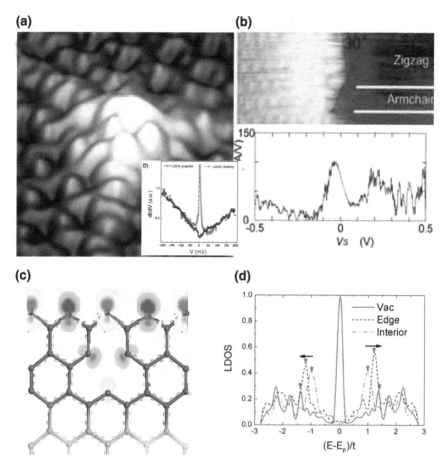

Fig. 7.1 STM protrusions and STS resonant current probed from **a** graphite surface atomic vacancies and from **b** GNR z-edge. **c, d** DFT derived edge states of asymmetric-dumbbell shaped, unpaired, and polarized electrons with spin up and down (color difference). The locally densely entrapped bonding and core electrons polarize and pin the dangling bond electrons [93]. Reprinted with permission from [43, 105]. Copyright 2010 and 2005 American Physical Society

However, such resonant states are absent from the flat skin, from graphene interior, or from graphene armchaired edges. The sharpness of the STM images and the resonant peaks are tip condition and probing temperature sensitive. These observations suggest that the point defects and the GNR z-edge are naturally the same in terms of bond relaxation, core electron entrapment, and nonbonding electron polarization [93].

DFT calculations [93] revealed that the DFPs with a high-spin-density create preferably at the z-edge or at an atomic vacancy of a GNR. The densely entrapped core electrons polarize the dangling σ-bond electrons of atoms of identical $\sqrt{3}d$

distance along the edge, see Fig. 7.1c. BOLS-TB and DFT derived edge LDOS for atomic vacancy and GNR edge manifest a sharp resonant peak at E_F, which is the same to that probed using STS from an atomic vacancy of graphite skin [43]. The locally and densely entrapped bonding electrons pin the DFPs through polarization. However, along the armchair-GNR edge and the reconstructed-zigzag-GNR edge, the quasi-triple-bond formation between the nearest edge atoms of d distance or less prevents the DFPs formation.

7.2.2 TEM: CN-Resolved C–C Bond Energy

Figure 7.2 shows the TEM image of a GNR with three types of edges: the zigzag- (I: ZGNR), the armchair- (II: AGNR), and the reconstructed- (III: rec-ZGNR) edges. Compared with a large graphene sheet or a single walled CNT, the ZGNR and vacancies share the hexagonal-sublattice possessing strongly-localized edge states [75], whereas the AGNRs have band gap (E_G). The E_G of the AGNR is proportional to the inverse width of the GNRs [51, 76–78, 82, 106]. The rec-ZGNR shares considerable similarity to that of the AGNR. The difference amongst these edges is only the distances between atoms along the outermost-edge. The atomic distance at the ZGNR edge is periodic with $\sqrt{3}d$ but at the AGNR the distance is d and $2d$ alternatively. A SWCNT is equivalent to a sheet of infinite large without edges though the rolling of the GNR may introduce some slight strains [92]. Edge only presents at the terminal ends or defects of the CNTs.

The TEM measurements revealed a 14.7% mean lattice contraction from 0.246 nm for graphite to 0.21 nm for the suspended GNR. For an 80-keV incident electron beam, the maximum energy that can be transferred to a carbon atom is 15.8 eV.

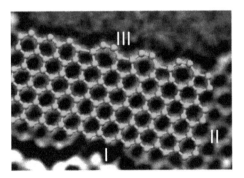

Fig. 7.2 TEM image of a GNR with three types of edges: I: ZGNR edge with identical ($\sqrt{3}d$) atomic distance, II: AGNR edge with the alternative d and $2d$ atomic distances, and III: rec-ZGNR with atomic distances being similar to that of the AGNR [49, 107]. The ZGNR edge generates Dirac-Fermions with magnetic and metallic nature while the AGNR and rec-ZGNR edge is the otherwise exhibiting semiconductor attributes Reprinted with permission from [49]. Copyright 2009 American Association for the Advancement of Science

The knock-on energy threshold for ejecting an in-lattice carbon atom with three bonds is 17.0 eV, corresponding to a beam-energy of 86 keV, and hence those atoms are not ejected by the 80-keV beams. However, this threshold drops below the maximum transfer energy to 15 eV for sites with a neighboring vacancy and may be even less, where atoms at the edge may have several vacant next-nearest-neighbor sites. The observation means that the minimal energy (15/2 = 7.50 eV/bond) required for breaking a bond between two-coordinated carbon atoms is 32% times higher than that (17/3 = 5.67 eV/bond) required for breaking a bond between three-coordinated carbon atoms in the suspended graphene [49].

7.2.3 XPS: Core Level and Work Function

Figure 7.3a shows the C 1s spectra measured using 635 eV photon energy from graphene flakes deposited on a SiO$_2$ substrate [88]. The well-resolved components at 285.97, 284.80, and 284.20 eV and their change of intensity with the number-of-layer confirmed the CN effect on the C 1s shift. These peaks counted from deeper to higher binding energy correspond to the GNR edge (E), monolayer GNR or skin (S), and the bulk graphite (B) in the layered graphene, respectively. The S and B components dominate the C 1s spectrum of the multilayered graphene while the E and S dominate the triple- and mono-layered graphene.

Figure 7.3b, c shows the number-of-layer resolved shift of the C 1s and the work function for the few-layer GNR grown on 6H-SiC(0001) substrate [89, 108–110]. The C 1s shifts positively from 284.42 to 284.83 eV associated with a work function reduction from 4.6 to 4.3 eV when the number-of-layer is decreased from ten to one

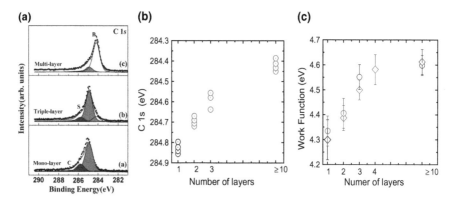

Fig. 7.3 Number-of-layer resolved **a** C 1s spectra [88] and the cooperative BE shift of the **b** C 1s energy and **c** work function [89] of GNR flakes indicate the coexistence of entrapment and polarization pertaining to undercoordinated C atoms. Reprinted with permission from [88, 89]. Copyright 2008 Wiley-VCH. Copyright 2009 American Physical Society

[89]. This observation follows the same trend demonstrated by C_{60}- deposited on CuPc substrate as detected using UPS, XPS, and synchrotron radiation [90]. Lin et al. [111] examined systematically the suspended and grounded graphene of different number of layers and confirmed the same trend of C 1s and work function change. The C 1s shift of other allotropes also show the same atomic-CN dependent trend [112]. Variation of the emission and azimuth angles [113] and surface termination affect the C 1s spectral features [114]. The cooperative relaxation of the C 1s and the work function indicates that quantum entrapment and polarization take place simultaneously and their extents increase as the atomic CN is reduced.

7.3 BOLS-TB Formulation and Quantification

Table 7.1 summarize the C–C bonding identities [115] derived from the measured stiffness and melting point of 1593 K for the terminal edge of the single-walled CNT [72]. The stiffness is the product of Young's modulus Y and the wall (bond) thickness t, $Yt = 0.3685$ TPa nm. The C–C bond of the SWCNT contracts by 18.5% and strengthens by 69% with respect to the C–C bond in a diamond (0.154 nm, 1.84 eV).

The C–C bond at the open edge of the CNT is 30% shorter and 152% stronger. The bond nature index for carbond is m = 2.56. The elastic modulus of CNT reaches 2.6 TPa compared with the value of 1.0 TPa for diamond [115]. The effective atomic CN of diamond is 12 instead of 4 as it is an interlock of two fcc-structured unit cells. The known C–C bond length of 0.154 nm for diamond and 0.142 nm for graphite result in the effective CN of 5.335 for graphite using the expression of bond contraction coefficient C_z [92].

The BOLS-TB notion correlates the XPS spectral components for carbon allotropes as [116],

Table 7.1 BOLS resolution of the C–C bond length, bond thickness, bond energy, the bond nature index m, the elastic modulus and the wall interior melting point with the measured $Yt_{z=3}$ and $T_m(2)$ data as input [92, 115]

$(Yt)_{z=3}$	0.3685 TPa nm
Tip-end melting point $T_m(2)$	1593 K
Bond nature index m	2.5585
Tube wall $T_m(3)$	1605 K
Elastic modulus Y	2.595 TPa
Effective wall thickness t(3)	0.142 nm
Bond length d(2) (c(2) = 0.6973)	0.107 nm
Bond length d(3) (c(3) = 0.8147)	0.126 nm
Relative bond energy, E(2)/E(12)	2.52
Relative bond energy, E(3)/E(12)	1.69

$$\frac{E_{1s}(z) - E_{1s}(0)}{E_{1s}(z') - E_{1s}(0)} = \frac{E_z}{E_{z'}} = \frac{C_z^{-2.56}}{C_{z'}^{-2.56}}; (z' \neq z) \qquad (7.1)$$

With the given C 1s values (for z = 2, 3, 5.335) of 285.97, 284.87, and 284.27 eV [88], one can easily calculate the values of $E_{1s}(0)$ and $\Delta E_{1s}(12)$. The mean value of $<E_{1s}(0)>$ is 282.57 ± 0.01 eV for an isolated C atom and the bulk shift $\Delta E_{1s}(12) = 1.321 \pm 0.001$ eV. Therefore, the following formulates the CN-resolved C 1s shift (z > 2):

$$E_{1s}(z) = E_{1s}(0) + \Delta E_E(12)C_z^{-2.56} = 282.57 \pm 0.01 + 1.32C_z^{-2.56}(eV) \qquad (7.2)$$

Figure 7.4 shows the BOLS-TB formulated C 1s shift of carbon allotropes. One can determine the effective CN of a graphene with the given number of layers. Conversely, the work function reduction arises from the elevation of E_F that is proportional to the density of charge centered at a specific energy, E, in the form of $[n(E)]^{2/\tau}$ [117] with τ being the dimensionality. Polarization of the dangling bond electrons [22, 91] will raise the DOS energies [118]. Hence, the observed work function reduction and the C 1s shift of the few-layer graphenes and the graphene flakes evidence the BOLS-NEP prediction of core level entrapment and nonbonding electron polarization.

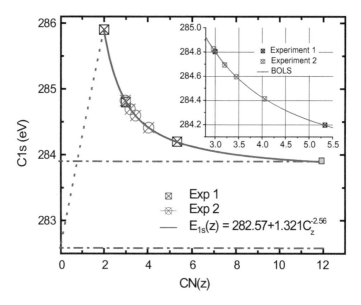

Fig. 7.4 Atomic CN dependence of the C 1s energy of carbon allotropes with scattered symbols representing measurements (Exp 1 [88] and Exp 2 [89]). Correlating the theoretical prediction to layer-resolved C 1s shift results in the effective CN for 1(z = 2.97), 2(3.20), 3(3.45) and 10(4.05) layer GNRs. Reprinted with permission from [116]. Copyright 2009 American Chemical Society

7.4   ZPS: Monolayer Skin Entrapment and Defect Polarization

Figure 7.5 shows the well-resolved XPS spectra collected from (a) the defect-free HOPG(0001) surface at different emission angles and (b) the surface of different defect densities at the emission angle of 50° [119]. The Ar^+ doses represent defect densities. One can control the density of the atomic vacancies by spraying the graphite surface using 0.5 keV Ar^+ ions along the surface normal with programmed time and current intensity. Ar^+ bombardment creates only vacancy defects without any chemical reaction or phase transition taking place in high vacuum [120].

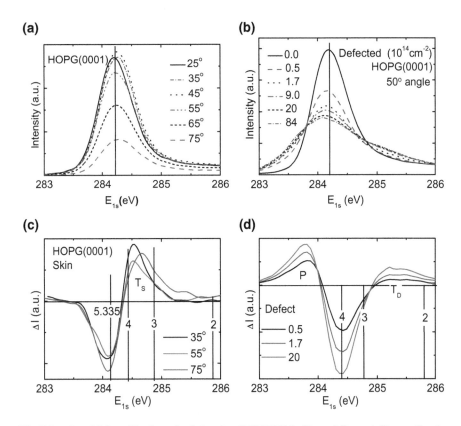

Fig. 7.5 **a** Raw XPS profiles for **a** the defect-free HOPG(0001) skin at different (off normal) polar angles and for **b** the defected skin at 50° polar angle of different defect densities generated by Ar^+ spraying. The ZPS distills **c** the entrapped monolayer skin states (T_S) that evolves from the bulk valley at z = 5.335 to z ~ 3.1 and **d** the entrapped defect to T_D, z ~ 2.2–2.4, and polarized (P) states of defects. The ZPS valleys correspond to the buk graphite (z = 5.35) and the mixture of skin and bulk (centered at z = 4) states. Reprinted with permission from [119]. Copyright 2012 The Royal Society of Chemistry

The angle-resolved C $1s$ spectra show a slight positive shift while the defect-resolved C $1s$ spectra exhibit tails at both spectral ends. The spectra collected at larger (off normal) emission angles or from higher defect densities weaken their overall intensities due to scattering loss [121].

The spectrum collected from the defect-free surface at the least emission angle (25°) serves as the reference for the ZPS processing upon all spectral peak area normalization and background correction. The skin ZPS is the difference between spectra collected at highest and lowest emission angles. The defect ZPS is the difference between a defected spectrum and this reference as well, as compared in Fig. 7.5c, d.

The ZPS in Fig. 7.5c, d purifies the energy states of the (a) defect-free monolayer skin and (b) the vacancy defects at graphite skin. The peaks above the x-axis are the DOS gain due to defects or the monolayer skin while the DOS loss under the axis is the bulk/skin components. According to the BOLS-TB notion, the separation between the specific spectral features and the $E_{1s}(0) = 282.57$ eV is proportional to the C–C bond energy at the particular atomic site.

$$[E_{1s}(z) - E_{1s}(12)]\big/[E_{1s}(5.335) - E_{1s}(12)] = (C_z/C_{5.335})^{-2.56}$$

The lateral axis is gridded with the effective atomic CN. The valley at 284.20 eV in (c) corresponds to graphite bulk ($z = 5.335$). The valley at 284.40 eV in (d) is a mixture of the bulk and the skin ($z = 4$). In addition to the spectral valleys, one entrapped peak (T_S) presents at the bottom edge of the C $1s$ band corresponding to the skin with $z \sim 3.1$. The T_S moves to energy even deeper and evolves into the T_D component with effective CN of 2.2–2.4, as defects generate. The shift from T_S to T_D is accompanied surprisingly by an emergence of both the P component at the upper edge of the C $1s$ band and the DFs at the E_F as probed using STM/S [43–45, 47–49, 87]. The T_D is deeper than the T_S means that the defect bonds are indeed shorter and stronger than that of the monolayer skin.

As the defect density is increased, the intensity of the T_D component grows but remains its energy. In contrast, the P component moves up in both energy and intensity. The atomic CN has reached and stabilized at the lowest value (2.2 for the nearest and 2.4 for the next nearest neighbors) and that the extent of polarization increases with defect density. The T_S energy depends only on the atomic CN but polarization on both the density and the CN of undercoordinated atoms. Only one neighbor short makes a great difference in the bond length and binding energy!

Figure 7.6 summarizes the ZPS spectra of graphite monolayer skin and vacancy defect. The skin ZPS differentiates two spectra collected at 75° and at 25°. The defect ZPS differentiates two spectra collected at 75° from the surface after and before high-density defect generation. Insets illustrate color zones contributing the excessive states in each case. The atomic CN for the skin is about 3.1, which is close to the ideal case of 3.0 of graphene interior. The atomic CN for the vacancy extends from 2.2 to 2.4, which indicates that the next nearest neighbors contribute to broadening the ZPS identity of the vacancy.

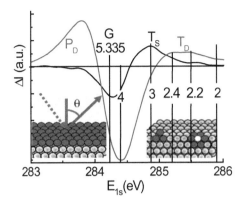

Fig. 7.6 ZPS purified C $1s$ spectra for the defected (9×10^{14} cm^{-2} dosed Ar$^+$ ion) and un-defected HOPG (0001) monolayer skin of graphite. The valley G centered at 284.20 eV ($z = 5.335$) corresponds to the bulk component and the 284.40 eV valley to a mixture of the bulk and the skin. The extra component T_S ($z \sim 3.1$) is the skin entrapment, and T_D ($z \sim 2.2$–2.4) the defect entrapment. The P component at the upper edge arises from the screening and splitting of the crystal potential by the Dirac-Fermion, see the STM/S data for graphite point defects in Fig. 7.1a. Insets illustrate the polar angle and the colored zones dominating the spectral signatures in each situation. Reprinted with permission from [119]. Copyright 2012 The Royal Society of Chemistry

Strikingly, the CNs of atoms annexed the vacancy defects are compatible to that of the GNR edge of 2.0. Based on Eq. (7.2), one can evaluate the length and strength of the C–C bonds and the C $1s$ shift associated with the undercoordinated atoms, as featured in Table 7.2. Consistency in the expected effective CN and the specific

Table 7.2 BOLS-TB-ZPS resolved z-dependent C–C bond length d_z, bond energy E_z, and C $1s$ BE of carbon allotropes in comparison to the documented C $1s$ shifts

	z	C_z	d_z(nm)	E_z(eV)	C $1s$ (eV)	Refs.	P (eV)
Atom	0	–	–	–	282.57	–	
Diamond	12.00	1.00	0.154	0.615	283.89	283.50–289.30 [124–126]	
GNR edge	2.00	0.70	0.107	1.548	285.89	285.97 [88]	283.85
Graphite vacancy	2.20	0.73	0.112	1.383	285.54		
	2.40	0.76	0.116	1.262	285.28	–	
GNR interior	3.00	0.81	0.125	1.039	284.80	284.80 [88]; 284.42 [127]; 284.90 [128]; 284.53–284.74 [129]	
Graphite skin	3.10	0.82	0.127	1.014	284.75	–	
Graphite	5.335	0.92	0.142	0.757	284.20	284.20 [88]; 284.30 [127, 128]; 284.35 [130]; 284.45 [131]	

Reprinted with permission from [119]. Copyright 2012 The Royal Society of Chemistry

energy between the present results and previous observations evidences sufficiently the accuracy and reliability of the BOLS-TB-ZPS derivatives.

The ratio of the energy shift between the 2- and the 3-coordinated atoms (252/169 = 149%) and the bond contraction of monolayer graphene (18.5%) agrees exceedingly well with that observed from monolayer graphene [49]. The C–C bond contracts by 14.7% from 0.246 to 0.207 nm and the minimal energy (7.50 eV/bond) required for breaking a bond between two-coordinated carbon atoms is 32% times higher than that (5.67 eV/bond) required for breaking a bond between three-coordinated carbon atoms in a suspended graphene. Information in Table 7.2 further confirms BOLS-NEP derivatives on the CN dependence of the Raman shift [122, 123]. With the known bond length, bond energy, effective CN, one is able to derive the local energy density E_{den} and atomic cohesive energy E_{coh} at different atomic sites of carbon allotropes.

Most strikingly, only one neighbor loss makes a great difference between C atoms at edges and C atoms in the monolayer skin. The defected P states of C are the same to Rh, Au, Ag, Cu and W adatoms or terrace edges and the skin entrapment is the same to Pt, Re, and Co adatoms or nanocrystals.

7.5 Summary

BOLS-NEP incorporation into the STM/S, TEM, and ZPS has enabled comprehensive information of the local bond length, bond energy, energy density, and cohesive energy at GNR edges, point defects, and monolayer skin of graphite. As compared in Fig. 7.1, STM protrusions and STS resonant peaks of vacancy defects at graphite surface are naturally the same to that of the graphene zigzag edge. One can therefore focus on the graphite surface vacancy more conveniently to mimic the GNR edge, as the latter is hardly accurately detectable in measurements. One neighbor loss differentiates greatly the edge carbon atom from the skin atom in the entrapment or the polarization dominance. Importantly, the practice clarifies the mechanism and dynamics of DFs creation. Polarization of the dangling-bond electrons by the locally densely entrapped bonding electrons dictates the unusual performance of GNRs.

References

1. S. Iijima, Helical microtubes of graphitic carbon. Nature **354**(6348), 56–58 (1991)
2. H.J. Dai, J.H. Hafner, A.G. Rinzler, D.T. Colbert, R.E. Smalley, Nanotubes as nanoprobes in scanning probe microscopy. Nature **384**(6605), 147–150 (1996)
3. P.G. Collins, A. Zettl, Unique characteristics of cold cathode carbon-nanotube-matrix field emitters. Phys. Rev. B **55**(15), 9391–9399 (1997)
4. W.A. Deheer, A. Chatelain, D. Ugarte, Acarbon nanotube field-emission electron cource. Science **270**(5239), 1179–1180 (1995)

5. A. Bachtold, P. Hadley, T. Nakanishi, C. Dekker, Logic circuits with carbon nanotube transistors. Science **294**(5545), 1317–1320 (2001)
6. J. Lee, H. Kim, S.J. Kahng, G. Kim, Y.W. Son, J. Ihm, H. Kato, Z.W. Wang, T. Okazaki, H. Shinohara, Y. Kuk, Bandgap modulation of carbon nanotubes by encapsulated metallofullerenes. Nature **415**(6875), 1005–1008 (2002)
7. A.C. Dillon, K.M. Jones, T.A. Bekkedahl, C.H. Kiang, D.S. Bethune, M.J. Heben, Storage of hydrogen in single-walled carbon nanotubes. Nature **386**(6623), 377–379 (1997)
8. C.Z. Wu, H.M. Cheng, Effects of carbon on hydrogen storage performances of hydrides. J. Mater. Chem. **20**(26), 5390–5400 (2010)
9. Y.Y. Xia, J.Z.H. Zhu, M.W. Zhao, F. Li, B.D. Huang, Y.J. Ji, X.D. Liu, Z.Y. Tan, C. Song, Y.Y. Yin, Enhancement of hydrogen physisorption on single-walled carbon nanotubes resulting from defects created by carbon bombardment. Phys. Rev. B **71**(7), 075412 (2005)
10. C. Liu, Y. Chen, C.Z. Wu, S.T. Xu, H.M. Cheng, Hydrogen storage in carbon nanotubes revisited. Carbon **48**(2), 452–455 (2010)
11. P.G. Collins, K. Bradley, M. Ishigami, A. Zettl, Extreme oxygen sensitivity of electronic properties of carbon nanotubes. Science **287**(5459), 1801–1804 (2000)
12. J. Kong, N.R. Franklin, C.W. Zhou, M.G. Chapline, S. Peng, K.J. Cho, H.J. Dai, Nanotube molecular wires as chemical sensors. Science **287**(5453), 622–625 (2000)
13. G. Centi, S. Perathoner, Problems and perspectives in nanostructured carbon-based electrodes for clean and sustainable energy. Catal. Today **150**(1–2), 151–162 (2010)
14. G.L. Che, B.B. Lakshmi, E.R. Fisher, C.R. Martin, Carbon nanotubule membranes for electrochemical energy storage and production. Nature **393**(6683), 346–349 (1998)
15. E.H.T. Teo, W.K.P. Yung, D.H.C. Chua, and B.K. Tay, A carbon nanomattress: a new nanosystem with intrinsic, tunable, damping properties. Adv. Mater., **19**(19), 2941–2945 (2007)
16. X. Chen, Z. Xie, W. Zhou, L. Tang, K. Chen, Thermal rectification and negative differential thermal resistance behaviors in graphene/hexagonal boron nitride heterojunction. Carbon **100**, 492–500 (2016)
17. X. Chen, Z. Xie, W. Zhou, L. Tang, K. Chen, Phonon wave interference in graphene and boron nitride superlattice. Appl. Phys. Lett. **109**(2), 023101 (2016)
18. B. Li, K. Chen, Huge inelastic current at low temperature in graphene nanoribbons. J. Phys.: Condens. Matter **29**(7), 075301 (2016)
19. C. Tan, Y. Zhou, C. Chen, J. Yu, K. Chen, Spin filtering and rectifying effects in the zinc methyl phenalenyl molecule between graphene nanoribbon leads. Org. Electron. **28**(28), 244–251 (2016)
20. K.S. Novoselov, A.K. Geim, S.V. Morozov, D. Jiang, Y. Zhang, S.V. Dubonos, I.V. Grigorieva, A.A. Firsov, Electric field effect in atomically thin carbon films. Science **306**(5696), 666–669 (2004)
21. K. Nakada, M. Igami, K. Wakabayashi, M. Fujita, Localized pi electronic edge state in nanographite. Mol. Cryst. Liq. Cryst. Sci. Technol. Sect. A-Mol. Cryst. Liq. Cryst. **310**, 225–230 (1998)
22. C.Q. Sun, S.Y. Fu, Y.G. Nie, Dominance of broken bonds and unpaired nonbonding pi-electrons in the band gap expansion and edge states generation in graphene nanoribbons. J. Phys. Chem. C **112**(48), 18927–18934 (2008)
23. O. Hod, V. Barone, J.E. Peralta, G.E. Scuseria, Enhanced half-metallicity in edge-oxidized zigzag graphene nanoribbons. Nano Lett. **7**(8), 2295–2299 (2007)
24. N. Levy, S.A. Burke, K.L. Meaker, M. Panlasigui, A. Zettl, F. Guinea, A.H.C. Neto, M.F. Crommie, Strain-induced pseudo-magnetic fields greater than 300 tesla in graphene nanobubbles. Science **329**(5991), 544–547 (2010)
25. R. Prasher, Graphene spreads the heat. Science **328**(5975), 185–186 (2010)
26. Y.M. Lin, C. Dimitrakopoulos, K.A. Jenkins, D.B. Farmer, H.Y. Chiu, A. Grill, P. Avouris, 100-GHz transistors from wafer-scale epitaxial graphene. Science **327**(5966), 662 (2010)
27. D. Hsieh, Y. Xia, D. Qian, L. Wray, J.H. Dil, F. Meier, J. Osterwalder, L. Patthey, J.G. Checkelsky, N.P. Ong, A.V. Fedorov, H. Lin, A. Bansil, D. Grauer, Y.S. Hor, R.J. Cava,

M.Z. Hasan, A tunable topological insulator in the spin helical dirac transport regime. Nature **460**(7259), 1101–1105 (2009)

28. A.K. Geim, Graphene: status and prospects. Science **324**(5934), 1530–1534 (2009)
29. T.W. Odom, J.L. Huang, P. Kim, C.M. Lieber, Atomic structure and electronic properties of single-walled carbon nanotubes. Nature **391**(6662), 62–64 (1998)
30. A.H. Castro Neto, F. Guinea, N.M.R. Peres, K.S. Novoselov, A.K. Geim, The electronic properties of graphene. Rev. Mod. Phys. **81**(1), 109–162 (2009)
31. M. Konig, S. Wiedmann, C. Brune, A. Roth, H. Buhmann, L.W. Molenkamp, X.L. Qi, S.C. Zhang, Quantum spin hall insulator state in HgTe quantum wells. Science **318**(5851), 766–770 (2007)
32. D. Hsieh, D. Qian, L. Wray, Y. Xia, Y.S. Hor, R.J. Cava, M.Z. Hasan, A topological dirac insulator in a quantum spin hall phase. Nature **452**(7190), 970–974 (2008)
33. R. Yu, W. Zhang, H.J. Zhang, S.C. Zhang, X. Dai, Z. Fang, Quantized anomalous hall effect in magnetic topological insulators. Science **329**(5987), 61–64 (2010)
34. T. Zhang, P. Cheng, X. Chen, J.F. Jia, X.C. Ma, K. He, L.L. Wang, H.J. Zhang, X. Dai, Z. Fang, X.C. Xie, Q.K. Xue, Experimental demonstration of topological surface states protected by time-reversal symmetry. Phys. Rev. Lett. **103**(26), 266803 (2009)
35. K.S. Novoselov, A.K. Geim, S.V. Morozov, D. Jiang, M.I. Katsnelson, I.V. Grigorieva, S.V. Dubonos, A.A. Firsov, Two-dimensional gas of massless Dirac fermions in graphene. Nature **438**(7065), 197–200 (2005)
36. N. Tombros, C. Jozsa, M. Popinciuc, H.T. Jonkman, B.J. van Wees, Electronic spin transport and spin precession in single graphene layers at room temperature. Nature **448**(7153), 571–574 (2007)
37. K.S. Novoselov, Z. Jiang, Y. Zhang, S.V. Morozov, H.L. Stormer, U. Zeitler, J.C. Maan, G.S. Boebinger, P. Kim, A.K. Geim, Room-temperature quantum hall effect in graphene. Science **315**(5817), 1379 (2007)
38. L. Brey, H. Fertig, Electronic states of graphene nanoribbons studied with the dirac equation. Phys. Rev. B **73**(23), 235411 (2006)
39. F.M.D. Pellegrino, G.G.N. Angilella, R. Pucci, Strain effect on the optical conductivity of graphene. Phys. Rev. B **81**(3), 035411 (2010)
40. H.B. Heersche, P. Jarillo-Herrero, J.B. Oostinga, L.M.K. Vandersypen, A.F. Morpurgo, Bipolar supercurrent in graphene. Nature **446**(7131), 56–59 (2007)
41. C. Girit, V. Bouchiat, O. Naamanth, Y. Zhang, M.F. Crommie, A. Zetti, I. Siddiqi, Tunable graphene do superconducting quantum interference device. Nano Lett. **9**(1), 198–199 (2009)
42. A.M. Black-Schaffer, S. Doniach, Possibility of measuring intrinsic electronic correlations in graphene using a d-wave contact Josephson junction. Phys. Rev. B **81**(1), 014517 (2010)
43. M.M. Ugeda, I. Brihuega, F. Guinea, J.M. Gómez-Rodríguez, Missing atom as a source of carbon magnetism. Phys. Rev. Lett. **104**, 096804 (2010)
44. T. Matsui, H. Kambara, Y. Niimi, K. Tagami, M. Tsukada, H. Fukuyama, STS observations of landau levels at graphite surfaces. Phys. Rev. Lett. **94**, 226403 (2005)
45. G. Li, E.Y. Andrei, Observation of landau levels of dirac fermions in g*raphite*. Nat. Phys. **3**(9), 623–627 (2007)
46. Y. Niimi, T. Matsui, H. Kambara, K. Tagami, M. Tsukada, H. Fukuyama, Scanning tunneling microscopy and spectroscopy of the electronic local density of states of graphite surfaces near monoatomic step edges. Phys. Rev. B **73**(8), 085421–085428 (2006)
47. Y. Niimi, H. Kambara, H. Fukuyama, Localized distributions of quasi-two-dimensional electronic states near defects artificially created at graphite surfaces in magnetic fields. Phys. Rev. Lett. **102**(2), 026803–026804 (2009)
48. T. Enoki, Y. Kobayashi, K.I. Fukui, Electronic structures of graphene edges and nanographene. Int. Rev. Phys. Chem. **26**(4), 609–645 (2007)
49. C.O. Girit, J.C. Meyer, R. Erni, M.D. Rossell, C. Kisielowski, L. Yang, C.H. Park, M.F. Crommie, M.L. Cohen, S.G. Louie, A. Zettl, Graphene at the edge: stability and dynamics. Science **323**(5922), 1705–1708 (2009)

50. D.L. Miller, K.D. Kubista, G.M. Rutter, M. Ruan, W.A. de Heer, P.N. First, J.A. Stroscio, Observing the quantization of zero mass carriers in graphene. Science **324**(5929), 924–927 (2009)

51. Y.W. Son, M.L. Cohen, S.G. Louie, *Energy gaps in graphene nanoribbons.* Phys. Rev. Lett. **97**(21), 216803 (2006)

52. M. Fujita, K. Wakabayashi, K. Nakada, K. Kusakabe, Peculiar Localized State at Zigzag Graphite Edge. J. Phys. Soc. Jpn. **65**(7), 1920–1923 (1996)

53. A.K. Geim, K.S. Novoselov, The rise of graphene. Nat. Mater. **6**(3), 183–191 (2007)

54. M.A.H. Vozmediano, M.P. Lopez-Sancho, T. Stauber, F. Guinea, Local defects and ferromagnetism in graphene layers. Phys. Rev. B **72**(15), 155121 (2005)

55. S.Y. Zhou, G.H. Gweon, J. Graf, A.V. Fedorov, C.D. Spataru, R.D. Diehl, Y. Kopelevich, D.H. Lee, S.G. Louie, A. Lanzara, First direct observation of dirac fermions in graphite. Nat. Phys. **2**(9), 595–599 (2006)

56. A. De Martino, L. Dell'Anna, R. Egger, Magnetic confinement of massless Dirac fermions in graphene. Phys. Rev. Lett. **98**(6), 066802 (2007)

57. K. Nomura, A.H. MacDonald, Quantum transport of massless dirac fermions. Phys. Rev. Lett. **98**(7), 076602 (2007)

58. J. Yan, Y.B. Zhang, P. Kim, A. Pinczuk, Electric field effect tuning of electron-phonon coupling in graphene. Phys. Rev. Lett. **98**(16), 166802 (2007)

59. G.H. Li, A. Luican, E.Y. Andrei, Scanning tunneling spectroscopy of graphene on graphite. Phys. Rev. Lett. **102**(17), 176804 (2009)

60. X.Z. Yan, C.S. Ting, Weak localization of Dirac fermions in graphene. Phys. Rev. Lett. **101**(12), 126801 (2008)

61. L. Yang, C.-H. Park, Y.-W. Son, M.L. Cohen, S.G. Louie, Quasiparticle energies and band gaps in graphene nanoribbons. Phys. Rev. Lett. **99**(18), 186801 (2007)

62. E. Hernandez, C. Goze, P. Bernier, A. Rubio, Elastic properties of C and BxCyNz composite nanotubes. Phys. Rev. Lett. **80**(20), 4502–4505 (1998)

63. M.F. Yu, O. Lourie, M.J. Dyer, K. Moloni, T.F. Kelly, R.S. Ruoff, Strength and breaking mechanism of multiwalled carbon nanotubes under tensile load. Science **287**(5453), 637–640 (2000)

64. M.F. Yu, B.S. Files, S. Arepalli, R.S. Ruoff, Tensile loading of ropes of single wall carbon nanotubes and their mechanical properties. Phys. Rev. Lett. **84**(24), 5552–5555 (2000)

65. M.M.J. Treacy, T.W. Ebbesen, J.M. Gibson, Exceptionally high Young's modulus observed for individual carbon nanotubes. Nature **381**(6584), 678–680 (1996)

66. J.P. Salvetat, A.J. Kulik, J.M. Bonard, G.A.D. Briggs, T. Stockli, K. Metenier, S. Bonnamy, F. Beguin, N.A. Burnham, L. Forro, Elastic modulus of ordered and disordered multiwalled carbon nanotubes. Adv. Mater. **11**(2), 161–165 (1999)

67. J.P. Salvetat, G.A.D. Briggs, J.M. Bonard, R.R. Bacsa, A.J. Kulik, T. Stockli, N.A. Burnham, L. Forro, Elastic and shear moduli of single-walled carbon nanotube ropes. Phys. Rev. Lett. **82**(5), 944–947 (1999)

68. E.T. Thostenson, Z.F. Ren, T.W. Chou, Advances in the science and technology of carbon nanotubes and their composites: a review. Compos. Sci. Technol. **61**(13), 1899–1912 (2001)

69. B.I. Yakobson, C.J. Brabec, J. Bernholc, Nanomechanics of carbon tubes: Instabilities beyond linear response. Phys. Rev. Lett. **76**(14), 2511–2514 (1996)

70. W. Liu, L.M. Jawerth, E.A. Sparks, M.R. Falvo, R.R. Hantgan, R. Superfine, S.T. Lord, M. Guthold, Fibrin fibers have extraordinary extensibility and elasticity. Science **313**(5787), 634 (2006)

71. Z.C. Tu, Z.C. Ou-Yang, Dimensional crossover of dilute neon inside infinitely long single-walled carbon nanotubes viewed from specific heats. Phys. Rev. B **68**, 153403 (2003)

72. A. Bai, F. Seiji, Y. Kiyoshi, Y. Masamichi, Surface superstructure of carbon nanotubes on highly oriented pyrolytic graphite annealed at elevated temperatures. Jpn. J. Appl. Phys. **37**(6B), 3809–3811 (1998)

73. P. Nikolaev, A. Thess, A.G. Rinzler, D.T. Colbert, R.E. Smalley, Diameter doubling of single-wall nanotubes. Chem. Phys. Lett. **266**(5–6), 422–426 (1997)

74. P.M. Ajayan, M. Terrones, A. de la Guardia, V. Huc, N. Grobert, B.Q. Wei, H. Lezec, G. Ramanath, T.W. Ebbesen, Nanotubes in a flash—Ignition and reconstruction. Science **296**(5568), 705 (2002)

75. K. Nakada, M. Fujita, G. Dresselhaus, M.S. Dresselhaus, Edge state in graphene ribbons: Nanometer size effect and edge shape dependence. Phys. Rev. B **54**(24), 17954–17961 (1996)

76. D. Gunlycke, C.T. White, Tight-binding energy dispersions of armchair-edge graphene nanostrips. Phys. Rev. B **77**, 115116 (2008)

77. M.Y. Han, B. Ozyilmaz, Y.B. Zhang, P. Kim, Energy band-gap engineering of graphene nanoribbons. Phys. Rev. Lett. **98**, 206805 (2007)

78. S.S. Yu, Q.B. Wen, W.T. Zheng, Q. Jiang, Electronic properties of graphene nanoribbons with armchair-shaped edges. Mol. Simul. **34**(10–15), 1085–1090 (2008)

79. S. Reich, J. Maultzsch, C. Thomsen, P. Ordejo, Tight-binding description of graphene. Phys. Rev. B **66**(3), 035412 (2002)

80. I. Zanella, S. Guerini, S.B. Fagan, J. Mendes, A.G. Souza, Chemical doping-induced gap opening and spin polarization in graphene. Phys. Rev. B **77**, 073404 (2008)

81. E. Rotenberg, A. Bostwick, T. Ohta, J.L. McChesney, T. Seyller, K. Horn, Origin of the energy bandgap in epitaxial graphene. Nat. Mater. **7**(4), 258–259 (2008)

82. Z.F. Wang, Q.X. Li, H.X. Zheng, H. Ren, H.B. Su, Q.W. Shi, J. Chen, Tuning the electronic structure of graphene nanoribbons through chemical edge modification: a theoretical study. Phys. Rev. B **75**, 113406 (2007)

83. S.Y. Zhou, D.A. Siegel, A.V. Fedorov, F. El Gabaly, A.K. Schmid, A.H.C. Neto, D.H. Lee, A. Lanzara, Origin of the energy bandgap in epitaxial graphene—Reply. Nat. Mater. **7**(4), 259–260 (2008)

84. S.Y. Zhou, G.H. Gweon, A.V. Fedorov, P.N. First, W.A. De Heer, D.H. Lee, F. Guinea, A.H.C. Neto, A. Lanzara, Substrate-induced bandgap opening in epitaxial graphene. Nat. Mater. **6**(10), 770–775 (2007)

85. G. Gui, J. Li, J.X. Zhong, Band structure engineering of graphene by strain: First-principles calculations. Phys. Rev. B **78**(7), 075435 (2008)

86. C.L. Kane, E.J. Mele, Quantum spin Hall effect in graphene. Phys. Rev. Lett. **95**, 226801 (2005)

87. T. Kondo, Y. Honma, J. Oh, T. Machida, J. Nakamura, Edge states propagating from a defect of graphite: Scanning tunneling spectroscopy measurements. Phys. Rev. B **82**(15), 153414 (2010)

88. K.J. Kim, H. Lee, J.H. Choi, Y.S. Youn, J. Choi, T.H. Kang, M.C. Jung, H.J. Shin, H.J. Lee, S. Kim, B. Kim, Scanning photoemission microscopy of graphene sheets on SiO_2. Adv. Mater. **20**(19), 3589–3591 (2008)

89. H. Hibino, H. Kageshima, M. Kotsugi, F. Maeda, F.-Z. Guo, Y. Watanabe, Dependence of electronic properties of epitaxial few-layer graphene on the number of layers investigated by photoelectron emission microscopy. Phys. Rev. B **79**, 125431 (2009)

90. H.Y. Mao, R. Wang, H. Huang, Y.Z. Wang, X.Y. Gao, S.N. Bao, A.T.S. Wee, W. Chen, Tuning of C[sub 60] energy levels using orientation-controlled phthalocyanine films. J. Appl. Phys. **108**(5), 053706 (2010)

91. C.Q. Sun, Dominance of broken bonds and nonbonding electrons at the nanoscale. Nanoscale **2**(10), 1930–1961 (2010)

92. W.T. Zheng, C.Q. Sun, Underneath the fascinations of carbon nanotubes and graphene nanoribbons. Energy Environ. Sci. **4**(3), 627–655 (2011)

93. X. Zhang, Y.G. Nie, W.T. Zheng, J.L. Kuo, C.Q. Sun, Discriminative generation and hydrogen modulation of the Dirac-Fermi polarons at graphene edges and atomic vacancies. Carbon **49**(11), 3615–3621 (2011)

94. Y.B. Zhang, Y.W. Tan, H.L. Stormer, P. Kim, Experimental observation of the quantum Hall effect and Berry's phase in graphene. Nature **438**(7065), 201–204 (2005)

95. T. Ohta, A. Bostwick, T. Seyller, K. Horn, E. Rotenberg, Controlling the electronic structure of bilayer graphene. Science **313**(5789), 951–954 (2006)

96. J.C. Meyer, A.K. Geim, M. Katsnelson, K. Novoselov, T. Booth, S. Roth, The structure of suspended graphene sheets. Nature **446**(7131), 60–63 (2007)
97. J. Caridad, F. Rossella, V. Bellani, M. Grandi, E. Diez, Automated detection and characterization of graphene and few-layer graphite via Raman spectroscopy. J. Raman Spectrosc. **42**(3), 286–293 (2011)
98. P.O. Lehtinen, A.S. Foster, Y. Ma, A.V. Krasheninnikov, R.M. Nieminen, Irradiation-induced magnetism in graphite: a density functional study. Phys. Rev. Lett. **93**(18), 187202 (2004)
99. J.J. Palacios, J. Fernandez-Rossier, L. Brey, Vacancy-induced magnetism in graphene and graphene ribbons. Phys. Rev. B **77**(19), 195428 (2008)
100. J. Červenka, M. Katsnelson, C. Flipse, Room-temperature ferromagnetism in graphite driven by two-dimensional networks of point defects. Nat. Phys. **5**(11), 840–844 (2009)
101. V.M. Pereira, F. Guinea, J.L. Dos Santos, N. Peres, A.C. Neto, Disorder induced localized states in graphene. Phys. Rev. Lett. **96**(3), 036801 (2006)
102. C. Soldano, A. Mahmood, E. Dujardin, Production, properties and potential of graphene. Carbon **48**(8), 2127–2150 (2010)
103. M. Acik, Y.J. Chabal, Nature of graphene edges: a review. Jpn. J. Appl. Phys. **50**(7), 070101 (2011)
104. S.S. Yu, W.T. Zheng, Effect of N/B doping on the electronic and field emission properties for carbon nanotubes, carbon nanocones, and graphene nanoribbons. Nanoscale **2**(7), 1069–1082 (2010)
105. Y. Kobayashi, K. Fukui, T. Enoki, K. Kusakabe, Y. Kaburagi, Observation of zigzag and armchair edges of graphite using scanning tunneling microscopy and spectroscopy. Phys. Rev. B **71**(19), 193406 (2005)
106. X. Zhang, J.L. Kuo, M.X. Gu, P. Bai, C.Q. Sun, Graphene nanoribbon band-gap expansion: broken-bond-induced edge strain and quantum entrapment. Nanoscale **2**(10), 2160–2163 (2010)
107. P. Koskinen, S. Malola, H. Hakkinen, Evidence for graphene edges beyond zigzag and armchair. Phys. Rev. B **80**, 073401 (2009)
108. U. Starke, C. Riedl, Epitaxial graphene on SiC(0001) and SiC(000(1)over-bar): from surface reconstructions to carbon electronics. J. Phys.-Condens. Matter **21**(13), 134016 (2009)
109. T. Filleter, K.V. Emtsev, T. Seyller, R. Bennewitz, Local work function measurements of epitaxial graphene. Appl. Phys. Lett. **93**(13), 133117 (2008)
110. K.V. Emtsev, F. Speck, T. Seyller, L. Ley, J.D. Riley, Interaction, growth, and ordering of epitaxial graphene on SiC{0001} surfaces: a comparative photoelectron spectroscopy study. Phys. Rev. B **77**(15), 155303 (2008)
111. C.-Y. Lin, H.W. Shiu, L.Y. Chang, C.-H. Chen, C.-S. Chang, F.S.-S. Chien, Core-level shift of graphene with number of layers studied by microphotoelectron spectroscopy and electrostatic force microscopy. J. Phys. Chem. C **118**(43), 24898–24904 (2014)
112. C.Q. Sun, Thermo-mechanical behavior of low-dimensional systems: the local bond average approach. Prog. Mater Sci. **54**(2), 179–307 (2009)
113. S. Lizzit, G. Zampieri, L. Petaccia, R. Larciprete, P. Lacovig, E.D.L. Rienks, G. Bihlmayer, A. Baraldi, P. Hofmann, Band dispersion in the deep 1 s core level of graphene. Nat. Phys. **6**, 345–349 (2010)
114. A. Stacey, B.C.C. Cowie, J. Orwa, S. Prawer, A. Hoffman, Diamond C 1s core-level excitons: Surface sensitivity. Phys. Rev. B **82**(12), 125427 (2010)
115. C.Q. Sun, H.L. Bai, B.K. Tay, S. Li, E.Y. Jiang, Dimension, strength, and chemical and thermal stability of a single C-C bond in carbon nanotubes. J. Phys. Chem. B **107**(31), 7544–7546 (2003)
116. C.Q. Sun, Y. Sun, Y.G. Nie, Y. Wang, J.S. Pan, G. Ouyang, L.K. Pan, Z. Sun, Coordination-resolved C–C bond length and the C 1 s binding energy of carbon allotropes and the effective atomic coordination of the few-layer graphene. J. Phys. Chem. C **113**(37), 16464–16467 (2009)
117. W.T. Zheng, C.Q. Sun, B.K. Tay, Modulating the work function of carbon by N or O addition and nanotip fabrication. Solid State Commun. **128**(9–10), 381–384 (2003)

118. C.Q. Sun, Size dependence of nanostructures: impact of bond order deficiency. Prog. Solid State Chem. **35**(1), 1–159 (2007)
119. C.Q. Sun, Y. Nie, J. Pan, X. Zhang, S.Z. Ma, Y. Wang, W. Zheng, Zone-selective photoelectronic measurements of the local bonding and electronic dynamics associated with the monolayer skin and point defects of graphite. RSC Adv **2**(6), 2377–2383 (2012)
120. K. Ostrikov, Colloquium: reactive plasmas as a versatile nanofabrication tool. Rev. Mod. Phys. **77**(2), 489–511 (2005)
121. D.Q. Yang, E. Sacher, s-p hybridization in highly oriented pyrolytic graphite and its change on surface modification, as studied by X-ray photoelectron and Raman spectroscopies. Surf. Sci. **504**(1–3), 125–137 (2002)
122. Y. Wang, X. Yang, J. Li, Z. Zhou, W. Zheng, C.Q. Sun, Number-of-layer discriminated graphene phonon softening and stiffening. Appl. Phys. Lett. **99**(16), 163109 (2011)
123. X.X. Yang, J.W. Li, Z.F. Zhou, Y. Wang, L.W. Yang, W.T. Zheng, C.Q. Sun, Raman spectroscopic determination of the length, strength, compressibility, Debye temperature, elasticity, and force constant of the C–C bond in graphene. Nanoscale **4**(2), 502–510 (2012)
124. G. Speranza, N. Laidani, Measurement of the relative abundance of sp(2) and sp(3) hybridised atoms in carbon based materials by XPS: a critical approach. Part I. Diam. Relat. Mater. **13**(3), 445–450 (2004)
125. S. Takabayashi, K. Motomitsu, T. Takahagi, A. Terayama, K. Okamoto, T. Nakatani, Qualitative analysis of a diamond like carbon film by angle-resolved x-ray photoelectron spectroscopy. J. Appl. Phys. **101**, 103542 (2007)
126. K.G. Saw, J. du Plessis, The X-ray photoelectron spectroscopy C 1 s diamond peak of chemical vapour deposition diamond from a sharp interfacial structure. Mater. Lett. **58**(7–8), 1344–1348 (2004)
127. T. Balasubramanian, J.N. Andersen, L. Wallden, Surface-bulk core-level splitting in graphite. Phys. Rev. B **64**, 205420 (2001)
128. Y.M. Shulga, T.C. Tien, C.C. Huang, S.C. Lo, V. Muradyan, N.V. Polyakova, Y.C. Ling, R.O. Loutfy, A.P. Moravsky, XPS study of fluorinated carbon multi-walled nanotubes. J. Electron Spectrosc. Relat. Phenom. **160**(1–3), 22–28 (2007)
129. A. Goldoni, R. Larciprete, L. Gregoratti, B. Kaulich, M. Kiskinova, Y. Zhang, H. Dai, L. Sangaletti, F. Parmigiani, X-ray photoelectron microscopy of the C 1s core level of freestanding single-wall carbon nanotube bundles. Appl. Phys. Lett. **80**(12), 2165–2167 (2002)
130. P. Bennich, C. Puglia, P.A. Bruhwiler, A. Nilsson, A.J. Maxwell, A. Sandell, N. Martensson, P. Rudolf, Photoemission study of K on graphite. Phys. Rev. B **59**(12), 8292–8304 (1999)
131. C.S. Yannoni, P.P. Bernier, D.S. Bethune, G. Meijer, J.R. Salem, NMR determination of the bond lengths in C60. J. Am. Chem. Soc. **113**(8), 3190–3192 (1991)

Chapter 8
Hetero-Coordinated Interfaces

Abstract Hetero-coordinated bond formation changes the local bond strength and charge distribution. Polarization dominance reduces the CL and makes the alloy a donor-like catalyst with weakened interface mechanical strength; entrapment dominance does it contrastingly. The ZPS offers such a unique yet straightforward means that diagnosis the interface bonding and electronic dynamics in alloys, compounds, impurities, and interfaces for devising functional substance.

Highlights

- Bond nature alteration shifts the interface CL positively or negatively.
- Cu/Pd, Cu/Sn, and Ge/Si show interface quantum entrapment dominance.
- Ag/Pd, Zn/Pd, Cu/Si, and Be/W show interface polarization dominance.
- Si/C and Ge/C show a mixture of C 1s polarization and Si(Ge) entrapment.

8.1 Observations

As the key components in alloys, compounds, dopants, impurities, glasses, super-lattices, hetero-coordination is ubiquitously important [1]. Bonds at the interface are completely different from those of the constituent parents [2, 3] because of the involvement of the A-B type exchange interactions. Formation of the interfacial bonds perturbs the Hamiltonian, BE density, atomic cohesive energy, and consequently the catalytic, electronic, dielectric, optic, mechanical, and thermal properties. The interface alloys have many industrial applications, such as in searching for new catalysts [4, 5] thermal barrier coatings [6], wear-resistance and joining strength [7, 8], optoelectronics [9], CMOS devices [10], and irradiation protection [11, 12].

Varying composition or thermally annealing upon continuous deposition of dissimilar metals has formed effective methods to modify interatomic strain and redistribute charge around the bonded atoms at the interface [13–15]. Controlling the performance of bonds, energies, and electrons are key concerns in mediating the macroscopic properties of a substance.

As the first wall material in a nuclear fusion device, Be/W alloying easily occurs in the processes of plasma interaction within the wall for radiation protection [11, 12]. However, understanding the energetics and electronics of the interface alloy is crucial to engineering such functional materials. Due to the electronic structures of W ($6s^2 5d^4$, delocalized d electrons dominate) and Be ($2s^2$ states dominate), a strong influence of the alloying on both the core and the valence electrons is apparent, but insofar, it remains poorly understood.

Another example is the catalytic nature and ability of alloys. Ag and Cu often form alloys with Pd because their interactions typically result in improved heterogeneous catalysts [16–18] that behave quite differently in reactions—reduction or oxidation [19, 20]. Cu/Pd is more active for the CO and alkene oxidation, CO, NO, benzene, toluene, and 1,3-butadiene hydrogenation and ethanol decomposition [21]. Ag/Pd is a good candidate for hydrogenation and permeation [22, 23]. Si, Ge, Sn, C and Cu are also used as constituent for the electrode of Li-ion batteries [24].

In the process of catalytic reaction, the direction and the ability of charge flow between the catalyst and the gaseous adsorbate are the key concern. The reactivity depends on the filling degree of the empty anti-bonding states of the catalyst by the reactant electrons. The reactivity also depends on the ability of the catalyst donating its valence electrons to the specimen. The catalyst and the adsorbate having orbits of similar energies should overlap for charge transportation during reaction [25, 26]. Although the catalytic behaviors of Cu/Pd and Ag/Pd alloys have been intensively investigated [14, 19, 27–31], laws governing the energetic behavior of the core and the valence electrons and their catalytic nature and ability are yet to be established.

Upon reacting with electronegative elements such as oxygen, nitrogen and fluorine with lone pair production, the core levels of metals also shift accordingly associated with excessive valence DOS features due to charge polarization and transportation [35, 36]. For instance, oxygen adsorption deepens the bulk and the skin $3d_{5/2}$ components of a clean Ru(0001) surface simultaneously by up to 1.0 eV [37]. Oxidation also deepens the Rh $3d_{5/2}$ level and its satellite by 0.40 eV further [38]. These observations confirm that both surface bond relaxation and new bond formation could shift the core-level positively in a superposition way by an amount that varies not only with the original $E_v(0)$ but also with the extent of reaction [39]. However, the polarization due to lone pair production is unapparent in the energy shift of deeper levels but the lone pairs create polarized energy states in the upper conduction and valence band. The polarized states may screen and split the local crystal potential, which in turn influence the CLS [40].

Hetero-junction bond formation shifts the XPS features positively or negatively, depending on the local potentials. For instances, Ag/Pd [41], Zn/Pd [42], and Be/W [32, 43] alloy formation shift the core and the valence bands upwardly but Cu/Pd [41, 44] alloy formation deepens all bands simultaneously, as summarized in Table 8.1. For alloy containing N constituent elements, there will be C(N, 2) terms of interactions such as the A-A, B-B, and A-B interaction for an AB alloy instance. The additional A-B type exchange interactions contribute to the overall potentials of an

Table 8.1 Summary of the component XPS peak energies and their intensity conversion upon Cu/Pd, Zn/Pd, Ag/Pd, Be/W alloy formation by annealing [32–34]

T(K)	Bulk	Interface	I_B/I_A	Bulk	Interface	I_B/I_A
Cu/Pd	Pd $3d_{5/2}$ (eV)			Cu $2p_{3/2}$ (eV)		
340	335.67	337.10	15.30	931.65	933.20	11.50
540	335.66	337.17	5.06	931.57	933.21	2.93
940	335.58	337.26	0.18	931.65	933.19	0.28
Mean	335.63	337.18	–	931.62	933.20	–
Zn/Pd	Pd $3d_{5/2}$ (eV)			Zn $3d_{5/2}$ (eV)		
540	335.43	334.75	4.0	9.6	8.8	4.0
940			0.2			0.2
Ag/Pd	Pd $3d_{5/2}$ (eV)			Ag $3d_{5/2}$ (eV)		
300	335.62	334.33	14.17	368.32	367.18	12.00
473	335.52		1.83	368.28	367.16	0.79
573	335.52		0.32	368.38	367.10	0.12
Mean	335.55	334.33	–	368.33	367.15	–
Be/W	Be $1s$ (eV)			W $4f_{7/2}$ (eV)		
300	111.11	110.48	8.47	31.07	30.66	9.19
970			0.19			1.02

amorphous state. Understanding the fundamental nature of the interface bond formation and its consequence on the electronic BE shifting as well as determination of the relevant energetics is a great challenge.

In order to verify the proposed mechanism of the interface quantum entrapment or polarization and to clarify their catalysis mechanism, Cu (2 nm) and Ag (2 nm) thin films were deposited, separately, onto Pd (10 nm) substrate using the physical vapor deposition method [15]. Both Ag and Cu grow on Pd in a layer-by-layer fashion at room temperature without alloy formation [14, 45, 46]. Heating up to 940 K for the Cu/Pd and up to 573 K for the Ag/Pd, alloys form completely, which result in transition of the XPS and UPS spectral peaks from the elemental bulk (B) dominance to the alloy interface (I) dominance [33].

8.2 BOLS-TB Formulation of the PES Attributes

One can derive the interface bond energy E_I and elucidate whether the entrapment or the polarization dictates the interface performance by ZPS analysis of the PES spectra. One can also calculate the interface BE density, atomic cohesive energy, and free energy with the known $E_v(12) - E_v(0)$ reference derived from elemental skin XPS analysis,

$$\gamma_I = \frac{E_v(x) - E_v(0)}{E_v(12) - E_v(0)} = E_I/E_B = \begin{cases} > 1 \ (T) \\ < 1 \ (P) \end{cases}$$

Figure 8.1a compares the normalized valence DOS of pure Ag (in the range of 4–7 eV), Cu (2–6 eV), and Pd (0–6 eV) with respect to E_F [27–29, 33]. The VB spectra clarify why Pd and Ag are so special for catalysis. The Pd valence DOS extends to energy across the E_F that is readily donating electrons to the reactant—donor-like. However, the valence DOS of Ag is far away from the E_F, which may allow Ag to capture electrons readily—acceptor-like. Therefore, Ag and Pd should perform differently in catalysis.

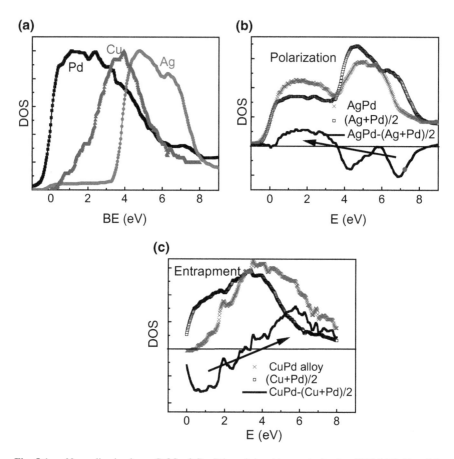

Fig. 8.1 **a** Normalized valence DOS of Cu, Pd, and Ag skins probed using UPS [41]. $E = 0$ is the E_F serving as the reference. ZPS reveals that **b** Ag/Pd alloy formation shifts the valence DOS negatively (polarization) and that **c** Cu/Pd alloying shifts the valence DOS positively (entrapment). The opposite directions (indicated by arrows) of polarization distinguish the Cu/Pd from the Ag/Pd in the catalytic nature. Ag/Pd readily donates electrons to benefit to reduction and Cu/Pd tends to capture electrons for oxidation. Reprinted with permission from [33]. Reproduced by permission of the PCCP Owner Societies

Subtracting the composed UPS spectrum, $I_{composed} = [xI_A + (1 - x)I_B]$, for instance, from the alloying UPS spectrum I_{alloy} upon spectral peak area normalization results in the valence ZPS, $\Delta I = I_{alloy} - I_{comosed}$, that distills the effect of alloy formation. The x is the concentration of A constituent in the AB alloy. The valence ZPS profiles in Fig. 8.1 reveal that (b) Ag/Pd alloy formation migrates the valence DOS upwardly and that (c) Cu/Pd alloy shifts the DOS to deeper energy. The opposite directions of valence DOS polarization distinguish the Cu/Pd (acceptor-like) from the Ag/Pd (donor-like) in their catalytic performance.

8.3   ZPS: Core Band Entrapment and Polarization

8.3.1   Ag/Pd, Cu/Pd, Zn/Pd and Be/W Interfaces

XPS spectra in Figs. 8.2 and 8.3 show the evolution of the Cu $2p_{3/2}$, Ag $3d_{5/2}$, Pd $3d_{5/2}$, Zn $3d_{5/2}$, Be $1s$, and W $4f_{7/2}$ bands upon annealing at transition temperatures [32, 33]. The intensity inversion of the I and the B component indicates the completeness of interface alloying as summarized in Table 8.1.

The ZPS features allow one to calibrate the potential depth, $\gamma = [\Delta E_v(I) - \Delta E_v(B)]/\Delta E_v(B)$, for the respective alloys, as summarized in Table 8.2. ZPS profiles in Figs. 8.2 and 8.3 reveal the following information:

1. Both the Cu $2p$ and the Pd $3d$ bands shift positively from B to I component upon Cu/Pd alloy formation, which is consistent with the valence band shift. Cu/Pd alloy formation strengthens the interface Cu–Cu and Pd–Pd bond, resulting in quantum entrapment.
2. Both the Ag $3d$ and the Pd $3d$ bands at the Ag/Pd interface shift upwardly from B to I, which also agrees with the valence band evolution. The Ag/Pd alloy formation weakens the interface Ag–Ag and Pd–Pd bond, resulting in polarization.
3. The Zn $3d$, Pd $3d$, Be $1s$ and W $4f$ bands shift up at the Zn/Pd and the Be/W interfaces, following the same polarization trend at the Ag/Pd interface.
4. The core and the valence bands evolve consistently in the same direction of the same alloy. One can infer that the Be/W and Zn/Pd valence bands follow the same polarization trend of their core bands though it is subject to experimental verification.

8.3.2   C/Si, C/Ge, Si/Ge, Cu/Si and Cu/Sn Interfaces

Figure 8.4 shows the C $1s$, Si $2p$, and Ge $3p$ ZPS for C/Si, C/Ge and Si/Ge alloys [54–56] and Table 8.3 summarizes the derive information on the interface entities. Results indicate the following:

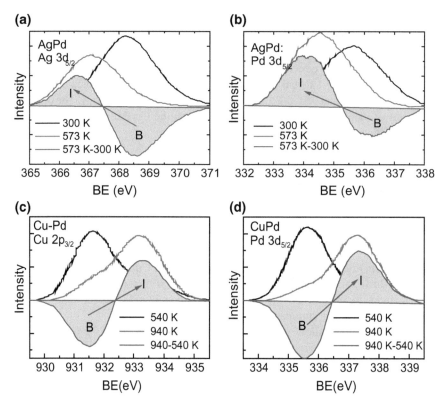

Fig. 8.2 Thermally driven core band XPS-intensity inversion in **a, b** Ag/Pd and **c, d** Cu/Pd alloys indicates alloy formation [41]. The ZPS reveals polarization dominance in the Ag/Pd interface and entrapment dominance in the Cu/Pd and SiGe alloying interfaces. Reprinted with permission from [33]. Reproduced by permission of the PCCP Owner Societies

1. For the Si/C interface, the C $1s$ BE shifts negatively by 1.46 eV from the bulk value of 284.20 eV to the interface of 282.74 eV with the ratio $\gamma = 0.11$, while the Si $2p$ BE shifts positively by 1.32 eV from the bulk value of 99.20 eV to the interface of 100.18 eV with $\gamma = 1.40$. Similarly, the γ values for C and Ge in C/Ge are 0.47 and 1.61, respectively. The C $1s$ in the C/Si and C/Ge interfaces become shallower than that in the bulk $V_{cryst}(r, B)$, while that of Si and Ge shows the entrapment dominance.

2. For the Si/Ge interface, both the Ge $3d$ and Si $2p$ level shift positively with respect to that of their respective bulk components. The γ values for Ge and Si in Si/Ge are 1.58 and 1.45, respectively, which indicates their crystal potential in the interface both become deeper than the respective bulk potential.

3. The BE shifts of the same component from different alloy are different. For example, the γ value for the C $1s$ in the C/Si is 0.11 and in the C/Ge is 0.47, and the γ value for the Si $2p$ in the C/Si is 1.40 and in the Si/Ge is 1.45.

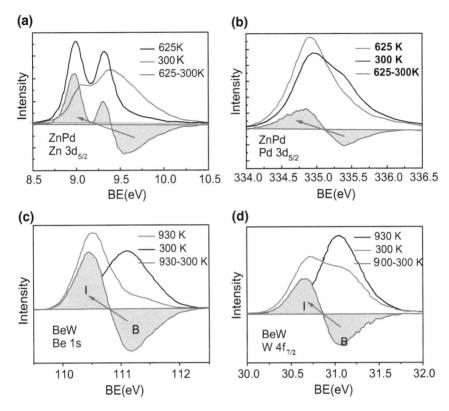

Fig. 8.3 Thermally driven XPS-intensity inversion in **a, b** Zn/Pd [42] and **c, d** Be/W [32] alloying interfaces. The ZPS reveals polarization dominance in all these interfaces. Reprinted with permission from [47]. Copyright 2011 Elsevier

Table 8.2 The relative BE shift $\gamma = [\Delta(I) - \Delta(B)]/\Delta(B)$ for the Cu/Pd, Ag/Pd, and Be/W alloy interfaces (unit in eV)

Alloy	Band	$E_v(0)$	$\Delta E_v(B)$	$\Delta E_v(I) - \Delta E_v(0)$	γ	
Cu/Pd	Cu $2p_{3/2}$ (eV) [49]	931.00	1.70	1.58	1.94	T
	Pd $3d_{5/2}$ (eV) [50]	330.26	4.36	1.55	1.36	
Ag/Pd				−1.22	0.72	P
	Ag $3d_{5/2}$ (eV) [51]	363.02	4.63	−1.18	0.75	
Be/W	Be $1s$ (eV) [52]	106.42	4.69	−0.63	0.86	
	W $4f_{7/2}$ (eV) [53]	28.91	2.17	−0.41	0.81	

The γ for Pd and Zn will be available provided with $E_{3d}(0)$ of Zn. Reprinted with permission from [34, 48]. Copyright 2013 and 2014 Elsevier

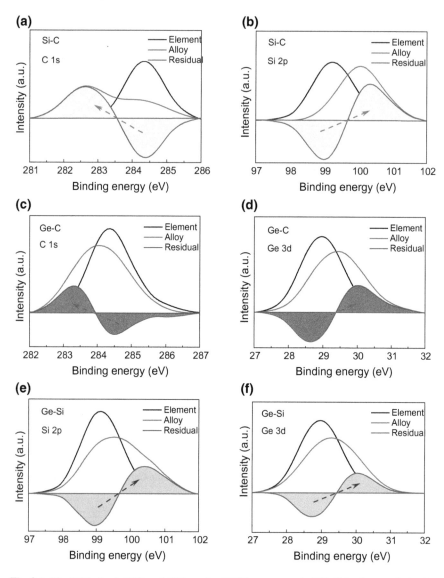

Fig. 8.4 The PZS of **a, b** C/Si, **c, d** C/Ge and **e, f** Si/Ge compounds [24]. C 1*s* shows polarization dominance in all specimens but Si 2*p* and Ge 3*d* show entrapment dominance at the interface

Figure 8.5 shows the ZPS for Cu/Si and Cu/Sn alloys and Table 8.4 summarizes the derived information. Both the Cu 2$p_{3/2}$ and the Si 2*p* levels shift negatively in the alloys, which indicates that polarization is dominant at the Cu/Si interface. However, Cu 2$p_{3/2}$ and Sn 3$d_{5/2}$ shift positively at the Cu/Sn interface, indicating that quantum entrapment dominates at the Cu/Sn interface. The migrated orientation of the charge in the Cu/Sn interface is opposite to that of Cu/Si alloy. That is, Cu/Si interface is mechanically weaker than either Cu or Si themselves and Cu/Sn interface become strengthened because of the formation of Cu/Sn alloy.

Table 8.3 Binding energies for an isolated atom $E_v(0)$, the bulk $E_v(B)$, the interface $E_v(I)$, and the respective shift $\Delta E_v(B)$ and $\Delta E_v(I)$ and the ratios of their relative shift γ for C $1s$, Si $2p$, Ge $3d$ in the C/Si, C/Ge, and Si/Ge interfaces (in eV unit)[a]

Interface	Energy levels	$E_v(0)$	$E_v(B)$	$E_v(I)$	$\Delta E_v(B)$	$\Delta E_v(I)$	γ
C/Si	C $1s$ [57]	282.57	284.20	282.74	1.63	0.17	0.11
	Si $2p$ [58]	96.74	99.20	100.18	2.46	3.44	1.40
C/Ge	C $1s$ [57]	282.57	284.20	283.33	1.63	0.76	0.47
	Ge $3d$ [59]	27.58	28.96	29.80	1.38	2.22	1.61
Si/Ge	Si $2p$ [58]	96.74	99.20	100.29	2.46	3.56	1.45
	Ge $3d$ [58]	27.58	28.96	29.76	1.38	2.18	1.58

[a] $\gamma > 1$ interface quantum entrapment dominance; otherwise, polarization dominance

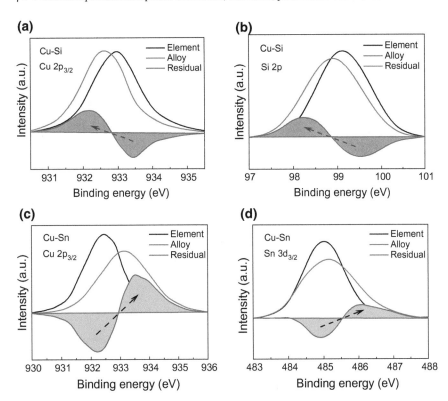

Fig. 8.5 The PZS of **a, b** Cu/Si and **c, d** Cu/Sn alloys [24]. Cu $2p$ and Si $2p$ show both polarization dominance in the Cu–Si allows but Si $2p$ and Ge $3d$ show entrapment dominance at the interface

Table 8.4 Binding energies for an isolated atom $E_v(0)$, the bulk $E_v(B)$, the interface $E_v(I)$, and the respective shift $\Delta E_v(B)$ and $\Delta E_v(I)$, and the ratios of their relative shift γ for Cu $2p_{3/2}$, Si $2p$, Sn $3d_{5/2}$ in the Cu/Si and Cu/Sn interfaces (in eV unit)[a]

Interface	CL	$E_v(0)$	$E_v(B)$	$E_v(I)$	$\Delta E_v(B)$	$\Delta E_v(I)$	γ
Cu/Si	Cu $2p_{3/2}$ [33]	931.00	932.70	932.00	1.70	1.00	0.59
	Si $2p$ [58]	96.74	99.20	98.46	2.46	1.72	0.70
Cu/Sn	Cu $2p_{3/2}$ [33]	931.00	932.70	933.82	1.70	2.82	1.66
	Sn $3d_{5/2}$ [24]	479.60	484.86	485.75	5.26	6.15	1.17

[a]$\gamma > 1$ interface quantum entrapment dominance; otherwise, polarization dominance

8.4 Energy Density, Cohesive Energy, and Free Energy

One can estimate the energy density, atomic cohesive energy, and the free energy in the interface region based on ZPS derived interface potential depth γ. The energy density is the sum of BE per unit cell. The cohesive energy is the sum of BE over all coordinates of an interface atom. Instead of the conventionally defined excessive energy required for creating a unit area of interface, the interface free energy equals the energy per unit cell divided by the cross-section area of the unit cell.

For simplicity, the interface is assumed an fcc structure with four atoms (N = 4) in a unit cell. Atoms in the interface region are fully coordinated with $z_I = 12$. The following determines the mean interface bond energy $\langle E_I \rangle$:

$$\frac{\langle E_I \rangle}{\langle E_b \rangle} = \frac{\Delta E_v(I)}{\Delta E_v(B)} = \gamma \tag{8.1}$$

The Vegard's notion expresses the mean interface bond energy $\langle E_{IS} \rangle$ and bond length $\langle d_{IS} \rangle$ with the involvement of the A-A, B-B and A-B type interactions [60]:

$$\begin{cases} \langle d_{IS} \rangle = x d_{IA} + (1-x) d_{IB} \\ \langle E_{IS} \rangle = x E_{IA} + (1-x) E_{IB} + x(1-x)\sqrt{E_{IA} E_{IB}} \end{cases} \tag{8.2}$$

The last term in the $\langle E_{IS} \rangle$ denotes the A-B exchange interaction and x the concentration of A. Tables 8.5 and 8.6 summarize the elucidated information regarding the interface energetics.

With the derived $\langle d_{IS} \rangle$ and $\langle E_{IS} \rangle$, we are able to determine the energy density E_{ID}, atomic cohesive energy E_{ID}, and the free energy γ_I at the interface:

$$\begin{cases} E_{IC} = z_I \langle E_{IS} \rangle & \text{(atomic cohesive energy)} \\ E_{ID} = \frac{E_{\text{sum\_cell}}}{V_{\text{cell}}} = \frac{N z_I \langle E_{IS} \rangle}{2 d_{IS}^3} & \text{(binding energy density)} \\ \gamma_I = \frac{E_{\text{sum\_cell}}}{A_{\text{sectional}}} = E_D d_{IS} = \frac{N z_I \langle E_{IS} \rangle}{2 d_{IS}^2} & \text{(interface free energy)} \end{cases} \tag{8.3}$$

Table 8.5 Interface mean bond energy (E_{IS}), mean lattice constant (d_{IS}), cohesive energy E_{IC}, energy density E_{ID}, and the free energy γ_I. E_b and d_b are the bond energy and bond length of the bulk constituent, respectively. E_I(A-A or B-B) is the bond energy of A or B in the interface; E_I(A-B) is the A-B exchange interaction energy [33, 34, 48]

Alloy	Atom	E_b (eV)	E_I (A-A) (eV)	E_I(A-B) (eV)	$\langle E_{IS}\rangle$ (eV)	d_b (nm)	$\langle d_{IS}\rangle$ (nm)	E_{IC} (eV)	E_{ID} (10^{10} J/m³)	γ_I (J/m²)
Be/W	Be	0.28	0.24	0.38	0.52	0.229	0.273	6.24	9.85	26.85
	W	0.74	0.60			0.316				
Cu/Pd	Cu	0.29	0.56	0.50	0.63	0.360	0.375	7.56	4.59	17.20
	Pd	0.32	0.45			0.389				
Ag/Pd	Pd	0.32	0.22	0.20	0.26	0.389	0.399	2.76	1.57	6.28
	Ag	0.25	0.19			0.409				

Table 8.6 Comparison of the interface lattice constant d_{IS} and energy E_I with respective bulk d_b and E_b values. E_I is the bond energy of specific atom in the interface, $\langle d_{IS} \rangle$ is the average interface lattice constant and $\langle E_{IS} \rangle$ is the average bond energy

Interface	Atom	E_b [61] (eV)	E_I (eV)	$\langle E_{IS} \rangle$ (eV)	d_b [61] (nm)	$\langle d_{IS} \rangle$ (nm)
Si/C	C	1.38	0.15	0.42	0.671	0.607
	Si	0.39	0.55		0.543	
Ge/C	C	1.38	0.65	0.73	0.671	0.618
	Ge	0.32	0.52		0.566	
Ge/Si	Si	0.39	0.56	0.66	0.543	0.555
	Ge	0.32	0.50		0.566	
Cu/Si	Cu	0.29	0.17	0.27	0.360	0.452
	Si	0.39	0.27		0.543	
Cu/Sn	Cu	0.29	0.48	0.48	0.360	0.472
	Sn	0.26	0.30		0.583	

Table 8.7 Interfacial atomic cohesive energy E_{coh}, binding energy density E_{den}, and free energy (γ_I (ε_I)) for the Si/C, Ge/C, Ge/Si, Cu/Si and Cu/Sn alloys [24]

Interface	E_{coh} (eV)	E_{den} (10^{10} J/m^3)	Γ (J/m^2)
Si/C	3.64	0.52	3.16
Ge/C	6.33	0.86	5.31
Ge/Si	7.92	1.48	8.21
Cu/Si	3.24	1.12	5.06
Cu/Sn	5.81	1.77	8.35

These quantities are different from those of the corresponding bulk constituent because of the involvement of interface exchange coupling. As given in Table 8.5, the E_{ID} of the Be/W interface is the highest among the three, which justifies that the Be/W becomes an important medium for radiation protection due to its higher energy density and interface polarization. Table 8.7 and Figure 8.6 feature the derivatives of the Si/C, Ge/C, Ge/Si, Cu/Si and Cu/Sn interface free energy.

8.5 Catalytic Nature, Toxicity, Radiation Protectivity and Mechanical Strength

ZPS revealed the quantum entrapment dominance in Cu/Pd alloy while polarization dominance in Ag/Pd, Zn/Pd, and the Be/W alloys. The entrapment generates holes at the upper edge of the valence band, thus the Cu/Pd, acting the same to Pt adatoms [63], serve as a charge acceptor in catalytic reaction. The entrapment enlarges the

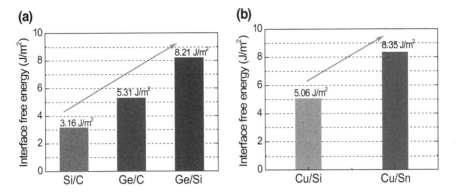

Fig. 8.6 The **a** C/Si, C/Ge and Si/Ge [54–56] and **b** Cu/Si and Cu/Sn interface free energy [62]

electroaffinity that is responsible for the toxicity of undercoordinated adatoms and hetero-coordinated impurities.

Comparatively, the polarization of Ag/Pd, Zn/Pd, and Be/W generates excessive electrons at the conduction band top edge, which makes Ag/Pd and Zn/Pd alloys, acting the same to Rh adatoms [63], serve as a charge donor in the process of catalytic reaction. The polarization of the valence electrons and the high interface energy density explain why the Be/W can protect nuclear radiation. Zn/Pd is an alternative for the donor type catalyst albeit its efficiency [64].

The Cu/Pd alloying entrapment seems disobey the rule of electronegativity difference between the Cu(1.9) and the Pd(2.2). The valence charge should flow partially from Cu to Pd, but results show that both tend to gain electrons from outside. This paradox suggests that interface quantum entrapment does occur due to the lattice strain and alloy bond formation; the initial electronegativity rule losses it effect in forming alloy.

Low-energy electron and photoelectron diffraction studies have revealed that the Cu-Pd distance contracts by up to $7.0 \pm 2.5\%$ at the interface [30, 65, 66] and the Ag-Pd bond contracts by 2.5% [31]. This fact explains that the hetero-coordinated Cu atom, having a half occupied $4s^1$ valence band, accepts charge from the Pd $4d^{10}$ valence band that is fully occupied, though the electronegativity of Cu (1.9) is lower than that of Pd (2.2) [2, 14]. However, the electronegativity of Ag(1.9) is the same to Cu, but the Ag/Pd shows charge polarization, instead. The possible reason for the polarity could be the number of electron shells or the radius of the atom. Electrons in the half-filled s-orbit of Ag($5s^1$) tend to be polarized more readily compared with those in the Cu($4s^1$) because of different radii of the s orbitals. The charge polarization also happens to Au($6s^1$) and Rh($5s^1$) but not to Pt ($6s^0$) and Co ($3d^7 4s^2$) at sites with localized strain [63]. Nevertheless, discrepancy in the charge transferring direction of Cu and Ag upon alloy formation with Pd is an interesting issue for further investigation.

The C $1s$ shows polarization dominance in the C/Si and C/Ge alloys, but the Si $2p$ and Ge $3p$ demonstrated entrapment dominance in the C/Si, C/Ge and Si/Ge alloys.

Both Cu $3d$ and Si $2p$ show polarization dominance in the Cu/Si alloy but the Cu $3d$ and Sn $3d$ show entrapment dominance in the Cu/Sn alloy. One can hardly predict how the energy levels change when they are subject to hetero-coordination without experimental verification.

8.6   Summary

An extension of the BOLS-NEP-ZPS spectrometrics has correlated the interface energetics to the interatomic bonding from the perspective of bond-energy-electron correlation and Hamiltonian perturbation. ZPS identifies readily quantum entrapment or polarization in the alloying interfaces. Interface entrapment makes the Cu/Pd a p-type catalyst while polarization makes Ag/Pd and Zn/Pd n-type catalysts. The high energy density and the polarization entitle the Be/W to protect nuclear irradiation. The combination of C, Si. Ge, Sn and Cu for alloys can modulate the interface stress unexpectedly. The following summarizes ZPS enabled findings of interfaces:

1. The concepts of energy-density-gain per unit volume and cohesive-energy-remnant per atom are essential to classify the interface energetics and their responsibilities.
2. Bond nature alteration and charge sharing determine the interfacial energetics, which follows unessentially the general rule of electronegativity difference.
3. BOLS-NEP-ZPS allows determination of the interface bond energy, energy density, cohesive energy, and free energy, at the atomic scale, which is beyond the scope of existing approaches.
4. The interface potential modulation due to bond order distortion and bond nature alteration perturbs essentially the Hamiltonian and hence leads to the BE shifts and the associated functionalities.

The accuracy of estimation is strictly subject to the measurement. Other factors such as materials purity, defect concentration, and testing techniques may render the accuracy of the derived $E_v(0)$ and $E_v(I)$ values. The concepts of interface quantum entrapment and polarization are essential for understanding the bonding and electronic behavior of hetero-coordinated atoms at the interface region or nearby impurities.

References

1. C.Q. Sun, Relaxation of the chemical bond. Springer Ser. Chem. Phys. **108**, 807pp (2014) (Heidelberg: Springer-Verlag)
2. J.A. Rodriguez, D.W. Goodman, The nature of the metal-metal bond in bimetallic surfaces. Science **257**(5072), 897–903 (1992)

3. P. Kamakoti, B.D. Morreale, M.V. Ciocco, B.H. Howard, R.P. Killmeyer, A.V. Cugini, D.S. Sholl, Prediction of hydrogen flux through sulfur-tolerant binary alloy membranes. Science **307**(5709), 569–573 (2005)

4. E.B. Fox, S. Velu, M.H. Engelhard, Y.H. Chin, J.T. Miller, J. Kropf, C.S. Song, Characterization of CeO_2-supported Cu-Pd bimetallic catalyst for the oxygen-assisted water-gas shift reaction. J. Catal. **260**(2), 358–370 (2008)

5. L.H. Bloxham, S. Haq, Y. Yugnet, J.C. Bertolini, R. Raval, trans-1,2-dichloroethene on Cu50Pd50(110) alloy surface: dynamical changes in the adsorption, reaction, and surface segregation. J. Catal. **227**(1), 33–43 (2004)

6. N.P. Padture, M. Gell, E.H. Jordan, Thermal barrier coatings for gas-turbine engine applications. Science **296**(5566), 280–284 (2002)

7. S. Veprek, M.G.J. Veprek-Heijman, The formation and role of interfaces in superhard nc-MenN/a-Si3N4 nanocomposites. Surf. Coat. Technol. **201**(13), 6064–6070 (2007)

8. S. Veprek, A.S. Argon, Towards the understanding of mechanical properties of super- and ultrahard nanocomposites. J. Vac. Sci. Technol., B **20**(2), 650–664 (2002)

9. N.T. Gabriel, J.J. Talghader, Optical coatings in microscale channels by atomic layer deposition. Appl. Opt. **49**(8), 1242–1248 (2010)

10. C.H. Tung, K.L. Pey, L.J. Tang, M. Radhakrishnan, W.H. Lin, F. Palumbo, S. Lombardo, Percolation path and dielectric-breakdown-induced-epitaxy evolution during ultrathin gate dielectric breakdown transient. Appl. Phys. Lett. **83**(11), 2223–2225 (2003)

11. A. Allouche, A. Wiltner, C. Linsmeier, Quantum modeling (DFT) and experimental investigation of beryllium–tungsten alloy formation. J. Phys.: Condens. Matter **21**(35), 355011 (2009)

12. R.P. Doerner, M.J. Baldwin, R.A. Causey, Beryllium-tungsten mixed-material interactions. J. Nucl. Mater. **342**(1–3), 63–67 (2005)

13. D.H. Zhang, W. Shi, Dark current and infrared absorption of p-doped InGaAs/AlGaAs strained quantum wells. Appl. Phys. Lett. **73**(8), 1095–1097 (1998)

14. G. Liu, T.P. St Clair, D.W. Goodman, An XPS study of the interaction of ultrathin Cu films with Pd(111). J. Phys. Chem. B **103**(40), 8578–8582 (1999)

15. M. Khanuja, B.R. Mehta, S.M. Shivaprasad, Geometric and electronic changes during interface alloy formation in Cu/Pd bimetal layers. Thin Solid Films **516**, 5435–5439 (2008)

16. M.A. Newton, The oxidative dehydrogenation of methanol at the CuPd[85: 15]{110} p(2x1) and Cu{110} surfaces: effects of alloying on reactivity and reaction pathways. J. Catal. **182**(2), 357–366 (1999)

17. A.M. Venezia, L.F. Liotta, G. Deganello, Z. Schay, D. Horvath, L. Guczi, Catalytic CO oxidation over pumice supported Pd-Ag catalysts. Applied Catalysis a-General **211**(2), 167–174 (2001)

18. C.L. Lee, C.M. Tseng, R.B. Wu, C.C. Wu, S.C. Syu, Catalytic characterization of hollow silver/palladium nanoparticles synthesized by a displacement reaction. Electrochim. Acta **54**(23), 5544–5547 (2009)

19. A.M. Venezia, L.F. Liotta, G. Deganello, Z. Schay, L. Guczi, Characterization of pumice-supported Ag-Pd and Cu-Pd bimetallic catalysts by X-ray photoelectron spectroscopy and X-ray diffraction. J. Catal. **182**(2), 449–455 (1999)

20. I.S. Choi, C.N. Whang, C.Y. Hwang, Surface-induced phase separation in Pd-Ag alloy: the case opposite to surface alloying. J. Phys.-Condens. Matter **15**(25), L415–L422 (2003)

21. J.P. Reilly, C.J. Barnes, N.J. Price, R.A. Bennett, S. Poulston, P. Stone, M. Bowker, The growth mechanism, thermal stability, and reactivity of palladium mono- and multilayers on Cu(110). J. Phys. Chem. B **103**(31), 6521–6532 (1999)

22. H. Amandusson, L.G. Ekedahl, H. Dannetun, Hydrogen permeation through surface modified Pd and PdAg membranes. J. Membr. Sci. **193**(1), 35–47 (2001)

23. I. Efremenko, U. Matatov-Meytal, M. Sheintuch, Hydrodenitrification with PdCu catalysts: catalyst optimization by experimental and quantum chemical approaches. Isr. J. Chem. **46**(1), 1–15 (2006)

24. Y. Wang, Y. Pu, Z. Ma, Y. Pan, C.Q. Sun, Interfacial adhesion energy of lithium-ion battery electrodes. Extrem. Mech. Lett. **9**, 226–236 (2016)

25. E. Roduner, Size matters: why nanomaterials are different. Chem. Soc. Rev. **35**(7), 583–592 (2006)
26. B. Hammer, J.K. Norskov, Why gold is the noblest of all the metals. Nature **376**(6537), 238–240 (1995)
27. N. Matensson, R. Nyholm, H. Cale, J. Hedman, B. Johansson, Electron-spectroscopic studies of the CuxPd1-x alloy system: chemical-shift effects and valence-electron spectra. Phys. Rev. B **24**(4), 1725 (1981)
28. A. Rochefort, M. Abon, P. Delichere, J.C. Bertolini, Alloying effect on the adsorption properties of Pd50Cu50{111} single crystal surface. Surf. Sci. **294**(1–2), 43–52 (1993)
29. T.D. Pope, K. Griffiths, P.R. Norton, Surface and interfacial alloys of Pd with Cu(100): structure, photoemission and CO chemisorption. Surf. Sci. **306**(3), 294–312 (1994)
30. C.J. Barnes, M. Gleeson, S. Sahrakorpi, M. Lindroos, Electronic structure of strained copper overlayers on Pd(110). Surf. Sci. **447**(1–3), 165–179 (2000)
31. S.K. Sengar, B. Mehta, Size and alloying induced changes in lattice constant, core, and valance band binding energy in Pd-Ag, Pd, and Ag nanoparticles: effect of in-flight sintering temperature. J. Appl. Phys. **112**(1), 014307 (2012)
32. A. Wiltner, C. Linsmeier, Surface alloying of thin beryllium films on tungsten. New J. Phys. **8**, 181 (2006)
33. C.Q. Sun, Y. Wang, Y.G. Nie, B.R. Mehta, M. Khanuja, S.M. Shivaprasad, Y. Sun, J.S. Pan, L.K. Pan, Z. Sun, Interface quantum trap depression and charge polarization in the CuPd and AgPd bimetallic alloy catalysts. Phys. Chem. Chem. Phys. **12**(13), 3131–3135 (2010)
34. Z.S. Ma, Y. Wang, Y.L. Huang, Z.F. Zhou, Y.C. Zhou, W.T. Zheng, C.Q. Sun, XPS quantification of the hetero-junction interface energy. Appl. Surf. Sci. **265**, 71–77 (2013)
35. C.Q. Sun, Oxidation electronics: bond-band-barrier correlation and its applications. Prog. Mater Sci. **48**(6), 521–685 (2003)
36. W.T. Zheng, C.Q. Sun, Electronic process of nitriding: mechanism and applications. Prog. Solid State Chem. **34**(1), 1–20 (2006)
37. G.A. Slack, S.F. Bartram, Thermal expansion of some diamondlike crystals. J. Appl. Phys. **46**(1), 89–98 (1975)
38. X.D. Pu, J. Chen, W.Z. Shen, H. Ogawa, Q.X. Guo, Temperature dependence of Raman scattering in hexagonal indium nitride films. J. Appl. Phys. **98**(3), 2006208 (2005)
39. H.B. Du, A. De Sarkar, H.S. Li, Q. Sun, Y. Jia, R.Q. Zhang, Size dependent catalytic effect of TiO_2 clusters in water dissociation. J. Mol. Catal. A-Chem. **366**, 163–170 (2013)
40. C.Q. Sun, Dominance of broken bonds and nonbonding electrons at the nanoscale. Nanoscale **2**(10), 1930–1961 (2010)
41. I. Coulthard, T.K. Sham, Charge redistribution in Pd-Ag alloys from a local perspective. Phys. Rev. Lett. **77**(23), 4824–4827 (1996)
42. J.A. Lipton-Duffin, J.M. Macleod, M. Vondracek, K.C. Prince, R. Rosei, F. Rosei, Thermal evolution of the submonolayer near-surface alloy of ZnPd on Pd(111). Phys. Chem. Chem. Phys. **16**, 4764–4770 (2014)
43. L. Schriver-Mazzuoli, A. Schriver, A. Hallou, IR reflection-absorption spectra of thin water ice films between 10 and 160 K at low pressure. J. Mol. Struct. **554**(2–3), 289–300 (2000)
44. Y. Nie, Y. Wang, Y. Sun, J.S. Pan, B.R. Mehta, M. Khanuja, S.M. Shivaprasad, C.Q. Sun, CuPd interface charge and energy quantum entrapment: a tight-binding and XPS investigation. Appl. Surf. Sci. **257**(3), 727–730 (2010)
45. C.L. Lee, Y.C. Huang, L.C. Kuo, High catalytic potential of Ag/Pd nanoparticles from self-regulated reduction method on electroless Ni deposition. Electrochem. Commun. **8**(6), 1021–1026 (2006)
46. R.J. Cole, N.J. Brooks, P. Weightman, S.M. Francis, M. Bowker, The physical and electronic structure of the Cu85Pd15(110) surface; Clues from the study of bulk CuxPd1-x alloys. Surf. Rev. Lett. **3**(5–6), 1763–1772 (1996)
47. Y. Wang, Y.G. Nie, L.K. Pan, Z. Sun, C.Q. Sun, Potential barrier generation at the BeW interface blocking thermonuclear radiation. Appl. Surf. Sci. **257**(8), 3603–3606 (2011)

48. Y. Nie, Y. Wang, X. Zhang, J. Pan, W. Zheng, C.Q. Sun, Catalytic nature of under- and hetero-coordinated atoms resolved using zone-selective photoelectron spectroscopy (ZPS). Vacuum **100**, 87–91 (2014)

49. D.Q. Yang, E. Sacher, Initial- and final-state effects on metal cluster/substrate interactions, as determined by XPS: copper clusters on Dow Cyclotene and highly oriented pyrolytic graphite. Appl. Surf. Sci. **195**(1–4), 187–195 (2002)

50. C.N.R. Rao, G.U. Kulkarni, P.J. Thomas, P.P. Edwards, Size-dependent chemistry: properties of nanocrystals. Chem-Eur. J. **8**(1), 29–35 (2002)

51. W. Qin, Y. Wang, Y.L. Huang, Z.F. Zhou, C. Yang, C.Q. Sun, Bond order resolved 3d(5/2) and valence band chemical shifts of Ag surfaces and nanoclusters. J. Phys. Chem. A **116**(30), 7892–7897 (2012)

52. Y. Wang, Y.G. Nie, J.S. Pan, L. Pan, Z. Sun, C.Q. Sun, Layer and orientation resolved bond relaxation and quantum entrapment of charge and energy at Be surfaces. Phys. Chem. Chem. Phys. **12**(39), 12753–12759 (2010)

53. Y.G. Nie, X. Zhang, S.Z. Ma, Y. Wang, J.S. Pan, C.Q. Sun, XPS revelation of tungsten edges as a potential donor-type catalyst. Phys. Chem. Chem. Phys. **13**(27), 12640–12645 (2011)

54. K. Miyoshi, D.H. Buckley, Tribological properties and surface chemistry of silicon carbide at temperatures to 1500 C. ASLE Trans. **26**(1), 53–63 (1983)

55. C. Jiang, J. Zhu, J. Han, Z. Jia, X. Yin, Chemical bonding and optical properties of germanium–carbon alloy films prepared by magnetron co-sputtering as a function of substrate temperature. J. Non-Cryst. Solids **357**(24), 3952–3956 (2011)

56. M. Arghavani, R. Braunstein, G. Chalmers, D. Shirun, P. Yang, XPS study of single crystal Ge-Si alloys. Solid State Commun. **71**(7), 599–601 (1989)

57. C.Q. Sun, Y. Sun, Y.G. Nie, Y. Wang, J.S. Pan, G. Ouyang, L.K. Pan, Z. Sun, Coordination-resolved C–C bond length and the C 1s binding energy of carbon allotropes and the effective atomic coordination of the few-layer graphene. J. Phys. Chem. C **113**(37), 16464–16467 (2009)

58. C.Q. Sun, L.K. Pan, Y.Q. Fu, B.K. Tay, S. Li, Size dependence of the 2p-level shift of nanosolid silicon. J. Phys. Chem. B **107**(22), 5113–5115 (2003)

59. X.J. Liu, M.L. Bo, X. Zhang, L. Li, Y.G. Nie, H. Tian, Y. Sun, S. Xu, Y. Wang, W. Zheng, C.Q. Sun, Coordination-resolved electron spectrometrics. Chem. Rev. **115**(14), 6746–6810 (2015)

60. L. Vegard, H. Schjelderup, Constitution of mixed crystals. Physik Z **18**, 93–96 (1917)

61. C. Kittel, *Introduction to Solid State Physics*, 8th edn. (Willey, New York, 2005)

62. F. Ringeisen, J. Derrien, E. Daugy, J. Layet, P. Mathiez, F. Salvan, Formation and properties of the copper silicon (111) interface. J. Vac. Sci. Technol., B **1**(3), 546–552 (1983)

63. C.Q. Sun, Y. Wang, Y.G. Nie, Y. Sun, J.S. Pan, L.K. Pan, Z. Sun, Adatoms-induced local bond contraction, quantum trap depression, and charge polarization at Pt and Rh surfaces. J. Phys. Chem. C **113**(52), 21889–21894 (2009)

64. A. Tamtögl, M. Kratzer, J. Killman, A. Winkler, Adsorption/desorption of H_2 and CO on Zn-modified Pd(111). J. Chem. Phys. **129**(22), 224706 (2008)

65. A. de Siervo, E.A. Soares, R. Landers, G.G. Kleiman, Photoelectron diffraction studies of Cu on Pd(111) random surface alloys. Phys. Rev. B **71**(11), 115417 (2005)

66. A. de Siervo, R. Paniago, E.A. Soares, H.D. Pfannes, R. Landers, G.G. Kleiman, Growth study of Cu/Pd(111) by RHEED and XPS. Surf. Sci. **575**(1–2), 217–222 (2005)

Chapter 9
Hybridized Bonding

Abstract Atoms of electronegative elements such as F, O, N and S bond to atoms of lower electronegativity form new chemical bonds with an association of positively-charged holes, nonbonding lone pairs, and antibonding dipoles, which modifies the valence band by introducing additionally four DOS features. The derived DOS features are detectable using STS (around E_F), IPES (inverse PES, $E > E_F$), and PES ($E < E_F$). New bond formation also modifies the crystal potential and hence results in further core-level entrapment. The polarization will in turn screen and split the potential to add excessive features to the core band. Atomic undercoordination and orbital hybridization enhance each other on the nonbonding electron polarization that fosters the monolayer high-T_C and topological insulator edge superconductivity.

Highlights

- The sp-orbital hybridization results in four excessive valence states.
- Adsorbate-host bond formation further deepens the core levels and adds bonding states.
- Electron-hole pair production opens the band gap; dipole formation lowers the work function.
- Spin-resolved polarization by atomic undercoordination and orbital hybridization foster the high-T_C and topological superconductivity.

9.1 STS and IPES: Antibonding and Nonbonding States

The antibonding and nonbonding states are located at energies near E_F, which can be detectable using STS and IPES (inverse photoemission). An STS probes the on-site DOS of a few atoms across at a surface [1, 2]. Features below E_F (sample negative bias) correspond to the occupied DOS of nonbonding states while features above E_F (sample positive) represent the allowed, yet to be occupied by the antibonding DOS [3]. Current flows across the barriers or the gap between the STS tip and sample from the negative to the positive direction. The IPES also probes the occupation of

© The Editor(s) (if applicable) and The Author(s), under exclusive license to Springer Nature Singapore Pte Ltd. 2020
C. Q. Sun, *Electron and Phonon Spectrometrics*,
https://doi.org/10.1007/978-981-15-3176-7_9

the unoccupied antibonding states by electrons of the surface dipoles above E_F, by integrating over morphological patches of a surface. Both STS and IPES measure consistently the presence of surface polarization but STS probes the nonbonding states that stem the surface dipoles.

Figure 3.2 compared the STS spectra collected from Cu(110) and O–Cu(110) surface at different atomic sites [1]. The clean surface shows an empty DOS at 0.8~1.8 eV above E_F with absence of any DOS features below E_F. The "O^{-2}:Cu^P:O^{-2}" chain formation reduces the original empty-DOS above E_F by occupation the states with the polarized electrons upon chemisorption. Chemisorption creates additional DOS features around $-(1.4–2.1)$ eV below the E_F with a sharp peak around -1.4 eV, as indication of the nonbonding lone pairs. The sharp feature has also been detected using the ARPES from O–Cu(110) and O–Cu(111) surfaces [4], the de-excitation spectroscopy of metastable atoms [5], and VLEED spectrometrics from O–Cu(001) surface [6, 7].

The STS profiles taken from O–Cu(110) surface between two Cu^P dipoles show more pronounced nonbond features than that from atop one Cu^P. Taking the tip-size effect of an STS (with ~2.5 Å lateral uncertainty) and the constant current mode into consideration, the intensity difference between these two sites, atop and between two Cu^P, results from that the STS collects more lone pairs information when its tip stands between two Cu^P dipoles. The occupation of the above-E_F feature results from the two neighboring dipole protrusions while the below-E_F information comes from the lone pair of the O^{-2} underneath. Therefore, the above-E_F features of the tip-between-Cu^P is stronger; the tip apt the Cu^P collects information of both the lone pair and the dipole but the signal is weaker because of the positive curvature at the dipole site.

The Cu 3d DOS are between -2 and -5 eV [4, 6–9], and the O–Cu bonding derivatives are around the 2p-level of oxygen, $-5.6 ~ - 7.8$ eV below E_F [10]. Both the Cu-3d and the O–Cu bond DOS features are outside the scope of the STS ($E_F \pm$ 2.5 eV).

Figure 9.1a shows the oxygen chemisorption induced STS spectra of the Nb(110) surface [11] and the IPES spectra of O–Cu(110) surfaces [12]. The STS spectra from the O–Nd(110) surface (Fig. 9.1a) show DOS features near the E_F is not so apparent as that from the O^{-2}:Cu^P:O^{-2} chain [11] (Fig. 9.10). The resonant peaks at the sample positive bias (unoccupied states) arise from a tunneling via quantized states in a potential well induced by the combination of the image states and the applied electrical field [13, 14]. Since the lowest image state is energetically tied to the vacuum level, the position of the first resonance can be used to estimate the work function [14]. The successive higher resonance at 6.4 and 7.7 eV are in accordance with resonances found for other transition metals such as Cu/Mo(110) [13] and Ni(001) surfaces [14, 15]. These resonances shift to lower energies on the O–Nb surfaces, independent of the tip positions, i.e. on or away from the O–Nb chains. The intensity reduction of the IPES 2.0 eV peak indicates the occupancy of the antibonding states by the surface dipole electrons induced by oxygenation. The rest two peaks remain constant because of the resonant and bulk tunneling [12].

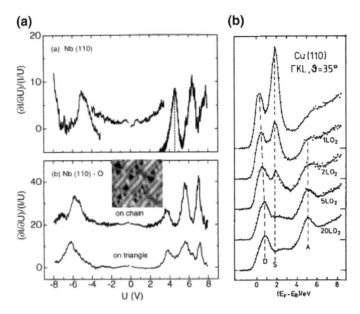

Fig. 9.1 Oxygen chemisorption derived **a** STS LDOS for the Nb(110) surface [11] and **b** the IPES (9.7 eV incident beam sources) for Cu(110) surface [12]. Inset a shows the STM image of O–Nb(110) surface with O–Nb chains and triangle-shaped atomic vacancies. Oxygen chemisorption attenuates the IPES S peak intensity at 2.0 eV as the occupancy of the empty surface states by antibonding dipoles. Features A and D are the O-induced and the direct transition in the bulk band structure. Reprinted with permission from [11, 12]

Room temperature oxygen chemisorption reduces the work function of Nb by $\Delta\phi = -0.45$ eV at less-than-one monolayer coverage and then is followed by an increase of $\Delta\phi = +0.8$ eV at higher oxygen coverages, which is even higher than that of the clean surface [16]. The shift of the second peak at 5.6 eV is consistent with the work function increase and the rest two peaks correspond to satellites of resonance. The occupied DOS features of the O–Nb bond are at -5.8 and at -6.2 eV, which is in the STM triangular vacancy positions. The clean Nb(110) surface DOS is at -5.0 eV. The energy shift of the occupied states was ever attributed to the overlap of the O $2p$ and the Nb $4d$ states of the reconstructed O–Nb(110) surface [11].

The obviously symmetric dispersion of this state with respect to the occupied oxygen $2p_{y-}$ states agrees with a two-level approximation for the O–Cu σ-bond. The band above E_F is an empty surface state, which means that the agreement of its dispersion with that of the O $2p_{y-}$ state below E_F is accidental [17]. No definite interpretation to this phenomenon has been available up to now. Nevertheless, it should be noted that the feature at $+2$ eV is in accordance with those probed using STS on the same O–Cu(110) surface. This empty feature decreases with increasing oxygen exposure.

Figure 9.1b shows the exposure dependence of the IPES unoccupied bands of the O–Cu(110) surface [12] exhibiting the resonant features similar to that observed

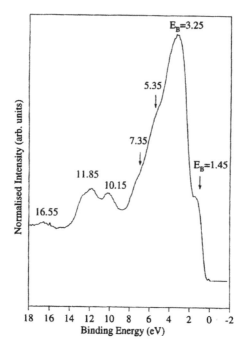

Fig. 9.2 O–Cu(001) surface valence DOS recorded at 70 eV photon energy, annoted are fine structures. Reprinted with permission from [18]

from O–Nb(110) using the STS. The states centered at 4 eV above E_F was attributed to the $Cu3d^{10}$ state, in keeping with the $Cu3d^9$ interpretation for the ground state, because with the IPES technique an additional electron is added to the O-Cu system [12].

Figure 9.2 shows the PES profile from an O–Cu(001) surface with three new DOS features around -1.45, -3.25 and -5.35 eV within the valence band (above -7.04 eV) compared to that of a clean Cu(001) surface [18]. The nonbonding states around -1.45 eV are anti-resonant, i.e., the intensity has no apparent change with varying incident beam energy. The anti-resonant DOS feature is the character of electrons that are strongly confined in one-dimension such as molecular chains [19]. The electron lone pairs zigzag the 'O^{-2}:Cu^p:O^{-2}' strings at the Cu(001)–O^{-2} surface [6, 7].

9.2 ARPES: Holes, Nonbonding and Bonding States

The ARPES with UV light source probes the holes, nonbonding and bonding states in the valence band and their anisotropic distribution in the real, two-dimensional surfaces. The ARPES from the O–Cu(110) surface, see Fig. 9.3, show three excessive

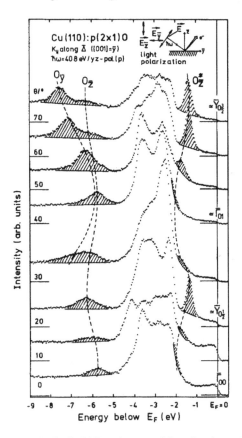

Fig. 9.3 ARPES profiles for the O–Cu(110) surface reveal the azimuth angular-resolved additional (shaded) nonbonding lone pair features around -1.4 eV and the O–Cu bonding states at ~ -6 eV [21]. Reprinted with permission from [20]

features due to oxidation. These features were previously assigned as Op_{y^-} and Op_{z^-} with two reasons [20]. Firstly, the strongest Cu–O interaction is assumed along the O–Cu–O chains and thus one will make the assignment Op_{y^-} to the structure with the largest dispersion. Secondly, this feature is only observable at large incident angles ϑ, which suggests that one is observing an orbital in the plane, and this could only be the Op_{y^-} orbital (along the O–Cu–O chain) because of the polarization dependence. The assignment of the $Op_{z^{\ast}}$ structure (the asterisk indicates an anti-bonding level) stems from the observed polarization dependence and the comparison with the dispersion of the Op_{z^-} orbital. It was expected [20] that the O 2p band located below the Cu 3d band corresponds to the O–Cu bonding band, and those above the Cu 2d band are the occupied O–Cu 'anti-bonding' bands.

However, UPS observations revealed three additional DOS features from the O–Cu/Ag(110)(2 × 2)p2mg phase surrounding -1, -3 and -6 eV and some of the features were ascribed as Cu 3d (-3 eV) and O 2p ($-1 \sim -2$ eV) states [22] without considering the charge transportation, *sp*-orbital hybridization, and polarization during chemisorption. This elemental orbital articulation changed completely in 1997 when the H_2O-like *sp*-orbital hybridization was firstly adopted for oxidation and nitridation with introduction of the antibonding, hole, nonbonding, and bonding states [21, 23–49].

Likewise, a set of ARPES spectra, shown in Fig. 9.4 displayed two adjacent DOS clusters populating below the E_F of the O–Pd(110) surface [50]. One is the nonbonding lone pair states around −2.0 eV and the other is O–Pd bonding states around −5.0 eV. These signatures arise at the expense of the weakening of the DOS close to the E_F due to hole production by transporting electron from Pd atoms to O adsorbates.

Figure 9.5 compares the ARPES spectra from the c(2 × 2)-O⁻ phase (radial reconstruction) and (2 × 2)p4g-O⁻² phase (clock and anti-clock rotation) on the

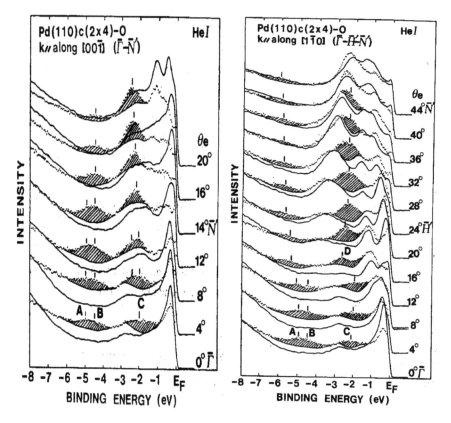

Fig. 9.4 ARPES profiles for the O–Pd(110) surface revealed the azimuth angular resolved (shaded) nonbonding lone pairs around −2 eV and O–Pd bonding states centered at −5 eV. The ARPES intensity reduction near E_F at some angles such as 0° and 44° arises from Pd⁺ hole production [21]. Reprinted with copyright permission from [21, 50]

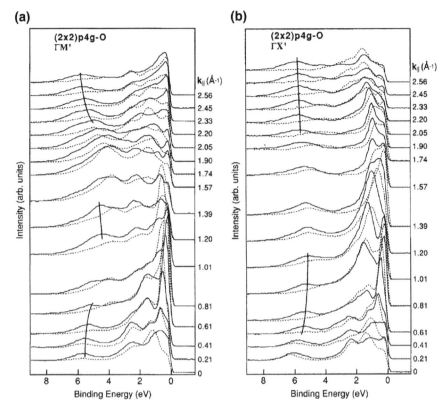

Fig. 9.5 ARPES in 4° steps (in the parallel momenta k$_{//}$) shows the azimuth angular resolved DOS for the clean Rh(001) (dotted curves) and Rh(001(2 × 2) p4g-O (solid curves) surfaces. The lines between adjacent spectra resolve the dispersion due to O adsorption. Data were collected **a** along ΓM′ <10> direction and **b** along the ΓX′ <11> direction [51]. Reprinted with permission from [51]

Rh(001) surface [51]. The PES spectra [51–53] showed that oxygen induces significant change in the energy states around −2 ~ −6 eV below E$_F$. DOS for holes below the E$_F$ can be resolved from the PES profiles of both phases [54]. Additional O-derivatives can be identified at around −5 eV for the first (2 × 2)-O⁻ phase. The feature around −5 eV shifts down a little with an additional peak at about −2.0 eV at the same azimuth angles for the O⁻²-induced 'p4g' phase.

The ARPES in Fig. 9.6 shows that S chemisorption results in the similar valence DOS for the O–Rh(001) surface. The similarity in the valence DOS identity suggests that the S may share the same 3sp-orbital hybridization creating the bonding, nonbonding, holes, and antibonding states though they are subject to further confirmation. Therefore, the four valence DOS features seemed common for surfaces with adsorption of electronegative elements having nonbonding lone pairs upon reaction.

(a)   **(b)**

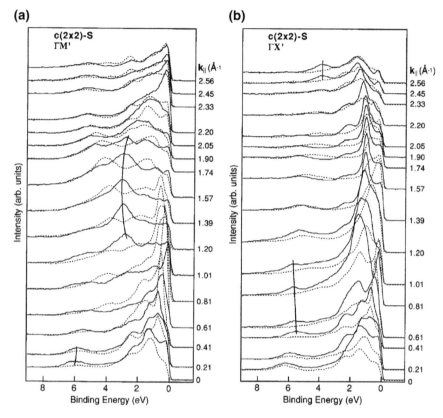

Fig. 9.6 ARPES from the clean Rh(001) (dotted curves) and Rh(001(2 × 2)-S (solid curves) surfaces. The parallel momenta $k_{//}$ shown are calculated for electrons emitted from the Fermi level. Data were collected **a** along the ΓM′ <10> and **b** the ΓX′ <11> direction. Reprinted with permission from [51]

9.3 Coverage Resolved O–Cu Valence DOS Evolution

The O-coverage-resolved PES (He-II, 21.22 eV), see Fig. 9.7, revealed that oxidation of a polycrystal copper surface proceeds in three stages [9]:

- Firstly (the lowest, 12 L, exposures), oxidation begins, which immediately raises a spectral shoulder at −1.5 eV for nonbonding states and a small peak at −6.0 eV for O–Cu bonding states. The emergence of the new DOS features is at the expense of a sharp fall of the DOS features between −3.0 eV and E_F as an indication of Cu^+ hole production.

- Secondly (12 ~ 1000 L), both the −6.0 eV DOS feature and the sharp-fall at $E_F >$ $E > −3.0$ eV become pronounced as the oxygen exposure increases as the number of the O–Cu bond grows with charge transporting from Cu to O.

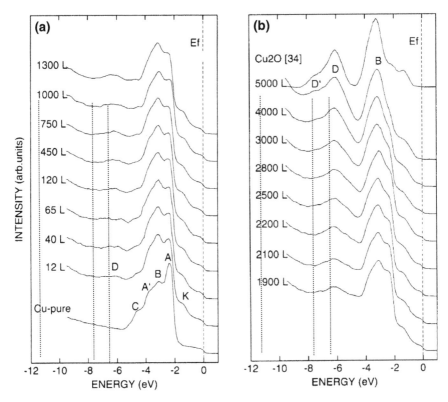

Fig. 9.7 Oxygen exposure-dependence of the PES at the successive stages of oxygen adsorption on polycrystal Cu surface. Spectra show the common features K (-1.5 eV, nonbonding) and D (-6 eV, bonding) and electron vacancies near E_F to oxide surfaces, as discussed above. Reprinted with permission from Ref. [9]

- Thirdly, at oxygen exposure of 5×10^3 L, a surface compound is formed, which exhibits semiconductive properties, and its electronic structure is very similar to that of the bulk copper protoxide Cu_2O. Emerging of the new features renders a sharp fall of the feature at -3.0 eV and above. The DOS at E_F fall to zero and produces a gap of ~1.0 eV, which is responsible for the semi-conductivity of the Cu_2O surface.

According to VLEED confirmation [28], these reaction steps agree well with the Cu_3O_2 bonding kinetics on the O–Cu(001) surface [21]. The feature around -6.0 eV corresponds to the O–Cu hybrid bonding and the feature centered at -1.5 eV to the oxygen nonbonding lone-pair states. The fall of the upper part of the d band corresponds to the process of electron transition from the outermost shell of the Cu atom to the deeper empty sp-hybrid orbitals of oxygen, or to the even higher empty levels of Cu to form diploes. The rise of the -1.5 eV feature is independent of the fall of the intensity near the E_F. The exposure-dependence of the valence DOS change provides valuable information about the dynamics of electron transportation [46].

Conventionally, the additional DOS features around -1.4 eV ~ -2.0 eV of the copper oxide were argued as: (i) O–Cu anti-bonding states [4, 18, 55]; (ii) O $2p$ anti-bonding states [4, 20, 55], (iii) oxygen $2s$ states [56, 57] and, (iv) the O $2p$ electrons with the spd hybridized electrons of Cu [9]. The additional DOS features around -5.5 eV were interpreted as O $2p$ states adding to the valence band of the host surface [4, 8, 20]. The sharp fall of the DOS features at $E_F > E > -3.0$ eV corresponds to the disappearance of the clean Cu surface states.

The DOS features appeared in the valence band or above of a chemisorbed surface, the PES features of the O–Pd(110) [58], O–Cu(110) [12, 18, 20], O–Cu(111) [4], O–Rh(001) [59] and S–Rh(110) and surfaces, as mentioned above, are substantially the same despite their surface crystal geometries and morphologies. In place of the conventional explanations in terms of individual orbital wise, the adsorbate-induced four DOS features result from the effect of sp-orbital hybridization and correspond to the antibond, nonbond, holes, and bonding states consistently.

9.4 DFT Derivatives

9.4.1 O–Ti(0001)

Figure 9.8 shows the optimal structural configuration of oxygen on the Ti(0001) surface and the hetero-coordination effect on its valence DOS. Oxygen chemisorption indeed results in four additional DOS features that modulate the band gap, work function, carrier lifetime for catalytic applications [60]. These features correspond to the O–Ti bonding, O lone pairs, Ti$^+$ electron holes, and Ti dipole anti-bonding states [61, 62].

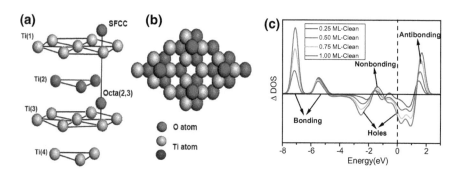

Fig. 9.8 a Side and **b** to views of O–Ti(0001) $-$ p(2 × 2) skin and **c** the calculated ZPS, n(Ti + O) $-$ n(Ti). Oxygen atoms occupy the surface fcc sites (SFCC) at 0.25 ML and then occupy both the fcc and the octahedral sites between the second and the third Ti layers (Octa(2, 3)) sites at 0.50 ML. Four DOS features correspond to the antibonding dipoles ($+1.6$ eV), nonbonding lone pairs (-1.6 ± 0.5 eV), holes (-1.5 ± 1.5), and bonding pair (-6.0 ± 1 eV) states. Reprinted with permission from [60]

9.4.2 N–Ti(0001)

Figure 9.9 shows the optimal N sites at the Ti(0001) surface and the coverage dependence of work function [63]. Nitrogen adsorbs to the Ti(0001) skin in a mixed monolayer and multilayer configuration rather than staying atop of the surface. According to LEED and AES measurements, an initial (1 × 1) monolayer and subsequent $(1 \times 1) + (\sqrt{3} \times \sqrt{3})$ pattern multilayer form in sequence. The monolayer and multilayer surface structures, i.e., (2 × 2) and (3 × 3) supercell used in DFT calculations [63] results in the work function changing from that for clean surface. The calculated work function for clean surface is 4.47 eV that agrees with previous DFT calculations of 4.45 eV [64] and measurements (4.45–4.60 eV) [65]. DFT Calculation revealed that the work function decreases initially and then reverses until saturation, agreeing with experimental observations [66]. The transition of work function with N coverage indicates the transformation from Ti^p to $Ti^{+/p}$ during reaction—hydrogen bond like formation—excessive N catch the polarized electrons to form bond stabilizing the surface.

Figure 9.10 shows the tetrahedral-coordination effect on the valence DOS of N–Ti(0001) surface as a function of N coverage. N chemisorption results in four additional DOS features that are the same appeared in the N–Ru(0001) [67] skin. These features correspond to the N–Ti bonding, N lone pairs, Ti^+ electron holes, and Ti^p dipole anti-bonding states [61, 62]. Panel b shows that the $((1 \times 1) + (\sqrt{3} \times \sqrt{3}))$-12 N reduces the antibonding states, which is responsible for the work function elevation.

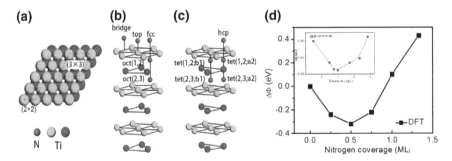

Fig. 9.9 **a** Top view of the Ti(0001) surface with indication of the (2 × 2) and (3 × 3) unit cells. **b, c** Perspective views of N adsorption sites. The symbol octan(i, j) denotes the octahedral sites between the i-th and $i + 1$ th layers, and the (i: aj(or bj)) means the tetrahedral site located directly below the i th layer. **d** Coverage dependence of the work function for the N-Ti(0001) surface, compared with experimental observations (inset). Reprinted with permission from [66]

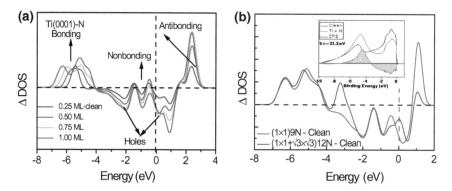

Fig. 9.10 Coverage dependence of the ZPS of n(Ti + N) − n(Ti) of the valence LDOS. **a** Monolayer adsorption that N occupied at oct(1, 2) sites in the (2 × 2) unit cell. **b** Multilayer adsorption for the Ti(0001)-(1 × 1)-9N and $(1 \times 1)\left(\sqrt{3} \times \sqrt{3}\right)$-12N in the (3 × 3) unit cell. Inset b compares the measured valence DOS with insignificance of the annihilated lone pair features [68]. Reprinted with permission from [63]

9.4.3 N–Ru(0001) and O–Ru(10$\bar{1}$0)

Figure 9.11 compares the residual DOS for both the N–Ru(0001) [67] and the O–Ru(10$\bar{1}$0) [69] surfaces with O and N at least 1.0 Å atop the surfaces without considering the tetrahedron bond formation. The DOS features are the same, corresponding to the bonding (−6.0 eV), non-bonding (−3.0), holes (−1.0) and anti-bonding (+3.0)

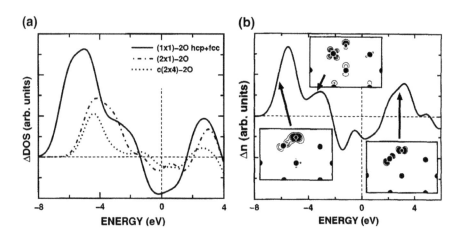

Fig. 9.11 DFT-derived differential valence DOS for N–Ru(0001) [67] and O–Ru(10$\bar{1}$0) [69] surfaces with O and N atop the surface. Both adsorbates create the same four valence DOS features. Reprinted with permission from [67, 69]. Copyright 1997 Elsevier and 1998 American Physical Society

states. N-induced lone pair states of nitrogen have been widely observed on nitrides surfaces [26, 41, 70, 71] and CNT nanotubes [72].

9.5 XPS and ZPS: Core Level Entrapment and Polarization

9.5.1 O–Ta(001) and O–Ta(111)

Figure 9.12 compares the calculated [73, 74] and measured ZPS profiles for the bcc-structured Ta(111) [75] and (b) Ta(001) [76] surfaces upon oxygen chemisorption. Consistency between experiments and calculations show that oxygen adsorption induces further entrapment of 4f core band. The valence ZPS shows four DOS features corresponding to the O–Ta bonding, O lone pairs, Ta^+ ions, and the antibonding dipoles of Ta^p due the lone pair induced polarization, with weak polarization attributes due to the $Ta^p \rightarrow Ta^{p/+}$ transition [21].

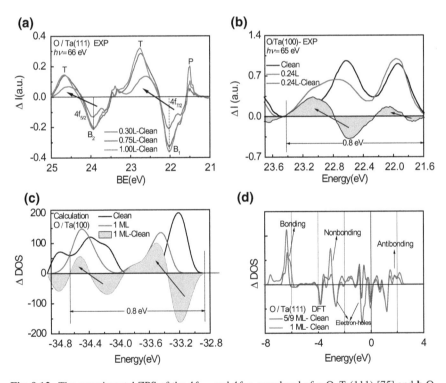

Fig. 9.12 The experimental ZPS of the $4f_{5/2}$ and $4f_{7/2}$ core band of **a** O–Ta(111) [75] and **b** O–Ta(001) [76] surfaces in comparison to the DFT derived 4f band of **c** Ta(100) and **d** the valence ZPS of Ta(111) upon oxygen adsorption. Core bands show entrapment dominance and the valence band shows the regular four-DOS features. Reprinted with permission from [74]

9.5.2 Monolayer High-T_C and Topological Edge Superconductivity

The spin-resolved polarization by processes of atomic undercoordination and sp^3-orbital hybridization may contribute to the high-T_C superconductors (HTSCs) and topological insulators (TIs) edge conductivity. Most HTSCs prefer the layered structure and the van der Waals gaps between layers serve as channels of charge transport. Dirac-Fermions associated with the even less coordinated edge atoms serve as the carriers transporting along the edges of the topological insulators. Figure 9.13 com-

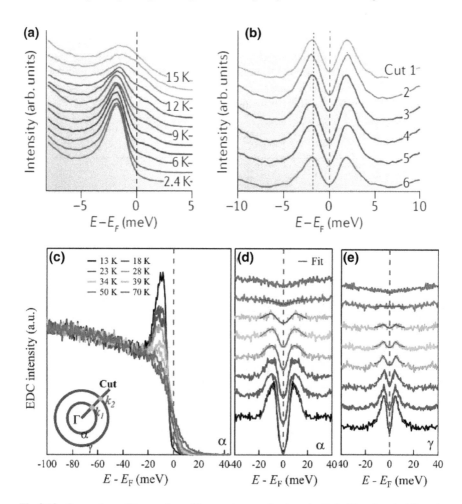

Fig. 9.13 Comparison of the angular and temperature resolved conductivity (Fermi pocket) for **a** the FeTe$_{0.55}$Se$_{0.45}$ ($T_C = 12$ K) topological insulator [77] and **b** the (Ba$_{0.6}$K$_{0.4}$)Fe$_2$As$_2$ superconductor ($T_C = 38.5$ K) [78]. The respective characteristic coherent peak energy at 2 and 20 meV is correated to the T_C value. Reprinted with permission from [77, 78]

pares the ARPES for the $FeTe_{0.55}Se_{0.45}$ topological insulator [77] and for the iron-based $(Ba_{0.6}K_{0.4})Fe_2As_2$ superconductor revealed that, while sharp superconducting coherence peaks emerge in the superconducting state on the hole-like Fermi surface sheets, no quasiparticle peak presents in the normal state. Its electronic behaviors deviate strongly from a Fermi liquid system. The superconducting gap of such a system exhibits an unusual temperature dependence that it is nearly a constant in the superconducting state and abruptly closes at T_C. In the superconducting state, the superconducting gap is extremely isotropic momentum dependence.

The TI and HTSC showed similar conductance but different T_C values or coherent peak energies. The $FeTe_{0.55}Se_{0.45}$ has a peak at 10 meV with a T_C of 38.5 K. Contrastingly, both the coherent peak energy and T_C of the TI are lower than those for the $FeTe_{0.55}Se_{0.45}$, which indicates that the TI and HTSC share the common yet different extents of interactions characterized by the splitting of the peak cross the E_F energy—Fermi pocket. The T_C corresponds to the temperature of the pocket gap disappearance.

Figure 9.14 shows the ARPES for the Bi-2212/3 HTSC states. The Bi-2212 has a peak at 30 meV with the T_C of 91 (up to 136) K. Contrastingly, both the coherent peak energy and T_C of the TI are one order lower compared with that for the Bi-2212. Apparently, the coherent peak energy and the T_C are cooperated proportionally.

Most strikingly, the conductivity of the HTSC Bi-2212 is thickness independence, see Fig. 9.15. At optimal doping, the monolayer Bi-2212 has $T_C = 105$ K [82] being identical to that of its bulk parent [79]. The monolayer refers to a half unit cell in the out-of-plane direction that contains two CuO_2 planes. The monolayers are

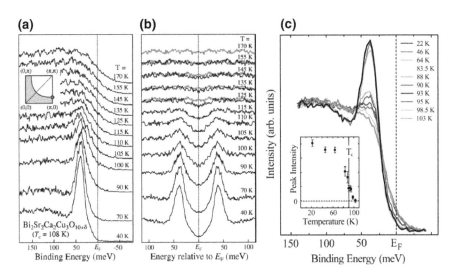

Fig. 9.14 Temperature dependent Fermi pocket conductance for the Bi-2223 ($Bi_2Sr_2Ca_2Cu_3O_{10+\delta}$, $T_C = 108$ K). **a** Raw spectra; **b** symmetrized spectra [79] and **c** Bi-2212 ($T_C = 91$ K). Inset: superconducting-peak intensity versus temperature. Reprinted with permission from [80, 81]

(a)   **(b)**   **(c)**

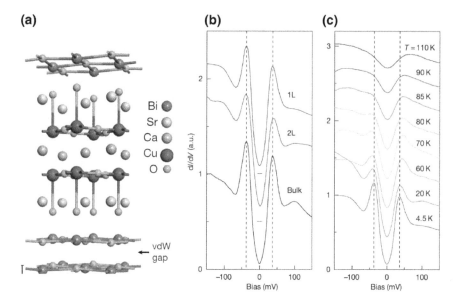

Fig. 9.15 a Atomic structure of Bi-2212. The 'monolayer' refers to a half unit cell in the out-of-plane direction that contains two CuO_2 planes. The monolayers are separated by van der Waals gaps in bulk Bi-2212. Spatially averaged **b** differential conductance of monolayer, bilayer and bulk and temperature dependence and superconductivity transition at $T \approx 105$ K for the monolayer [82], which is identical to that for the bulk Bi-2212 at optimal doping [79]. Reprinted with permission from [82]

separated by van der Waals gaps in bulk Bi-2212. This observation reveals not only the skin dominance of the Bi-2212 superconductivity but also the impact of the atomic undercoordination to the HTSC superconductivity.

It is even amazing, an array of tiny holes (~100 nm across) turns the yttrium barium copper oxide (YBCO) HTSC into a regular conductor having resistance to carrier transportation [83]. Rather than moving in concert, electron pairs conspire to stay in place, stranded on tiny islands and unable to jump to the next island, in a very thin HTSC. When the material has a current running through it and is exposed to a magnetic field, charge carriers in the YBCO will orbit the pores like water circling a drain in a fashion of Bosonic cyclotronic ocsilasion.

In conventional superconductors like niobium or lead, the normal state is a Fermi liquid with a well-defined Fermi surface and quasiparticles move along the surface. Superconductivity is realized by the Fermi surface instability in the superconducting state and the formation and condensation of the electron pairs (Cooer pairing). The original Bardeen-Cooer-Schrieffer (BCS) theory of superconductivity adequately describes the origination and behavior of conventional superconducting metals and alloys, whose critical temperatures of superconductivity transition are below 30 K.

According to the BCS theory, a large Bose-Einstein condensation resulting from the coupling of electron pairs near the Fermi surface, which are known as Cooer pairs, governs the superconductivity. Resistance is created when electrons rattle around in the atomic lattice of a material as they move. But when electrons join together to become Cooer pairs, they undergo a remarkable transformation. Electrons by themselves are fermions, particles that obey the Pauli exclusion principle, which means each electron tends to keep its own quantum state. Cooer pairs, however, act like bosons, which can happily share the same state. That bosonic behavior allows Cooer pairs to coordinate their movements with other sets of Cooer pairs in a way that reduces resistance to zero [84].

The high temperature cuprate superconductors, represent another extreme case that superconductivity can be realized in the underdoped region where there is neither well-defined Fermi surface due to the pseudogap formation nor quasiparticles near the antinodal regions in the normal state. The superconductivity is realized in a system with well-defined Fermi surface but without quasiparticles along the surface in the normal state.

However, the HTSC and the effect of atomic undercoordination in the monolayer and defect are beyond the description of the BCS theory.

One may consider first the elemental selectivity in the HTSC and TI substance. The fact that some compounds of B, C, N, O, F and elements surrounding them, particularly in group V and VI in the periodic table, form HTSC and TIs, albeit the T_C and the coherent energy, implies an underlying similarity in these elements. It has been certain that N, O and F could generate nonbonding and antibonding states near the Fermi surface upon their sp^3-orbital hybridization. C and N can undergo sp^2-orbital hybridization as well with creation of the unpaired and paired electrons [85]. In turn, these localized lone pairs and the associated antibonding electrons may have a high chance of forming Copper pairs dominating the character of the HTSCs and TIs. When an external electric field is applied, these localized pairs of electrons are easily excited and hence become highly conducting, in a concerted fashion, given suitable channels of transportation. Compared with the findings of graphene edge Dirac Fermion states, the effective mass of these electrons is very small and their group velocity is extremely high. An important characteristics is that these HTSCs all assume a two-dimensional layered structure, such as "Cu^p:O^{-2}:Cu^p:" chains or CuO_2 planes, on which superconductivity relays.

As a plausible mechanism governing the HTSC and TI, the strong correlation of electronic spins has attracted much attention. The presence of the nonbonding and the antibonding states near Fermi surface should play at least a role of competence. If the $1s$ electrons of B and C are excited to occupy the hybridized $2sp^3$ orbits, B and C would likely form valence band structures similar to those of N and O and hence result in the superconductivity. Atoms of group V and VI elements should maintain features of weak sp-orbital hybridization, and therefore, the nonbonding lone pairs could be a factor of dominance in both HTSC and TI conductivities. The exposition of the mechanism of HTSC from the perspective of antibond and nonbond formation and the corresponding electronics and energetics would be an approach culminating new knowledge. The spin-spin coupling may determine the coherent peak energy and thermal decoupling of the spins may correspond to the critical temperatures. From this perspective, the spin coupling in the TI is weaker than it is in the HTSC because the former has much lower T_C and coherent energy and the conductivity proceeds along the even undercoordinated edge atoms.

On the other hand, atomic undercoordination enhances the polarization due to *sp*–orbital hybridization, which discriminates the skin dominance of HTSC conductivity and Dirac-Fermion generation TI edge conductivity. Because of the localization and entrapment of the polarized states, the artificially holed HTSC transits into a regular conductor. Although it is subject to further justification, the dual process of spin-resolved nonbonding electron polarization by atomic undercoordination and the intrinsic *sp*-orbital hybridization may provide a feasible mechanism for the HTSC and TI superconductivity.

9.5.3 Other Surfaces

It is common that chemisorption deepens the core band of metals because stronger bond formation. Oxygen adsorption deepens the Pd $3d_{5/2}$ level by ~0.6 eV [58]. The O $1s$ level (-529.5 eV) shifts 0.6 eV to -530.1 eV when the oxygen reacts with the Cu(001) surface [86]. The Rh $3d$ binding energy increases about 0.3 eV per O–Rh bond on the O–Rh(111) surface [87]. With increasing oxygen coverage on the Ru(0001) surface, the Ru $3d_{5/2}$ core-level peaks shift by up to -1.0 eV [88].

Chemisorption is a kinetic process in which the valences of the bonding constituents change and hence the sizes and positions of surface atoms change. Electrons transport from the valence band of the host to the empty *p*-orbital of oxygen for the bonding, and then the oxygen hybridizes with the production of lone pairs. The lone pairs polarize in turn their surrounding neighbors and the electrons of the host dipoles move from the original energy level to the higher energy levels. These sequential processes will redistribute electrons in the valence band and above of the host with four additional DOS features.

Experimental databases for PES, IPESS, and XPS are available for Pd-O [89, 90], O–Cu(110) [91], O–Nb(110) [92], AgO [93], and $Bi_2Sr_2CaCu_2O_8$ [94]. These spectra share considerable DOS similarities in the valence band and above. Au nanoclusters deposited on $TiO_2(110)$ substrate also exhibits a weak feature at -1.0 eV due to the interfacial oxidation [95].

9.6 Summary

Electron spectroscopic observations confirm the generality of the creation and evolution dynamics of the valence states during chemisorption and its effect on the core level shift. In addition to the bonding states, the nonbonding lone pair of the adsorbate, the antibonding dipole states of the host, and the electron hole of the host atom can never be neglected, which modify the physical properties of a chemisorbed surface. These observations shall be general to the chemisorption of light electronegative elements to other surfaces. Table 9.1 summarized the observations of O, N, S adsorption on typical examples. The energy states of nonbonding lone pairs and positively-charged holes exit but they may not be detectable because they may share the same binding energy. Transition from M^p to $M^{+/p}$ due to hydrogen bond like formation may attenuate the antibonding states.

Topological insulators edge superconductivity and high-T_C monolayer superconductivity share the same identities of spin-resolved polarization by sp-orbital hybridization and atomic undercoordination but different coupling strengths that discriminate their coherent peak energy, critical temperatures and channels of superconductivity. The thickness independence of the Bi-2212 superconductivity evidence the skin dominance of HTSC.

Table 9.1 Adsorbate-derived valence DOS features (unit in eV)

Refs.	Guest	Host (X)	Anti-bond M^p dipole $> E_F$	Lone pair $< E_F$	M^+ (hole) $< R_F$	X-M bond $\ll E_F$
[18, 96]	O	Cu(001)		-1.5 ± 0.5	-3.0 ± 1.0	-6.5 ± 1.5
			1.2	-2.1		
[55]	O	Cu(001)		-1.37; -1.16		
[97]	O	Ni(001)				$E_F \sim -6.0$
[1, 4, 8, 12, 98, 99]	O	Cu(110)	~2.0	-1.5 ± 0.5	-3.0 ± 1.0	-6.5 ± 1.5
			1.3 ± 0.5	-2.1 ± 0.5		
[9]	O	Cu(poly)		-1.5	-3.0 ± 1.0	-6.5 ± 1.5
[22]	O	Cu/Ag(110)		-1.5		-3.0; -6.0
[100]	O	Rh(001)	1.0	-3.1		-5.8
[58]	O	Pd(110)		-2.0 ± 0.5	$-0.5, 3.0$	-4.5 ± 1.5
[101]	O	Al(poly)	1.0			
[102]	O	Gd(0001)		-3.0	$-1.0, -8.0$	-6.0
[103]	O	Ru(0001)		-1.0 ± 1.0		-5.5 ± 1.5

(continued)

Table 9.1 (continued)

Refs.	Guest	Host (X)	Anti-bond M^P dipole $> E_F$	Lone pair $< E_F$	M^+ (hole) $< R_F$	X-M bond $\ll E_F$
[104]				−0.8		−4.4
[105]			1.5	−4		−5.5, −7.8
[106]			1.7	−3.0		−5.8
[69]	O	Ru(10$\bar{1}$0)	2.5	−2 ~ − 3.0		−5.0
[107]		MgO/Ag(001)		−3.0 ± 1.0		
[108]	O	Co(Poly)		−2.0	−0.7	−5.0
[109]	O	C(diamond)		−3.0		
[110]	O	CNT	0.8			
[111]	O, S	Cu(001)P		−1.3		−6.0
[112]	O, S, N	Ag(111)		−3.4		−8.0
[97, 111]	N	Cu(001)	3.0	−1.2		−5.6
				−1.0	−4.0	−5.5
[67]	N	Ru(0001)	3.0	−3.0		−6.0
[111]	N	Ag(111)		−3.4		−8.0
[113]		TiCN		0.0 ± 1.0		−5.7
[114]		a-CN		−4.5		−7.1
[115]		CN		−2.3		

Reprinted with permission from [46]

References

1. F.M. Chua, Y. Kuk, P.J. Silverman, Oxygen chemisorption on Cu(110): an atomic view by scanning tunneling microscopy. Phys. Rev. Lett. **63**(4), 386–389 (1989)
2. J. Tersoff, D. Hamann, Theory of the scanning tunneling microscope, in *Scanning Tunneling Microscopy* (Springer, Heidelberg, 1985), pp. 59–67
3. J. Wintterlin, R.J. Behm, *In the Scanning Tunnelling Microscopy I*, ed. H.J. Günthert, R. Wiesendanger (Springer, Berlin, 1992)
4. W. Jacob, V. Dose, A. Goldmann, Atomic adsorption of oxygen on Cu (111) and Cu (110). Appl. Phys. A **41**(2), 145–150 (1986)
5. W. Sesselmann et al., Probing the local density of states of metal-surfaces by deexcitation of metastable noble-gas atoms. Phys. Rev. Lett. **50**(6), 446–450 (1983)
6. C.Q. Sun, O-Cu(001): II. VLEED quantification of the four-stage Cu_3O_2 bonding kinetics. Surf. Rev. Lett. **8**(6), 703–734 (2001)
7. C.Q. Sun, O-Cu(001): I. Binding the signatures of LEED, STM and PES in a bond-forming way. Surf. Rev. Lett. **8**(3–4), 367–402 (2001)
8. R. DiDio, D. Zehner, E. Plummer, An angle-resolved UPS study of the oxygen-induced reconstruction of Cu (110). J. Vac. Sci. Technol., A **2**(2), 852–855 (1984)
9. V.P. Belash et al., Transformation of the electronic structure of Cu into Cu_2O in the adsorption of oxygen. Surf. Rev. Lett. **6**(3–4), 383–388 (1999)

10. R. Courths et al., Dispersion of the oxygen-induced bands on Cu (110)—an angle-resolved UPS study of the system p(2x1)O/Cu(110). Solid State Commun. **63**(7), 619–623 (1987)
11. C. Surgers, M. Schock, H. von Lohneysen, Oxygen-induced surface structure of Nb(110). Surf. Sci. **471**(1–3), 209–218 (2001)
12. V. Dose, Momentum-resolved inverse photoemission. Surf. Sci. Rep. **5**, 337–378 (1986)
13. G. Binnig et al., Tunneling spectroscopy and inverse photoemission—image and field states. Phys. Rev. Lett. **55**(9), 991–994 (1985)
14. T. Jung, Y.W. Mo, F.J. Himpsel, Identification of metals in scanning-tunneling-microscopy via image states. Phys. Rev. Lett. **74**(9), 1641–1644 (1995)
15. M. Portalupi et al., Electronic structure of epitaxial thin NiO(100) films grown on Ag(100): towards a firm experimental basis. Phys. Rev. B **64**(16), 165402 (2001)
16. R. Pantel, M. Bujor, J. Bardolle, Continuous measurement of surface-potential variations during oxygen-adsorption on (100), (110) and (111) faces of niobium using mirror electron-microscope. Surf. Sci. **62**(2), 589–609 (1977)
17. C.T. Chen, N.V. Smith, Unoccupied surface-states on clean and oxygen-covered Cu(110) and Cu(111). Phys. Rev. B **40**(11), 7487–7490 (1989)
18. S. Warren et al., Photoemission studies of single crystal CuO (100). J. Phys.: Condens. Matter **11**(26), 5021 (1999)
19. E.G. Emberly, G. Kirczenow, Antiresonances in molecular wires. J. Phys.-Condens. Matter **11**(36), 6911–6926 (1999)
20. S. Hüfner, *Photoelectron Spectroscopy: Principles and Applications* (Springer Science & Business Media, 2013)
21. C.Q. Sun, Oxidation electronics: bond-band-barrier correlation and its applications. Prog. Mater Sci. **48**(6), 521–685 (2003)
22. D. Sekiba et al., Electronic structure of the Cu-O/Ag(110)(2x2)p2mg surface. Phys. Rev. B **67**(3), 035411 (2003)
23. C.Q. Sun, Time-resolved VLEED from the O-Cu(001): atomic processes of oxidation. Vacuum **48**(6), 525–530 (1997)
24. C.Q. Sun, Nature of the O-fcc(110) surface-bond networking. Mod. Phys. Lett. B **11**(25), 1115–1122 (1997)
25. C.Q. Sun, Oxygen-reduced inner potential and work function in VLEED. Vacuum **48**(10), 865–869 (1997)
26. C.Q. Sun, A model of bond-and-band for the behavior of nitrides. Mod. Phys. Lett. B **11**(23), 1021–1029 (1997)
27. C.Q. Sun, Angular-resolved VLEED from O-Cu(001): valence bands, chemical bonds, potential barrier, and energy states. Int. J. Mod. Phys. B **11**(25), 3073–3091 (1997)
28. C.Q. Sun, Exposure-resolved VLEED from the O-Cu(001): bonding dynamics. Vacuum **48**(6), 535–541 (1997)
29. C.Q. Sun, Coincidence in angular-resolved VLEED spectra: brillouin zones, atomic shifts and energy bands. Vacuum **48**(6), 543–546 (1997)
30. C.Q. Sun, Spectral sensitivity of the VLEED to the bonding geometry and the potential barrier of the O-Cu(001) surface. Vacuum **48**(5), 491–498 (1997)
31. C.Q. Sun, What effects in nature the two-phase on the O-Cu(001)? Mod. Phys. Lett. B **11**(2–3), 81–86 (1997)
32. C.Q. Sun, C.L. Bai, Modelling of non-uniform electrical potential barriers for metal surfaces with chemisorbed oxygen. J. Phys.-Condens. Matter **9**(27), 5823–5836 (1997)
33. C.Q. Sun, C.L. Bai, A model of bonding between oxygen and metal surfaces. J. Phys. Chem. Solids **58**(6), 903–912 (1997)
34. C.Q. Sun, C.L. Bai, Oxygen-induced nonuniformity in surface electrical-potential barrier. Mod. Phys. Lett. B **11**(5), 201–208 (1997)
35. C.Q. Sun et al., Spectral correspondence to the evolution of chemical bond and valence band in oxidation. Mod. Phys. Lett. B **11**(25), 1103–1113 (1997)
36. C.Q. Sun, On the nature of the O-Co(1010) triphase ordering. Surf. Rev. Lett. **5**(5), 1023–1028 (1998)

37. C.Q. Sun, Driving force behind the O-Rh(001) clock reconstruction. Mod. Phys. Lett. B **12**(20), 849–857 (1998)
38. C.Q. Sun, O-Ru(0001) surface bond and band formation. Surf. Rev. Lett. **5**(2), 465–471 (1998)
39. C.Q. Sun, Origin and processes of O-Cu(001) and the O-Cu(110) biphase ordering. Int. J. Mod. Phys. B **12**(9), 951–964 (1998)
40. C.Q. Sun, Nature and dynamics of the O-Pd(110) surface bonding. Vacuum **49**(3), 227–232 (1998)
41. C.Q. Sun, A model of bonding and band-forming for oxides and nitrides. Appl. Phys. Lett. **72**(14), 1706–1708 (1998)
42. C.Q. Sun, On the nature of the O-Rh(11O) multiphase ordering. Surf. Sci. **398**(3), L320–L326 (1998)
43. C.Q. Sun, On the nature of the triphase ordering. Surf. Rev. Lett. **5**(05), 1023–1028 (1998)
44. C.Q. Sun, Mechanism for the N-Ni(100) clock reconstruction. Vacuum **52**(3), 347–351 (1999)
45. C.Q. Sun, P. Hing, Driving force and bond strain for the C-Ni(100) surface reaction. Surf. Rev. Lett. **6**(1), 109–114 (1999)
46. C.Q. Sun, S. Li, Oxygen-derived DOS features in the valence band of metals. Surf. Rev. Lett. **7**(3), 213–217 (2000)
47. C.Q. Sun, Oxygen interaction with Rh(111) and Ru(0001) surfaces: bond-forming dynamics. Mod. Phys. Lett. B **14**(6), 219–227 (2000)
48. C.Q. Sun et al., Preferential oxidation of diamond {111}. J. Phys. D-Appl. Phys. **33**(17), 2196–2199 (2000)
49. W.T. Zheng, C.Q. Sun, Electronic process of nitriding: mechanism and applications. Prog. Solid State Chem. **34**(1), 1–20 (2006)
50. K. Yagi, K. Higashiyama, H. Fukutani, Angle-resolved photoemission study of oxygen-induced c (2 × 4) structure on Pd (110). Surf. Sci. **295**(1), 230–240 (1993)
51. J. Mercer et al., Angle-resolved photoemission study of half-monolayer O and S structures on the Rh (100) surface. Phys. Rev. B **55**(15), 10014 (1997)
52. M. Zacchigna et al., Photoemission from atomic and molecular adsorbates on Rh(100). Surf. Sci. **347**(1–2), 53–62 (1996)
53. C.W. Tucker, Oxygen faceting of rhodium (210) and (100) surfaces. Acta Metall. **15**(9), 1465–1474 (1967)
54. J. Wintterlin et al., Atomic motion and mass-transport in the oxygen induced reconstructions of Cu(110). J. Vac. Sci. Technol., B **9**(2), 902–908 (1991)
55. F. Pforte et al., Wave-vector-dependent symmetry analysis of a photoemission matrix element: the quasi-one-dimensional model system Cu(110)(2X1)O. Phys. Rev. B **63**(16), 165405 (2001)
56. C. Benndorf et al., The initial oxidation of Cu (100) single crystal surfaces: an electron spectroscopic investigation. Surf. Sci. **74**(1), 216–228 (1978)
57. C. Benndorf et al., Oxygen interaction with Cu (100) studied by AES, ELS, LEED and work function changes. J. Phys. Chem. Solids **40**(12), 877–886 (1979)
58. V.A. Bondzie, P. Kleban, D.J. Dwyer, XPS identification of the chemical state of subsurface oxygen in the O/Pd(110) system. Surf. Sci. **347**(3), 319–328 (1996)
59. E. Schwarz et al., The interaction of oxygen with a rhodium (110) surface. Vacuum **41**(1–3), 167–170 (1990)
60. L. Li et al., Oxygenation mediating the valence density-of-states and work function of Ti (0001) skin. Phys. Chem. Chem. Phys. **17**, 9867–9872 (2015)
61. C.Q. Sun, *Relaxation of the Chemical Bond* (Springer Ser. Chem. Phys.), vol 108 (Springer, Heidelberg, 2014), 807 pp
62. L. Li et al., Defects improved photocatalytic ability of TiO$_2$. Appl. Surf. Sci. **317**, 568–572 (2014)
63. L. Li et al., Nitrogen mediated electronic structure of the Ti (0001) surface. RSC Adv. **6**(18), 14651–14657 (2016)
64. M. Huda, L. Kleinman, Density functional calculations of the influence of hydrogen adsorption on the surface relaxation of Ti (0001). Phys. Rev. B **71**(24) (2005)

65. D. Hanson, R. Stockbauer, T. Madey, Photon-stimulated desorption and other spectroscopic studies of the interaction of oxygen with a titanium (001) surface. Phys. Rev. B **24**(10), 5513–5521 (1981)

66. Y. Fukuda, W.T. Elam, R.L. Park, Nitrogen, oxygen, and carbon monoxide chemisorption on polycrystalline titanium surfaces. Appl. Surf. Sci. **1**, 278–287 (1978)

67. S. Schwegmann et al., The adsorption of atomic nitrogen on Ru (0001): geometry and energetics. Chem. Phys. Lett. **264**(6), 680–686 (1997)

68. D. Eastman, Photoemission energy level measurements of sorbed gases on titanium. Solid State Commun. **10**(10), 933–935 (1972)

69. S. Schwegmann et al., Oxygen adsorption on the Ru(10(1)over-bar0) surface: anomalous coverage dependence. Phys. Rev. B **57**(24), 15487–15495 (1998)

70. Y.Q. Fu et al., Crystalline carbonitride forms harder than the hexagonal Si-carbonitride crystallite. J. Phys. D-Appl. Phys. **34**(9), 1430–1435 (2001)

71. C.Q. Sun et al., Bond contraction and lone pair interaction at nitride surfaces. J. Appl. Phys. **90**(5), 2615–2617 (2001)

72. M. Terrones et al., N-doping and coalescence of carbon nanotubes: synthesis and electronic properties. Appl. Phys. A-Mater. Sci. Process. **74**(3), 355–361 (2002)

73. Y. Guo et al., Tantalum surface oxidation: lattice reconstruction, bond relaxation, energy entrapment, and electron polarization, in *Applied Surface Science*, 2016. in press

74. Y. Guo et al., Tantalum surface oxidation: bond relaxation, energy entrapment, and electron polarization. Appl. Surf. Sci. **396**, 177–184 (2017)

75. J. Van der Veen, F. Himpsel, D. Eastman, Chemisorption-induced 4 f-core-electron binding-energy shifts for surface atoms of W (111), W (100), and Ta (111). Phys. Rev. B **25**(12), 7388 (1982)

76. C. Guillot et al., Core-level spectroscopy of clean and adsorbate-covered Ta (100). Phys. Rev. B **30**(10), 5487 (1984)

77. B. Lv, T. Qian, H. Ding, Angle-resolved photoemission spectroscopy and its application to topological materials. Nat. Rev. Phys. **1**(10), 609–626 (2019)

78. Z. Lin et al., Multiple nodeless superconducting gaps in (Ba0. 6K0. 4) Fe2As2 superconductor from angle-resolved photoemission spectroscopy. Chin. Phys. Lett. **25**(12), 4402 (2008)

79. T. Sato et al., Low Energy Excitation and Scaling in $Bi_2Sr_2Ca_{n-1}Cu_nO_{2n+4}$ (n = 1–3): angle-resolved photoemission spectroscopy. Phys. Rev. Lett. **89**(6), 067005 (2002)

80. A. Fedorov et al., Temperature dependent photoemission studies of optimally doped $Bi_2Sr_2CaCu_2O_8$. Phys. Rev. Lett. **82**(10), 2179 (1999)

81. A. Damascelli, Z. Hussain, Z.-X. Shen, Angle-resolved photoemission studies of the cuprate superconductors. Rev. Mod. Phys. **75**(2), 473 (2003)

82. Y. Yu et al., High-temperature superconductivity in monolayer $Bi_2Sr_2CaCu_2O_8+\delta$. Nature (2019)

83. C. Yang et al., Intermediate bosonic metallic state in the superconductor-insulator transition. Science (2019), p. eaax5798

84. K. Stacy, Research reveas new state oof matter: a Cooper pair metal, in *Brown University* (2019)

85. L. Zhang et al., Stabilization of the dual-aromatic Cyclo-N_5^- anion by acidic entrapment. J. Phys. Chem. Lett. **10**, 2378–2385 (2019)

86. H. Tillborg et al., O/Cu(100) studied by core level spectroscopy. Surf. Sci. **270**, 300–304 (1992)

87. M.V. Ganduglia-Pirovano, M. Scheffler, Structural and electronic properties of chemisorbed oxygen on Rh(111). Phys. Rev. B **59**(23), 15533–15543 (1999)

88. S. Lizzit et al., Surface core-level shifts of clean and oxygen-covered Ru(0001). Phys. Rev. B **63**(20), 205419 (2001)

89. T. Pillo et al., The electronic structure of PdO found by photoemission (UPS and XPS) and inverse photoemission (BIS). J. Phys.-Condens. Matter **9**(19), 3987–3999 (1997)

90. K. Yagi, H. Fukutani, Oxygen adsorption site of Pd(110)c(2x4)-O: analysis of ARUPS compared with STM image. Surf. Sci. **412–13**, 489–494 (1998)

91. R. Ozawa et al., Angle-resolved UPS study of the oxygen-induced 2x1 surface of Cu(110). Surf. Sci. **346**(1–3), 237–242 (1996)

92. Y.S. Wang et al., An AES, UPS and HREELS study of the oxidation and reaction of NB(110). Surf. Sci. **372**(1–3), L285–L290 (1997)

93. A.I. Boronin, S.V. Koscheev, G.M. Zhidomirov, XPS and UPS study of oxygen states on silver. J. Electron Spectrosc. Relat. Phenom. **96**(1–3), 43–51 (1998)

94. A.M. Aprelev et al., UPS (8.43-ev and 21.2-ev) data on the evolution of DOS spectra near E(f) of $Bi_2Sr_2CaCu_2O_8$ under thermal and light treatments. Physica C **235**, 1015–1016 (1994)

95. A. Howard et al., Initial and final state effects in photoemission from Au nanoclusters on $TiO_2(110)$. Surf. Sci. **518**(3), 210–224 (2002)

96. C.Q. Sun et al., Solution certainty in the Cu(110)-(2x1)-2O(2-) surface crystallography. Int. J. Mod. Phys. B **16**(1–2), 71–78 (2002)

97. H. Tillborg et al., Electronic-structure of atomic oxygen adsorbed on Ni(100) and Cu(100) studied by soft-x-ray emission and photoelectron spectroscopies. Phys. Rev. B **47**(24), 16464–16470 (1993)

98. A. Spitzer, H. Luth, The adsorption of oxygen on copper surfaces.1. Cu(100) and Cu(110). Surf. Sci. **118**(1–2), 121–135 (1982)

99. A. Spitzer, H. Luth, The adsorption of oxygen on copper surfaces. 2. Cu(111). Surf. Sci. **118**(1–2), 136–144 (1982)

100. D. Alfe, S. de Gironcoli, S. Baroni, The reconstruction of Rh(001) upon oxygen adsorption. Surf. Sci. **410**(2–3), 151–157 (1998)

101. A.C. Perrella et al., Scanning tunneling spectroscopy and ballistic electron emission microscopy studies of aluminum-oxide surfaces. Phys. Rev. B **65**(20), 201403 (2002)

102. J.D. Zhang et al., Angle-resolved photoemission-study of oxygen-chemisorption on Gd(0001). Surf. Sci. **329**(3), 177–183 (1995)

103. A. Bottcher, H. Niehus, Oxygen adsorbed on oxidized Ru(0001). Phys. Rev. B **60**(20), 14396–14404 (1999)

104. A. Bottcher, H. Conrad, H. Niehus, Reactivity of oxygen phases created by the high temperature oxidation of Ru(0001). Surf. Sci. **452**(1–3), 125–132 (2000)

105. G. Bester, M. Fahnle, On the electronic structure of the pure and oxygen covered Ru(0001) surface. Surf. Sci. **497**(1–3), 305–310 (2002)

106. C. Stampfl et al., Catalysis and corrosion: the theoretical surface-science context. Surf. Sci. **500**(1–3), 368–394 (2002)

107. S. Altieri, L.H. Tjeng, G.A. Sawatzky, Electronic structure and chemical reactivity of oxide-metal interfaces: MgO(100)/Ag(100). Phys. Rev. B **61**(24), 16948–16955 (2000)

108. R. Mamy, Spectroscopic study of the surface oxidation of a thin epitaxial Co layer. Appl. Surf. Sci. **158**(3–4), 353–356 (2000)

109. J.C. Zheng et al., Oxygen-induced surface state on diamond (100). Diam. Relat. Mater. **10**(3–7), 500–505 (2001)

110. L.W. Lin, The role of oxygen and fluorine in the electron-emission of some kinds of cathodes. J. Vacuum Sci. Technol. A-Vacuum Surf. Films **6**(3), 1053–1057 (1988)

111. G.G. Tibbetts, J.M. Burkstrand, J.C. Tracy, Electronic properties of adsorbed layers of nitrogen, oxygen, and sulfur on copper (100). Phys. Rev. B **15**(8), 3652–3660 (1977)

112. G.G. Tibbetts, J.M. Burkstrand, Electronic properties of adsorbed layers of nitrogen, oxygen, and sulfur on silver (111). Phys. Rev. B **16**(4), 1536–1541 (1977)

113. G.G. Fuentes, E. Elizalde, J.M. Sanz, Optical and electronic properties of TiC_xN_y films. J. Appl. Phys. **90**(6), 2737–2743 (2001)

114. S. Souto et al., Electronic structure of nitrogen-carbon alloys (a-CN_x) determined by photoelectron spectroscopy. Phys. Rev. B **57**(4), 2536–2540 (1998)

115. Z.Y. Chen et al., Valence band electronic structure of carbon nitride from x-ray photoelectron spectroscopy. J. Appl. Phys. **92**(1), 281–287 (2002)

Chapter 10
Hetero- and Under-Coordination Coupling

Abstract A combination of the hetero- and under-coordination forms a promising means of mediating the bonding and electronic dynamics and properties of a substance as the hetero- and under-coordination enhance each other on the charge entrapment and polarization. However, at a critical size, polarization may compensate or override quantum entrapment because of the polarization screens and splits the interatomic potentials.

Highlights

- Hetero- and under-coordination enhances each other on the CL shift and valence states.
- The enhanced polarization screens and splits the interatomic potential and offsets the CLS.
- The joint effect mediates the band gap, electroaffinity, carrier life, and work function.
- Coordination engineering forms powerful means of modulating materials performance.

10.1 Ti(0001) Skin and TiO_2 Nanocrystals

10.1.1 Photoactivity of Defected TiO_2

The photoactivity of the nanoscaled or the highly defected TiO_2 has received much attention owing to its tunable band gap and work function [1–6]. The hetero-coordinated bulk TiO_2 is only active under UV irradiation exciting electrons to overcome the band gap of 3.2 eV for the anatase and 3.0 eV for the rutile phase [7, 8]. Although the defected, or undercoordinated TiO_2 that contains the Ti^{3+} ions or oxygen vacancies absorbs visible light [9, 10], the mechanism for the defect modulation of the band gap, affinity, carrier life, and work function is very important for devising materials having derived functions.

C. Q. Sun, *Electron and Phonon Spectrometrics*, https://doi.org/10.1007/978-981-15-3176-7_10

205

It has been proposed that the oxygen-derived Ti $3d$ states located ~0.85 eV below the E_F narrows the band gap [7, 11]. Such Ti $3d$ states arise from oxygen (O_{br}) vacancies that bridge two Ti^{3+} ions across [12–14]. Two excessive electrons per O_{br} vacancy transfer to the neighboring Ti atoms in an ionic route. However, based on their UPS, STM, and DFT investigations of $TiO_2(110)$ surfaces, Martinez et al. [1] suggested that the Ti $3d$ defect states were primarily due to Ti^{3+} interstitials in the near-surface region rather than the surface O_{br} vacancies. The defected black TiO_2 exhibits substantial activity in photocatalytic production of hydrogen from water under sunlight radiation [15]. Both the valence and the conduction bands of the black TiO_2 shift upwards because of the defect induced band bending [16].

10.1.2   XPS: Ti(0001) Skin 2p Band Shift

Figure 10.1 shows the Ti $2p_{3/2}$ XPS spectrum collected from a well-faceted Ti(0001) surface [17]. The BOLS-TB decomposition of the spectrum results in the $E_{2p3/2}(0)$, $\Delta E_{2p3/2}(12)$, and the CN-resolved $E_{2p3/2}(z)$ for calibrating the BE shift of the Ti $2p_{3/2}$ upon oxidation and defect formation: $E_{2p_{3/2}}(z) = 451.47 \pm 0.003 + 2.14C_z^{-4.6}$.

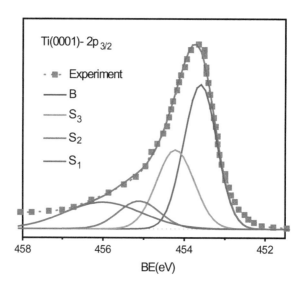

Fig. 10.1 BOLS-TB decomposition of the Ti(0001) $2p_{3/2}$ spectrum [18] with derived information featured in Table 10.1. The optimized atomic CN ($z_1 = 3.50$, $z_2 = 4.36$, $z_3 = 6.48$ and $z_b = 12$ for the hcp structure sublayers) is identical to those derived from the same hcp skins [19]. The refinement leads to the bond nature index m = 4.6 for Ti. Reprinted with permission from [17]

10.1.3 ZPS: Defect-Induced Entrapment and Polarization

The ZPS in Fig. 10.2 resolves the Ti $2p_{3/2}$ and the O $1s$ defect states. Ar⁺ bombard-ment creates defects of controllable concentrations [20]. Subtracting the spectrum collected from an un-defected $TiO_2(110)$ surface from that of the defected-$TiO_2(110)$ surface reveals the following:

(1) The valley at 458.41 eV in (a) corresponds to the $2p_{3/2}$ bulk component and the valley at 529.83 eV in (b) to the O $1s$ in the bulk TiO_2. Ti–O interaction substantially magnifies the crystal potential, which shifts the Ti $2p_{3/2}$ states positively by 4.8 eV with respect to that of bulk Ti metal.

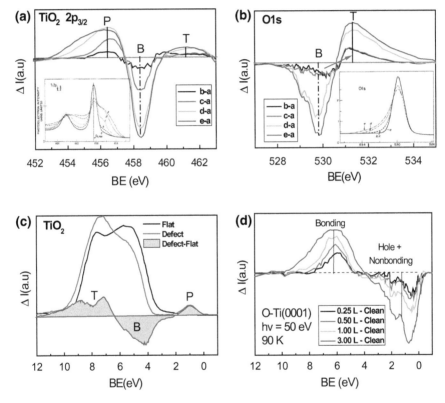

Fig. 10.2 ZPS of **a** Ti $2p_{3/2}$, **b** O $1s$, and **c** the valence band of the defected TiO_2 at different defect concentrations with respect to the un-defected TiO_2. Insets are the raw data for a—flat surface; b—thermally and Ar⁺ bombardment produced defects; for c—10 min; d—30 min; and e—50 min durations [20]. **d** Oxygen coverage (in Langmuir) dependence [22] of the valence ZPS of TiO_2 with respect to that of clean Ti(0001) surface. Both the core and the valence bands show coexistence of the entrapment and polarization due to the combination of hetero- and undercoordination. The deeper O $1s$ BE shows only entrapment. Reprinted with permission from [17]

(2) Defects create both entrapment (T = 461.14 eV) and polarization states (P = 456.41 eV) to the TiO_2 in addition to the B valley. However, the O $1s$ shows only entrapment (T = 531.33 eV) without polarization. The O $1s$ orbit, 528.83 − 458.41 = 71.42 eV deeper than that of the TiO_2, seems insensitive to the polarization that screens and splits the interatomic potential acting on oxygen.

(3) The peak intensities of the entrapment and the polarization increase with defect concentration. The effective CN of the defected-TiO_2 is lower than that of the ideal surface of 3.5 for the Ti(0001) skin. The ZPS gives polarization coefficient:

$$p = \left[E_{2p_{3/2}}(P) - E_{2p_{3/2}}(0)\right]/\left[E_{2p_{3/2}}(12(TiO_2)) - E_{2p_{3/2}}(0)\right] = 0.71.$$

(4) Because the CLS is proportional to the equilibrium bond energy, one can obtain the TiO_2 bulk bond energy $E_b(TiO_2) = 1.51$ eV/bond and defected bond energy $E_b(\text{defect}) = 2.11$ eV/bond in comparison to that of Ti bulk bond energy $E_b(Ti) = 0.41$ eV/bond [21].

The valence ZPS in Fig. 10.2c shows co-existence of entrapment and polarization. Therefore, the valence and the core band shift simultaneously in the same direction because of the screening effect on the core charge, which are the same to Ag/Pd alloy and Rh adatoms. The ZPS in Fig. 10.2d for different oxygen coverages matches DFT derivatives in energy below E_F. Table 10.1 summarizes the oxidation and defect effect on the E $2p$ and the VB attributes of Ti and TiO_2.

Table 10.1 Layer-order resolved E $2p_{3/2}$ for Ti(0001), the VB for O–Ti(0001), and the defect-induced entrapment and polarization of the core and the VB for TiO_2

	z	Ti(0001) E $2p_{3/2}$	Defected TiO_2	O-adsorbed Ti(0001) [22]
	m	4.6 [17]	5.34 [23, 24]	
Atom	0	451.47	–	–
Bulk	12.00	453.61	458.41(B)	–
S_3	6.48	454.22	–	–
S_2	4.36	455.11	–	–
S_1	3.50	456.00	–	–
E $2p_{3/2}$			461.14(T)	–
			456.41(P)	–
O $1s$				529.8(B)
				531.3(T)
VB			1.0 (P)	1.6 (antibond)
			4.5 (B)	−1.6 ± 0.5 (nonbond)
				−1.5 ± 1.5 (hole)
			8 ± 1(T)	−6.0 ± 1 (bond)

10.1.4 Defect Enhanced Photocatalytic Ability

When a TiO$_2$ surface is illuminated by light with energy equaling to that of the band gap, the radiation excites electrons in the ground VB to the upper conduction band (CB, e$^-$), leaving a hole (h$^+$) behind the VB. The excited carriers are highly reactive to radicals with robust reducing and oxidizing capacity. The carriers may recombine or be trapped by metastable surface states, or react with suitable electron acceptors/donors pre-adsorbed on the catalyst surface.

Major concerns in improving the photocatalytic ability of TiO$_2$ include:

(i) reducing the band gap to match the wavelength of visible light for electron excitation;
(ii) lowering the work function to ease migration of the excited electrons; and,
(iii) prolonging the lifetime of the carriers for slowing electron-hole recombination.

Currently, the band gap matches only UV range that accounts only 4% of solar light. During photocatalytic reaction, when the reduction and oxidation do not proceed simultaneously, there is an electron accumulation in the CB, thereby causing a fast recombination of the e–h pairs. Therefore, improve the utilization rate of sunlight by modulating band gap and work function and by raising carrier lifetime and electroaffinity via locally pinning the polarized electrons is a feasible means.

Atomic undercoordination and hetero-coordination are useful in this situation. At first, the entrapment of the valence electrons deepens the energy states and enlarges the electroaffinity, which polarizes the conduction electrons shifting up in energy to lower the work function. Meanwhile, the polarization screens and splits the local potential to create the polarized states to raise the upper edge of the VB. This process narrows the band gap. The strong localization of the defect dipoles by pinning prolongs the carrier life effectively. Therefore, undercoordinated defects modulate the band gap, carrier life, and work function is a promising means.

It is also clear now why metal atoms added to a defected TiO$_2$ and graphene surface could improve the efficiency of CO oxidation at room temperature compared with the otherwise fully covered surface, like Au [25]. TiO$_2$ defect dipoles polarize the undercoordinated metal adatoms further to lowers the work function and hence improves the reactivity of the supported metallic adatom dipoles [26].

10.2 ZnO Nanocrystals Passivated with H, N, and O

10.2.1 ZPS: Size-Induced Entrapment-Polarization Transition

Figure 10.3a shows the joint under- and hetero-coordination effects on the Zn 2p energy. XPS revealed that a transition from the positive to the negative CLS happens at 8.5 nm [27]. When the ZnO crystal reduces its size from 200 to 8.5 nm, entrapment

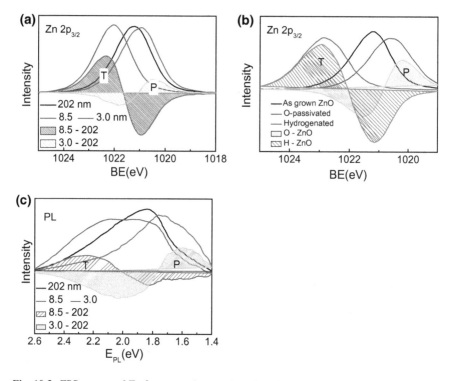

Fig. 10.3 ZPS spectra of Zn $2p_{3/2}$ reveal **a** transition from entrapment (T) to polarization (P) at 8.5 nm size [27]; **b** annealing under 0.21 O_2 + 0.79 N_2 ambient enhances polarization and annealing under 0.03 H_2 + 0.97 Ar ambient enhances entrapment [28, 29]. H-passivation annihilates the surface dipoles and weakens the screening effect on the crystal potential [30]. **c** Photoluminescence spectra show the same size trend of (**a**) band gap transition from blueshift (T) to redshift (P) at 8.5 nm [27]

dominates. The extent of entrapment is proportional to the size reduction. Further size reduction from 8.5 to 3.0 nm reverses, however, the CLS direction, which evidences that further CN-reduction enhances the polarization that becomes dominance at ultra-low atomic CNs.

Figure 10.3b demonstrates the hetero-coordination effect on the CLS of 200 nm sized ZnO. Both oxidation and nitrogenation enhance the polarization through lone pair interaction. However, hydrogenation attenuates the polarization by annihilating skin dipoles, which deepens the CLS. Hydrogen annihilation of skin dipoles attenuates the magnetism of Pt clusters [30]. Photoluminescence spectra show the same size trend of (a) transition from blueshift (T) to redshift (P) at the critical size 8.5 nm for ZnO [27]. The photoluminescence energy shift involves the processes of intrinsic band gap relaxation, electron-phonon coupling, and bond relaxation, which is proportional to the bond energy as well [31, 32].

10.2.2 Band Gap, Work Function, and Magnetism

Annealing the specimen up to 900 °C under the ambient pressure of $0.03 H_2 + 0.97$ Ar (type I) and $0.21 O_2 + 0.79 N_2$ (type II) for 24 h create two kinds of defected ZnO [29]. Type I (H-induced) shows the PL energy at 2.46 eV and type II (O and N induced) at 2.26 eV (Fig. 10.3c). Annealing under pure O_2 at ambient pressure lowers the PL peak energy to an even lower value of 2.15 eV [29].

N and O passivation reduces the band gap of ZnO rather than enlarges it because of the potential weakening by lone pair induced polarization. The band gap and the CLS as well, is proportional to the bond energy. Dipoles formation narrows the band gap by screening the crystal potential and band tail creation. For both I and II types, the VB maximum moves down, and the VB expands slightly as the relative intensity of the green emission to that of UV emission increases. However, H passivation removes the dipoles of unpaired electrons by hydrogen termination, which removes the band tails and widens the band gap.

The skin dipoles are critical to the dilute magnetism, electron emission, and hydrophobicity of ZnO. The effects of under- and hetero-coordination do enhance each other in modulating the band gap, work function, electroaffinity, and the density of surface dipoles that determine surface hydrophobicity. These attributes make the sharp edge of ZnO structures hydrophobic, magnetic, and easy to emit electrons and sensitive in photocatalysis.

10.3 Scratched $SrTiO_3$ Skin: Defect States

An *in situ* XPS measurement confirmed the defect-enhanced entrapment and polarization at the $SrTiO_3$ skin shown in Fig. 10.4 [33]. The ZPS of two VB spectra collected from the $SrTiO_3$ skin before and after 3-keV Ar^+ bombarding for 20 min creating the additional polarization at 1 eV and entrapment at 12 eV. O $1s$ band exhibits a 0.6 eV entrapment without presence of polarization because of its low sensitivity. These observations confirmed the BOLS-NEP expectation that defect creation by surface roughening enhances the entrapment and polarization. The valence ZPS shows the features of polarization, electron-hole pair production, and the entrapment, which is the same to the defected TiO_2 (see Fig. 10.4a).

Table 10.2 summarizes the irregular-coordination coupling effect on the electronic structures of ZnO and $SrTiO_3$ nanocrystals. The photoluminescence energy correlates to the CLS with involvement of electron-phonon coupling as an addition [34, 35]. Atomic CN reduction by creating defect enhances the BOLS-NEP effect. The chemical passivation and undercoordination enhances each other on the charge entrapment and polarization.

Mechanical scratching of the SiO_2 surface [36] and O-plasma etching of the HOPG [37] also induce the local charge entrapment for the same reason of under- and hetero-coordination combination. Using the ZPS technique, one can readily

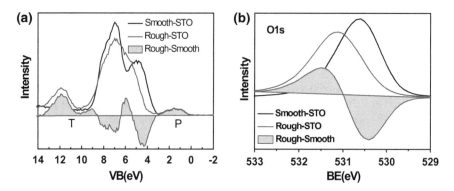

Fig. 10.4 ZPS of the **a** valence and the **b** O $1s$ band for SrTiO$_3$ with and without defects creation by 3-keV Ar$^+$ beam bombardment [33]. The DOS shows polarization at 1 eV and entrapment at 12 eV. The O $1s$ band shows entrapment without presence of polarization because its energy is too far away from the energy of the polarized states. Reprinted with permission from [33]

Table 10.2 Hetero- and under-coordination coupling effect on the energy states of ZnO and SrTiO$_3$. The bulk valley for ZnO should remain constant with reference to large bulk. Energy is in eV unit. O $1s$ is insensitive of the polarization effect

ZnO		Polarization (P)	Entrapment (T)	Bulk valley (B)
8.5–200 nm	Zn $2p_{3/2}$ (eV)	–	1022.5	1021
	PL (eV)		2.3	1.8
3.5–200 nm	Zn $2p_{3/2}$ (eV)	1020.7	–	1021.8
	PL (eV)	1.6		2.0
O-passivated 200 nm	Zn $2p_{3/2}$ (eV)	1020.3	–	1021.4
H-passivated 200 nm	Zn $2p_{3/2}$ (eV)	–	1023.0	1021.2
SrTiO$_3$	VB (eV)	1	12	4–8
	O $1s$ (eV)	–	531.5	530.4

identify the effect of metal skin nanometerization followed by passivation so clarify the origin for metal skin functionalization such as Fe thin films [38].

10.4   Summary

A combination of the BOLS-NEP notion and the ZPS distillation has enabled identification of the coupling effect of hetero- and under-coordination on the electron binding energy shift of TiO$_2$, ZnO, and SrTiO$_3$. Coupling of the irregular-coordination effect may form efficient means to tune the band gap, carrier life, electroaffinity, and work function, which determine the catalytic ability, electron and photon emissivity, hydrophobicity, magnetic property, and toxicity of the undercoordinated compounds.

Narrowing band gap allows for extension of light adsorption to the visible and lower work function promote electronic consumption and prevent the recombination of photogenerated carriers.

References

1. U. Martinez, J.Ø. Hansen, E. Lira, H.H. Kristoffersen, P. Huo, R. Bechstein, E. Lægsgaard, F. Besenbacher, B. Hammer, S. Wendt, Reduced step edges on Rutile TiO_2 (110) as competing defects to oxygen vacancies on the terraces and reactive sites for ethanol dissociation. Phys. Rev. Lett. **109**(15), 155501 (2012)
2. S. Jin, Y. Li, H. Xie, X. Chen, T. Tian, X. Zhao, Highly selective photocatalytic and sensing properties of 2D-ordered dome films of nano titania and nano Ag^{2+} doped titania. J. Mater. Chem. **22**(4), 1469–1476 (2012)
3. A. Borodin, M. Reichling, Characterizing TiO_2 (110) surface states by their work function. Phys. Chem. Chem. Phys. **13**(34), 15442–15447 (2011)
4. M. Kong, Y. Li, X. Chen, T. Tian, P. Fang, F. Zheng, X. Zhao, Tuning the relative concentration ratio of bulk defects to surface defects in TiO_2 nanocrystals leads to high photocatalytic efficiency. J. Am. Chem. Soc. **133**(41), 16414–16417 (2011)
5. J. Tao, M. Batzill, Role of surface structure on the charge trapping in TiO_2 photocatalysts. J. Phys. Chem. Lett. **1**(21), 3200–3206 (2010)
6. C.M. Yim, C.L. Pang, G. Thornton, Oxygen vacancy origin of the surface band-gap state of TiO_2(110). Phys. Rev. Lett. **104**(3), 036806 (2010)
7. U. Diebold, The surface science of titanium dioxide. Surf. Sci. Rep. **48**(5), 53–229 (2003)
8. R. Daghrir, P. Drogui, D. Robert, Modified TiO_2 for environmental photocatalytic applications: a review. Ind. Eng. Chem. Res. **52**(10), 3581–3599 (2013)
9. K. Kollbek, M. Sikora, C. Kapusta, J. Szlachetko, K. Zakrzewska, K. Kowalski, M. Radecka, X-ray spectroscopic methods in the studies of nonstoichiometric TiO_{2-x} thin films. Appl. Surf. Sci. **281**, 100–104 (2013)
10. F. Zuo, L. Wang, T. Wu, Z. Zhang, D. Borchardt, P. Feng, Self-doped Ti^{3+} enhanced photocatalyst for hydrogen production under visible light. J. Am. Chem. Soc. **132**(34), 11856–11857 (2010)
11. K. Mitsuhara, H. Okumura, A. Visikovskiy, M. Takizawa, Y. Kido, The source of the Ti 3d defect state in the band gap of rutile titania (110) surfaces. J. Chem. Phys. **136**(12), 124707 (2012)
12. S. Chrétien, H. Metiu, Electronic structure of partially reduced Rutile TiO_2 (110) surface: where are the unpaired electrons located? J. Phys. Chem. C **115**(11), 4696–4705 (2011)
13. P. Krüger, S. Bourgeois, B. Domenichini, H. Magnan, D. Chandesris, P. Le Fèvre, A.M. Flank, J. Jupille, L. Floreano, A. Cossaro, A. Verdini, A. Morgante, Defect states at the TiO_2 surface probed by resonant photoelectron diffraction. Phys. Rev. Lett. **100**(5), 055501 (2008)
14. Z. Zhang, S.-P. Jeng, V.E. Henrich, Cation-ligand hybridization for stoichiometric and reduced TiO_2 (110) surfaces determined by resonant photoemission. Phys. Rev. B **43**(14), 12004 (1991)
15. X. Chen, L. Liu, Y.Y. Peter, S.S. Mao, Increasing solar absorption for photocatalysis with black hydrogenated titanium dioxide nanocrystals. Science **331**(6018), 746–750 (2011)
16. Z. Zhang, J.T. Yates Jr., Band bending in semiconductors: chemical and physical consequences at surfaces and interfaces. Chem. Rev. **112**(10), 5520–5551 (2012)
17. L. Li, H.-W. Tian, F.-L. Meng, X.-Y. Hu, W.-T. Zheng, C.Q. Sun, Defects improved photocatalytic ability of TiO_2. Appl. Surf. Sci. **317**, 568–572 (2014)
18. M. Kuznetsov, A. Tel minov, E. Shalaeva, A. Ivanovskii, Study of adsorption of nitrogen monoxide on the Ti (0001) surface. Phys. Metals Metallogr. c/c fizika metallov i metallovedenie **89**(6), 569–580 (2000)

19. Y. Wang, Y.G. Nie, L.L. Wang, C.Q. Sun, Atomic-layer- and crystal-orientation-resolved $3d_{5/2}$ binding energy shift of Ru(0001) and Ru(1010) surfaces. J. Phys. Chem. C **114**(2), 1226–1230 (2010)

20. W. Göpel, J. Anderson, D. Frankel, M. Jaehnig, K. Phillips, J. Schäfer, G. Rocker, Surface defects of $TiO_2(110)$: a combined XPS, XAES AND ELS study. Surf. Sci. **139**(2), 333–346 (1984)

21. C. Kittel, *Intrduction to Solid State Physics*, 8th edn. (Willey, New York, 2005)

22. D. Hanson, R. Stockbauer, T. Madey, Photon-stimulated desorption and other spectroscopic studies of the interaction of oxygen with a titanium (001) surface. Phys. Rev. B **24**(10), 5513–5521 (1981)

23. X.J. Liu, L.W. Yang, Z.F. Zhou, P.K. Chu, C.Q. Sun, Inverse Hall-Petch relationship of nanostructured TiO_2: skin-depth energy pinning versus surface preferential melting. J. Appl. Phys. **108**, 073503 (2010)

24. X.J. Liu, L.K. Pan, Z. Sun, Y.M. Chen, X.X. Yang, L.W. Yang, Z.F. Zhou, C.Q. Sun, Strain engineering of the elasticity and the Raman shift of nanostructured TiO_2. J. Appl. Phys. **110**(4), 044322 (2011)

25. M.S. Chen, D.W. Goodman, The structure of catalytically active gold on titania. Science **306**(5694), 252–255 (2004)

26. J. Xie, H. Zhang, S. Li, R. Wang, X. Sun, M. Zhou, J. Zhou, X.W. Lou, Y. Xie, Defect-rich MoS_2 ultrathin nanosheets with additional active edge sites for enhanced electrocatalytic hydrogen evolution. Adv. Mater. **25**(40), 5807–5813 (2013)

27. Y.Y. Tay, S. Li, C.Q. Sun, P. Chen, Size dependence of Zn 2p 3/2 binding energy in nanocrystalline ZnO. Appl. Phys. Lett. **88**(17), 173118 (2006)

28. Y. Tay, T. Tan, M. Liang, F. Boey, S. Li, Specific defects, surface band bending and characteristic green emissions of ZnO. Phys. Chem. Chem. Phys. **12**(23), 6008–6013 (2010)

29. Y.Y. Tay, T. Tan, F. Boey, M.H. Liang, J. Ye, Y. Zhao, T. Norby, S. Li, Correlation between the characteristic green emissions and specific defects of ZnO. Phys. Chem. Chem. Phys. **12**(10), 2373–2379 (2010)

30. C.Q. Sun, Dominance of broken bonds and nonbonding electrons at the nanoscale. Nanoscale **2**(10), 1930–1961 (2010)

31. L. Pan, S. Xu, X. Liu, W. Qin, Z. Sun, W. Zheng, C.Q. Sun, Skin dominance of the dielectric electronic-phononic-photonic attribute of nanoscaled silicon. Surf. Sci. Rep. **68**(3–4), 418–445 (2013)

32. J.W. Li, S.Z. Ma, X.J. Liu, Z.F. Zhou, C.Q. Sun, ZnO meso-mechano-thermo physical chemistry. Chem. Rev. **112**(5), 2833–2852 (2012)

33. C.Q. Sun, Y. Sun, Y.G. Ni, X. Zhang, J.S. Pan, X.H. Wang, J. Zhou, L.T. Li, W.T. Zheng, S.S. Yu, L.K. Pan, Z. Sun, Coulomb repulsion at the nanometer-sized contact: a force driving superhydrophobicity, superfluidity, superlubricity, and supersolidity. J. Phys. Chem. C **113**(46), 20009–20019 (2009)

34. L. K. Pan, Chang Q. Sun, Coordination imperfection enhanced electron-phonon interaction. J. Appl. Phys. **95**(7), 3819–3821 (2004)

35. L. Pan, S. Xu, X. Liu, W. Qin, Z. Sun, W. Zheng, C. Q. Sun, Skin dominance of the dielectric–electronic–phononic–photonic attribute of nanoscaled silicon. Surf. Sci. Rep. **68**(3–4), 418–445 (2013)

36. M. Hasegawa, T. Shimakura, Observation of electron trapping along scratches on SiO_2 surface in mirror electron microscope images under ultraviolet light irradiation. J. Appl. Phys. **107**(8), 084107 (2010)

37. J.I. Paredes, A. Martinez-Alonso, J.M.D. Tascon, Multiscale Imaging and tip-scratch studies reveal insight into the plasma oxidation of graphite. Langmuir **23**(17), 8932–8943 (2007)

38. W.P. Tong, N.R. Tao, Z.B. Wang, J. Lu, K. Lu, Nitriding iron at lower temperatures. Science **299**(5607), 686–688 (2003)

Chapter 11
Liquid Phase

Abstract Supersolidity of water ice, proposed in 2013 and intensively verified since then, refers to those water molecules being polarized by molecular undercoordination pertained to the skin of bulk water ice, nanobubbles, and nanodroplets (often called confinement) or by electrostatic fields of ions in salt solutions or a capacitor. From the perspective of hydrogen bond (O:H–O or HB with ":" being the lone pairs on O^{2-}) cooperative relaxation and polarization, this section features the recent progress and recommends future trends in understanding the bond-electron-phonon correlation in the supersolid phase. The supersolidity is characterized by the shorter and stiffer H–O bond and the longer and softer O:H nonbond, deeper $O1s$ energy band, and longer photoelectron and phonon lifetime. The supersolid phase is hydrophobic, less dense, viscoelastic, mechanically and thermally more stable. The O:H–O bond cooperative relaxation offsets boundaries of structural phases and raises the melting point and meanwhile lowers the freezing and evaporating temperatures of water ice—known as supercooling and superheating.

Highlights

- Molecular undercoordination and ionic hydration effect the same on O:H–O relaxation.
- XPS O $1s$ and K–edge absorption energy shifts in proportional to the H–O bond energy.
- Electron hydration probes the site– and size–resolved bounding energy and the electron lifetime.
- DPS, SFG, and calculations confirm the site–resolved H–O contraction and thermal stability.

C. Q. Sun, *Electron and Phonon Spectrometrics*,
https://doi.org/10.1007/978-981-15-3176-7_11

11.1 Wonders of H₂O Molecular Undercoordination and Salt Hydration

As independent degrees of freedom, water molecular undercoordination and electro-static polarization by ions in salt solutions or an parallel field in a capacitor make the mysterious water ice even more fascinating [1]. Undercoordinated water molecules are referred to those with fewer than four nearest neighbors (CN < 4) as they occur in the bulk interior of water and ice. Molecular undercoordination takes place in the terminated hydrogen bonded network, in the skin of a large or small volume of water and ice, molecular clusters, ultrathin films, snowflakes, clouds, fogs, nanodroplets, nanobubbles, and water in the vapor phase.

Such a kind of water molecules shows the extraordinary features of hydrophobic, less dense, elastoviscous, melting point (T_m) elevation, freezing temperature (T_N) and evaporating temperature (T_V) depression, and superfluidity when travelling in microchannels [2–6]. A few molecular-layer of ice can form at the room tempera-ture on SiO_2 substrate [7]. Nanobubbles are long lived, mechanically stronger and thermally more stable. The T_N drops from 258 K to even 150 K when the droplet size is reduced—called supercooling or "no man's land" [4, 8]. Likewise, nuclear magnetic resonance and differential scanning calorimetry measurements revealed that the melting of ice in porous glass having different pores sizes proceeds inho-mogeneously. There exists a 0.5 nm thick interface liquid layer between the pores surface and the ice crystal [9]. The superfluidity occurs at most six-layer molecular thickness sandwiched between two graphene sheets [10].

Excessive properties also include the longer O–O distance, stiffer H–O phonons and softer O:H phonons, deeper O $1s$ core level, and longer photoelectron and H–O phonon lifetimes [11–16]. The dynamics of water molecules in the confined geome-tries and near different types of surfaces are substantially slower than in the bulk (about one order of magnitude) [8, 17]. These features become more pronounced as the molecular coordination number decreases.

Salt solvation differs the local physical–chemical properties of solutions in the hydration shells from those of the ordinary bulk water. Intensive pump–probe spectro-scopic investigations have been conducted to pursue the mechanism behind molecular performance in the spatial and temporal domains. For instance, the sum frequency generation (SFG) spectroscopy resolves information on the molecular dipole orien-tation or the skin dielectrics, at the air–solution interface [18], while the ultrafast two–dimensional infrared absorption (2DIR) probes the solute or water molecular diffusion dynamics in terms of phonon lifetime and the viscosity of the solutions [19].

Salt solutions demonstrate the Hofmeister effect [20] on regulating the solution surface stress and the solubility of proteins with debating mechanisms of structural maker and breaker [21], ionic specification [22], quantum dispersion [23], skin induc-tion [24], quantum fluctuation [25], and solute–solvent interactions [26]. Increasing the chloride, bromide and iodide solute concentration stiffens the H–O stretching vibration mode to higher frequencies. The H–O phonon blueshifts are usually referred to that the Cl^-, Br^-, and I^- ions weaken their surrounding O:H nonbond, known as water structure breaking [27, 28].

An external electric field in the 10^9 V/m order slows down water molecular motion and even crystallizes the solution matrix. The field generated by a Na^+ ion acts rather locally to reorient and even hydrolyze its neighboring water molecules according to molecular dynamics (MD) computations [29]. A cation can form a stiffener cylindrical volume between the cation and the graphite-oxide (negatively charged) defects of the adjacent layers in a "$(-) \sim (+) \sim (-)$" fashion perpendicular to the graphene sheets. This stiffer hydration volume expands the graphene-oxide interlayer spacing up to 1.5 nm and the modulated layer separation varies with the type of cations [30, 31]. Strikingly, charge injection in terms of ions separation by salt solvation shares the same effect of molecular undercoordination on the hydrogen bond (O:H–O or HB) segmental length and stiffness and the H–O phonon lifetime [32–34].

Salt hydration and water molecular undercoordination have been intensively investigated using the following multiscale approaches:

(1) Classical continuum thermodynamics [35–37] embraced the dielectrics, diffusivity, surface stress, viscosity, latent heat, entropy, nucleation, and liquid/vapor phase transition in terms of free energy, though this approach has faced difficulties in dealing with solvation dynamics and the properties of water and ice.

(2) MD computations and the ultrafast phonon spectroscopies [38–40] are focused on the spatial and temporal performance of water and solute molecules as well as the proton and lone pair transportation behavior. Information includes the phonon relaxation or the molecular residing time at sites surrounding the solute or under different coordination conditions or perturbations.

(3) Nuclear quantum interactions [41, 42] simulation has enabled visualizing the concerted quantum tunneling of protons within the water clusters and quantify the impact of zero–point motion on the strength of a single hydrogen bond at a water/solid interface. An interlay of STM/S and the *ab initio* path–integral molecular dynamics (PIMD) verifed unambiguously that the sp^3–orbital hybridization takes place at 5 K temperature. The proton quantum interaction elongates the longer part and shortens the shorter part of the O:H–O bond.

(4) O:H–O bond cooperativity [1, 43–45] enables resolution to multiple mysteries of water and ice. A combination of the Lagrangian mechanics, MD and density functional theory (DFT) computations with the static phonon spectrometrics has enabled quantification of O:H–O transition from the mode of ordinary water to the conditioned states. Obtained information includes the fraction, stiffness, and fluctuation order transition upon perturbation and their consequence on the solution viscosity, surface stress, phase boundary dispersity, and the critical pressure and temperature for phase transition.

However, knowledge insufficiency about the O:H–O bond cooperativity and polarizability [46] has hindered largely the progress in understanding the effect of molecular undercoordination and salt solvation in terms of bonding dynamics, solute capabilities, and inter- and intramolecular interactions. It has been hardly possible to resolve the network O:H–O bond segmental cooperative relaxation induced by salt solvation and molecular undercoordination. It is yet to be known how the cation and anion mediate the O:H–O network and properties of salt solutions such as the surface stress, solution viscosity, solution temperature, and critical pressures and temperatures for phase transition [27, 47]. It is also unclear how the undercoordinated water molecules define the supercooling and superheating processes and the unusual performance of nanodroplets, nanobubbles, and the skins of water ice. Fine–resolution detection and consistently deep insight into the intra- and intermolecular interactions and their consequence on the solution and undercoordination derived properties of water have been an area of increasingly active study.

According to Pauling [48], the nature of the chemical bond bridges the structure and property of a crystal and molecule. Therefore, bond formation and relaxation and the associated energetics, localization, entrapment, and polarization of electrons mediate the macroscopic performance of substance accordingly [32]. O:H–O bond segmental disparity and O–O repulsivity form the soul dictating the extraordinary adaptivity, cooperativity, recoverability, and sensitivity of water and ice [43]. One must focus on the chemical bond relaxation [48] and the valence electron polarization [32, 43] in the skin region or in the hydration volume with the afore–mentioned multiscale approaches for improved knowledge.

This section shows the electronic and phononic spectrometric evidence for the bond–electron–phonon correlation of the undercoordinated and the hydrated O:H–O bond network and the supersolid state. Ions in salt solutions form each a center of electric field that clusters, stretches and polarizes its neighboring H_2O molecules that in turn screen the electric field of the ions to form the hydration shell of limited sizes [33]. Molecular undercoordination shortens the H–O bond spontaneously and lengthens the O:H nonbond associated with the polarization [33]. Molecular undercoordination and salt solvation share the same effect of O:H–O relaxation and polarization on the bonding network and properties such as the prolonged H–O phonon lifetime and the mediated critical conditions of phase transition.

11.2 O:H–O Bond Oscillator Pair

11.2.1 Basic Rules for Water

Water prefers the statistic mean of the tetrahedrally–coordinated, two–phase structure in a core–shell fashion of the same geometry but different O:H–O bond lengths [1, 43]. Figure 11.1a illustrates the $2H_2O$ unit cell having four oriented O:H–O bonds bridging oxygen anions. Transiting from the V-shaped H_2O motif of C_{2v} symmetry to the $2H_2O$ unit cell of C_{3v} symmetry aims to resolving the O:H–O cooperativity. As the basic structure and energy storage unit, the O:H–O bond integrates the intermolecular weaker O:H nonbond (or called van der Waals bond with ~0.1 eV energy) and the intramolecular stronger H–O polar–covalent bond (~4.0 eV).

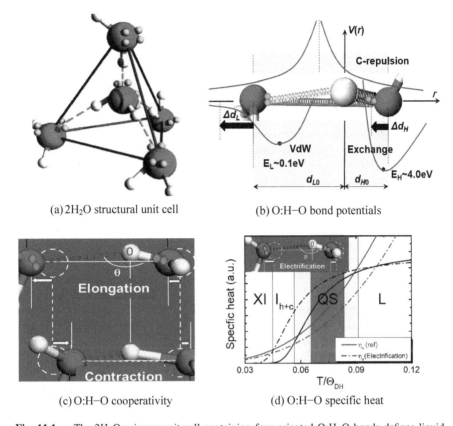

(a) 2H₂O structural unit cell

(b) O:H–O bond potentials

(c) O:H–O cooperativity

(d) O:H–O specific heat

Fig. 11.1 **a** The $2H_2O$ primary unit cell containing four oriented O:H–O bonds defines liquid water as a crystal with molecular and proton motion restrictions. **b** The asymmetrical, short–range, coupled three–body potentials for the segmented O:H–O bond. **c** Cooperative relaxation of the segmental O:H–O bond proceeded by elongating one part and contracting the other with respect to the H^+ coordination origin. **d** Superposition of the segmental specific heats η_x defines the phases from high temperature downward of the Vapor ($\eta_L = 0$, not shown), Liquid and Ice I_{h+c} ($\eta_L/\eta_H < 1$), Quasisolid (QS) ($\eta_L/\eta_H > 1$), XI ($\eta_L \cong \eta_H \cong 0$), and the QS boundaries ($\eta_L/\eta_H = 1$) closing to T_m and T_N, respectively [2]. Electrification (ionic polarization) [33] or molecular undercoordination [34] disperses the QS boundaries outwardly. Reprinted with permission from [2, 34, 56]

The asymmetrical and short–range interactions of the segmented O:H–O are coupled with the ever-overlooked Coulomb repulsion between electron pairs on adjacent oxygen ions, which dictates the extraordinary adaptivity, cooperativity, recoverability sensitivity of water ice when subjecting to perturbation [1]. O:H–O bond angle and length determines the crystal geometry and mass density of water ice. The segmental stretching vibration frequencies ω_x determine the respective Debye temperatures Θ_{Dx} of their specific heats through the Einstein relation, $\Theta_{Dx} \propto \omega_x$, (X = L and H denotes, respectively, the O:H and the H–O interactions). The segmental binding energy E_x correlated to the thermal integration of the specific heat $\eta_x(T/\Theta_{Dx})$ in Debye approximation. O:H–O bond length and its containing angle relaxation changes the system energy, but fluctuation contributes little to the system energy on average.

It is essential to treat water as a crystalline–like structure with well–defined lattice geometry, strong correlation, and high fluctuation. For a specimen containing N oxygen atoms, there will be 2N numbers of protons H^+ and lone pairs ":" to form the O:H–O bonds uniquely, except for the dangling H–O radicals and the dangling lone pairs at the network terminals. The 2N numbers and the O:H–O bond configuration conserve regardless of structural phase [49] unless excessive H^+ or ":" is introduced. The H^+ or the ":" cannot stay alone of moving freely but attach to a H_2O.

Excessive H^+ injection by acid solvation will bond to a H_2O to form the H_3O^+ hydronium and the H↔H anti–HB [50]. Base or H_2O_2 solvation introduces excessive number of ":" in the form of HO^- hydroxide that form the O:⇔:O super–HB with one of its neighboring H_2O [51]. The H_3O^+ or the HO^- replaces the central H_2O in Fig. 11.1a but the neighboring H_2O remain their orientations because of the lattice geometry and interactions with other neighboring H_2O molecules. The H_3O^+ or the HO^- may undergo Brownian motion or drift diffusion under a graded electric or thermal field.

The motion of a H_2O molecule or the proton H^+ transportation is subject to restriction. If the central molecule rotate above 60° around the C_{3v} symmetrical axis of the $2H_2O$ unit cell, an H↔H and O:⇔:O repulsion will come into play, which is energetically forbidden. Because of the H–O bond energy of ~4.0 eV, translational tunneling of the H^+ between adjacent H_2O molecules is also forbidden. In fact, only can 121.6 nm wavelength laser radiation breaks the H–O or the D–O bond in the vapor phase [52, 53]. The 121.6 nm wavelength corresponds to 5.1 eV energy that is greater than the 4.0 eV estimated by resolving the $T_C − P$ profiles of ice Regelation [54, 55]. The least molecular coordination number of the gaseous monomer shortens and stiffens the H–O bond most.

11.2.2   O:H–O Bond Potentials and Cooperativity

Figure 11.1b illustrates the asymmetrical, short–range, coupled three–body potentials for the segmented O:H–O bond [44, 45]. The proton serves as the coordination origin. The left–hand side is the O:H van der Walls (vdW) interaction and the right–hand side is the H–O polar-covalent bond. The Coulomb repulsion between electron pairs on neighboring O^{2-} couple the O:H–O bond to be an oscillator pair.

The O:H nonbond and the H–O bond segmental disparity and the O–O coupling dislocate the O^{2-} of the segmented O:H–O bond in the same direction but by different amounts under an external stimulus, see Fig. 11.1c. The softer O:H nonbond always relaxes more than the stiffer H–O bond with respect to the H^+ as the coordination origin. The ∠O:H–O angle θ relaxation contributes mainly to the geometry and mass density. The O:H–O bond bending has its specific vibration mode that does not interfere with the H–O and the O:H stretching vibrations [1]. The O:H–O bond cooperativity determines the properties of water and ice under external stimulus such as molecular undercoordination [5, 6, 57–59], mechanical compression [27, 47, 54, 60], thermal excitation [2, 61], solvation [62] and determines the molecular behavior such as solute and water molecular thermal fluctuation, solute drift motion dynamics, or phonon relaxation.

11.2.3 Specific Heat and Phase Transition

Figure 11.1d shows the superposition of the specific heat $\eta_x(T/\Theta_{Dx})$ of Debye approximation for the O:H and the H–O segments [2]. The segmental specific heat meets two conditions. One is the Einstein relation, $\Theta_{Dx} \propto \omega_x$, which directly correlates the shape of the specific heat curve to the measured phonon frequency ω_x. The specific heat curves vary with external perturbation through Einstein's relation. The other is the thermal integration that is proportional to the bond energy E_x. The (ω_x, E_x) are $(200 \text{ cm}^{-1}, \sim 0.1\text{–}0.2 \text{ eV})$ for the O:H nonbond and they are $(3200 \text{ cm}^{-1}, \sim 4.0 \text{ eV})$ for the H–O bond. The superposition of the segmental specific heats η_x defines the phases from high temperature downward of Vapor (not shown), Liquid, Quasisolid (QS), I_{h+c} ice, XI, and the QS boundaries of extreme densities that are closing to the T_m and the T_N [2, 4].

The thermodynamics of water ice is subject to its segmental specific heat difference in terms of the η_L/η_H ratio. The segment having a lower specific heat follows the regular rule of thermal expansion, but the other segment responds to thermal excitation contrastingly. The thermodynamics and density oscillation of the bulk water ice under the ambient pressure proceed in the following phases [2, 4]:

(1) In the Vapor phase (≥ 373 K), $\eta_L \cong 0$, the O:H interaction is negligibly weak though the H$_2$O motifs still hold.

(2) In the Liquid phase (277, 373 K), $\eta_L/\eta_H < 1$, O:H cooling contraction and H–O elongation take place, but the O:H contracts more than H–O contracts, cooling contraction of water take place, reaching to the value of $d_{OO} = d_H + d_L = 1.0004 + 2.6946$ Å.

(3) In the QS phase (258, 277 K), $\eta_L/\eta_H > 1$, H–O cooling contracts less that the O:H expands; O:H–O expands, which triggers ice floating.

(4) At the QS boundaries (258, 277 K), $\eta_L/\eta_H = 1$, the density drops from its maximum 1.0 to its minimum of 0.92 g cm^{-3}. No apparent singularity presents to the specific heat or to the density profile, so the 277 K is recommended be the temperature for phase transition from Liquid to the QS phase. The 258 K corresponds to the T_N for the homogeneous ice nucleation.

(5) The I_{c+h} ice (258, 100 K), $\eta_L/\eta_H < 1$, repeats the thermodynamics of the Liquid phase at a lower transition rate, transiting the density from 0.92 to 0.94 g cm^{-3}.

(6) In the XI phase, $\eta_L \cong \eta_H \cong 0$, neither O:H nor H–O responds sensitively to thermal excitation, so the density remains almost constant except for the slight $\angle$O:H–O angle expansion at cooling.

One should note that any relaxation of the O:H–O bond length and energy under perturbation will offset the phase boundaries through Einstein's relation: $\Theta_{DX} \propto \omega_X$; and therefore, the O:H–O bond oscillation and relaxation dictate the thermodynamics of water ice and aqueous solutions.

11.3 Supersolidity and Quasisolidity

11.3.1 Signatures

The concept of supersolidity was initially extended from the ^{4}He fragment at mK temperatures, demonstrating elastic, repulsive and frictionless between the contact motion of ^{4}He segments [63]. Atomic undercoordination induces local densification of charge and energy and the associated electron polarization at the fragment surfaces [64]. The supersolidity features the behavior of water and ice under polarization by undercoordination or electrostatic polarization. When the molecular CN is less than four, the H–O bond contracts spontaneously associated with O:H elongation and strong polarization. At the surface, the H–O bond contracts from 1.00 to 0.95 Å and the O:H expands from 1.70 to 1.95 Å associated with the O:H vibration frequency transiting from 200 to 75 cm^{-1} and the H–O from 3200 to 3450 cm^{-1} [65]. The shortened H–O bond raises its vibration frequency to a higher value that increases again with further reduction of the molecular CN.

The quasisolidity describes phase transition from Liquid density maximum of 1.0 g cm^{-3} at 4 °C to the Solid density minimum of 0.92 g cm^{-3} at −15 °C. The unusual property of the QS phase is its $\eta_L/\eta_H < 1$ defined cooling expansion and the QS boundaries tunability. Cooling the QS phase shortens the H–O bond and lengthens the O:H and enlarges the $\angle$O:H–O angle from 160° to 165° for bulks specimen [2].

The O:H–O bond relaxation under perturbation shifts its segmental phonon in opposite directions, which offsets the phase boundaries accordingly, including the T_C for the I_c–XI transition dropping from 100 K to 60 K with the droplet size [66, 67]. The outward dispersion of the QS phase boundaries by molecular undercoordination or electrostatic polarization depresses the T_N for homogenous ice nucleation and raises the T_m for melting, which accounts for the room-temperature thin ice formation [7] and mechanical and thermal stability of nanobubbles [68]. The supercooling of water nanodroplets to the "no man's land" could arise from the QS boundary dispersion due to the high fraction of undercoordinated molecules [1]. However, compression has a contrasting effect on the QS boundary dispersion, raising the T_N and lowering the T_m, resulting in the ice Regelation—compression depresses the T_m that reverse when and the pressure is relieved [55].

Because of the strong polarization and O:H–O relaxation, the skin of water and ice is offered with a supersolid skin that is elastoviscous, less dense (0.75 unit), and mechanically and thermally more stable. The high elasticity of the softer O:H phonons, $\Delta\omega_L < 0$, ensures the high adaptivity and the densely-packed surface dipoles secures the contacting interface repulsivity, resulting in the slipperiness of ice [69], nanobubble endurability [68], and the toughest water skin [70]. It is the supersolid skin's lower density that raises the skin thermal diffusivity for the heat transport in the Mpemba effect [71].

Salt solvation derives cations and anions dispersed in the solution [33]. Each of the ions serves as a source center of electric field that aligns, clusters, stretches and polarizes the neighboring O:H–O bonds, resulting the same supersolidity within the hydration shell whose size is subject to the screening of the hydrating H_2O dipoles and the ionic charge quantity and volume size.

11.3.2 Supercooling of Supersolid Phase

Figure 11.2 shows the T_N depression by droplet size reduction and by salt solvation. As illustrated in Fig. 11.1d, the QS boundaries offset outwardly by the phonon frequency shift $\Delta\omega_H > 0$ and $\Delta\omega_L < 0$ that disperses the Debye temperatures Θ_{Dx}, resulting in the supercooling at freezing and superheating at melting, as one observes as the "no man's land". XRD, Raman, and MD observations show that 1.2 nm sized droplet freezes at 173 K [5], and the $(H_2O)_{3-18}$ clusters do not form ice even at 120 K [6].

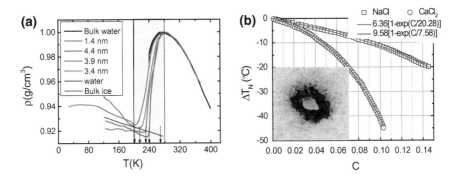

Fig. 11.2 Ice nucleation temperature T_N depression **a** by H_2O droplet size-reduction from 258 K for the bulk to 205 K for 1.4 nm droplet [2–6] and **b** by concentrated NaCl and $CaCl_2$ solvation [72] with inset showing the salt anti–icing. Reprinted with permission from [2, 72, 73]

11.4   Electron Spectrometrics

11.4.1   STM and STS: Strong Polarization

Figure 11.3 shows the orbital images and the dI/dV spectra of a H_2O monomer and a $(H_2O)_4$ tetramer deposited on a NaCl(001) surface probed using STM/S at 5 K temperature [74]. The highest occupied molecular orbital (HOMO) below the E_F of the monomer appears as a double—lobe structure with a nodal plane in between, while the lowest unoccupied molecular orbital (LUMO) above E_F appears as an ovate lobe developing between two HOMO lobes. STS spectra at different depths discriminate the tetramer from the monomer in the density of states (DOS) crossing E_F.

These STM/S observations [74] confirmed the occurrence of sp^3—orbit hybridization of oxygen in H_2O monomer occurs at 5 K or lower and the intermolecular interaction involved in $(H_2O)_4$. Therefore, the 2N number conservation of protons and

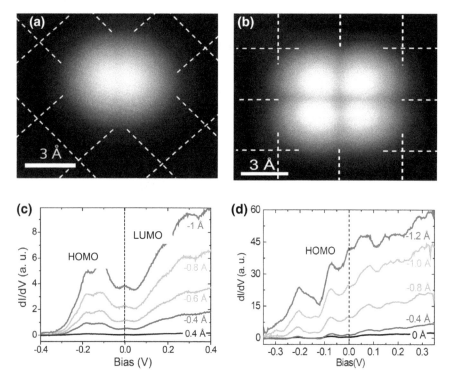

Fig. 11.3 STM images of **a** a H_2O monomer and **b** a $(H_2O)_4$ tetramer, and **c, d** the respective dI/dV spectra obtained under conditions of V = 100 mV, I = 100 pA, and dI/dV collected at 50 pA of different heights at 5 K temperature. Grid in images denotes the Cl^- lattice of the NaCl(001) substrate. The LUMO (>E_F) and HOMO (<E_F) indicated in **c** denote the orbital energy states. Reprinted with permission from [74]

":" holds regardless of temperature. Even at extreme conditions of 2000 K temperature and 2 TPa pressure, when the $2H_2O$ transits into H_3O^+:HO^- the superionic state
[75], the 2N number conservation of protons and lone pairs remains. According to the
bond–band–barrier correlation [32, 76], the HOMO located below E_F corresponds to
the energy states occupied by electron lone pairs of oxygen, and the LUMO to states
yet to be occupied by electrons of antibonding dipoles. The image of the monomer
showing the directional lone pairs that point into the open end of the surface. As the
H^+ protons share its unpaired electron with oxygen, the Cl^- ion in the NaCl substrate
interacts with the H^+ only electrostatically.

11.4.2 Water Skin: Entrapment and Polarization

Following the same size trend of "normal" materials, molecular undercoordination imparts to water local charge densification [13, 14, 77–80], binding energy
entrapment [77, 81–83], and nonbonding electron polarization [79]. XPS shown in
Fig. 11.6a confirms that the O $1s$ level shifts more deeply from the bulk value of
536.6–538.1 eV and 539.7 eV when move from bulk water to its skin and monomer
in gaseous phase [84–86]. The O $1s$ binding energy shift is a direct measure of the
H–O bond energy and the contribution from the O:H nonbond is negligibly small
[87].

The nonbonding electrons are subject to dual polarization when the molecular
CN is reduced [1]. Firstly, H–O bond contraction deepens the H–O potential well
and entraps and densifies electrons in the H–O bond and those in the core orbitals
of oxygen. This locally and densely entrapped electrons polarize the lone pair of
oxygen from the net charge of −0.616 to −0.652 eV according to DFT calculation
for ice skin [65]. The increased charge of O ions further enhances the O–O repulsion
as the second round of polarization. This dual polarization raises the valence band
energy up, from the bulk value of 3.3 eV, as shown in Fig. 11.6b. The bound energy
of solvated electrons in the skin and in the bulk reduces further with the number n
of $(H_2O)_n$ clusters toward zero [88–90] n = 2 − 11 [88]. (Fig. 11.4).

11.4.3 Ultrafast PES: Nonbonding Electron Polarization

Molecular undercoordination induced skin polarization have been detected using
an ultrafast pump—probe liquid—jet ultra-violet photoelectron spectroscopy (UPS)
[79]. A free electron injected into water [91] will be trapped by locally oriented solvent molecules and transiently confined within a roughly spherical cavity defined by
H–O bonds oriented toward the hydrated electron [92]. The hydrated electron serves

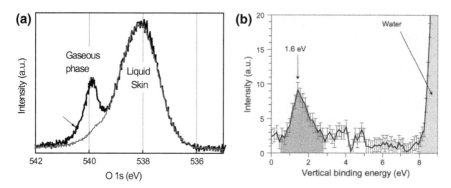

Fig. 11.4 Molecular undercoordination resolved **a** O 1s level entrapment from the liquid 536.6 eV to the skin at 538.1 eV and to the gaseous phase at 539.9 eV. (Reprinted with permission from [86]). This entrapment is associated with polarization of the **b** nonbonding solvent electrons by lowering the vertical bound energy for solvated electrons from 3.3 eV in the bulk to **b** 1.6 eV in the skin of the liquid water. Reprinted with copyright permission from [83]

as a probe to the local environment without changing the solvent geometry. Using the ultrafast pump–probe photoelectron spectroscopy, Verlet et al. [13] discovered that an excess electron can bound to the surface of a water cluster and to the ambient water/air interface. The internally solvated electron bound energy for $(D_2O)_{50}^-$ is centered at -1.75 eV and the surface localized states are centered at -0.90 eV. These two states vary with the cluster size and from $(D_2O)_{50}^-$ to $(H_2O)_{50}^-$ slightly. The vertical bound energies (being equivalent to work function) of the hydrated electrons is 1.6 eV in the skin and 3.2 eV in the bulk interior of pure water. The bound energy decreases with the number n of the $(H_2O)_n$ clusters toward zero [89, 90].

The hydrated electrons live longer than 100 ps near the surface compared with those solvated inside the bulk interior. The unexpectedly long lifetime of solvated electrons bound at the water surface is attributed to a free-energy barrier that separates surface and interior states [79]. Observations evidence that molecular undercoordination substantially enhances nonbonding electron polarization [34], which increases the viscoelasticity and hence lowers the skin molecular mobility. The anchored skin dipoles allow nanodroplet interacting with other substance through electrostatic, van der Waals, and hydrophobic interactions without exchanging electrons or bond formation, named non-additivity [93].

The nonbonding electrons are subject to dual polarization when the molecular CN is reduced [1]. Firstly, H–O bond contraction deepens the H–O potential well and entraps and densifies electrons in the H–O bond and those in the core orbitals of oxygen. This locally and densely entrapped electrons polarize the lone pair of oxygen from the net charge of -0.616 e to -0.652 eV according to DFT calculations for ice skin [65]. The increased charge of O^{2-} further enhances the O–O repulsion as the second round of polarization. This dual polarization raises the valence band energy up.

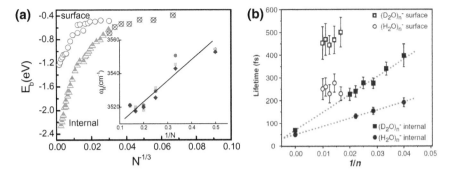

Fig. 11.5 a $(H_2O)_n^-$ (n = 2 − 11) cluster size and molecular site resolved bound energy of the hydrated electron [89, 90] with inset showing the size-resolved H–O phonon frequency [94, 95]; **b** The lifetime of hydrated electrons [13]). The relaxation time scales are given after electronic excitation of the surface isomers at 0.75 eV (open symbols) and internal isomers at 1.0 eV (solid symbols) of $(H_2O)_n^-$ (circles) and $(D_2O)_n^-$ (squares). Reprinted with permission from [79]

Further reduction of cluster size, or the molecular CN, enhances this dual polarization, resulting observations in Fig. 11.5a—cluster trend of the solvate electron polarization. Therefore, electronic dipoles formed on the flat and the curved skins enhances the polarization, which creates the repulsive force, making liquid water hydrophobic and ice slippery. Nonbonding electron polarization notion indicates that molecular undercoordination polarizes nonbonding electrons in two rounds by the densely entrapped H–O bonding electrons and by the repulsion between electron pairs on adjacent oxygen anions [1].

The hydrated electron can reside at the water/air interface, but remain below the dividing surface, within the first nanometer, shown in Fig. 11.5a [90, 96]. Comparatively, the neutral $(H_2O)_N$ size–resolved H–O vibration frequency (inset a [19]) and photoelectron lifetime (Fig. 11.5b) shifts linearly with the inverse of cluster size [94, 95, 97]. The longer lifetime of the surface phonons and electrons than the internal states indicates the slower electron/phonon energy dissipation in the supersolidity phase. These cluster size and molecular site resolved electron bound energy, phonon stiffness, and electron phonon lifetimes confirm consistently the molecular undercoordination induced supersolidity.

11.4.4 XAS: Supersolid Thermal Stability

Figure 11.6a, b compares the near-edge X-ray fine structure adsorption spectroscopy (NEXFAS) profiles of nanobubbles [98], vapor, liquid skin, and bulk water [99]. The spectra show three majors at 535.0–536.8 and 540.9 eV corresponding, respectively, to the molecular coordination resolved bulk interior, skin, and H–O dangling bond radicals.

(a)   **(b)**

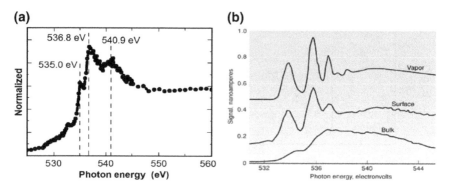

Fig. 11.6 NEXFAS spectra of **a** nanobubbles [98] and **b** vapor, liquid skin, and bulk liquid [99]. The spectra resolve discrete peaks that correspond to the bulk, skin, and H–O dangling bond radicals. Reprinted with permission from [99, 100]

The NEXFAS energy conservation mechanism is different from that of the XPS [87]. The NEXFAS involves the shift of both the valence band and the O1s core band but the XPS involves the O 1s band only. The NEXFAS pre–edge shift is the relative shift of the O1s core level ΔE_{1s} and its valence band shift ΔE_{vb} (occupied $4a_1$ orbital) from their energy levels of the isolated O atom: $\Delta E_{edge} = \Delta E_{vb} - \Delta E_{1s} \propto \Delta E_H > 0$ (H–O bond energy) [34]. The ΔE_{vb} is always greater than the ΔE_{1s} because of the shielding of electron in the outer electronic orbitals [87].

11.4.4.1 Thermal XAS and DPS: H–O Bond Stability

Figure 11.7 compares the thermal NEXAFS spectra for bulk water and 5M LiCl solutions. Thermal heating shifts the pre-edge peak for the liquid water negatively faster than that for the LiCl solutions as the O:H undergoes thermal elongation at different rates for the deionized water and LiCl solutions [101]. At 25 °C, the cation effect on the pre-edge shift from the value of 534.67 eV to deeper in alkali chlorides is remarkable: Li$^+$ (0.27 eV), Na$^+$ (0.09 eV), and K$^+$ (0.00 eV). The energy shift of Li$^+$ in the 5 M LiCl solution (0.30 eV) is close to that happened to the 3M LiCl/H$_2$O solution. On the other hand, in the NaCl solution, the anion effect is small: Cl$^-$ (0.09 eV), Br$^-$ (0.04 eV), and I$^-$ (0.02 eV). The energy trend of the pre-edge shifts is the same as the differencial phonon spectroscopy (DPS) ω_H shifting positively for the solutions and the skin of water [102].

The contributions from the Li:O polarization or the O:H binding energy are negligibly small. At polarization, the H–O bond becomes shorter and stiffer, the stiffer H–O bond is thermally more stable than those in the bulk of deionized water, as it does in the water skin [102].

Figure 11.8 compares the XAS pre-edge thermal shift for the bulk water and the supersolid hydrating water in the Li$^+$ hydration shell and the Raman H–O phonon thermal shift for the concentrated KCl/H$_2$O solutions encapsulated in a silica capillary tube [103]. The XAS peak of the hydration oxygen shifts negatively slower than it is in the bulk water. The Raman peak 1 at 3200 cm^{-1} corresponds to the ω_H for

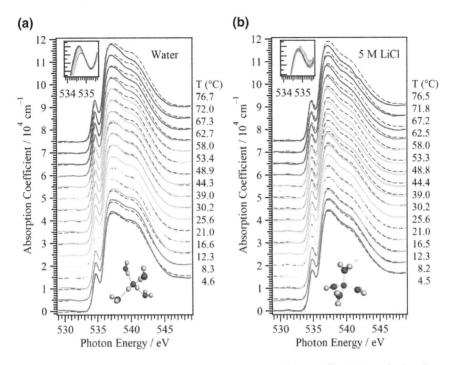

Fig. 11.7 Temperature resolved XAS for **a** bulk water and **b** 5M LiCl solutions. Broken lines are respective references collected at 25 °C. Insets show structure models for pure water and the hydrating molecules and the pre-edge negative shift due to H–O thermal contraction. Reprinted with permission from [101]

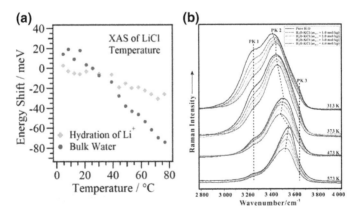

Fig. 11.8 Thermal stability of the supersolid **a** Li$^+$ hydrating volume probed using NEXAFS [101] and **b** the H–O components of concentrated KCl/H$_2$O solutions probed using Raman spectroscopy [102]. The O E$_{edge}$ peak shifts less negatively in the supersolid hydration volume than that in the bulk water. The thermal reversion at 373–473 K further proves the thermal stability of the KCl supersolid hydrating H–O bonds, see context for detailed discussion. Reprinted with permission from [101, 103]

ordinary bulk water and peak 2 at 3450 cm^{-1} to the skin ω_H mode. At the ambient temperature, ionic polarization shifts the 3200 cm^{-1} up to overlap the 3450 cm^{-1}. Heating and polarization enhance each other to shorten and stiffen the H–O bond at different rates, depending on the solute concentration. In the liquid phase below 373 K, the ω_H shifts with temperature and more with the solute concentration. The ω_H shifts further as the temperature increases, but the increasing rate reverts with solute concentration. At 373 K and above, the liquid partially becomes vapor up to the saturation pressure of 100 MPa. The thermal reversion of the ω_H shift between 373 and 473 K shows that the hydrating H–O bonds are thermally more stable than that of the deionized water in liquid and vapor phases. The polarization distorted H–O bonds are hardly further deformed.

11.4.4.2   Solution XAS: Polarization Dominance

Figure 11.9 compares the XAS for the concentrated LiCl/H$_2$O, 3M YCl/H$_2$O and NaX/H$_2$O solutions. Salt solvation shifts the pre-edge peak to the positive direction in contrasting to the effect of heating that shifts the peak in the negative direction, which appeared conflicting to the fact that heating and ionic polarization share the same effect of H–O bond contraction. Bond contraction dictates the energy shift of all the core level and valence band [87], as the O:H binding energy contributes insignificantly to the energy level shift.

To clarify the inconsistency of pre-edge peak shift induced by heating and ionic polarization, one must consider the competition between H–O bond contraction and electrostatic polarization on the energy level shift [87]. Firstly, electrons transiting from O 1s (K) level to the upper edge (the occupied 4a$_1$ orbital) of the valence band absorbs energy equals to the XAS pre-edge peak energy, E$_{edge}$. Secondly, separation between the energy shifts of the valence band and the O 1s level from their respective energy level of an isolated O atom determines the XAS pre-edge shift, $\Delta E_{edge} = \Delta E_{1s} - \Delta E_{vb}$. Thirdly, the involved energy levels undergo positive shift because of the addition of the local crystal potential to the intra-atomic potential, which

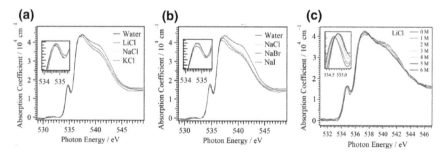

Fig. 11.9 NEXAFS oxygen spectra for the 3M **a** YCl (Y=Li, Na, K), **b** NaX (X=Cl, Br, I), and **c** the concentrated LiCl solutions collected at 25 °C. Insets show magnification of the pre-edge positive shift due to polarization dominance. Reprinted with permission from [101, 105]

is proportional to the binding energy of bond in the first coordination neighbors, according to the tight-binding approximation [104]. A contraction of the neighboring bonds shifts the energy levels further—deepens the energy levels further—quantum entrapment occurs [87]. The amount of energy shift varies from level to level because of the screening of the potential by electrons in the outer orbitals. Therefore, the energy of the valence band composed of electrons in the outermost orbital shifts more than an inner energy level. One has thus, $\Delta E_{vb} > \Delta E_{1s}$, and $\Delta E_{edge} = \Delta E_{1s} - \Delta E_{vb} < 0$, negative shift due to bond contraction.

On the other hand, electrostatic polarization changes the situation contrastingly. Charge polarization screens and splits the local potential and then shifts a proportion of electrons up in energy bands causing their negative shift, then, $-\Delta E_{vb} \gg -\Delta E_{1s}$, and thus $\Delta E_{edge} = \Delta E_{1s} - \Delta E_{vb} > 0$, positive shift takes place due to the polarization. In fact, bond contraction by molecular undercoordination or salt solvation is associated with polarization. Therefore, H–O bond thermal contraction without polarization results in the pre-edge energy negative shift. For the salt solution, ionic polarization becomes dominance, which overweight the effect of H–O bond contraction on the binding energy shift. The polarization shifts the energy levels up and thus, $\Delta E_{edge} = \Delta E_{1s} - \Delta E_{vb} > 0$. Therefore, the XAS pre-edge shift is very sensitive to the local energetic environment, being capable of discriminating the effect of entrapment by bond contraction and charge polarization. If the entrapment and polarization is compatible, no shift will occur to the XAS pre-edge energy peak. This discrimination could be the advantage of the XAS that probes the effect of valence band polarization while an XPS detects only the O 1s level shift subjecting to weak perturbation of the polarization.

11.4.4.3 XAS Capability: Entrapment and Polarization

NEXFAS measurements [101] revealed that Li^+, Na^+, and K^+ cations shift the pre-edge component peak energy more than the Cl^-, Br^-, and I^- anions and the pre–edge component associated with the first hydration shell of Li^+ ion is thermally more stable than those beyond, see Fig. 11.9. At 25 °C, the cation effect on the pre–edge shift from the value of 534.67 eV in alkali chlorides is remarkable: Li^+(0.27 eV), Na^+ (0.09 eV), and K^+ (0.00 eV). The energy shift of Li^+ ion in 5 M LiCl solution (0.30 eV) is close to that in its 3 M solution. On the other hand, in sodium halides, the anion effect is small: Cl^-(0.09 eV), Br^-(0.04 eV), and I^- (0.02 eV). The energy trend of the pre–edge shifts is the same as the DPS $\Delta\omega_H$ of the solutions and the skin of water [102].

Contributions from the Li:O polarization or the O:H binding energy change are negligibly small. At polarization, the H–O bond becomes shorter and stiffer; the stiffer H–O bond is thermally more stable than those in the bulk of ordinary water, as it does in the skin of deionized water [102]. The identical O_{fw}–O_{bw} and O_{bw}–O_{bw} thermal expansion in both the pure water and in the 5M LiCl solution indicates the invariance of the Li^+ hydration shell size, which does not interfere the O–O thermal behavior between the hydrated and non–hydrated oxygen anions.

Table 11.1 The O:Y^+ and $O^{2-} \leftrightarrow X^-$ distances (Å) in the YX solutions [101, 107, 108]

	Li^+:O^{2-} (5 M LiCl)	Na^+:O^{2-} (Å)	K^+:O^{2-} (Å)
Y^+:O^{2-} distance MD [101]	1.99 (25 °C)	2.37	2.69
Neutron diffraction [101, 107, 108]	1.90	2.34	2.65
MD [101] $X^- \leftrightarrow O^{2-}$ distance	$Cl^- \cdot H^+ - O^{2-}$	$Br^- \cdot H^+ - O^{2-}$	$I^- \cdot H^+ - O^{2-}$
	3.26	3.30	3.58
DFT(acid) [50][a]	$Cl^- \cdot (H - O:H)$	$Br^- \cdot (H - O:H)$	$I^- \cdot (H - O:H)$
1st (ε_H; ε_L)%	− 0.96; +26.1	− 1.06; +30.8	− 1.10; + 41.6
2nd (ε_H; ε_L)%	− 0.73; +19.8	− 0.78; +22.8	− 0.83; +28.6
XAS (5 M LiCl) [101]	Li^+:(first O:H–O)	Li^+:(next O:H–O)	O:H–O (H_2O)
5 °C	$d_{O-O} = 2.71$ (at 4 °C, $d_{O-O} = d_H + d_L = 1.0004 + 1.6946 = 2.695$ [56])		
80 °C	2.76 [$\Delta d_L > -\Delta d_H (< 0)$]		

[a](ε_H; ε_L)% is the DFT derived segmental strain for the first and the second O:H–O bonds radially surrounding X^- anions in acid solutions. The strain is refereed the standard values of $d_H = 1.0004$ and $d_L = 1.6946$ Å for 4 °C water [56]. $X^- \cdot H$ represents the anions and H^+ Coulomb interaction

Table 11.1 lists the MD estimation and neutron diffraction resolved the first O^{2-}:Y^+ and $O^{2-} \leftrightarrow X^-$ hydration shell sizes of the solutes [101], which agree with the DFT derived segmental strains of the O:H–O bonds surrounding X^- solutes [50]. The consistency between Raman and XAS observations confirmed the thermal stability of the supersolid hydration volume and clarifies the origins of the pre-edge energy shift due to competition between thermal H–O bond contraction and charge polarization. Observations verified the following:

(1) The XAS pre-edge shift features the energy difference between the O $1s$ core level shift ΔE_{1s} and its valence band shift ΔE_{vb} from their energy levels of an isolated O atom: $\Delta E_{edge} = \Delta E_{1s} - \Delta E_{vb} < 0$ [34] when the H–O contraction, $\Delta E_H > 0$, is dominant [87]. In contrasting, $\Delta E_{edge} = \Delta E_{1s} - \Delta E_{vb} > 0$ when polarization becomes dominant, as polarization shifts all energy levels upwardly. Competition between H–O bond contraction and polarization dictates the pre-edge energy shift that is in a contrasting manner of the XPS O $1s$ level shift.

(2) The H–O bond undergoes energy gain in liquid water heating [28], skin molecular undercoordination [34], and polarization [106]. The shortened H–O bonds are thermally and mechanically more stable because the stiffened bonds are less sensitive to perturbation. It is harder to further deform an already deformed H–O bond by stimulations such as heating in the present case: $(d|E_H|/dT)_{supersolid}/(d|E_H|/dT)_{regular} < 1$ and $(d\omega_H/dT)_{supersolid}/(d\omega_H/dT)_{regular} < 1$.

(3) The QS phase upper boundary is at 4 °C for regular water and it seems at 25 °C for the supersolid states according to the slopes of the XAS profile in Fig. 11.8.

The supersolidity disperses outwardly the QS phase. Outside the QS phase, O:H–O bond follows the regular thermodynamics: $dd_L/dT > 0$, $dd_H/dT < 0$.

(4) The local electric field of a small cation is stronger than that of a larger anion because of the $X^- \leftrightarrow X^-$ repulsion due to the insufficient number of the hydrating dipoles with the ordered molecular structure of the supersolid hydration shells [33].

11.5 Proton Capture-Ability and Electron Emissibility of Halide Anions

Why does the X^- follow the Hofmeister polarizability series order and why does HF not polarize water molecules? Discoveries using a combination of *ab initio* calculations and the anion photodetachment PES measurements of $HSO_4^- \cdot (HX)$ complexes [109] could clarify this issue crisply. Figure 11.10 shows the electron emissibility

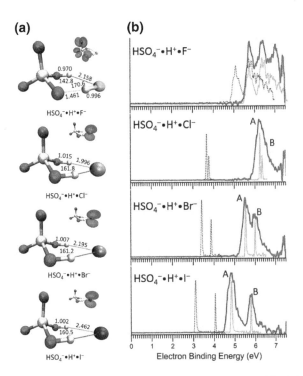

Fig. 11.10 **a** The optimal configurations of $HSO_4^- \cdot (HX)$ complexes and **b** their photoelectron spectra measured at 20 K using 157 nm laser excitation. The doted and grey curves represent, respectively, the original and shifted spectra of HSO_4^- and X^-. The spectra of X^- are adapted from [111]. Panel **a** also plots their highest occupied molecular orbitals (HOMOs) (insets). Denoted are bond lengths (Å) and bond angles (o). Reprinted with permission from [109]

and the proton capture-ability of halide and oxygen anions. The optimal $HSO_4^- \cdot (HX)$ configuration shows that HCl, HBr, and HI dissolve into H^+ and X^- and then the H^+ binds to the neighboring O^{2-}. Ions have a higher electronegativity tend to capture the proton that initially bonds to the less negative ions. The HF is not dissolvable and keeps itself the contacted ion pair. Ions of highly negative are hard to emit electrons under the excitation of 157 nm laser under the same temperature of 20 K.

Results also show the X:H–O cooperativity in its segmental length and the polarizability of the anions. The X in the $HSO_4^- \cdot (HX)$ complexes forms each two identical O–H:X hydrogen bonds with its two neighboring O for X = Cl, Br, and I. The O–H:X segmental lengths vary cooperatively—if the O–H is shorter, the H:X will be longer [43]. As the X moves from Cl to Br and I, the anion polarization shortens the O–H bond from 1.015 to 1.007 and to 1.002 Å, and lengthens the X:H from 1.996 to 2.155 and to 2.462 Å, correspondingly, evidencing the O–X repulsive coupling. The strongest O–I repulsion lengthens the O–H:X bond most. However, according to the $HSO_4^- \cdot (HF)$ configuration, the HF retains and the F forms one O–H:F and one O:H–F without HF being separated (called contact ion pair). The segmental lengths and containing angles for the O:H–F bond are different from those of the O–H:F bond.

Figure 11.10b compares the PE spectra for the complexes. The spectral signature of $HSO_4^- :(HF)$ resembles that of an isolated HSO_4^- but shifting positively by ~0.65 eV (dotted and grey curves) upon the complex formation. The electrons bind more strongly in the complexes than they do to the HSO_4^- (~4.75 eV) [110] or the X^- (~3.06 − 3.61 eV) standing alone. On the other hand, the spectral signatures of the $HSO_4^- \cdot (HX)$ shift deeper by ~2.6, 2.2, and 1.8 eV for X = Cl, Br, and I, respectively. Two spectral bands A and B resolve the spin-orbit (SO) splitting states ($\Omega = 3/2$ and 1/2) more pronounced for Br and I than that of Cl [111–114]. The energy states of these complexes show that HSO_4^- carries most electron cloud for X = F and the X^- carries most when X = Cl, Br, and I. The separations between the vacuum energy ($E = 0$) and the upper band edges, show in Table 11.1, coincide with the ionic order of polarizability that shifts the H–O phonon frequency positively. The I^- polarizes water molecules most [118] .

Table 11.2 compares the adiabatic detachment energy (ADE) and vertical detachment energy (VDE) of the complexes derived from photoelectron spectral measurements, and computations using the B3LYP functional [115, 116] and coupled cluster

Table 11.2 Ionic radius R, electronegativity, polarizability $\alpha(\text{Å}^3)$ [117], and electron EBE of $HSO_4^- \cdot HX$ complexes

X	R	η	α	O–H	H:X	B3LYP	CCSD(T)	Exptl.
F	1.33	4.0	0.952	0.970	2.158	5.71	5.95	5.74
Cl	1.81	3.0	3.475	1.015	1.996	5.94	6.04	6.22
Br	1.96	2.8	4.821	1.007	2.195	5.56	5.49	5.52
I	2.20	2.5	7.216	1.002	2.462	4.93	4.84	4.84

Reprinted with permission from [109]

including singlet, doublet, and triplet excitation (CCSD(T)) levels in DFT calculations. Observation may challenge that if the H frustration hopping happens in water and solutions, as the O in the O:H–O has the identical electronegativity. The answer from the author is negative because of the high H–O bonding energy. The exchange location between the lone pair and the proton needs similar amount energy to break the H–O bond.

11.6 Perspectives

A combination of the STM/S, XPS, XAS, DPS, and ultrafast UPS and FTIR observations and quantum calculations reveals consistently the bond—electron—phonon correlation of the supersolidity of the confined and the hydrating water. Molecular undercoordination and charge dispersion by salt solvation share the same effect on the O:H–O bond relaxation and nonbonding electron polarization, which modulate the local hydrogen bonding network and the water properties through H–O bond contraction and O:H nonbond elongation. The O $1s$ energy shifts deeper, electronic vertical bound energy decreases but the electron and phonon lifetimes become longer. The supersolid phase is less dense, elastoviscous, mechanically and thermally more stable with the hydrophobic and frictionless surface. The O:H–O bond cooperative relaxation disperses outwardly the quasisolid phase boundary to raise the melting point and meanwhile lower the freezing point of water ice.

   From what we have learnt in the described exercises, one can be recommended the following ways of thinking and approaching to complement the available premises:

(1) The nonbonding electron lone pairs pertained to N, O, F and their neighboring elements in the Periodic Table form the primary element being key to our life, which should receive deserved attention. The lone pair is related to DNA folding and unfolding, regulating, and messaging. NO medication and CF_4 anticoagulation in synthetic blood are realized through lone pairs interaction with living cells. The lone pair forms the O:H and the O:⇔:O interactions together with the H ↔ H determines the molecular interactions. Without lone pairs, neither O:H–O bond nor oxidation could be possible; molecular interaction equilibrium could not be realized. Extending the knowledge about lone pairs and their functionality of polarization to catalysis, solution-protein, drug-cell, liquid-solid, colloid-matrix, interactions and even energetic explosives and other molecular crystals would be even more fascinating.

(2) The key to the O:H–O bond is the O–O coupling. Without such a coupling none of the cooperative relaxation or the mysterious of water ice and aqueous solutions such as ice floating, ice slipperiness, regelation, supercooling/heating or the negative thermal expansion, warm water cooling faster. Unfortunately, the O–O coupling has been long oversighted in practice. It is necessary to think

about water and solvent matrix as the highly ordered, strongly correlated, and fluctuating crystals, particularly, the supersolid phase caused by salt solvation and molecular undercoordination rather than the amorphous or multiphase structures. Water holds the two-phase structure in the core-shell configuration, rather than the randomly domain-resolved mixture of density patches. Liquid water and the matrix of aqueous solutions must follow the conservation rules for the 2N number of protons and lone pairs and for the O:H–O configuration despite its segmental length and energy relaxation unless excessive H^+ or lone pairs are introduced.

(3) As a degree of freedom, atomic and molecular undercoordination forms the foundation of and reconciles the defect and surface science, nanoscience and engineering. The undercoordination derived bond contraction and the associated electron and energy entrapment and local polarization govern the performance of the undercoordinated systems. One can consider the solvation as a process of charge injection with multiple interactions. Charge injection in terms of hydrated electrons, ions, protons, lone pairs, and even molecular dipoles mediate the HB network and properties of a solution. Salt solvation and molecular undercoordination share the same effect of polarization on the viscosity, surface stress, phonon stiffness and lifetime transition. The quasisolid (or quasiliquid) of negative thermal extensity due to O:H–O bond segmental specific disparity, the supersolidity due to molecular undercoordination and electric polarization are both important to water, ice and aqueous solutions. The quasisolid phase boundary dispersion by perturbation determines the solution O:H–O bond network and thermodynamic behavior such critical pressures and temperatures for phase transition.

(4) Focusing on the bond-electron-phonon-property correlation and interlaying the spatially- and temporarily-resolved electron/phonon/photon spectrometrics would substantiate the advancement of related studies. Combining the spatially resolved electron/phonon DPS and the temporarily resolved ultrafast pump-probe spectroscopies not only distill the phonon abundance-stiffness-lifetime-fluctuation due to liquid conditioning but also fingerprint the electron/phonon energy dissipation and the ways of interactions. Molecular residing time or drift motion under a certain coordination environment fingerprints the way of energy dissipation but these processes could hardly give direct information of energy exchange under perturbation. Polarization, entrapment, and absorption determine the energy dissipation. Embracing the emerged O:H–O bond segmental disparity and cooperativity and the specific heat difference would be even more revealing.

Understanding may extend to water-protein interaction, biochemistry, environmental and pharmaceutical industries. As the primary functional and structural unit, lone pair and proton play the key role in molecular interactions. Hydrophobic interface is the same to free surface. Charge injection by salt and other solute solvation provides the local electric fields. As the important degrees of freedom, molecular undercoordination and electric polarization are ubiquitous to our daily life and living conditions. It would be very promising for one to keep mind open and be on the way to developing experimental strategies and innovating theories toward resolution to the wonderful world.

References

1. Y.L. Huang, X. Zhang, Z.S. Ma, Y.C. Zhou, W.T. Zheng, J. Zhou, C.Q. Sun, Hydrogen-bond relaxation dynamics: resolving mysteries of water ice. Coord. Chem. Rev. **285**, 109–165 (2015)
2. C.Q. Sun, X. Zhang, X. Fu, W. Zheng, J.-L. Kuo, Y. Zhou, Z. Shen, J. Zhou, Density and phonon-stiffness anomalies of water and ice in the full temperature range. J. Phys. Chem. Lett. **4**, 3238–3244 (2013)
3. M. Erko, D. Wallacher, A. Hoell, T. Hauss, I. Zizak, O. Paris, Density minimum of confined water at low temperatures: a combined study by small-angle scattering of X-rays and neutrons. PCCP **14**(11), 3852–3858 (2012)
4. F. Mallamace, C. Branca, M. Broccio, C. Corsaro, C.Y. Mou, S.H. Chen, The anomalous behavior of the density of water in the range 30 K < T < 373 K. Proc. Natl. Acad. Sci. U. S. A. **104**(47), 18387–18391 (2007)
5. F.G. Alabarse, J. Haines, O. Cambon, C. Levelut, D. Bourgogne, A. Haidoux, D. Granier, B. Coasne, Freezing of water confined at the nanoscale. Phys. Rev. Lett. **109**(3), 035701 (2012)
6. R. Moro, R. Rabinovitch, C. Xia, V.V. Kresin, Electric dipole moments of water clusters from a beam deflection measurement. Phys. Rev. Lett. **97**(12), 123401 (2006)
7. J. Hu, X.-D. Xiao, D. Ogletree, M. Salmeron, Imaging the condensation and evaporation of molecularly thin films of water with nanometer resolution. Science **268**(5208), 267–269 (1995)
8. S. Cerveny, F. Mallamace, J. Swenson, M. Vogel, L.M. Xu, Confined water as model of supercooled water. Chem. Rev. **116**(13), 7608–7625 (2016)
9. J. Rault, R. Neffati, P. Judeinstein, Melting of ice in porous glass: why water and solvents confined in small pores do not crystallize? Eur. Phys. J. B **36**(4), 627–637 (2003)
10. S. Chen, A.P. Draude, X. Nie, H. Fang, N.R. Walet, S. Gao, J. Li, Effect of layered water structures on the anomalous transport through nanoscale graphene channels. J. Phys. Commun. **2**(8), 085015 (2018)
11. R. Ludwig, Water: from clusters to the bulk. Angew. Chem. Int. Ed. **40**(10), 1808–1827 (2001)
12. A. Michaelides, K. Morgenstern, Ice nanoclusters at hydrophobic metal surfaces. Nat. Mater. **6**(8), 597–601 (2007)
13. J.R.R. Verlet, A.E. Bragg, A. Kammrath, O. Cheshnovsky, D.M. Neumark, Observation of large water-cluster anions with surface-bound excess electrons. Science **307**(5706), 93–96 (2005)
14. N.I. Hammer, J.W. Shin, J.M. Headrick, E.G. Diken, J.R. Roscioli, G.H. Weddle, M.A. Johnson, How do small water clusters bind an excess electron? Science **306**(5696), 675–679 (2004)
15. C. Perez, M.T. Muckle, D.P. Zaleski, N.A. Seifert, B. Temelso, G.C. Shields, Z. Kisiel, B.H. Pate, Structures of cage, prism, and book isomers of water hexamer from broadband rotational spectroscopy. Science **336**(6083), 897–901 (2012)
16. T. Ishiyama, H. Takahashi, A. Morita, Origin of vibrational spectroscopic response at ice surface. J. Phys. Chem. Lett. **3**, 3001–3006 (2012)
17. J. Swenson, S. Cerveny, Dynamics of deeply supercooled interfacial water. J. Phys. Condens. Matter **27**(3), 032101 (2014)
18. Y.R. Shen, V. Ostroverkhov, Sum-frequency vibrational spectroscopy on water interfaces: polar orientation of water molecules at interfaces. Chem. Rev. **106**(4), 1140–1154 (2006)

19. S.T. van der Post, C.S. Hsieh, M. Okuno, Y. Nagata, H.J. Bakker, M. Bonn, J. Hunger, Strong frequency dependence of vibrational relaxation in bulk and surface water reveals sub-picosecond structural heterogeneity. Nat Commun **6**, 8384 (2015)
20. F. Hofmeister, Concerning regularities in the protein-precipitating effects of salts and the relationship of these effects to the physiological behaviour of salts. Arch. Exp. Pathol. Pharmacol. **24**, 247–260 (1888)
21. W.M. Cox, J.H. Wolfenden, The viscosity of strong electrolytes measured by a differential method. Proc. Roy. Soc. Lond. A. **145**(855), 475–488 (1934)
22. K.D. Collins, Charge density-dependent strength of hydration and biological structure. Biophys. J. **72**(1), 65–76 (1997)
23. K.D. Collins, Why continuum electrostatics theories cannot explain biological structure, polyelectrolytes or ionic strength effects in ion–protein interactions. Biophys. Chem. **167**, 43–59 (2012)
24. X. Liu, H. Li, R. Li, D. Xie, J. Ni, L. Wu, Strong non-classical induction forces in ion-surface interactions: general origin of Hofmeister effects. Sci. Rep. **4**, 5047 (2014)
25. H. Zhao, D. Huang, Hydrogen bonding penalty upon ligand binding. PLoS ONE **6**(6), e19923 (2011)
26. W.B. O'Dell, D.C. Baker, S.E. McLain, Structural evidence for inter-residue hydrogen bonding observed for cellobiose in aqueous solution. PLoS ONE **7**(10), e45311 (2012)
27. Q. Zeng, T. Yan, K. Wang, Y. Gong, Y. Zhou, Y. Huang, C.Q. Sun, B. Zou, Compression icing of room-temperature NaX solutions (X = F, Cl, Br, I). Phys. Chem. Chem. Phys. **18**(20), 14046–14054 (2016)
28. X. Zhang, T. Yan, Y. Huang, Z. Ma, X. Liu, B. Zou, C.Q. Sun, Mediating relaxation and polarization of hydrogen-bonds in water by NaCl salting and heating. Phys. Chem. Chem. Phys. **16**(45), 24666–24671 (2014)
29. M. Druchok, M. Holovko, Structural changes in water exposed to electric fields: a molecular dynamics study. J. Mol. Liq. **212**, 969–975 (2015)
30. J. Abraham, K.S. Vasu, C.D. Williams, K. Gopinadhan, Y. Su, C.T. Cherian, J. Dix, E. Prestat, S.J. Haigh, I.V. Grigorieva, P. Carbone, A.K. Geim, R.R. Nair, Tunable sieving of ions using graphene oxide membranes. Nat Nano **12**(6), 546–550 (2017)
31. L. Chen, G. Shi, J. Shen, B. Peng, B. Zhang, Y. Wang, F. Bian, J. Wang, D. Li, Z. Qian, G. Xu, G. Liu, J. Zeng, L. Zhang, Y. Yang, G. Zhou, M. Wu, W. Jin, J. Li, H. Fang, Ion sieving in graphene oxide membranes via cationic control of interlayer spacing. Nature **550**(7676), 380–383 (2017)
32. C.Q. Sun, Relaxation of the chemical bond. Springer Ser. Chem. Phys. **108**, 807 (2014)
33. C.Q. Sun, J. Chen, Y. Gong, X. Zhang, Y. Huang, (H, Li)Br and LiOH solvation bonding dynamics: molecular nonbond interactions and solute extraordinary capabilities. J. Phys. Chem. B **122**(3), 1228–1238 (2018)
34. C.Q. Sun, X. Zhang, J. Zhou, Y. Huang, Y. Zhou, W. Zheng, Density, elasticity, and stability anomalies of water molecules with fewer than four neighbors. J. Phys. Chem. Lett. **4**, 2565–2570 (2013)
35. J.C. Araque, S.K. Yadav, M. Shadeck, M. Maroncelli, C.J. Margulis, How is diffusion of neutral and charged tracers related to the structure and dynamics of a room-temperature ionic liquid? Large deviations from Stokes-Einstein behavior explained. J. Phys. Chem. B **119**(23), 7015–7029 (2015)
36. K. Amann-Winkel, R. Böhmer, F. Fujara, C. Gainaru, B. Geil, T. Loerting, Colloquiu: water's controversial glass transitions. Rev. Mod. Phys. **88**(1), 011002 (2016)
37. Z. Zhang, X.-Y. Liu, Control of ice nucleation: freezing and antifreeze strategies. Chem. Soc. Rev. **47**(18), 7116–7139 (2018)
38. M. Thämer, L. De Marco, K. Ramasesha, A. Mandal, A. Tokmakoff, Ultrafast 2D IR spectroscopy of the excess proton in liquid water. Science **350**(6256), 78–82 (2015)
39. Z. Ren, A.S. Ivanova, D. Couchot-Vore, S. Garrett-Roe, Ultrafast structure and dynamics in ionic liquids: 2D-IR spectroscopy probes the molecular origin of viscosity. J. Phys. Chem. Lett. **5**(9), 1541–1546 (2014)

40. S. Park, M. Odelius, K.J. Gaffney, Ultrafast dynamics of hydrogen bond exchange in aqueous ionic solutions. J. Phys. Chem. B **113**(22), 7825–7835 (2009)
41. J. Guo, X.-Z. Li, J. Peng, E.-G. Wang, Y. Jiang, Atomic-scale investigation of nuclear quantum effects of surface water: experiments and theory. Prog. Surf. Sci. **92**(4), 203–239 (2017)
42. J. Peng, J. Guo, P. Hapala, D. Cao, R. Ma, B. Cheng, L. Xu, M. Ondráček, P. Jelínek, E. Wang, Y. Jiang, Weakly perturbative imaging of interfacial water with submolecular resolution by atomic force microscopy. Nat. Commun. **9**(1), 122 (2018)
43. C.Q. Sun, Y. Sun, The attribute of water: single notion, multiple myths. Springer Ser. Chem. Phys. **113**. 494 (2016)
44. Y.L. Huang, X. Zhang, Z.S. Ma, G.H. Zhou, Y.Y. Gong, C.Q. Sun, Potential paths for the hydrogen-bond relaxing with (H_2O)(N) cluster size. J. Phys. Chem. C **119**(29), 16962–16971 (2015)
45. Y. Huang, X. Zhang, Z. Ma, Y. Zhou, G. Zhou, C.Q. Sun, Hydrogen-bond asymmetric local potentials in compressed ice. J. Phys. Chem. B **117**(43), 13639–13645 (2013)
46. Editorial, So much more to know. Science, **309**(5731), 78–102 (2005)
47. Q. Zeng, C. Yao, K. Wang, C.Q. Sun, B. Zou, Room-Temperature NaI/H_2O Compression Icing: solute–solute interactions. PCCP **19**, 26645–26650 (2017)
48. L. Pauling, *The Nature of the Chemical Bond*, 3 edn. (Cornell University Press, Ithaca, NY, 1960)
49. X. Zhang, P. Sun, Y. Huang, T. Yan, Z. Ma, X. Liu, B. Zou, J. Zhou, W. Zheng, C.Q. Sun, Water's phase diagram: from the notion of thermodynamics to hydrogen-bond cooperativity. Prog. Solid State Chem. **43**, 71–81 (2015)
50. X. Zhang, Y. Zhou, Y. Gong, Y. Huang, C. Sun, Resolving H(Cl, Br, I) capabilities of transforming solution hydrogen-bond and surface-stress. Chem. Phys. Lett. **678**, 233–240 (2017)
51. C.Q. Sun, J. Chen, X. Liu, X. Zhang, Y. Huang, (Li, Na, K)OH hydration bondin thermodynamics: solution self-heating. Chem. Phys. Lett. **696**, 139–143 (2018)
52. S.A. Harich, X. Yang, D.W. Hwang, J.J. Lin, X. Yang, R.N. Dixon, Photodissociation of D2O at 121.6 nm: a state-to-state dynamical picture. J. Chem. Phys. **114**(18), 7830–7837 (2001)
53. S.A. Harich, D.W.H. Hwang, X. Yang, J.J. Lin, X. Yang, R.N. Dixon, *Photodissociation of H_2O at 121.6 nm: a state-to-state dynamical picture*. J. Chem. Phys. **113**(22), 10073–10090 (2000)
54. C.Q. Sun, X. Zhang, W.T. Zheng, Hidden force opposing ice compression. Chem. Sci. **3**, 1455–1460 (2012)
55. X. Zhang, Y. Huang, P. Sun, X. Liu, Z. Ma, Y. Zhou, J. Zhou, W. Zheng, C.Q. Sun, Ice regelation: hydrogen-bond extraordinary recoverability and water quasisolid-phase-boundary dispersivity. Sci. Rep. **5**, 13655 (2015)
56. Y. Huang, X. Zhang, Z. Ma, Y. Zhou, J. Zhou, W. Zheng, C.Q. Sun, Size, separation, structure order, and mass density of molecules packing in water and ice. Sci. Rep. **3**, 3005 (2013)
57. K. Sotthewes, P. Bampoulis, H.J. Zandvliet, D. Lohse, B. Poelsema, Pressure induced melting of confined ice. ACS Nano **11**(12), 12723–12731 (2017)
58. H. Qiu, W. Guo, Electromelting of confined monolayer ice. Phys. Rev. Lett. **110**(19), 195701 (2013)
59. B. Wang, W. Jiang, Y. Gao, B.K. Teo, Z. Wang, Chirality recognition in concerted proton transfer process for prismatic water clusters. Nano Res. **9**(9), 2782–2795 (2016)
60. H. Bhatt, A.K. Mishra, C. Murli, A.K. Verma, N. Garg, M.N. Deo, S.M. Sharma, Proton transfer aiding phase transitions in oxalic acid dihydrate under pressure. Phys. Chem. Chem. Phys. **18**(11), 8065–8074 (2016)
61. D. Kang, J. Dai, H. Sun, Y. Hou, J. Yuan, Quantum simulation of thermally-driven phase transition and oxygen K-edge x-ray absorption of high-pressure ice. Sci. Rep. **3**, 3272 (2013)
62. F. Li, Z. Li, S. Wang, S. Li, Z. Men, S. Ouyang, C. Sun, Structure of water molecules from Raman measurements of cooling different concentrations of NaOH solutions. Spectrochim. Acta Part A Mol. Biomol. Spectrosc. **183**, 425–430 (2017)

63. J. Day, J. Beamish, Low-temperature shear modulus changes in solid He-4 and connection to supersolidity. Nature **450**(7171), 853–856 (2007)
64. C.Q. Sun, Size dependence of nanostructures: impact of bond order deficiency. Prog. Solid State Chem. **35**(1), 1–159 (2007)
65. X. Zhang, Y. Huang, Z. Ma, Y. Zhou, W. Zheng, J. Zhou, C.Q. Sun, A common supersolid skin covering both water and ice. Phys. Chem. Chem. Phys. **16**(42), 22987–22994 (2014)
66. C. Medcraft, D. McNaughton, C.D. Thompson, D.R.T. Appadoo, S. Bauerecker, E.G. Robertson, Water ice nanoparticles: size and temperature effects on the mid-infrared spectrum. Phys. Chem. Chem. Phys. **15**(10), 3630–3639 (2013)
67. C. Medcraft, D. McNaughton, C.D. Thompson, D. Appadoo, S. Bauerecker, E.G. Robertson, Size and temperature dependence in the far-ir spectra of water ice particles. Astrophys. J. **758**(1), 17 (2012)
68. X. Zhang, X. Liu, Y. Zhong, Z. Zhou, Y. Huang, C.Q. Sun, Nanobubble skin supersolidity. Langmuir **32**(43), 11321–11327 (2016)
69. X. Zhang, Y. Huang, Z. Ma, L. Niu, C.Q. Sun, From ice supperlubricity to quantum friction: electronic repulsivity and phononic elasticity. Friction **3**(4), 294–319 (2015)
70. X. Zhang, Y. Huang, S. Wang, L. Li, C.Q. Sun, Supersolid skin mechanics of water and ice. Procedia IUTAM **21**, 102–110 (2017)
71. X. Zhang, Y. Huang, Z. Ma, Y. Zhou, J. Zhou, W. Zheng, Q. Jiang, C.Q. Sun, Hydrogen-bond memory and water-skin supersolidity resolving the Mpemba paradox. Phys. Chem. Chem. Phys. **16**(42), 22995–23002 (2014)
72. D.R. Lide, *CRC Handbook of Chemistry and Physics*, 80th edn. (CRC Press, Boca Raton, 1999)
73. A.K. Metya, J.K. Singh, Nucleation of aqueous salt solutions on solid surfaces. J. Phys. Chem. C **122**(15), 8277–8287 (2018)
74. J. Guo, X. Meng, J. Chen, J. Peng, J. Sheng, X.Z. Li, L. Xu, J.R. Shi, E. Wang, Y. Jiang, Support: real-space imaging of interfacial water with submolecular resolution. Supp. Nat. Mater. **13**, 184–189 (2014)
75. Y. Wang, H. Liu, J. Lv, L. Zhu, H. Wang, Y. Ma, High pressure partially ionic phase of water ice. Nat. Commun. **2**, 563 (2011)
76. C.Q. Sun, Oxidation electronics: bond-band-barrier correlation and its applications. Prog. Mater Sci. **48**(6), 521–685 (2003)
77. O. Marsalek, F. Uhlig, T. Frigato, B. Schmidt, P. Jungwirth, Dynamics of electron localization in warm versus cold water clusters. Phys. Rev. Lett. **105**(4), 043002 (2010)
78. S. Liu, J. Luo, G. Xie, D. Guo, Effect of surface charge on water film nanoconfined between hydrophilic solid surfaces. J. Appl. Phys. **105**(12), 124301–124304 (2009)
79. K.R. Siefermann, Y. Liu, E. Lugovoy, O. Link, M. Faubel, U. Buck, B. Winter, B. Abel, Binding energies, lifetimes and implications of bulk and interface solvated electrons in water. Nat. Chem. **2**, 274–279 (2010)
80. D.H. Paik, I.R. Lee, D.S. Yang, J.S. Baskin, A.H. Zewail, Electrons in finite-sized water cavities: hydration dynamics observed in real time. Science **306**(5696), 672–675 (2004)
81. R. Vacha, O. Marsalek, A.P. Willard, D.J. Bonthuis, R.R. Netz, P. Jungwirth, Charge transfer between water molecules as the possible origin of the observed charging at the surface of pure water. J. Phys. Chem. Lett. **3**(1), 107–111 (2012)
82. F. Baletto, C. Cavazzoni, S. Scandolo, Surface trapped excess electrons on ice. Phys. Rev. Lett. **95**(17), 176801 (2005)
83. L. Turi, W.S. Sheu, P.J. Rossky, Characterization of excess electrons in water-cluster anions by quantum simulations. Science **309**(5736), 914–917 (2005)
84. M. Abu-Samha, K.J. Borve, M. Winkler, J. Harnes, L.J. Saethre, A. Lindblad, H. Bergersen, G. Ohrwall, O. Bjorneholm, S. Svensson, The local structure of small water clusters: imprints on the core-level photoelectron spectrum. J. Phys. B **42**(5), 055201 (2009)
85. K. Nishizawa, N. Kurahashi, K. Sekiguchi, T. Mizuno, Y. Ogi, T. Horio, M. Oura, N. Kosugi, T. Suzuki, High-resolution soft X-ray photoelectron spectroscopy of liquid water. Phys. Chem. Chem. Phys. **13**, 413–417 (2011)

86. B. Winter, E.F. Aziz, U. Hergenhahn, M. Faubel, I.V. Hertel, Hydrogen bonds in liquid water studied by photoelectron spectroscopy. J. Chem. Phys. **126**(12), 124504 (2007)

87. X.J. Liu, M.L. Bo, X. Zhang, L. Li, Y.G. Nie, H. Tian, Y. Sun, S. Xu, Y. Wang, W. Zheng, C.Q. Sun, Coordination-resolved electron spectrometrics. Chem. Rev. **115**(14), 6746–6810 (2015)

88. J. Kim, I. Becker, O. Cheshnovsky, M.A. Johnson, Photoelectron spectroscopy of the 'missing' hydrated electron clusters $(H_2O) - n$, $n = 3, 5, 8$ and 9: isomers and continuity with the dominant clusters $n = 6, 7$ and $\geqslant 11$. Chem. Phys. Lett. **297**(1–2), 90–96 (1998)

89. J.V. Coe, S.M. Williams, K.H. Bowen, Photoelectron spectra of hydrated electron clusters vs. cluster size: connecting to bulk. Int. Rev. Phys. Chem. 27(1), 27–51 (2008)

90. A. Kammrath, G. Griffin, D. Neumark, J.R.R. Verlet, Photoelectron spectroscopy of large (water)[sub n][sup −] ($n = 50$–200) clusters at 4.7 eV. J. Chem. Phys. **125**(7), 076101 (2006)

91. E.J. Hart, J. Boag, Absorption spectrum of the hydrated electron in water and in aqueous solutions. J. Am. Chem. Soc. **84**(21), 4090–4095 (1962)

92. L. Kevan, Solvated electron structure in glassy matrixes. Acc. Chem. Res. **14**(5), 138–145 (1981)

93. C.A. Silvera Batista, R.G. Larson, N.A. Kotov, Nonadditivity of nanoparticle interactions. Science, **350**(6257), 1242477 (2015)

94. J. Ceponkus, P. Uvdal, B. Nelander, On the structure of the matrix isolated water trimer. J. Chem. Phys. **134**(6), 064309 (2011)

95. V. Buch, S. Bauerecker, J.P. Devlin, U. Buck, J.K. Kazimirski, Solid water clusters in the size range of tens-thousands of H_2O: a combined computational/spectroscopic outlook. Int. Rev. Phys. Chem. **23**(3), 375–433 (2004)

96. D. Sagar, C.D. Bain, J.R. Verlet, Hydrated electrons at the water/air interface. J. Am. Chem. Soc. **132**(20), 6917–6919 (2010)

97. J. Ceponkus, P. Uvdal, B. Nelander, Water tetramer, pentamer, and hexamer in inert matrices. J. Phys. Chem. A **116**(20), 4842–4850 (2012)

98. L.-J. Zhang, J. Wang, Y. Luo, H.-P. Fang, J. Hu, A novel water layer structure inside nanobubbles at room temperature. Nucl. Sci. Tech. **25**, 060503 (2014)

99. K.R. Wilson, B.S. Rude, T. Catalano, R.D. Schaller, J.G. Tobin, D.T. Co, R.J. Saykally, X-ray spectroscopy of liquid water microjets. J. Phys. Chem. B **105**(17), 3346–3349 (2001)

100. L. Belau, K.R. Wilson, S.R. Leone, M. Ahmed, Vacuum ultraviolet (VUV) photoionization of small water clusters. J. Phys. Chem. A **111**(40), 10075–10083 (2007)

101. M. Nagasaka, H. Yuzawa, N. Kosugi, Interaction between water and Alkali metal ions and its temperature dependence revealed by oxygen K-edge X-ray absorption spectroscopy. J. Phys. Chem. B **121**(48), 10957–10964 (2017)

102. Y. Zhou, Y. Zhong, Y. Gong, X. Zhang, Z. Ma, Y. Huang, C.Q. Sun, Unprecedented thermal stability of water supersolid skin. J. Mol. Liq. **220**, 865–869 (2016)

103. Q. Hu, H. Zhao, Understanding the effects of chlorine ion on water structure from a Raman spectroscopic investigation up to 573 K. J. Mol. Struct. **1182**, 191–196 (2019)

104. M.A. Omar, *Elementary Solid State Physics: Principles and Applications* (Addison-Wesley, New York, 1993)

105. M. Nagasaka, H. Yuzawa, N. Kosugi, Development and application of in situ/operando soft X-ray transmission cells to aqueous solutions and catalytic and electrochemical reactions. J. Electron. Spectrosc. Relat. Phenom. **200**, 293–310 (2015)

106. C.Q. Sun, Supersolidity of the undercoordinated and the hydrating water (perspective). Phys. Chem. Chem. Phys. **20**, 30104–30119 (2018)

107. R. Mancinelli, A. Botti, F. Bruni, M.A. Ricci, A.K. Soper, Hydration of sodium, potassium, and chloride ions in solution and the concept of structure maker/breaker. J. Phys. Chem. B **111**, 13570–13577 (2007)

108. N. Ohtomo, K. Arakawa, Neutron diffraction study of aqueous ionic solutions. I. aqueous solutions of lithium chloride and cesium chloride. Bull. Chem. Soc. Jpn. **52**, 2755–2759 (1979)

109. G.-L. Hou, X.-B. Wang, Spectroscopic signature of proton location in proton bound HSO4-·H+·X-(X = F, Cl, Br, and I) clusters. J. Phys. Chem. Lett. **10**(21), 6714–6719 (2019)
110. G.-L. Hou, W. Lin, S.H.M. Deng, J. Zhang, W.-J. Zheng, F. Paesani, X.-B. Wang, negative ion photoelectron spectroscopy reveals thermodynamic advantage of organic acids in facilitating formation of bisulfate ion clusters: atmospheric implications. J. Phys. Chem. Lett. **4**, 779–785 (2013)
111. M. Cheng, Y. Feng, Y. Du, Q. Zhu, W. Zheng, G. Czako, J.M. Bowman, Communication: probing the entrance channels of the $X+CH_4^-\to HX+CH_3$ (X=F, Cl, Br, I) reactions via photodetachment of X^-CH_4. J. Chem. Phys. **134**, 191102 (2011)
112. H. Haberland, On the spin-orbit splitting of the rare gas-monohalide molecular ground state. Z. Phys. A Atoms Nuclei **307**, 35–39 (1982)
113. Y. Zhao, C.C. Arnold, D.M. Neumark, Study of the $I\cdot CO_2$ van der waals complex by threshold photodetachment spectroscopy of $I\cdot CO_2$. J. Chem. Soc. Faraday Trans. **89**, 1449–1456 (1993)
114. D.W. Arnold, S.E. Bradforth, E.H. Kim, D.M. Neumark, Study of halogen-carbon dioxide clusters and the fluoroformyloxyl radical by photodetachment of $X-(CO_2)$ (X=I, Cl, Br) and FCO_2^-. J. Chem. Phys. **102**, 3493–3509 (1995)
115. A.D. Becke, Density-functional thermochemistry. III. The role of exact exchange. J. Chem. Phys. **98**, 5648–5652 (1993)
116. C. Lee, W. Yang, R.G. Parr, Development of the Colle-Salvetti correlation-energy formula into a functional of the electron density. Phys. Rev. B **37**, 785–789 (1988)
117. K. Fajans, Polarizability of alkali and halide ions, especially fluoride ion. J. Phys. Chem. **74**(18), 3407–3410 (1970)
118. Chang Q. Sun, (2018) Aqueous charge injection: solvation bonding dynamics, molecular nonbond interactions, and extraordinary solute capabilities. International Reviews in Physical Chemistry 37 (3-4):363-558

Chapter 12
Perspectives

Abstract A combination of the BOLS-NEP-LBA theory and its enabled ZPS and APECS have enabled ever-deep and consistent insight into the irregularly-coordination effect on the bonding and electronic dynamics and associated performance of materials. This set of strategies derive the atomistic, local, quantitative information on the bond length and energy, binding energy density, atomic cohesive energy, energy levels of an isolated atoms and its shift upon coordination environment change, which are crutial to functional material devising. It is emphasized that the premise of "coordination bonding and electronic dynamics" could feature the performance of atoms, elelctrons, and molecules in the energetic-spatial-temporal domains effectively and the electron spectrometrics offers information for genomic engineering of matter and life.

Highlights

- BOLS-NEP-LBA notion enables ZPS to refine atomistic bonding and electronic dynamics.
- APECS, STM/S, PES, and AES complement one another to probe comprehensive information.
- Irregular coordination results in bond relaxation and electron localization, entrapment, and polarization.
- Bond length and energy, energy density, atomic cohesive energy, charge distribution are critical.

12.1 Advantages and Attainments

The following summarizes major progress advanced from the presented strategies:

(1) Bond formation and relaxation and the associated energetics, localization, densification, entrapment, and polarization mediate the structure and property of a substance.

(2) Perturbation to the Hamiltonian by undercoordination-induced bond contraction, by hetero-coordination-induced bond nature alteration, and the associate subjective polarization of the nonbonding electrons shift intrinsically all energy levels of a substance cooperatively in the same direction but at different extents. The highest energy level, or the outermost electronic orbit, shifts most.

(3) BOLS-NEP notion formulates adequately the bond-energy-electron attribute of the irregularly-coordinated systems based on the Tight-Binding convention. One needs only care the $V_{cryst}(r)$ perturbation at equilibrium without bothering the Bloch wavefunction or the particular shape of the $V_{cryst}(r)$ in examination.

(4) The energy level of an isolated atom, $E_\nu(0)$, is the reference for the CLS shift. Local densification and quantum entrapment shifts globally positively energy levels while polarization of the nonbonding electrons by the densely entrapped bonding electrons screens and splits the crystal potential and hence offsets the entrapped states negatively.

(5) The conventional "initial-final states" relaxation and "surface charging" exist throughout the course of measurement, which could be minimized in numerical calibration particularly in the ZPS processing.

(6) The BOLS-NEP enhanced capabilities of APECS, VLEED, ZPS, STS/M, UPS, APECS and XAS for quantitative information on the local bonding and electronic dynamics.

Most strikingly, being extremely sensitive to a tiny change of atomic CN or chemical condition, ZPS resolves directly the desired information without needing any approximation or assumption, which is beyond the scope of available approaches. ZPS discriminates the DOS gain from its loss due to interface and skin conditioning with high sensitivity. The ZPS is particular of use in purifying energy states in the following situations:

(1) Surface reconstruction exhibits slight CN difference between different patterns such as the Rh(110)-(1 × 2) and the (1 × 1) + (1 × 2) skins. The former corresponds to situation of every other row missing and the latter every other pairing-raw missing;

(2) Surface contamination, chemisorption, and catalytic reaction (O on Re skins in, O, N, H contamination and cluster size variation of ZnO) enhances the coordination effect on surface charge distribution;

(3) Surface roughening (such as $SrTiO_3$ skin, HOPG plasma etching and SiO_2 mechanical etching) has an important effect on polarization of surface charge;

(4) N and O chemisorption results in four distingct valence states of bonding (~6 eV), nonbonding lone pairs (~2 eV), ionic holes (~1–3 eV), and antibonding dipoles (>E_f), which reduces the work function and creates band gap of conductors;

(5) Under- and hetero-coordination effect jointly on the electronic structure and the catalytic performance of TiO_2 and the band gap and core level shift of ZnO;

(6) Electron and phonon spectroscopic confirmed the supersolid states of the confined and hydrating water. Polarization by molecular undercoordination and

charge injection of salt salvation effect the same on the O:H nonbond elongation, H–O contraction and nonbonding electron polarization.

(7) ZPS resolves the spin-degenerated sublevel relative shift and correction of the charging effect (Sects. 6.3 and 6.8);

(8) ZPS distills energy states for the monolayer skin, point defects, adatoms, and terrance and nanoribbon edges.

The ZPS requires no critical background correction but spectral peak area normalization. Inaccurate background correction results in only slight quantitative deviation without rendering the nature and trend behind observations. Spectral area normalization compensates for the anisotropy of photoemission caused by photoelectron diffraction as the anisotropy affects only the peak intensity instead of the peak energy.

APECS and XAS resolve simultaneously the energy shifts of two energy levels. One is the upper valence band and the other is a core band. The energy shift of the Auger parameter equals twice that of the upper band instead of the average shift of both the valence and the core band, as conventionally thought. An extended Wagner plot reveals information of the screening effect on the two levels and the chemical effect on the respective shift.

On the base of electron spectrometrics, we have learnt the following:

(1) The skin of a solid consisting of at most three atomic layers forms such a phase that the interatomic bonds are shorter and stronger and the charge and energy are denser while the atomic cohesive energy is generally lower compared to the bulk interior. The energy density and BE shift determine the performance of the under-coordinated and the hetero-coordinated systems.

(2) Defects, adatoms, terrace edges, grain boundaries, and nanostructures share the common origin of atomic undercoordination. One atomic neighbor shortage makes a great difference in performance for the even undercoordinated atoms—polarization dominance at sites with fewer nearest neighbors than that in the flat skin, which is of use in modulating band gap, work function, electroaffinity, and the emissibility of electrons and photons.

(3) Pt adatom and Cu/Pd alloy serve as acceptor-type catalysts because of the dominance of entrapment while Rh adatoms and Ag/Pd alloy as donor-type catalysts due to the dominance of polarization. Under coordinated Co, Ni, and Re atoms exhibit entrapment dominance while W, Au, Ag, and Cu manifest polarization dominance. Dominance of entrapment or polarization may provide guidelines for catalysts design and diagnosis.

(4) Graphitic Dirac-Fermi polaritons result from the isolation and polarization of the dangling σ bond electrons by the densely locally entrapped bonding electrons surrounding the defects and at the zigzag edges of nanoribbon, which may extend to single-atom catalysis, thermal electronics, and superconductors with the dominance of strong local polarization.

(5) The high interface energy density and charge polarization entitles Be/W to protect fusion radiation, which may help in searching and fabricating such devices for nuclear industry.

(6) Defect formation improves the photocatalytic ability of TiO_2 by band gap and workfunction reduction and carrier life extension through the undercoordination-induced quantum entrapment and polarization—pining dipoles. Transition from entrapment to polarization dominance occurs at a critical size or surface curvasture.

(7) Water is the simplest among the molecular crystals as it contains equal number of dangling protons and lone pairs. Lone pair and dangling proton are the primary functional and structural units of molecular crystals. Molecular undercoordoina-tion and electrostatic polarization shorten the H–O bond and lengthen the O:H nonbond associated with strong polarization, which leads to a supersolid phase that is elastoviscus, hydrophobic, less dense, mechanically and thermally more stable. The supersolidity disperses the quansisolid phase boundaries to raise the melting temperature and lowers the freezing point and evaporation of water ice.

(8) Spin-resolved polarization by atomic undercoordination and sp-orbital hybridization could reconcile the skin and monolayer high-T_C superconductivity and topological insulator edge superconductivity. The strength of spin-spin coupling determines the T_C and the conduction bandgap width; the localized polarization determines the edge or skin paths of carrier travelling.

Findings not only demonstrate the power of the BOLS-NEP-ZPS and the APECS strategies in complementing STM/S and PES but also the essentiality of the perspective of bonding and electronic dynamics associated with irregular atomic and molecular coordination and orbital hybridization. Controlling the process of bond and nonbond formation, dissociation, vibration, and relaxation and the associated dynamics and energetics of charge repopulation, polarization, densification and localization forms the effective means engineering bonds and electrons towards devising functional materials.

12.2 Limitations and Precautions

In using this set of BOLS-NEP notion and the APECS and ZPS strategies, the following precautions are necessary to ensure the accuracy and reliability in determining the local bond relaxation and electronic BE shift:

(1) The CLS is proportional to the sum of the exchange and the overlap integrals in the TB approach. The contribution of the overlap integral is 3% or less that of the exchange integral, particularly for the deeper core levels. Involving such overlap integral may improve the numerical accuracy of determination but it is more complicated.

(2) Unlike the core band of a particular element, the valence band is more complicated and it includes the convoluted features of charge transport between constituent elements in alloy or compound. The resultant valence DOS shows states of bonding, nonbonding, holes, and antibonding dipoles.

(3) BOLS-TB derivatives from the skin PES analysis are more reliable than that from analyzing the size PES of nanocrystals as the uniformity of crystal size and shape can hardly be certain. The reliability and accuracy also depends on the volume of skin PES database. The larger the database is, the more reliable and higher accuracy of the analysis will be. For example, PES profiles collected from the layered and oriented (hkl) skins determine the energy levels of an isolated atom and its z-dependent shift with a standard deviation that is inversely proportional to the square root of the number of date sets.

(4) In analyzing the APECS profiles, the relative shift of the spin-orbit degenerated levels is negligible particularly for the deeper energy levels.

(5) ZPS applies to monitor coordination and chemical effect on the performance of bonds and electrons in conductors and semiconductors but insulators. One may appeal to phonon and photon spectrometrics for detecting the bonding energetic behavior of insulators and liquids particularly under mechanical and thermal stimuli without needing high-vacuum environment.

(6) The surface registry and sublayer dependent atomic CN conserves for specific crystal geometry regardless chemical composition, which enabled the standard of atomic CN calibration.

12.3 Prospects and Perspectives

The high sensitivity of the ZPS to a tiny change of the chemical and coordination environment makes it of particular use in monitoring *in situ* trace element adsorption both statically and dynamically. The ZPS process keeps the meaningfully intrinsic information by removing the general background. For instance, in purifying the adatom or defect states, the ZPS keeps features due to the least atomic CN as new peaks and the features due to the highest atomic CN in the bulk as a valley. The ZPS removes the energy states belong to those of intermediate atomic CNs. The ZPS should be able to resolve a bimetallic system with surface enrichment by one component, change of surface composition of alloys caused by adsorption or catalytic reaction, etc. The energy and intensity of the peak may change with the richness of the excessive skin element.

Besides the chemical and coordination modulation, the ZPS should be sensitive to any change of electric, magnetic, mechanical or thermal fields applied to the substance examined. With the establishment of the BOLS-NEP-LBA notion and ZPS strategy, one is able to gain quantitative information of the local bond length and energy, charge distribution in various bands, BE density, and atomic cohesive energy, which form the key to mediating the macroscopic properties of a substance at the atomic scale in a way of bond-by-bond, and electron-by-electron engineering.

This atomistic, CN-resolved electron spectrometrics, or the BOLS-NEP-ZPS strategy, may extend to spectrometrics in more general such as phonons and photons to resolve the multiple-field effect on the collective performance of bond, electron, phonon, and photon at a specific atomic site. One can imagine what will happen if we turn our current focusing point from the bond energy at equilibrium to the phonon frequency or to the photoemission wavelength. Bond energy is the zeroth and the phonon vibration energy is the second derivative of the interatomic potential in Taylor series. Energies of photon emission and photon absorption depend on the band gap and electron-phonon coupling. The band gap is proportional to the bond energy at equilibrium. Therefore, atomistic spectrometrics for bond, electron, phonon, and photon is useful. This extension could be extremely appealing, fascinating, promising, and rewarding, which will advance surely profoundly the under-coordination Physics and the hetero-coordination Chemistry.

Part II
Electron Diffraction:
Bond-Band-Barrier Transition

VLEED resolves the bond geometry, Brillouin zones, band structure, inner-potential constant, surface potential, work function, and four-stage Cu_3O_2 bonding and four-state oxidation valence banding dynamics at the outermost two atomic layers.

Chapter 13
Introduction

Abstract Extending the very low-energy electron diffraction (VLEED) crystallography to the bond-band-barrier forming dynamics has been proven to offer comprehensive information on bond geometry from the outermost two atomic layers and on electron performance from the valence band and above, and on surface potential barrier evolution upon chemisorption, which are consistent with STM and PES observations.

Highlights

- The impact of extending LEED crystallography to bond-band-barrier forming dynamics is enormous.
- VLEED monitors the behavior of atoms, valence electrons, and chemisorption bonding dynamics.
- VLEED probes bond geometry, energy states, potential function in the outermost two atomic layers.
- VLEED resolves work function, muffin-tin inner potential, and Brillouin zone boundaries.

13.1 LEED, VLEED, STM/S, and PES

LEED crystallography is an important technique for studying the geometric arrangement of atoms in the sublayers of crystalline surfaces [1, 2]. In normal LEED, the scattering of electron beams with energies greater than 30 eV is dominated by interaction between the incident electron beams and the ion cores of a number of stacking layers of atoms [3, 4]. The LEED pattern portrays the two-dimensional structure of the surface lattice. By studying how the diffraction intensities change with increasing the energy of the incident electron beam at different azimuth angles, it is possible to infer details about the vertical location of the atoms and to measure the distance of the surface interlayer spacing, as does the X-ray diffraction for bulk crystals. The analysis of LEED patterns and LEED I-E profiles is very important, and LEED is so established as a prestigious tool in surface science for analyzing the structure of

C. Q. Sun, *Electron and Phonon Spectrometrics*,
https://doi.org/10.1007/978-981-15-3176-7_13

chemisorbed overlayers and the reconstructed clean surfaces. LEED pattern analysis forms the basis for structure models of a chemisorption process and it is vital to an understanding of catalysis and corrosion.

However, pattern analysis only tells us the relative size, the symmetry and the orientation of the unit cell in the overlayer; it tells us little in quantity about the spacing between the overlayer and the substrate. Spectra analysis can find the adsorption site but the accuracy of bond lengths and surface layer spacings is often subject to accuracy [5].

In contrast, LEED at very low energies (VLEED) below plasma excitation or ionization (~15 eV below E_F) holds rich and profound information not only about the crystal geometry of the outermost atomic layer of a crystallite but also the behavior of surface electrons described using the real and the imaginary part of the surface potential barrier (SPB). The conjunction of the SPB with the multiple scattering dynamics gives rise to the interesting phenomena of surface states, surface resonance or VLEED fine structures [2, 6].

Interaction between the incident electron beams and the surface electrons defines the fine-structure features in the VLEED I-E spectra. The so-called surface electrons tie closely to the positions and valence states of the atoms at the surface. VLEED spectrum integrates the following information [7, 8]:

- diffraction from the ion cores of a very small number of surface atomic layers and the number of diffraction layer depends on the incident beam energy,
- scattering and interference by the elastic SPB; and,
- attenuation by the imaginary SPB, or inelastic damping, because of energy exchange with surface electrons and excitation of phonons.

The mechanism for VLEED is more complicated than that for the normal LEED but it is much more comprehensively revealing. Because of the difficulties arising from the effects of stray electronic and magnetic fields on slowly moving electrons, VLEED technique needs to screen the stray field by installing Halmholtz coils [53]. Theory calculations usually do not extend to these low energies because of the approximation limitation in the calculation package. The use of an optical potential independent of electron energy in convention is valid at higher energies but it becomes unreliable at energies below that for the plasmon-excitation. Furthermore, the spectral fine structures are sensitive to the shape of the SPB. In conventional calculations the SPB scattering is treated in a very simple uniform fashion, which is insufficient for the chemisorbed systems.

It seemed conventionally impossible to use LEED to simultaneously determine the crystal geometry and the SPB as one was unable to identify the spectral feature arising from the change of either atomic positions or from the shape of the SPB. In circumstances where the shape of the SPB is unknown, the energy dependence of the muffin-tin inner potential constant is unknown, and the spatial decay and energy dependence of the inelastic potential is unknown, it would be hardly possible to derive comprehensive information on the structural and energetic configurations from the VLEED fine-structure features alone. Appropriate modeling including all the contributions and their nature links in decoding the VLEED data is necessary.

However, VLEED possesses more advantages than LEED in obtaining information about the SPB and the energy band structure as demonstrated for several mostly close-packed surfaces of transition metals [9–16] and ZnO(0001) surface [17]. The scattering from light adsorbates is often stronger, even comparable to the host substrate at very low energies, and this can lead to knowledge of the adsorbate positions [18] as well as the valence states of individual surface atoms with the assistance of an appropriate structural model [7, 8, 19].

In fact, bond formation modifies the electronic structures of both the negative ions of the light adsorbate and the host valence states [20]. Moreover, in VLEED calculations, fewer beams and fewer phase shifts are considered and spectral peaks occur with a greater density in the energy range. Hence, computation time is considerably reduced. Furnished with a proper calculation code and suitable structural models, the high-resolution VLEED I-E spectra can be analyzed for comprehensive information about the behavior of atoms and electrons at a surface simultaneously.

There are two kinds of characteristic features on a measured VLEED I-E spectrum. One is the Rydberg series that relate to the interference or resonance effect due to the SPB [21]. The Rydberg features come from the interference between the measured beam and the pre-emergent beams that reflect repeatedly between the substrate lattice and the SPB [1]. The Rydberg series are used to be known as "threshold effects", or "SPB resonance or interference" [22], since they converge from below the emergence thresholds of diffracted beams.

The other kind of sharp features comes from the band gap Bragg diffraction at the Brillouin zone boundaries [23]. The positions of the narrow, sharp, solitary and violent peaks rely both on the crystal geometry of the surface and on the incident and azimuth angles of the electron beams [24, 25]. The sharp peaks converge into the thresholds of emerging new beams inside the crystal, as found by many researchers from the O-Ru(001) [26], Cu(111) [27, 28], and Ni(111) and the ZnO(0001) [17] surfaces. These sharp features result from the energy-band structure. The troughs relate to the intensity variation caused by the wave-like energy dependence of the additional sextets in VLEED patterns [29].

The relationship between the band structure and the elastic reflection coefficient in VLEED can be obtained through the matching method [30, 31]. The matching between the vacuum wave function with the superposition of Bloch waves excited in the solid determines the elastic reflection coefficient. The energy location of a trough in the spectrum corresponds to the rapid change of reflection, coinciding with the location of the band-structure critical positions at the boundaries of Brillouin zones [32, 33]. Therefore, it is possible to determine the connection between the VLEED profiles and the band structure by directly measuring the energy positions of the critical points.

Many researchers have devoted their efforts to using VLEED to characterize the energy states above vacuum level of the surfaces—unoccupied states. Strocov et al. [10–13] demonstrated that VLEED measurements are ideally suited for accurate determination of the desired upper states. Knowledge about the excited states could be substantially improved than could the band mapping by photoelectron spectroscopy. From the band structure point of view, VLEED is very sensitive to modifications of the upper band formed by the variations of geometric structure, which encourages one to use VLEED as a tool to investigate which way these two entities, i.e., atomic geometry and band structures, are interdependent. Furthermore, Bartos et al. [27, 34] found that corresponding theoretical I-E curves could be obtained in good agreement with experimental data from the densely packed Cu(111) surface when the anisotropy of the electron attenuation (inelastic potential) was taken into account. Therefore, VLEED gives information about the anisotropy of the SPB.

Unfortunately, little progress had been made until 1995 [35] in the structure determination of surfaces with large unit meshes or surfaces with adsorbed molecules using VLEED. No attempts had even been given using VLEED to simultaneously determine the bond geometry, energy states, the three-dimensional (3D) SPB and the bond forming dynamics of systems with chemisorbed oxygen for that time being [7, 8]. Situation was changed when a multi-atom and multiple diffraction code was developed to deal with the dynamics of oxygen chemisorption on Cu(001) surface [35].

In the multiple scattering calculation code [35], a complex unit cell containing five atoms treats the SPB as an additional layer of interference. However, the parameter space used to describe the elastic ReV(z) and the inelastic ImV(z, E) part of the SPB was too large. At least nine variables in a one-dimensional system were required to describe the spatial variation and energy dependence of the SPB [37]. If the strongly correlated SPB parameters were treated independently the one-dimensional approximation did not work for the chemisorbed surface besides the information miss on the localized nature of the valence states and the anisotropy of the SPB.

Such an approximation worked fairly well for the clean Cu(001) surface but it faced severe difficulties in decoding the VLEED data from the chemisorbed O-Cu(001) surface. Spectra near the <11> azimuth of the O-added Cu(001) surface could not be fitted despite various structure models and SPB parameters being used [35]. Three-dimensional SPB attributes are very important for the adsorbate-involved system, which is local on atomic scale in many properties [37].

As a matter of fact, correct SPB parameters must depend on energy E, lateral momentum $K_{//}$, and surface atomic coordinates (x, y). However, the primary effect of the 3D terms in the SPB is to generate off-diagonal matrix elements in the SPB scattering matrix, which leads to too much longer computer running. One way to include the 3D effects in a simpler manner is to use a hybridized one-dimensional model that incorporates some of the more important 3D features. As the multiple scattering is often less important in the VLEED energies, the hybridization of one-dimensional model with 3D contribution is acceptable.

When the high-energy photons, X-ray or hard ultraviolet ionizing radiation, slam into the sample in vacuum chamber, they evict some of its electrons, launching them out of the material. Detectors then count these homeless electrons and measure their energies and directions of travelling. UPS, which examines electrons ejected from valence shells, is more suited to establishing the bonding characteristics and the details of valence shell electronic structures of substances on the surface. Its usefulness is its ability to reveal which orbitals of the adsorbate are involved in the bond to the substrate.

Consequently, if photoelectrons ejected from filled levels will have kinetic energies (KEs) that fall within this structured region, the observed intensity will be a convolution of the filled and empty DOS together with the matrix of transition probabilities. This is the case in UPS and gives rise to the strong dependence of valence spectra on photon energy. In the case of XPS, the KE of the valence photoelectrons is such that the final states are quite devoid of structure; thus, the observed DOS closely reflects the initially filled density of states. The complete spectrum at high resolution shows the band structure and the sharp cut-off in electron density at E_F.

The invention of STM/S has led to enormous progress in revealing direct and qualitative information on an atomic scale about the surface but the images challenge proper physical interpretation [36]. VLEED, PES and STM/S can be used as complementary tools to get such information that integrates both real-space and k-space, on an atomic scale, and over macroscopic areas of the surface. A combination of VLEED with STM/S reduces certainly the limitations inherent in each of these techniques used separately. STM/S observation provides one with a direct vision of the surface topography; decoding VLEED spectra rewards the quantitative details, while a logic thinking of models in real and energy domain will link the morphology and spectroscopy. STM imaging aids establishment the models of chemical bond formation and electronic states evolution, which is able to discriminate individual atomic states and the non-uniformity of the SPB. Derivations of these models from STM images can be used as input to be justified in simulating VLEED spectra.

The VLEED simulation results, in turn, improve the understanding of the STM/S and PES observations. Thus one can extract comprehensive information from such a combination about the behavior of atoms and electrons at surfaces, and hence, we can obtain the common senses about the nature and dynamics of the chemisorption bonding and band forming dynamics, which can generalize other chemisorbed surfaces [19, 20]. Under the basic principles, one may be able to specify the individual atomic states and thereby to derive the reaction formula for surface reaction from the profiles of VLEED, STM/S and PES. If one could specify first the STM signatures coming from whether ionic, polarized, or vacancy of missed atoms, then elucidation of identities due to chemisorption becomes much easier [37].

13.2   O–Cu(001) Surface Reaction

As one of the prototypes of oxygen chemisorption, O-Cu(001) surface has been intensively investigated both experimentally and theoretically. Since 1956, when Young et al. [38] found that the Cu(001) surface is easier to be oxidized than other planes of copper single crystal, numerous scientists have been devoted their efforts to the understanding of oxygen chemisorption. There have been many conflicting opinions regarding the oxygen-induced reconstruction of Cu(001) surface. Different atomic superstructures have been derived with various experimental techniques [39] and theoretical approaches [40, 41]. However, the observed structures vary from researcher to researcher even though they used the same approach.

Most researchers agreed on the missing-row (MR) type Cu(001)-(2$\sqrt{2}$ × $\sqrt{2}$)R45°-O reconstruction first proposed by Zeng and Mitchell in 1990 [42]. A combination of STM observations [43, 44] and VLEED computations [7, 8, 19] confirmed that an off-centered pyramid Cu(001)-c(2 × 2)-O^{-1} forms first, at oxygen exposure lower than 25 L, and then the MR (2$\sqrt{2}$ × $\sqrt{2}$)R45°-2O^{-2} superstructure follows.

Figure 13.1 shows the STM images [44, 45] of the two reconstructed phases with oxygen adsorption, in terms of the MR rigid-sphere model description [46]. Based on the effective-medium theory approach, Jacobsen [47] suggested that the missing-row type reconstruction is most stable for both the O-Cu(001) and the O-Cu(110) surfaces. On the O-Cu(001) surface, *"the O atoms go underneath the first layer that is then shifted out by 0.3–0.5 Å. At the same time there is a pairing of the Cu atoms over the missing row"* with unclear mechanism.

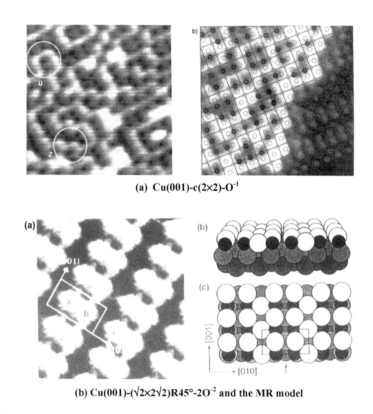

(a) Cu(001)-c(2×2)-O^{-1}

(b) Cu(001)-($\sqrt{2}$×2$\sqrt{2}$)R45°-2O^{-2} and the MR model

Fig. 13.1 STM images of **a** the nanometric c(2 × 2)-O domains with zigzag- and U-shaped protruding boundaries [from Ref. [43, 50]], which was obtained under 25 L oxygen exposure. Guidelines of ($\sqrt{2}$ × 2$\sqrt{2}$)R45° mesh indicate the ideal positions of original Cu substrate atoms. The open and filled circles represent the Cu and O adsorbate. **b** The fully-developed Cu(001)-($\sqrt{2}$ × 2$\sqrt{2}$)R45°-2O structure, which was obtained by exposing the sample to 1000 L oxygen at 300 °C and followed by an anneal at 300 °C for 5 min [from Ref. [45]] **c** shows the side and top view of the missing-row type rigid-sphere reconstruction model [From Ref. [19]]

In 1993, Lederer et al. [44], Arvanitis et al. [48], Yokoyama et al. [49], asserted that there are two phases during the O-Cu(001) reconstruction. One corresponds to an unreconstructed state and the other a reconstructed phase. In the former case, oxygen atom locates 0.8 Å above the first Cu layer to form an off-centered pyramid, or a c(2×2)-O structure with oxygen shifting 0.1 Å away from the symmetry site center. Upon reconstruction, the adsorbate locates 0.2 Å above the first Cu layer with a MR ($\sqrt{2} \times 2\sqrt{2}$)R45°-O structure. In the two sequential phases, oxygen adsorbate retains at the apical site of an off-centered pyramid and then at the apical site of a distorted tetrahedron. The nature of the O-Cu bond was suggested as "*predominantly ionic for the first phase and more covalent for the reconstructed second phase.*"

It became certain in 1996 when Fujita et al. [43] confirmed the bi-phase structures with STM observations. In the first phase, oxygen sits in the near-center of the next-nearest-neighboring hollow site to form a pyramid at an oxygen exposure of 25 L or lower. The precursor phase is composed of nanometric c(2×2)-O domains with protruding boundaries. Upon increasing oxygen exposures, the c(2×2)-O evolves into the ($2\sqrt{2} \times \sqrt{2}$)R45°-O structure by every fourth row of Cu atoms missing. In 1998, Tanaka et al. [50] demonstrated that the two phases are reversible under the bombardment of energetic Ar$^+$ beams.

In 1997, Sun [19], Sun and Bai [51] clarified that the MR formation results from the process of oxide tetrahedron bond formation that breaks all bonds of the MR atoms being evaporated without needing other perturbation. The two sequential phases result from the transition from O^{-1} to O^{-2} under the specific Cu(001) geometric condition [52].

Figure 13.1a shows the zigzag and the U-shape protruding domain boundaries of the first O^{-1} derived phase. Comparatively, the O^{-2} derived "dumb-bell" shaped protrusions in Fig. 13.1b bridge over the missing rows. Jensen et al. [45] interpreted these protrusions as: "*Pairing of Cu-O-Cu chains forms by displacing the Cu and/or O atoms next to the missing row by about 0.35 Å towards the missing row; the O-Cu-O chain is linked by the delocalized antibonding states*". However, Sun affirmed that the nonbonding electron lone pair ":" connects the O^{-2}:Cu^p:O^{-2} and the pairing Cu^p ↔ Cu^p dipoles (p) crosses the missing row. The "↔" represents the repulsive force between the dipoles [19].

From the STM image of Fig. 13.1b, the separation of the paired rows was estimated at 2.9 ± 0.3 Å. The length of the bright spot was estimated at 5.1 Å; the height of the bright spot was 0.45 Å; whereas, for the pure Cu(001) surface the protrusion was about 0.3 Å in height [38]. These STM observations founded the H$_2$O-like tetrahedral models to specify the bond nature and its consequence on the individual atomic states.

Having overcome the experimental difficulties, Hitchen, Thurgate, and Jennings [53] collected high-resolution VLEED spectra from O-Cu(001) surfaces in energy range from 6.0 to 16.0 eV, far lower than the energy of plasmon excitation. The VLEED O-Cu(001) spectra displayed a dramatic intensity change that occurs at oxygen exposure of 200–300 L. This dramatic change was related to the transition from a c(2×2) to a ($\sqrt{2} \times 2\sqrt{2}$)R45° phase. Unfortunately, with either or the combination of the two superstructures acceptable fit of the VLEED data from the

O-Cu(001) surface could not be realized by using the uniform one-dimensional SPB model.

13.3 Objectives

The main aim of this part is to show that an extension of the H_2O molecule to oxide formation, a model of the 3D SPB model, and the VLEED decoding strategies has enabled the combination of VLEED, STM, and PES to provide comprehensive information about the following:

- Quantification of the bond geometry, bond length, and the $CuO–Cu_3O_2$ transition four-stage bonding dynamics
- Specification of surface atomic valance states and the four signatures of valence DOS
- Description of the anisotropy and nonuniformity of the SPB
- Determination of the reduced work function and inner potential constant due to oxidation
- Derivation of the deformed 2D Brillouin zones and the effective electron masses at zone boundaries
- Clarification of factors controlling bond formation and forces driving the reconstruction
- Confirmation of the tetrahedron bond formation and atomic undercoordination derived surface bond contraction
- Specification of the STM/S, PES and VLEED spectral signatures

13.4 Scope

VLEED is the unique technique that collects information from the skin region about the networking bond geometry of the outermost atomic layer and the electronic energy distribution in real and momentum domains. Interplaying with STM and PES, VLEED at $E \leq 16.0$ eV can reveal the bond breaking and making dynamics and the associated variation of the valence DOS and the surface potential barrier (SPB) for chemisorbed surfaces. An examination of the VLEED sensitivity to calculation parameters indicates that the bond geometry and the elastic potential define the VLEED fine-structure features while the inelastic damping dominated by electrons located in the second atomic layer and above determines the spectral intensity.

Decoding the azimuth angular-resolved VLEED profiles yielded statistic information about the deformation of the Brillouin zones, bond geometry, and the reduction of both the muffin-tin inner potential constant and the work function of oxygen chemisorbed surface. It is essential to take the SPB anisotropy and the nonuniformity of valence charge distribution in real space into consideration. Excitingly, the reaction upon oxygen exposure increase demonstrates four-step transition from the CuO_2

pairing off-centered-pyramid into the Cu_3O_2 pairing-tetrahedron on the Cu(001) surface, and the corresponding change of the valence DOS. The annealing at ~550 K de-hybridizes the oxygen of the chemisorbed system and hence demotes oxidation.

It is appropriate to describe a process of oxidation in terms of tetrahedron bond making rather with evolution of energy states in place of dislocation of an individual atom in a certain direction at a time, which represents essentially the true situation and correlates observations in terms of crystallography, microscopy, and electron spectroscopy. Besides, the sp^3-orbital hybridization with creation of nonbonding lone pairs and metal dipoles as well as the bond contraction are essential events for chemisorption. Therefore, combining with STM and PES, VLEED provides such a unique means that it collects comprehensive, nondestructive, and quantitative information from outside the second atomic layer of a surface and covers the valence band energies and above.

This part is arranged as follows. Chapter 1 introduces briefly the VLEED spectrometrics and a survey on the knowledge acquired insofar about O-Cu(001) surface reaction. Chapter 2 describes the VLEED calculation code for multi-atom and multiple beam diffraction developed by Thurgate [35] from the LEED package of Van Hove and Tong's [4], and a bond-band-barrier (3B) model of Sun [19, 20] addressing the oxide tetrahedron bonding, the valence DOS evolution, and the change of the SPB in general, for the specific O-Cu(001) surface in particular [19]. Chapter 3 justifies numerically the models, the decoding strategies and the reliability of VLEED technique. Chapter 4 demonstrates outcomes of decoding a series of dynamic VLEED I–E spectra from the O-Cu(001) surface. Quantitative information is gained about the change of bond geometry, Brillouin zones, valence DOS, SPB, work function and inner potential due to the reaction [54]. Evolutions of the bond geometry and the valence DOS, particularly the DOS features of nonbonding states demonstrate consistently four-stage dynamics of Cu_3O_2 forming on Cu(001) surface upon exposure increase. The last Chap. 5 presents a perspective of further extension of the knowledge gained in this part.

References

1. M.A. VanHove, W.H. Weinberg, C.-M. Chan, *Low-Energy Electron Diffraction: Experiment, Theory and Surface Structure Determination*, vol. 6 (Springer Science & Business Media, 2012)
2. R.O. Jones, P.J. Jennings, LEED fine structure: origins and applications. Surf. Sci. Rep. **9**(4), 165–196 (1988)
3. J. Pendry, G.P. Alldredge, Low energy electron diffraction: the theory and its application to determination of surface structure. Phys. Today **30**, 57 (1977)
4. M.A. Van Hove, S.Y. Tong, *Surface Crystallography by LEED: Theory, Computation and Structural Results*, vol. 2 (Springer Science & Business Media, 2012)
5. M. Van Hove, G. Somorjai, Adsorption and adsorbate-induced restructuring: a LEED perspective. Surf. Sci. **299**, 487–501 (1994)
6. E. McRae, Electron diffraction at crystal surfaces: I. generalization of Darwin's dynamical theory. Surf. Sci. **11**(3), 479–491 (1968)

7. C.Q. Sun, O-Cu(001): II. VLEED quantification of the four-stage Cu_3O_2 bonding kinetics. Surf. Rev. Lett. **8**(6), 703–734 (2001)

8. C.Q. Sun, O-Cu(001): I. Binding the signatures of LEED, STM and PES in a bond-forming way. Surf. Rev. Lett. **8**(3–4), 367–402 (2001)

9. E. McRae, C. Caldwell, Absorptive potential in nickel from very low energy electron reflection at Ni (001) surface. Surf. Sci. **57**(2), 766–770 (1976)

10. V. Strocov, H. Starnberg, P. Nilsson, H. Brauer, L. Holleboom, New method for absolute band structure determination by combining photoemission with very-low-energy electron diffraction: application to layered VSe_2. Phys. Rev. Lett. **79**(3), 467 (1997)

11. V. Strocov, P. Blaha, H. Starnberg, M. Rohlfing, R. Claessen, J.-M. Debever, J.-M. Themlin, Three-dimensional unoccupied band structure of graphite: very-low-energy electron diffraction and band calculations. Phys. Rev. B **61**(7), 4994 (2000)

12. V. Strocov, Intrinsic accuracy in 3-dimensional photoemission band mapping. J. Electron Spectrosc. Relat. Phenom. **130**(1), 65–78 (2003)

13. V.N. Strocov, R. Claessen, G. Nicolay, S. Hüfner, A. Kimura, A. Harasawa, S. Shin, A. Kakizaki, P. Nilsson, H. Starnberg, Absolute band mapping by combined angle-dependent very-low-energy electron diffraction and photoemission: application to Cu. Phys. Rev. Lett. **81**(22), 4943 (1998)

14. R. Jaklevic, L. Davis, Band signatures in the low-energy-electron reflectance spectra of fcc metals. Phys. Rev. B **26**(10), 5391 (1982)

15. L.R. Bedell, H. Farnsworth, A study of the (00) LEED beam intensity at normal incidence from CdS (0001), Cu (001), Cu (111), and Ni (111). Surf. Sci. **41**(1), 165–194 (1974)

16. R. Feder, P. Jennings, R. Jones, Spin-polarization in LEED: a comparison of theoretical predictions. Surf. Sci. **61**(2), 307–316 (1976)

17. P.J. Møller, S. Komolov, E. Lazneva, VLEED from a ZnO (0001) substructure. Surf. Sci. **307**, 1177–1181 (1994)

18. H. Pfnür, M. Lindroos, D. Menzel, Investigation of adsorbates with low energy electron diffraction at very low energies (VLEED). Surf. Sci. **248**(1–2), 1–10 (1991)

19. C.Q. Sun, Oxidation electronics: bond-band-barrier correlation and its applications. Prog. Mater Sci. **48**(6), 521–685 (2003)

20. C.Q. Sun, *Relaxation of the chemical bond*. Spr. Ser. Chem. Phys. **108**, 807 (Springer-Verlag, Heidelberg, 2014)

21. E. McRae, Electronic surface resonances of crystals. Rev. Mod. Phys. **51**(3), 541 (1979)

22. S. Papadia, M. Persson, L.-A. Salmi, Image-potential-induced resonances at free-electron-like metal surfaces. Phys. Rev. B **41**(14), 10237 (1990)

23. C.Q. Sun, Angular-resolved VLEED from O-Cu(001): valence bands, chemical bonds, potential barrier, and energy states. Int. J. Mod. Phys. B **11**(25), 3073–3091 (1997)

24. G. Hitchen, S. Thurgate, Azimuthal angular dependence of LEED fine structure from Cu (001). Surf. Sci. **197**(1–2), 24–34 (1988)

25. G. Hitchen, S. Thurgate, Determination of azimuth angle, incidence angle, and contact-potential difference for low-energy electron-diffraction fine-structure measurements. Phys. Rev. B **38**(13), 8668 (1988)

26. M. Lindroos, H. Pfnür, D. Menzel, Theoretical and experimental study of the unoccupied electronic band structure of Ru (001) by electron reflection. Phys. Rev. B **33**(10), 6684 (1986)

27. I. Bartoš, M. Van Hove, M. Altman, Cu (111) electron band structure and channeling by VLEED. Surf. Sci. **352**, 660–664 (1996)

28. W. Jacob, V. Dose, A. Goldmann, Atomic adsorption of oxygen on Cu (111) and Cu (110). Appl. Phys. A **41**(2), 145–150 (1986)

29. J.-M. Baribeau, J.-D. Carette, P. Jennings, R. Jones, Low-energy-electron-diffraction fine structure in W (001) for energies from 0 to 35 eV. Phys. Rev. B **32**(10), 6131 (1985)

30. E. Tamura, R. Feder, J. Krewer, R. Kirby, E. Kisker, E.L. Garwin, F. King, Energy-dependence of inner potential in Fe from low-energy electron absorption (target current). Solid State Commun. **55**(6), 543–547 (1985)

31. J. Baribeau, J. Carette, Observation and angular behavior of Rydberg surface resonances on W (110). Phys. Rev. B **23**(12), 6201 (1981)
32. H.-J. Herlt, R. Feder, G. Meister, E. Bauer, Experiment and theory of the elastic electron reflection coefficient from tungsten. Solid State Commun. **38**(10), 973–976 (1981)
33. S.A. Komolov, *Total Current Spectroscopy of Surfaces* (CRC Press, 1992)
34. I. Bartoš, P. Jaroš, A. Barbieri, M. Van Hove, W. Chung, Q. Cai, M. Altman, Cu (111) surface relaxation by VLEED. Surf. Rev. Lett. **2**(04), 477–482 (1995)
35. S.M. Thurgate, C. Sun, Very-low-energy electron-diffraction analysis of oxygen on Cu(001). Phys. Rev. B **51**(4), 2410–2417 (1995)
36. M. Van Hove, J. Cerda, P. Sautet, M.-L. Bocquet, M. Salmeron, Surface structure determination by STM vs LEED. Prog. Surf. Sci. **54**(3), 315–329 (1997)
37. C.Q. Sun, C.L. Bai, Modelling of non-uniform electrical potential barriers for metal surfaces with chemisorbed oxygen. J. Phys. -Condensed Matter **9**(27), 5823–5836 (1997)
38. F.W. Young, J.V. Cathcart, A.T. Gwathmey, The rates of oxidation of several faces of a single crystal of copper as determined with elliptically polarized light. Acta Metall. **4**(2), 145–152 (1956)
39. D. Woodruff, T. Delchar, *Modern Techniques of Surface Analysis* (Cambridge University Press, New York, 1986)
40. P.S. Bagus, F. Illas, Theoretical analysis of the bonding of oxygen to Cu (100). Phys. Rev. B **42**(17), 10852 (1990)
41. J. Nørskov, Theory of adsorption and adsorbate-induced reconstruction. Surf. Sci. **299**, 690–705 (1994)
42. H. Zeng, K. Mitchell, Further LEED investigations of missing row models for the Cu (100)-(2√2×√2) R45°-O surface structure. Surf. Sci. **239**(3), L571–L578 (1990)
43. T. Fujita, Y. Okawa, Y. Matsumoto, K.-I. Tanaka, Phase boundaries of nanometer scale c (2× 2)-O domains on the Cu (100) surface. Phys. Rev. B **54**(3), 2167 (1996)
44. T. Lederer, D. Arvanitis, G. Comelli, L. Tröger, K. Baberschke, Adsorption of oxygen on Cu (100). I. local structure and dynamics for two atomic chemisorption states. Phys. Rev. B **48**(20), 15390 (1993)
45. F. Jensen, F. Besenbacher, E. Laegsgaard, I. Stensgaard, Dynamics of oxygen-induced reconstruction on Cu(100) studied by scanning tunneling microscopy. Phys. Rev. B **42**(14), 9206–9209 (1990)
46. F. Besenbacher, J.K. Nørskov, Oxygen chemisorption on metal surfaces: general trends for Cu, Ni and Ag. Prog. Surf. Sci. **44**(1), 5–66 (1993)
47. K.W. Jacobsen, Theory of the oxygen-induced restructuring of Cu (110) and Cu (100) surfaces. Phys. Rev. Lett. **65**(14), 1788 (1990)
48. D. Arvanitis, G. Comelli, T. Lederer, H. Rabus, and K. Baberschke, Characterization of two different adsorption states for O on Cu (100). Ionic versus covalent bonding. Chem. Phys. Lett. **211**(1), 53–59 (1993)
49. T. Yokoyama, D. Arvanitis, T. Lederer, M. Tischer, L. Tröger, K. Baberschke, G. Comelli, Adsorption of oxygen on Cu (100). II. Molecular adsorption and dissociation by means of O K-edge x-ray-absorption fine structure. Phys. Rev. B **48**(20), 15405 (1993)
50. K.-I. Tanaka, T. Fujita, Y. Okawa, Oxygen induced order–disorder restructuring of a Cu (100) surface. Surf. Sci. **401**(2), L407–L412 (1998)
51. C.Q. Sun, C.L. Bai, A model of bonding between oxygen and metal surfaces. J. Phys. Chem. Solids **58**(6), 903–912 (1997)
52. C.Q. Sun, Exposure-resolved VLEED from the O-Cu(001): bonding dynamics. Vacuum **48**(6), 535–541 (1997)
53. C. Hitchen, S. Thurgate, P. Jennings, A LEED fine structure study of oxygen adsorption on Cu (001) and Cu (111). Aust. J. Phys. **43**(5), 519–534 (1990)
54. C.Q. Sun, Spectral sensitivity of the VLEED to the bonding geometry and the potential barrier of the O-Cu(001) surface. Vacuum **48**(5), 491–498 (1997)

Chapter 14
Principles: Bond-Band-Barrier Correlation

Abstract Oxide tetrahedral bond formation with orbital occupation by the shared bonding and nonbonding electron pairs determine uniquely the bond geometry, valence density of states, and the surface potential barrier. Parameterization of all involved parameters as a function of the bond angle and length and the origin of the SPB not only simplified the calculations but also importantly ensured the solution approaching true situations.

Highlights

- VLEED sums the Bragg and the SPB elastic and inelastic resonant diffractions.
- Bond geometry, valence states, and complex SPB dictate collectively the VLEED fine structures.
- Proper modelling and parameterization of the bond geometry and E-resolved 3D SPB is necessary.
- Dynamic VLEED resolves the bond-band-barrier evolution dynamics at reaction.

14.1 VLEED: Multibeam Resonant Diffraction

The earliest VLEED code with double-diffraction scheme [1] calculates the SPB by integrating from infinitely far away of the surface to a point z_c in the bulk where the potential has become equal to the atomic muffin-tin inner potential constant. Good agreement in simulating the VLEED spectra collected from pure Cu(001) surfaces had been realized in the mono-atom scheme [2]. However, the double-diffraction and single-atom wise could not work properly for the chemisorbed surfaces so the multi-order diffraction on the reflectance intensities is necessary [3], which made the calculation for the O–Cu(001) surface possible. The following highlight some essential details of the multi-atom and multiple diffraction scheme.

© The Editor(s) (if applicable) and The Author(s), under exclusive license to Springer Nature Singapore Pte Ltd. 2020
C. Q. Sun, *Electron and Phonon Spectrometrics*,
https://doi.org/10.1007/978-981-15-3176-7_14

14.1.1   Multi-Beam Diffraction

To model the substrate scattering it is assumed that the substrate forms a semi-infinite stack of atomic layers. As a representative of each atomic layer, a scattering matrix describes how each incident beam transforms into beams leaving the bulk. The scattering from the SPB is represented by a set of matrices describing the repeated transmission and reflection of a set of beams by the SPB. These matrices are added to the substrate lattice reflection matrix using the layer-doubling formalism. This model has all possibly multiple reflections between the substrate lattice and the SPB. Thus, the scattering amplitude from the crystal, with the SPB in place, forms a geometrical series summing over infinite number of internal reflections [4]. Figure 14.1 illustrates the multiple diffraction mechanism [3] for the LEED patterns and Fig. 14.2 shows the VLEED (00) beam intensity-energy spectroscopy [5].

The scattering by the SPB is represented by a set of matrices representing the transmission and reflection of beams. These metrics are added to the substrate reflection matrix using the layer-doubling formulism. This accounts for all possible multiple reflections between the substrate and the SPB. In the formulism, the scattered amplitude from the crystal, with the SPB in place, can be written as,

$$R_T^{-+} = r^{-+} + t^- R^{-+}\left(1 - r^{+-} R^{-+}\right)t^{++}$$

With

$$r = \frac{ik_\perp \psi + \psi'}{ik_\perp \psi - \psi'}\exp(-2ik_\perp z)$$

$$t = \frac{2ik_\perp}{ik_\perp \psi - \psi'}\exp(ik_\perp z) \tag{14.1}$$

Here the convention used is that the substrate occupies the positive half space $z > 0$ and the superscripts refer to the directions of the electrons, with the second superscript indicating the initial direction and the first indicating the final direction. Hence the total reflection matrix is then R_T^{-+}, while R^{-+} is the substrate reflection matrices without the SPB and r^{-+}, r^{+-}, t^-, and t^{++} are, respectively, the reflection and transmission matrices of the SPB. In the r and t matrices, ψ and ψ' be the wave function at point z and $k_\perp$ is the perpendicular momentum of the electron (Fig. 14.1).

14.1.2   Beam Interference

The interference between the measured (usually the specular) beams and a pre-emergent beam varies with the crystal geometry and the SPB, which determines the VLEED fine-structure features. When an electron beam impinges on to the crystal it

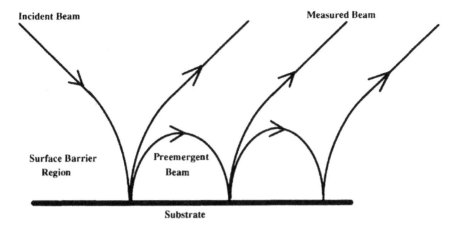

Fig. 14.1 SPB multi-reflection produces the VLEED fine structures. Part of the amplitude of the incident beam is diffracted into the direction of the measured beam and part of the beam into the preemergent beam. The amplitude in the preemergent beam is turned back by the SPB. Part of this amplitude diffracts back into the direction of the preemergent beam. The measured intensity sums the squares of the repeatedly amplitudes. Reprinted with permission from [3]

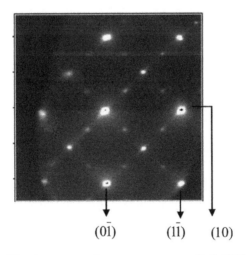

$$(0\bar{1}) \qquad (1\bar{1}) \qquad (10)$$

Fig. 14.2 A LEED diffraction pattern of a room temperature O–Cu(001) surface after 480 L showing a well ordered ($\sqrt{2} \times 2\sqrt{2}$) superstructure. The (00) pattern intensity-energy (I-E) profiles in the 6.0–16.0 eV range form the VLEED spectroscopy [3, 5]

is diffracted into the specular direction and then scattered into high-order beams in all directions, depending on the incident conditions. If the high-order beams do not have sufficient energy perpendicular to the surface to escape from the SPB confinement they are totally internally reflected. They can then be re-scattered by the substrate back into the specular direction.

As the energy of the beam increases, the perpendicular momentum of the diffracted beam increases, so the beam travels further from the substrate, increasing the phase change. When the beam energy is close to the emergence energy, a small increase causes a large change in phase, so the peaks cluster together towards the fixed energy. The same physical process is responsible for the image-like states seen in the inverse PES spectra [6, 7]. Each pre-emergent beam contributes to the observed fine structure. The interference will produce visible modulations in the observed intensity. The interference between the pre-emergent and the specular beams must meet the following conditions:

- Same direction. The incident beam is partially diffracted into other beams that are then diffracted by the substrate and the SPB repeatedly. These diffracted beams are partially diffracted into the (00) direction being responsible for the detection.
- Compatible amplitudes. The pre-emergent beam and the specular beam must have similar amplitudes for them to be any observable modulation and interfere in the intensity.
- Fixed phase difference. Same wavelengths of the diffracted beams are vital to the interference and the difference in phase must be an integer multiple of 2π.

The emergence energy of the pre-emergent beam meets Bragg-diffraction condition. The emergence energy depends on the two-dimensional crystal geometry only. The dependence of the emergence energy on the two-dimensional crystal geometry can be assessed from the Ewald construction. The emergence energy corresponds to the point where the sphere first touches the rod in the reciprocal space. The fine-structure features converge in a Rydberg-like series and the height and shape of the SPB modulates the beam energies slightly.

14.1.3 Scattering Matrices and Phase Shift

The barrier-scattering matrices were calculated by integrating the complex Schrödinger's equation from the position of the top atomic layer to a point some distance away from the surface where the influence of the potential has become insignificant. To create all the required reflection and transmission coefficients, the integration was started from inside the crystal to large distances in the negative z-direction pointing outward of the surface. The integration range depends on the energy of the electron compared to the inner potential. The proper reflection (r) and transmission (t) coefficients can be found from the value of the wave function and its derivative following this integration, using the expressions given in Eq. (14.1). The choice of the integration limits depends on which part of coefficients is being evaluated. If r^{-+} and t^{-} are being evaluated, then the integration of the wave function is started at some large distance from the surface and continued into the point z, which is the terminal of the surface $z = 0$. The choice of where the integration starting from depends on the energy of the electron compared to the potential.

For electrons close to the binding energy, the path is about 150 atomic unit long. However, if the energy of electron is small or large compared with the barrier height, then the path is shortened to around 50 atomic unit. The other pair of reflection and transmission coefficients was found by reversing the direction of integration. The Runga-Kutta technique was used to integrate through the barrier. This approach provides both ψ and ψ'. The initial wave function was taken to be a plane wave. The partial-wave phase shifts were determined by numerically integrating the radial Schrödinger's equation for the ion cores and matching the solutions at the muffin-tin radius.

14.1.4 Multi-Atom Calculation Code

The code used previously to do VLEED calculation was developed for a single atom per unit cell and so it was unsuitable for systems where the surface had reconstructed or where there was more than one atom per unit cell. To extend the analysis to such complex systems, a multi-atom code is necessary with the following modifications to the previous one-atom code [3]:

• Based on the code of Malmstrom and Rundgren [8], Lindroos had developed a code to calculate the effect of the SPB on the I-E curves by counting in the reflection and transmission coefficients of the SPB. A modification of the code of Lindroos et al. [9] by dealing with the layer-scattering matrices to make it suitable for such low energies. The Kambe's method was used to sum the scattering from atoms in the layers.

• The Runga-Kutta method was used to integrate the Schrödinger's equation from the surface layer out to distance from the substrate. The package by Malmstrom and Rundgren calculated the reflection and transmission coefficients of a SPB by integrating the image-like part of the potential to large distances from the substrate. All these attempts to find the effect of the SPB assumed that the SPB could be represented by a one-dimensional function of distance from the substrate.

• A subroutine sums the effect of the SPB interference to the substrate lattice scattering. Once the reflection and transmission coefficients are calculated, reflection and transmission matrices for the SPB layer are produced. The scattering of a single layer from all incident beams into all exciting beams is represented by a (n × n) matrix, where n is the number of active beams. In this convention, the barrier matrices must be diagonal, as the barrier is one-dimensional so a beam incident in one direction cannot be scattered into a beam in other directions. To save space, Thurgate represented the reflection and transmission matrices as vectors and wrote a subroutine to add the SPB effect to the substrate scattering. This was done by summing to infinity all multiple-scattering effects between the substrate and the SPB.

14.2 Oxide Tetrahedron Bond Formation

14.2.1 Observations

Table 14.1 summarizes the known behaviour of oxygen at metal surfaces and its consequences on various observations. These experimental observations and the corresponding explanations represent up-to-date understanding of oxygen-metal interaction, which provide foundations and justifications and provoked the current efforts

Table 14.1 Summary of the distinct observations and the conclusions on oxygen chemisorption on metals, which provides experimental foundations and also justifications for the proposed model

Microscopy—Spatial electron distribution	• Intensive contrast between STM protrusions and depressions [10] • Patches of protrusions detected with PEEM [11]
Spectroscopy—Variation of DOS	$E > E_F$ • Occupation of empty surface states [12] • Reduction of work function ϕ [13] • Presence of extra DOS above E_F [14]
	$E < E_F$ • Creation of new occupied DOS at 1.4–2.1 eV [15] • Up-shift of Cu $3d$-band and O-p states [16, 17]
Crystallography	• Lateral reconstruction and interlayer spacing relaxation [18] • Formation of O–M–O chains and missing row vacancies [17]
Characteristic properties	• Strongly localized energy states and bright patterned STM protrusions • Non-Ohmic rectification
Explanations and predictions	• Oxygen adsorbate affects STM/S current predominantly by polarization of metal electrons [19] • Surface dipole formation lowers the work function [20] • The "strong O–M bond" formation drives the relaxation and reconstruction [21] • The bond between adsorbate and substrate increases the extent of relaxation as O-exposure increases [22]
Motivation	• Valence states and atomic sizes should change when reaction takes place • Atoms displace collectively other than a certain atom moves in one direction at a time • Bond formation stems all phenomena in terms of microscopy, spectroscopy and crystallography, as well as mass transportation and structural phase formation • Reaction is a dynamic process associated with electron entrapment and polarization, which should beyond the description of static location of atoms

towards a compact model of chemical bond, energy band, and SPB for metal surfaces with chemisorbed oxygen.

14.2.2 Rules for Surface Bond Formation

Three principles govern the nature, number, and geometry (angle and length) of the oxygen chemisorption bond formation [23, 24]:

14.2.2.1 Electronegativity Specificity

The difference of the electronegativity ($\Delta\chi$) between two elements describes the ability of the atom of higher χ capturing electron from the other partner. Comparatively, electro-affinity being separation between the vacuum level and the bottom edge of the conduction band, describes the ability of the element holding the electrons caught from other elements. The electronegativity is elementally intrinsic but the electroaffinity refers to bulk solid and it is an adjustable quantity. Pauling [25] pointed out that if atoms differ sufficiently (by about two units) in electronegativity, they would form bonds that are mainly ionic. If the $\Delta\chi$ is much less than this, the bonds are mainly covalent, or polar-covalent.

The χ of oxygen is 3.5. The χ is around 1.8 for transition metals. For noble metals, the χ is about 2.2. The high χ value of oxygen indicates a great tendency for oxygen to form compounds with ionic or polar-covalent bonds to metals by catching electrons. Reactions with elemental oxygen give oxide products in which the oxidation state of oxygen is -2. The net charge transportation from metal to oxygen can be estimated by introducing a coefficient [26]:

$$\varepsilon = \varepsilon_c + \Delta\chi(\varepsilon_i - \varepsilon_c)/2, \quad \text{for } \Delta\chi \leq 2$$

This means that the lower χ element has a net charge loss:

$$q = \varepsilon e \leq e$$

where $\varepsilon_c = 0.5$ (for $\Delta\chi = 0$) and $\varepsilon_i = 1.0$ (for $\Delta\chi = 2$) correspond to ideally covalent and ionic states, respectively. According to Pauling, if $\Delta\chi \geq 2$ the bond is ideally ionic, the net charge contribution of a metal to oxygen is $q = e$. For H_2O, the χ for H is 2.2, and the H–O bond is polar-covalent ($\Delta\chi = 1.3$) and the charge transfer is about 0.62–0.65 e [27].

14.2.2.2   Intrinsic *sp*-Orbital Hybridization of Oxygen

The *sp*-orbital hybridization is intrinsic to oxygen, which generates four tetrahedrally directional orbits. The *sp*-orbital hybridization is more stable than the original $2s$, $2p_x$, $2p_y$ and $2p_z$ orbital configurations though the hybridization requires a small amount of energy. The hybridization of *sp* orbitals is fairly insensitive to its partners that donate electrons. The geometrical arrangement of the electron pairs in the valence shell of the central atom A for any molecule AB_n [28].

The AB_2 molecule is a common structure for the element A, like oxygen, possessing two lone-pair nonbonding orbitals and two bonding orbitals. In the H_2O molecule, the $2s$ and $2p$ orbits of an oxygen atom hybridize to form sp^3-type orbits. There are two electron pairs "−" in the bonding orbits (BP) and two lone pairs ":" in the non-bonding orbits (LP). The interaction system consisting of lone pair and polar-covalent bond is widely known as hydrogen bond (O^{-2}: H^+–O^{-2}) being the key element to water and ice [29].

Hypothesis can be made that the interaction of oxygen atom with metal atoms arranges in the general AB_2 structure and then the bond configuration is M_2O tetrahedron. Similar to a H_2O molecule, the charge cloud in the lone-pair orbit is under the influence of only one (oxygen) nucleus, so this orbital is larger than a bonding orbital occupied by sharing pair of electrons between two positive nuclei (O and H). The repulsion between the lone pairs (H^+: O^{-2}:H^+) occupied orbits, increases the angle between them to $109°28'$ or more.

One needs to note that one oxygen atom can never capture two electrons simultaneously from a certain atom because of the directional nature of the sp^3-hybridized bond geometry [30]. Meanwhile, their repulsions towards the electron pairs in the bonding orbitals (H^+–O^{-2}–H^+) push the bonding orbitals closer together, thereby reducing the H^+–O^{-2}–H^+ bond angle from the standard tetrahedral value to $104.5°$ or less. The repulsion energies between the tetrahedral orbits are in the order of $LP \leftrightarrow LP > BP \leftrightarrow BP > LP \leftrightarrow BP$.

14.2.2.3   Surface Bond Contraction

Atomic radii are no longer constant when atoms alter their states from metallic to either ionic or polarized. For instance, the radius of Cu reduces from 1.27 to 0.53 Å when the Cu becomes Cu^+ while the radius of O increases from 0.64 to 1.32 Å when the O turns to be O^{-2} [31].

On the other hand, atomic radius is not constant but changes with the coordination environment. The BOLS correlation [28] described in Part 1 applies to the chemisorbed surfaces and also to the intramolecular H–O bond of H_2O molecules, which results in the supersolid phase of nanoclusters, nanodroplets, nanobubbles and the skins of water and ice [29].

According to Pauling, the Cu radius will change from 1.276 to 1.173 Å (about 8% contraction) if the CN changes from 12 to one (only one neighbor). Jorgensen [32] and Kamimura and Suwa [33] noticed that the Cu-apical oxygen (at the apex of the octagonal CuO_6 structure) bond length is shortened with the increase of dopant-hole concentration. The reduction of the O–Cu distance is as high as 0.26 Å (about 14% contraction) in La–Sr–Cu–O superconducting materials. Hence, besides the atomic radius alternation due to the change of valence state, atomic undercoordination further shortens the surface bonds, which is independent of the bond nature [31]. Compared with Pauling's notion that is specific atomic radius dependent, Goldschmidt scheme is more general. The following equation formulates the BOLS correlation that is universally true [34]:

$$R(z) = R(12)[1 - Q(z)] = \frac{2R(12)}{1 + \exp\left[\frac{12-z}{8z}\right]}$$

This BOLS notion provides an understanding of the commonly-known fact that the first interlayer spacing contracts [35, 36] and the tiny protrusions appeared in the STM images of pure metal surfaces, showing the effect of polarization as a result of the interlayer local charge densification [15]. The CN-reduction induced bond contraction has been found as origin of the change of many properties of nanometric materials [28, 31].

14.2.3 The Primary M_2O Structure

Observations discussed above laid the ground for the primary M_2O bond structure, as shown in Fig. 14.3 [30]. Occupation of the hybridized orbitals by the valence electrons of oxygen (6e) and metal atoms (2e) generates two lone pairs (4e) and two contracting ionic bonds (4e). Each of the metal ions, 1 and 2, donates one electron to the central oxygen to form the ionic bonds. The radii of atoms 1, 2 and O^{-2} change with the alternation of their atomic states and sizes.

Further, the nonbonding lone pairs are apt to polarize metal atoms on which the lone pairs are acting. Atoms 3 are the lone-pair induced metal dipoles with expansion of their dimensions and elevation of their energy states, which are responsible for the protrusions in the STM images and the reduction of the local work function. This event agrees with Lang's tunneling theory [37] which predicted that the oxygen adsorbate affects the tunnel current predominantly by electronic polarization of metal atoms.

The M_2O primary tetrahedron is distorted for reasons: (i) the repulsion varies the bond angles [BAij ($\angle iOj$), i, j = 1, 2, 3; BA12 $\leq$ 104.5°, BA33 > 109.5°] and, (ii) the CN difference adjusts individual bond length [BLi = ($R_{M^+} + R_{O^{-2}}$) × (1 – Q_i), i = 1, 2; Q_i is the effective contracting coefficient. BL3 and BA33 vary with the coordination environment in a real system.

The coordination surroundings of a specific system—crystal geometry and the scale of lattice constant facilitate the Cu_2O tetrahedron [30]. The ideal coordination environment is that, as denoted in the diagram, the distances of 1–2 and 3–3 matches

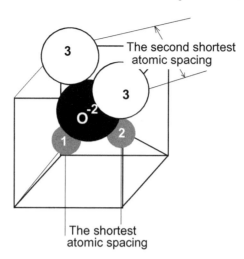

Fig. 14.3 Quasi-tetrahedron oxide bond model (from Ref. [38]). Each of the two ions, 1 and 2, donates one electron to the central oxygen to form the ionic bonds. Atoms labeled 3 are the lone-pair-induced metal dipoles with expansion of sizes and elevation of energy states. The tetrahedron is distorted for the repulsion effect on the bond angles and the CN effect on the bond lengths. The interaction of 3-O is much weaker than that of (1, 2)-O

closely the first and the second shortest atomic spacings. The plane composed of 1O2 is perpendicular to the plane of 3O3. In reality, atomic dislocation and bond angle distortion happen. Moreover, oxygen always seeks four neighbors for a stable tetrahedron formation. It should be noted that the lone pair induced dipoles tend to directing into the open end of a surface due to the strong repulsion between the lone pair-induced dipoles. Therefore, atomic dislocations during the reaction are determined by the bonding dynamics and the bond geometry.

With the adaptation of the general AB_2 molecule to the M_2O system, the M_2O tetrahedron contains three valence states, namely, O^{-2}-hybrid with bonding and nonbonding orbitals, the lone-pair-induced metal dipoles and metal cations with electronic positive holes. Interaction between the dipole and another oxygen (O^{-2}: $M^{+/p}$–O^{-2}) is similar to the *hydrogen bond* (O^{-2}: H^+–O^{-2}) of water and ice [29]. The interaction between oxygen and the lone pair induced dipole (O^{-2}: $M^{+/p}$) is stronger than the ordinary Van der Waals bond but it is much weaker than the ionic or the covalent O–M bond.

Conveniently, such interaction (O^{-2}: $M^{+/p}$–O^{-2}) can be termed as "*hydrogen bond like*" because the H^+ is simply replaced by the $M^{+/p}$. Besides the bonding between oxygen and metals, nonbonding lone pairs of oxygen, antibonding metal dipoles and hydrogen bond like present at reaction. It is worth emphasizing that the oxide bonding creates inhomogeneous electronic structures surrounding a certain atom in the compound. This local feature provides the basis for granularly anisotropy of ceramics such as colossal magneto-resistance. Because focus was more on electron transportation between oxygen and metals, the lone pairs and hydrogen-bond like formation in oxygen-chemisorbed systems has received little attention.

At the initial stage of oxidation, oxygen molecule dissociates and interacts with metal atoms through one bond. The O^{-1} specifies a position where the O^{-1} bonds directly to one of its neighbors. For the transition metals of lower electronegativity ($\chi < 2$) and smaller atomic radius (<1.3 Å), such as Cu and Co, O often bonds to a neighbor at the surface; while for noble metals of higher electronegativity ($\chi > 2$) and larger atomic radius (>1.3 Å), such as Rh and Pd, O tends to sink into the hollow site and forms the first bond to its neighbor underneath [30]. The O^{-1} also polarizes other neighbors and pushes the metal dipoles radially away the adsorbate. This process also leads to the STM protrusions and creates antibonding dipole states but different morphologies.

14.2.4 O–Cu(001) Bond Geometry and Atomic Valency

For the particular O–Cu(001) system, two phases present during the reaction. One is the nanometric Cu(001)-c(2 × 2) − $2O^{-1}$ domain with protruding boundary that occurs when the surface exposes to an oxygen exposure lower than 25 L, the other is the MR type Cu $(\sqrt{2} \times 2\sqrt{2})$ R45° − $2O^{-2}$ structure. Figure 13.1 showed the corresponding STM images of the two phases, which provide the experimental ground for the bond models discussed below.

14.2.4.1 Precursor: CuO_2 Pairing Pyramid

Figure 14.4 illustrates a single Cu(I)O($Cu^{+1} + O^{-1}$) pyramid structure, part of the c(2 × 2) − $2O^{-1}$ complex unit cell. The unit cell can also be illustrated as a CuO_2 paring pyramid ($Cu^{+2} + 2O^{-1} + 6Cu^p$). In the precursor state, each of the two oxygen atoms catches one electron from the same Cu neighbor, or separately from two, to form one contracting ionic bond, BL1. Meanwhile, the O^{-1} polarizes its rest neighbors that form the protruding domain boundaries. If $DO_x = 0$, the O will be located at the apex of a centered pyramid to form *four* identical O–Cu bonds, which is forbidden. Therefore, the first phase is an off-centered pyramid. The geometrical parameters (DO_x, DO_z, BL1, and BL2) of the pyramid can be determined by DO_z, and BL1.

14.2.4.2 Cu_3O_2 Pairing Tetrahedron

Upon increasing oxygen exposure, the pairing-pyramid evolves into a pairing-tetrahedron, as shown in Fig. 14.5. Each of the O^{-1} ions forms another bond with its neighboring atom underneath and then the Cu_3O_2 pairing-tetrahedron forms, being a case in which "two oxygen get four electrons from three coppers" [39]. Atoms labeled

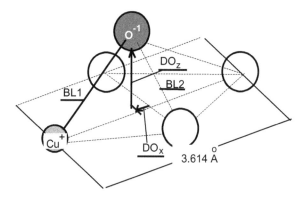

Fig. 14.4 An off-centered pyramid describes the precursor Cu(001)-c(2 × 2) − 2O^{-1} reaction (Ref. [30]):

O_2 (adsorbate) + 4Cu (surface) + 2 Cu (substrate)

$\Rightarrow 2O^{-1} + Cu^{+2}$ (surface)   (CuO$_2$ bonding)

$+3\,Cu^p$ (buckled up) + 2Cu (substrate)   (bonding effect)

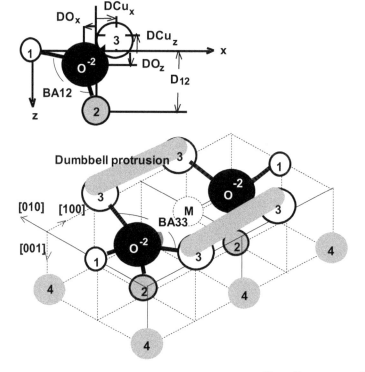

Fig. 14.5 Cu$_3$O$_2$ pairing tetrahedron models the Cu(001) − (2√2 × √2) R45 − 2O^{-2} recon-struction (Ref. [30]):

O_2 (adsorbate) + 4 Cu (surface) + 2 Cu (substrate)

$2\,O^{-2}$ (hybrid) + Cu^{+2} (surface) + 2 Cu$^+$ (substrate)   (Cu$_3$O$_2$ bonding)

$+2\,Cu^p$ (buckled up and crossing the MR) + Cu (MR vacancy) (bonding effect)

1 is Cu^{2+} and 2 is Cu^+. Atoms 3 are the lone pair induced Cu^p dipoles. This configuration gives rise to a unit cell of the $Cu(001) - \left(\sqrt{2} \times 2\sqrt{2} \right) R45° - 2O^{-2}$ structure that contains one O^{-2}, one Cu vacancy, and two diploes in the top layer. For the $Cu(001)$ surface, the first and the second shortest atomic spacings are 2.555 and 3.614 Å, respectively. Such a surrounding accommodates the Cu_3O_2 pairing-tetrahedron in the way to form the complex unit cell.

The O^{-2} prefers the center of a quasi-tetrahedron. Atoms 1 and 2 are Cu^{+2} and Cu^+. Atom 3 is Cu^p and M is the vacancy of the missing Cu arisen from the isolation of this atom. Atom 4 is a metallic Cu atom. The oppositely coupled 3 ↔ 3 dipoles bridge over the MR vacancy (Fig. 14.5).

14.2.5 Bond Geometry Versus Atomic Position

14.2.5.1 Parameters Required in Calculation

In the multi-atom VLEED calculation code, the geometrical variables are the layer spacing D_{12}, and the atomic positions in the complex unit cell of the topmost plane which added to the normal lattice matrix of the $Cu(001)$ crystal. For the O–$Cu(001)$ system, there are no y-directional (along the missing row) displacements for all the atoms due to the lattice periodicity. The D_{12} and the x, z-directional displacements of oxygen (DO_x, DO_z) and Cu dipoles (DCu_x, DCu_z) are used as input used in calculations. Atom 1 was taken as the coordination origin. The pairing dipoles dislocate by the same (DCu_x, DCu_z) amount but in opposite direction. The MR vacancy is in the symmetric central position. There are five independent parameters in terms atomic dislocations if considers the symmetry and periodicity.

14.2.5.2 Bond Variables

Instead of the conventional dislocation of individual atoms, the following Cu_3O_2 bond geometry for the $Cu(001) - (\sqrt{2} \times 2\sqrt{2}) R45° - 2O^{-2}$ phase employs:

BL1 - distance between the O^{-2} and the Cu^{+2} that was taken as coordination origin

BL2 - distance between the O^{-2} and the Cu^+ in the second layer

BL3 - distance between the O^{-2} and the Cu^p that bulked up of the surface

BAij($\angle iOj$) - angle between atoms i-O^{-2}-j(i, j = 1, 2, 3).

BL1 and BL2 are subject to the undercoordination-induced contraction:

$$BLi = \left(R_{O^{-2}} + R_{Cu^+} \right) \times (1 - Q_i) \quad (i = 1, 2)$$
$$Q_i (i = 1, 2) \text{ is the CN-resolved contraction coefficient.}$$

Setting the radii of O^{-2} and Cu^+ equal to the standard Goldschmidt radii of 1.32 Å and 0.53 Å, respectively, then the ionic bond length equals the standard bulk value of 1.85 Å for the Cu_2O. The effective CNs of Cu^{+2} and Cu^+ are taken as 4 and 6, respectively. So, the corresponding contracting coefficients are $Q_1 = 0.12$ and $Q_2 = 0.04$, respectively. Thus, the lengths of the contracting ionic bonds are:

$$BL1 = 1.85 \times 0.88 = 1.628(\text{Å})$$
$$BL2 = 1.85 \times 0.96 = 1.776(\text{Å})$$

The bond angle BA12 is constrained to be 104.5° or less, because of the smaller repulsion between the bonding orbitals. BA33 can be any value greater than 109.5° due to the strong repulsion between the lone pair induced dipoles. In calculations using this model for the dynamic processes, the contracting coefficient Q_2 is taken as an adjustable variable, while the Q_1 ($= 0.12$) is always assumed as a constant because it forms immediately upon oxygen molecule dissociates. Thus, the variables of Q_2, DCu_x, and BA12 are independent variables in determining the collective motion of the atoms in the complex unit cell during the reaction.

Change of the variables (setting $DCu_x = 0.25 \pm 0.25$ Å, BA12 $\leq 104.5°$, $Q_2 = 0.04 \pm 0.04$) is independent and restricted to finite intervals. This differs from the atomic-dislocation wise in which one must consider the atomic dislocation of a certain atom once in one direction.

The advantage of such a set of variables is that the number of the adjustable variables is reduced from the conventional five to two (for stable system) or three (for dynamic system), and the bond geometry is constrained by the principle of tetrahedron formation. As will be demonstrated, any single variation of these variables dislocates almost all the atoms in the unit cell. These variables also reconcile all the simultaneously geometrical variations due to bond formation without dislocating individual atoms independently. In a real reacting system, any individual atomic-shift, in principle, will affect other atoms of the entire system. In other words, the bond geometry is more realistic and convenient than the atomic-disposition wise. It never happens that atoms dislocate independently during reaction.

14.2.5.3 Bond Relaxation Versus Atomic Dislocation

Variation of the bond relaxation in geometry and length ($Q_1 = 0.12$, Q_2, DCu_x, and BA12) determines all the structural parameters (D_{12}, DO_x, DO_z, DCu_x, and DCu_z) used in calculation. DCu_x, together with DCu_z, determines the orientation and the distance between Cu^P and O^{-2}. The parameter spaces of atomic-position and bond-geometry are interchangeable as illustrated in Fig. 14.6.

• Coordination System and Constraints

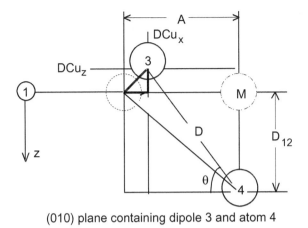

(010) plane containing dipole 3 and atom 4

Fig. 14.6 Illustration of the relation between DCu$_x$ and DCu$_z$ [23, 24]

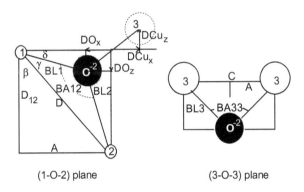

(1-O-2) plane   (3-O-3) plane

Fig. 14.7 Geometrical relationship between atomic-displacement and bond-geometry. BA12, BA33, β, γ and δ are angles [23, 24]

Atom denoted 1 in Fig. 14.5 is the origin of the coordination system with z-axis directing into the substrate. The x-axis is along the [100] direction, perpendicular to the MR. As the origin of the coordination, atom 1 only has z-directional displacement. Atom 2, as part of the tetrahedron, fixes at the usual site in the non-reconstructed substrate second layer. As the VLEED probes nondestructive information dominated by the second atomic layer and above. Calculations revealed little geometrical change of layers deeper than the second. DCu$_z$ and DCu$_x$, describing the dislocation of dipole 3, are in the same order and they are much smaller than the distance D, from dipole 3 to atom 4 (underneath the missing row). The DCu$_z$ is constrained by (A = 1.807 Å is the nearest row spacing):

$$D^2 \approx A^2 + D_{12}^2$$

As a parameter for structural sensitivity testing, D takes the value of normal atomic spacing D_1 so as to keep the usual atomic distance between dipole 3 and atom 4, D, and the value D_2 by taking into the relaxation D_{12} into consideration:

$$D = D_1 = 2.555\text{Å, or}$$
$$= D_2 = \left[A^2 + D_{12}^2\right]^{1/2}$$

• Parameter transformation and initialization

The bond geometrical change defines the atomic shifts, which is under the following constrains:

$$(D_{12}, DCu_x, DCu_z, DO_x, DO_z) \Leftrightarrow$$
$$\left(Q_1 = 0.12, Q_2 = 0.04 \pm 0.04, DCu_x = 0.25 \pm 0.25\text{Å}, BA12 \leq 1040.5°\right).$$

The DCu_x and D determine the DCu_z (Fig. 14.5):

$$DCu_z = \left[D^2 - (A - DCu_x)^2\right]^{1/2} - D_{12}$$

14.2.6 Mechanism of Surface Reconstruction

14.2.6.1 Surface Atomic Valence States

The adaptation of the primary AB_2 model to the O–Cu(001) system improves the following understandings [30]:

• Bond length and geometry relaxation determines atomic collective dispositions. The variation of the BL1, BL2 and BA12 determines the first interlayer distance. The valence states alternation of atoms 2 from metallic to Cu^+ in the second layer shortens the second layer spacing. Interaction between charged ions and the neutral metal atoms in the third layer is stronger than it is between the initially neutral metal atoms.
• Atomic sizes and valence states change during the reaction. Oxygen adsorption is the process of bond transforming that gives rise to Cu^{+2}, Cu^+, O^{-1} and O^{-2} and the lone-pair-induced Cu^p as well as the MR vacancy. The Cu_3O_2 structure creates three sublayers and six different rows in a complex unit cell. The top layer is divided into three sublayers. The first sublayer is composed of the protruding Cu^p with dimension expansion; the second sublayer is of Cu^{+2} ions with radii contraction and lowered energy states; the third sublayer is composed of O^{-2}. Both Cu^{+2} and O^{-2} ions are detected as depressions with an STM due to the relatively lower occupied energy states. Along the [100] direction, on each side of the missing row, there is a Cu^p row, an O^{-2} row and a Cu^{+2} row. The Cu^p row displaces outwardly and

closely to the missing row. The Cu^{2+} row pairs two Cu^P rows. The ($Cu^P \leftrightarrow Cu^P$) bridges over the missing row vacancies.

- O^{-2} prefers the central position of the M_2O tetrahedron rather than at an apical site of the tetrahedron. Therefore, O^{-2} locates underneath the top layer and moving close to atom 1, and away from the MR, due to bond contraction.
- The O–Cu–O string is zigzagged by electron lone pairs (Cu^P: O^{-2}: Cu^P) rather than any other kinds of bonding or antibonding states. The $Cu^P \leftrightarrow Cu^P$ antibonding quadruple is responsible for the "dumbbell" protrusion in the STM image. Missing row atom is produced by the isolation of this atom from other neighbors during the bond forming. All neighbors of the MR atom have bonded to the adsorbate and thus the MR atoms are "evaporated" from the surface.

14.2.6.2   Reaction Dynamics

The present model describes logically beautifully the dynamic process of oxygen bonding to the Cu(001) surface, in which O^{-1} forms first and then turns to O^{-2} with sp-orbital hybridization and lone pair production:

The precursor phase Cu(001) $- (2 \times 2) - 2O^{-1}$ can be simply described as a pairing CuO_2 pyramid formation:

$$O_2(\text{adsorbate}) + 4\,Cu(\text{surface}) + 2\,Cu(\text{substrate})$$
$$\Rightarrow 2O^{-1} + Cu^{+2}(\text{surface}) \qquad\qquad (CuO_2\text{bonding})$$
$$+3\,Cu^P(\text{buckled up}) + 2\,Cu(\text{substrate}) \;(\text{bonding effect});$$

The MR type Cu(001) $- (\sqrt{2} \times 2\sqrt{2})\,R45 - 2O^{-2}$ structure is a consequence of the pairing CuO_2 pyramid evolves into a novel pairing tetrahedron Cu_3O_2:

$$\Rightarrow 2\,O^{-2}(\text{hybrid}) + Cu^{+2}(\text{surface}) + 2\,Cu^+(\text{substrate}) \;(Cu_3O_2\text{bonding})$$
$$+2\,Cu^P(\text{buckled up}) + Cu(\text{MR vacancy}) \qquad\qquad (\text{bonding effect})$$

As the bonding effect on reconstruction and charge transformation, only the MR vacancies and the buckled Cu^P are within the scope of experimental observation while the origin of the phenomena, dynamics of bonding and electron polarizing, is within the capacity of logic imagination.

14.2.6.3   Surface Bond Network and STM Morphology

Repeatedly packing the complex unit cell forms the bond network of the O–Cu(001) surface. The STS profiles measured along the O–Cu chain at different sites from Cu(110) surface verified the electronic configurations of the tetrahedron [40]. According to the Fourier transformation, the performance of atoms and electrons in the unit cell represents all possible events happened during reaction. Figure 14.8 gives an analogue of the STM image for the O^{-2} induced Cu(001) surface reconstruction.

STM imaging [41] revealed no overlap of electron cloud on a clean Cu(001) surface although the atomic spacing of 2.555 Å is much shorter than that between the pairing dipole chains of 2.9 ± 0.3 Å separation for the O-added Cu(001) surface. However, the pronounced "dumb-bell" protrusions bridge over the missing row. In the constant-current mode, the protrusions are maps of spatial DOS sampled at a potential related to the voltage applied between the tip and sample. The rigid-sphere scheme is unable to rationalize the underlying mechanism of such a big protrusion. This kind of protrusion arisen from the bond formation and electron polarization, which also applies to STM imaging from the polarized states induced by atomic undercoordination such as graphene zigzag edges, monatomic Ag chains, and Ag adatoms [28].

From a spatial point of view, the displacement of the ion-core position and the shift of the charge centers of the lone-pair-induced dipoles determine the dimension of the STM protrusions. The polarization of the metal electrons results in the strong localization of the surface charges and the pronounced protrusion of the polarized DOS. On the other hand, from an energy point of view, interaction between the lone pair and Cu^P, even further the repulsion of $Cu^P \leftrightarrow Cu^P$ along the [100] direction will further raise the energy levels and the protrusions of the dipoles, namely, the antibonding states.

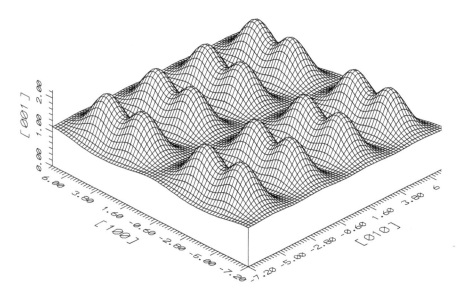

Fig. 14.8 Analogue of $z_0(x, y)$ counter plot to the STM images of $Cu(001) - (2\sqrt{2} \times \sqrt{2}) R45° - 2O^{-2}$ surface. The dumbbell protrusions correspond to the oppositely paired metal dipoles that cross over the missing row. The depressions are missing row vacancies and the Cu^{+2} ions. The O^{-2}: Cu^P: O^{-2} string is along the [010] direction [23, 24]

This physical picture agrees with that probed with STS from the O–Cu(110) system showing the energy levels of the occupied DOS increases relative to the pure Cu system. These results evidence for the polarization of metal electrons. On another hand, the strong localization of the electron density resulting from the polarization of metals, formation of the missing rows and ionization of metals cause the nonuniformity of the contact-potential difference from one site to another on the surface.

14.2.6.4   The Force Driving Reconstruction

Unlike bonding and anti-bonding states, the energy levels of the non-bonding electron pairs change little relative to an unpaired electrons in its isolated atomic orbital. However, the lone pair is capable of polarizing the neighboring metal atoms, raising their electronic energy above E_F [28]. The centers of the negative and positive charges of the dipoles shift oppositely along the resultant direction at which the two lone pairs are acting. The strength of interaction for a "O^{-2}: Cu^p: O^{-2}" system is twice that of a nonbonding interaction, which is in the order of ~0.1 eV [29].

Transformation from the metallic bond of ~1.0 eV to the contracting ionic bonds gains ~3.0 eV energy. Part of the amount of energy contributes to the hybridization of the sp-orbitals of oxygen. The hybridization of the sp-orbitals further lowers the system energy. These energies due to ionic bond formation provides forces driving the reconstruction and missing-row formation.

14.3   Valence Density-of-States

14.3.1 O^{-1} Derived Three DOS Features

Because of oxide bond formation, the valence band of the host metal exhibits additional features. Figure 14.8 illustrates the evolution of the DOS in the valence band, or above the vacuum level E_0, of the host. Arrows represent the dynamic processes of electron transportation between (a) the energy bands of an arbitrary metal and (b) the energy levels of oxygen adsorbate. Initially, the energy states below Fermi level of metals are fully occupied. The work function ϕ_0, the Fermi energy E_F, and the E_0 follows the relationship: $E_0 = \phi_0 + E_F$. For Cu example, $E_0 = 12.04$ eV, $\phi_0 = 5.0$ eV and $E_F = 7.04$ eV. The Cu-3d band is located at energies from -2.0 to -5.0 eV below the E_F. For oxygen, the O-2p level is at -5.5 eV with respect to the E_F of Cu.

At the initial stage of reaction, one electron transports from the outmost shell of a metal atom to the unfilled O-2p states of oxygen. The O^{-1} polarizes its rest neighbors. Figure 14.8c represents the resultant effect of O^{-1} formation on metal surface, which is not simply a superimposition of (a) and (b) but a resultant of the DOS features:

- The resultant band shows an extra DOS feature as indicated as O-p states charge transportation from metal to the O^{-1}.
- The O^{-1} polarizes its nearest neighbors to produce dipoles that form sub-band above the E_F, which lowers the local work function from original ϕ_0 to ϕ_1.
- The processes of bond and dipole formation create electronic holes right below the E_F, which creates a band gap E_G to the metal or widens the band gap of semiconductor from E_{g0} to E_{g1}.

14.3.2 O^{-2} Derived Four DOS Features

Upon sp-orbital hybridization, the band configuration of the host metal evolves from Fig. 14.8c, d. Besides the holes and the antibonding dipoles appeared in the O^{-1} precursor, the O-p sub-band divided into nonbonding (lone pair) and sp-bonding sub-bands. The antibonding states sustained now by the lone-pair induced dipoles instead of that induced by O^{-1}. [23] The sp-hybrid nonbonding (lone pair) states of oxygen locate somewhere (normally ~1.5 eV) below E_F and above the sp-hybrid bonding states that shift slightly towards energy lower than the $2p$-level of the oxygen because hybridization lowers system energy. For the Cu example, the $4s$ electrons (in conduction band, CB) either contribute to the sp-bonding or jump to its own outer empty-shell (for $4p$ orbital example) with energy even higher than the E_F. Such a process empties the states below the E_F, which yields the Cu-oxide to be a semiconductor. However, the lone pair states may overlap the electronic holes in energy domain so some time observation cannot recognize their presence.

14.3.3   Anomalous H-Bond like

Upon overdosing of oxygen, H-bond like forms. The dipoles contribute the polarized electrons to the bonding orbitals of an additional oxygen adsorbate. The arrow in Fig. 14.8d from the antibonding states to the sp-bonding subband of oxygen represents the process of the H-bond like formation. STS and VLEED revealed that the antibonding states of the O-Cu system range over 1.3 ± 0.5 eV and the nonbonding states -2.1 ± 0.7 eV around E_F. The photoemission electron microscopy (PEEM) studies of O–Pt surfaces, [43–45] detected the conversion of the dark islands, in the scale of 10^2 μm, into very bright ones with work functions ~1.2 eV lower than that of the clean surface, which evidence the localization nature of polarization.

14.4 Energy-Dependent 3D-SPB

14.4.1 Initiatives

The SPB describes the charge distribution both in real space and in energy space, which links to the change of valence states and geometrical arrangement of atoms of the surface [46]. The SPB is a complex function. The real part describes the image elastic potential and the imaginary part describes the charge distribution in the real and energy domain. The former charges the direction and phase shift of the reflected electron beams and the latter causes inelastic damping of the incident beam energy. The inelastic damping has been assumed as monotonically energy-dependent. For clean metals, such as Cu(001) [47], W(001) [1, 48, 49], Ru(110) [9, 50], and Ni [51], the SPB approaches a uniform layer of thin-film interference.

STM/S observations [41, 52] support the uniform-SPB approximation for clean metals. For example, STM reveals that the ion cores with small protrusions (0.15–0.30 Å) arrange regularly in the homogeneous background or Fermi sea; STS studies of the Cu(110) surface [15] confirmed the uniformity of the DOS below E_F. Hence, clean metal surfaces can be described as nearly-ideal Fermi systems and the uniform-SPB approximation is acceptable within the error of detection.

Surfaces with chemisorbed oxygen differ much from those of the pure metals in that the chemisorbed surfaces have many "rather local" features—varying site from site around an atom. The chemisorption of oxygen results not only in the dislocation of ion cores but also in the alternation of atomic sizes and atomic valence states by charge transportation and polarization. More importantly, as consequences of oxygen-metal bonding, the formation of the dipole layer and, in some cases, the removal of atoms roughens the surface largely.

Even at very low exposure to oxygen, the Cu(001) surface is roughened by the protruding domain boundaries, see Fig. 13.1a. The STM scale differences of the systems with chemisorbed oxygen are much higher (0.45–1.1 Å) compared with pure metals of 0.3 Å or lower. The STS profiles from the O–Cu chain region on the O–Cu(110) surface show that there is a general elevation of energy states. The above-E_F empty surface state is occupied and a new DOS feature is generated below E_F.

Furthermore, in decoding VLEED data from the O–Cu(001) surface, Thurgate and Sun [3] found that the spectrum collected at azimuth closing to the <11> direction (perpendicular to the missing-row) could not be simulated with the uniform-SPB by using either the Cu(001)-c(2×2) $- 2O^{-1}$ or the ($\sqrt{2} \times 2\sqrt{2}$) R45°$- 2O^{-2}$ structure, or even their combination with various SPB parameters. Therefore, it is implausible with VLEED to determine the SPB and the crystal structure without proper modeling parameterization.

The atomic-scale localization and the on-site variation of energy states of the systems with chemisorbed oxygen suggest that it is necessary to consider the electron distribution on the surface site by site. The 3D effect, the variation of energy states

of the surface and the correlation among the parameters used in calculations consti-
tute the complexity in de-coding VLEED data from the systems with chemisorbed
oxygen.

Therefore, besides the M_2O bond geometry, a proper nonuniform-SPB for systems
with chemisorbed oxygen is necessary to:

- reduce the number of independent SPB variables to properly correlate the intrinsic
  entities of the surface and to simplify VLEED optimization ensuring the certainty
  of solution.
- allow the calculation code to automatically optimize the image plane $z_0(E)$ to
  reproduce the experimental data.
- let the $z_0(E)$ profiles vary on-site with the crystal structure to reflect appropriately
  the interdependence between the atomic geometry and the electronic structures;
  and, eventually
- produce the essential DOS features in the energy range covered by VLEED.

14.4.2   Complex Form of the Surface Potential Barrier (SPB)

Electrons with energy E traversing the surface region can be described as moving in
a complex optical potential [53, 54]:

$$V(\mathbf{r}, E) = ReV(\mathbf{r}) + iImV(\mathbf{r}, E)$$
$$= ReV(\mathbf{r}) + iIm[V(\mathbf{r}) \times V(E)] \qquad (14.2)$$

The $V(\mathbf{r}, E)$ satisfies the following constraints.

14.4.2.1   Elastic and Inelastic Correlation

The elastic potential $ReV(\mathbf{r})$ correlates to the electric field $\varepsilon(\mathbf{r})$, charge density $\rho(\mathbf{r})$,
and the imaginary $ImV(\mathbf{r}, E)$ in the following form:

$$\nabla[ReV(r)] = -\varepsilon(r)$$
$$\nabla^2[ReV(r)] = -\rho(r) \propto ImV(r, E) = ImV(r) \times ImV(E)$$

The $ReV(\mathbf{r})$ satisfies Poisson equation and the gradient of the $ReV(\mathbf{r})$ equals the
electric field $\varepsilon(\mathbf{r})$. If $\rho(\mathbf{r}) = 0$, then the $ReV(\mathbf{r})$ corresponds to a conservation field
without charging source being included; that is, the moving electrons will suffer no
energy loss and the spatial variation of the inelastic damping potential $ImV(\mathbf{r}) \propto \rho(\mathbf{r})$
$= 0$. Therefore, the $ImV(\mathbf{r})$ and the $ReV(\mathbf{r})$ are correlated each other uniquely through
the electron distribution $\rho(\mathbf{r})$.

14.4.2.2 Inelastic Electron Dissipation

The energy dependence of the inelastic potential, ImV(E), represents all the dissipative processes that are dominated by phonon and single-electron excitation at energies below that for plasma excitation. This means that the single-electron excitation occurs in the electron-occupied space, described by $\rho(\mathbf{r})$, and with any incident energy E greater than the work function ϕ.

As remarked by Pendry and Alldredge [55], no contribution to ImV(E) from a particular loss mechanism can come about until the incident electron has enough energy to excite this mechanism. Plasma excitation needs energies around 15 eV below E_F [56] in metals with a free-electron conduction band. The contribution of plasma excitation to ImV(E) does not come into play below this value. Even in non-free-electron metals there is a more general cluster of excitations usually at around the equivalent free-electron plasmon energy. Below the plasmon energy, single electrons can still be excited from the conduction band, though they produce a smaller ImV(E) features than plasma excitation.

In metals, single electron can be excited by an incident electron that has any energy greater than the work function, $(E \geq \phi)$, but in insulators the situation is different. On the other hand, phonon excitation and photon excitation may occur at energies that are much lower than the E_F. Therefore, at energies between the work function and the plasma excitation threshold, single electron excitation is dominant, which adds "humps" to the inelastic damping showing the DOS features in this energy regime.

14.4.2.3 Amplitude and Phase Shift of Diffracted Beams

The z-directional integration of the ImV(z, E) and the ReV(z) corresponds, respectively, to the amplitude loss ΔA and the phase change $\Delta \Phi$ of the reflected electron beams that are described with plane waves:

$$\varphi = A\exp(-ik \cdot r + \Phi), \, and,$$

$$\Delta A \propto \int_a^{-\infty} \varphi \text{Im} V(z, E)\varphi^* dz,$$

$$\Delta \Phi \propto \int_a^{-\infty} \varphi \text{Re} V(z)\varphi^* dz \qquad (14.3)$$

where k is the plane wave vector. Integration starts from a certain point, a, inside the crystal, to infinitely far away of the surface. The parameter a varies depending on the penetration depth of the incident beams. This constraint provides a leeway for the mathematical expressions of the specific forms of ImV(z), ImV(E), and ReV(z). Therefore, the exact values of the strongly correlated parameters are much less important than the integration is in the physics, see Eq. (14.2). The correlation among the

SPB parameters leads to finite numerical solutions that corresponds to reality and be physically meaningful.

Most importantly, the variation in bond length and geometry and the behavior of electrons, both in real space (represented by $ImV(r)$) and in energy space (described by $ImV(E)$ or DOS), depend on each other, as they are consequences of the surface bond forming and the electrons connecting to the sites, sizes and the chemical states of the atoms.

14.4.3 One-Dimensional SPB

Jones, Jennings and Jepsen formulated the $ReV(z)$ [1] that approximated the results of jellium and density functional theory calculations of the SPB. This model has widely been used for the fitting both of the VLEED fine-structure features and of the inverse-photoemission image states, and has the form [4]:

$$
\begin{aligned}
ReV(z) &= \frac{-V_0}{1 + A\exp[-B(z-z_0)]}, \; z \geq z_0 \, (\text{a pseudo-Fermi-z function}) \\
&= \frac{1 - \exp[\lambda(z-z_0)]}{4(z-z_0)}, \; z < z_0 \, (\text{the image potential})
\end{aligned}
\tag{14.4}
$$

where A and B are constants given by $B = V_0/A$ and $A = -1 + 4V_0/\lambda$. The z-axis is directed into the crystal. V_0 is the muffin-tin inner potential constant of the crystal and z_0 is the origin of the image potential. The degree of saturation is described by the λ parameter.

The $ReV(z)$ transforms at $z = z_0$ from the pseudo-Fermi z function to the $1/(z - z_0)$ type classical image potential with a spatial decay term. The Poisson equation correlates the $ReV(z)$ and the charge density,

$$
\nabla^2 ReV(z) \begin{cases} -\rho(z) \; (z > z_0) \\ 0 \qquad\;\; (z \leq z_0) \end{cases}
\tag{14.5}
$$

The origin of the image-plane, z_0, acts as the boundary of the region occupied by surface electrons. If we permit z_0 to vary with the surface coordinates, then $z_0(x, y)$ provides a contour of spatial electron distribution being mimic of the STM image. The SPB features are thus characterized by the z_0 and this effect allows us to choose z_0 as the key character in the subsequent single-variable parameterization of the nonuniform-SPB.

There are several models for the energy-dependence of the inelastic damping, $ImV(E)$, for pure metal. An $E^{1/3}$ dependence of the inelastic damping worked well in the LEED calculations of Ni surfaces [51]. An alternative was proposed by McRae and Caldwell [57] in 1976 from their VLEED investigations of Ni surfaces, which was widely used in dealing with other metal surfaces. The damping varies with energy monotonically:

$$\mathrm{Im}V(E) = \gamma\left(1 + E/\phi\right)^{\delta}, \quad (\gamma = -0.26, \ \delta = 1.7)$$

the ϕ is the work function of the crystal of concern.

The spatial-decay of the inelastic damping, $\mathrm{Im}V(z)$, was often expressed typically with a step and a Gaussian-type function:

$$\mathrm{Im}V(z) = \beta \times \exp[-\alpha|z - z_1|], \quad (z < z_1)$$
$$= \eta \qquad\qquad\qquad (z < z_1 < z_{SL}) \qquad (14.6)$$

where β, α, η and z_1 are adjustable parameters to be fixed in calculations. β and η were employed to modify the intensity of damping at various regions and z_{SL} is the atomic positions of the second atomic layer.

The $\mathrm{Re}V(z)$, $\mathrm{Im}V(z)$, and $\mathrm{Im}V(E)$ are traditionally independent from one another. The independent treatment is indeed convenient in examining contributions made by the individual component but renders the correlation among the parameters. In the one-dimensional approach, there are all together seven independent parameters to be optimized in VLEED calculations with z being the variable of integration:

$$\mathrm{Re}V(z; \lambda, z_0, V_0)\text{and,}$$
$$\mathrm{Im}V(z, E; z_1, \alpha, \beta, \eta, \gamma, \delta)$$

The independent treatment of such a huge number of correlated parameters causes troublesome in the uniqueness of solutions. Figure 14.9 shows the schematic diagrams of the (a) $\mathrm{Re}V(z)$, (b) the pseudo-Fermi z function, (c) the $\mathrm{Im}V(z)$ in a step-Gaussian wise and, (d) the Fermi-z decay $\rho(z)$ defined in [42].

14.4.4 Energy Dependent 3D-SPB

VLEED integrates over large areas of a surface, which effects $z_0(x, y)$ to be $z_0(E)$. As it is known that at $z = z_0$, the Poisson equation approaches zero. Therefore, z_0 is the boundary of the region occupied by electrons. If we permit z_0 to vary as a function of the surface coordinates, then $z_0(x, y)$ provides a contour describing the spatial distribution of electrons. The surface coordinate (x, y) relates to the $k_{//}$ in reciprocal lattice ($k_x \propto 1/x$). In reality, the SPB parameters are functions of energy E, lateral wave vector $k_{//}$ and surface coordinates (x, y) but the resultant $z_0(E, k_{//}(x, y))$ is so complicated that it is impractical to attempt to generalize the functional dependence of a mixture of the real and reciprocal spaces in the theory models.

On the other hand, the term $k_{//}(x, y)$ is the contribution of multiple beams in the VLEED calculation integrated over large areas as each beam has a specific $k_{//}$. VLEED calculation can only provide the average effect of the $z_0(E)$, which is the

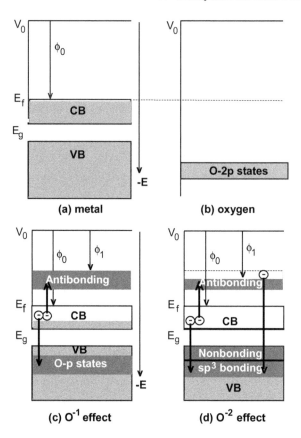

Fig. 14.9 Valence DOS modification of metal surface with chemisorbed oxygen. Panel (**a**) and (**b**) corresponds to the valence band of pure metal and oxygen, respectively. O-$2p$ level is much lower relative to the Fermi level of the metal. Panel (**c**) shows the resultant of O adding to metal at the initial stage of reaction. O^{-1} formation produces three features of bonding states, holes, and antibonding dipoles. Panel (**d**) corresponds to the effect of O^{-2} formation, which gives rise to four characteristic sub-band features that exhibit all localized nature. Reprinted with copyright permission from [38, 42]

joint contribution of the energy and the surface coordinates. For the effect of spatial integration, the energy effects dictate the $z_0(E)$ features. The multi- or high-order diffraction can only modify the surface-coordinate contribution to the shape of the $z_0(E)$. Therefore, we may treat the image plane as functional dependent of energy $z_0(E)$. The $z_0(E)$ profile should be in any form other than a constant or monotonically energy dependent exhibiting joint features of valence DOS and surface morphology, as revealed by STM and STS.

14.4.5 Parameterization and Functionalization

14.4.5.1 Local DOS and Work Function

Instead of the complex form of $\rho(z)$ derived from the Poisson equation, we may define a Fermi z function to describe the spatial decay $\mathrm{Im}V(z)$ and electron spatial distribution (Fig. 14.9d) [42]:

$$\rho(z) = \frac{V_0}{1 + \exp\left[-\frac{(z-z_1)}{\alpha}\right]} \qquad (14.7)$$

$\rho(z)$, characterized by z_1 and α, is constrained by $\rho(z_0) \approx 0$. The z-directional integration of $\rho(z)$ outside a certain atomic layer of the lattice is therefore proportional to the occupied local DOS, $n(x, y)$. The region of integration was determined in VLEED as just within one atomic layer on the basis that the inelastic damping dominates in this region [58].

The work function changes with oxygen adsorption in a localization manner. STM images of oxygen-metal surfaces show pronounced corrugations on the atomic scale, whereas, the measured work function is an average over large areas. Thus, it is necessary to introduce the concept of local work function $\phi_L(x, y)$. The $\phi_L(x, y)$ depends on $[n(x, y)]^{2/3}$. The $n(x, y)$ is an integration of the Fermi z function (14.6).

Although the Fermi decay of $\mathrm{Im}V(E, z)$ represents the spatial distribution of electrons it is difficult to calibrate the integration because the $\mathrm{Im}V(z, E)$ varies with energy. Fortunately, the local work function is such a convenient variable that is used to link the $\mathrm{Re}V(z)$ and the $\mathrm{Im}V(z, E)$. The surface local DOS is proportional to the integration of the Fermi z function $\rho(z)$ from a position inside the crystal to infinitely far away. Letting $n(x, y)$ and n_0 be the DOS for oxygen-added and clean metal surfaces, respectively, then

$$n(x, y) = \int_{D_{12}}^{-\infty} \rho(z, V_0, z_0(x, y), \lambda(z_0))dz$$

$$n_0 = \int_{D_{12}}^{-\infty} \rho(z, 11.56, -2.5, 0.9)dz \qquad (14.8)$$

Therefore, the localized work function varies with atomic coordinates and energy in the form,

$$\phi_L(x, y) = E_0 - E_F \left(\frac{n(x, y)}{n_0}\right)^{2/3} \qquad (14.9)$$

where $E_0 = 12.04$ eV and $E_F = 7.04$ eV are the vacuum and Fermi level of a pure Cu surface. For calibration n_0 was given by the data for the Cu(001) surface ($V_0 = 11.56$ eV, $z_1 = z_0 = -2.5$ Bohr radii, $1/\alpha = \lambda = 0.9$) [59]. Different calibrations merely offset the $\phi_L(x, y)$ value.

The work function depends on the occupied DOS and it is independent of the dimensions of whatever sample is being considered. The concept of local work function ϕ_L has been employed to explain variations on the scale of patches of unit cells with chemisorbed oxygen [30]. This concept can be extrapolated to the atomic scale so that variations occur over the dimensions of a single atom.

For metal systems with chemisorbed oxygen, the usual concept of ϕ is no longer valid due to the strongly "localized" features. It is even unlikely that the strongly localized electrons with low mobility move from the site of "lower" ϕ_L to the site of "higher" ϕ_L on the same surface described with $\phi_L(x, y)$. Since the VLEED integrates over a large area of surface, all the quantities depending on surface coordinates (x, y) become energy-dependent. Accordingly, the $n(x, y)$ in ϕ_L becomes $n(E)$. The $n(E)$ is precisely the occupied DOS that is characterized by $z_0(E)$. In the current modeling approach, the $\phi_L(E)$ becomes E dependent and it can also be extended to large surface areas over which VLEED integrates for the DOS.

14.4.5.2   Parameterization of the 3D-SPB

One may define an inelastic potential to unify the effect that damping occurs in the electron-occupied space (Fermi z decay) with any energy greater than the work function, which depends on the occupied DOS [42]:

$$\begin{aligned}
\mathrm{Im}V(z, E) &= \mathrm{Im}[V(z) \times V(E)] \\
&= \gamma \times \rho(z) \times \exp\left[\frac{E - \phi_L(E)}{\delta}\right] \\
&= \frac{\gamma \times \exp\left[\frac{E - \phi_L(E)}{\delta}\right]}{1 + \exp\left[-\frac{z - z_1(z_0)}{\alpha(z_0)}\right]}
\end{aligned} \qquad (14.10)$$

where γ and δ are constants depending on the calibration of the measured spectral intensities. The $z_1(z_0)$ and the $\alpha(z_0)$ in the Fermi z function characterizes the electron distribution.

Because single electron can be excited by an incident electron beam that has any energy greater than the work function, i.e., $E \geq \phi$, we choose the form of $[E - \phi_L(E)]$ in the inelastic damping. It is to be noted that the surface electron density $\rho(z)$ is so important that it ties all the identities for surface electrons together meeting the basic requirement. These identities are local work function $\phi_L(x, y)$, elastic potential $\mathrm{Re}V(z)$, spatial decay $\mathrm{Im}V(z)$ and energy dependent of the inelastic damping $\mathrm{Im}V(E)$ of the electronic system.

In order to reduce numerical efforts and ensure the uniqueness of solutions, one can define the SPB parameters as functional dependents on z_0 in the following form [42]:

$$z_1(z_0) = z_0 \times \exp\left[-\left(\frac{z_0 - z_{0M}}{\tau_1}\right)^2\right],$$

$$\alpha(z_0) = 1/\lambda(z_0) \times \exp\left[-\left(\frac{z_0 - z_{0m}}{\tau_2}\right)^2\right], \qquad (14.11)$$

where the constants τ_1 and τ_2 are the full width at half maximum (FWHM) of the Gaussian functions and optimized to be 0.75 and 1.50, respectively, by minimizing the Δz_0 in the calculations. The z_{0m} is estimated to be -1.75 Bohr radii.

The λ in Eq. (14.11) increases monotonically with the outward shift of z_0 [42]:

$$\lambda(z_0) = \lambda_{0M}\left\{x + (1 - x) \times \exp\left[-\left(\frac{z_0 - z_{0M}}{\lambda_z}\right)^2\right]\right\}, \quad (x = 0.4732) \quad (14.12)$$

where the $\lambda_{0M} = 1.275$ is the maximum of λ corresponding to $z_{0M} = -3.425$ Bohr radii and $\lambda_z = 0.8965$. Constants for O–Cu(001) surface were obtained by least-square simulation of a $z_0(E) - \lambda(E)$ curve.

Equations (14.10) and (14.11) represent not only correlation among SPB parameters but also hypotheses that, at the dipole site, $z_{1M} \approx z_{0M}$, $\alpha \approx \lambda^{-1}$, while in the missing-row or ion position, $z_{1m} \ll z_{0m}$ and that the ImV(z) is much less saturated than is ReV(z) in the depressed sites of STM images. The SPB increases its degree of saturation with the outward shift of the image plane z_0 by polarization. The z_0-dependent of the SPB parameters is illustrated in Fig. 14.10.

14.4.5.3 Physical Indication of the SPB

Figure 14.11 shows an interface between the bulk and the vacuum for typical O-metal surfaces to illustrate the coordinate-resolved z_0 and z_1. The z-axis originating from the top layer (z_{0L}) directs into the bulk. The two typical broken curves, at the sites of the dipole and the missing-row vacancy, are ReV(z) of Fig. 14.9a to illustrate the difference in z_0 and λ from site to site on the surface. That $z_0(x, y)$ is usually farther away from the surface than the $z_1(x, y)$, which results from the contribution of the surrounding electrons to the image potential. It is also to be noted that:

$$\rho(z_1) = 0.5\rho(bulk, z \geq z_{SL}) > \rho(z < z_0) \approx 0,$$

as defined by the Fermi-z function and the correlation between the ReV(z) and the ImV(z). In the vacant position, the smallest z_{1m} is much lower than z_{0m} because it is

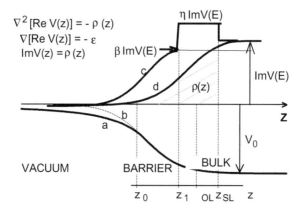

Fig. 14.10 An illustration of the non-uniform SPB model [42]. Curve (a) is ReV(z) and the broken curve (b) the quasi-Fermi z function. Curve (c) is the conventional step-Gaussian decay of the inelastic damping, in which β and η are independent parameters used to modulate the intensities in different regions. Curve (d) is the Fermi-z function, $\rho(z)$, proposed to model the spatial electron distribution. ImV(E) is the energy-dependent bulk damping and the V_0 the inner potential constant, respectively. The z_{OL} and z_{SL} are the positions of the overlayer and the second layer of lattice, respectively

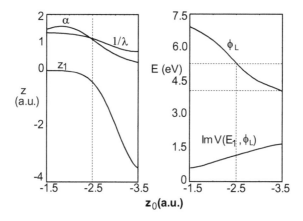

Fig. 14.11 Z_0-dependence of the single-variable parameterization of the barrier functions [42]. E_1 and E_2 are particular (terminal of the VLEED window) energies at 6.3 and 16.0 eV, respectively. Left panel shows the z_0-dependent of λ, z_1 and α. Right panel illustrates that $\phi_L(x, y)$ reduces its value with the outwards shift of $-z_0$. Inelastic damping ImV(E) increases with $-z_0$. If z_0 kept constant, the approximation will degrade to the one-dimensional uniformity

assumed that no free electrons flow into the missing-row vacancy. At the dipole site, the largest $z_{1M} \cong z_{0M}$ (Fig. 14.12).

The formation of metal dipoles results in the outward-shift and the saturation of electron clouds. Hence, the higher the protrusion in the STM image or z_0 is, the

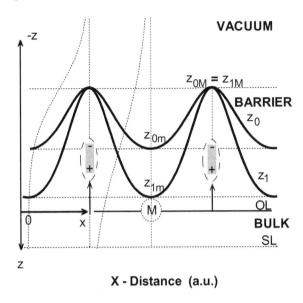

Fig. 14.12 Coordinate (viewed in the x-z plane) dependence of the z_0 and z_1 that characterize the 3D-SPB [42]. Regions of bulk ($z > z_{OL}$), barrier and vacuum are indicated. M is the vacancy of missing atom. Displacement and vertical component of dipoles are also denoted. That the z_0 in ReV(z) is usually higher than the z_1 in ImV(z) results from the contribution of the surrounding electrons to the image potential characterized by z_0. At distances sufficiently far away from the surface, the SPB approaches uniformity. The broken lines are the ReV(z) corresponding to the locations at the dipole and the missing-row vacancy, showing the difference in saturation degree

higher the saturation degree of the SPB will be, and the lower the work function will be. The gradient of ReV(z), or the intensity of the electric field at the surface, should also be site-dependent (see Fig. 14.9). At the dipole site, the electric field is much stronger than that at a clean surface or in the STM depressions. At distances sufficiently far away from the surface, the nonuniform SPB degenerates into the conventional uniform type.

In the VLEED energy range, single-electron excitation processes dominate the damping, which may occur at energy near top edge of a fully occupied band. Phonon scattering and other minor processes may cause the incident beam to decay in a monotonic way. However, single electron excitation will change the damping function by adding a "hump" to the ImV(E). On the other hand, from the density point of view, the denser the electrons are, the higher the damping will be. Inelastic damping will never occur in the region without electrons. The sum of monotonic decay of incident wave and "humps" caused by single excitation as well as the density effect implies that the real forms of inelastic damping are more complicated that beyond the description of the ImV(E) in a constant or a monotonic way.

14.4.6   Significance and Limitations of the 3D-SPB

Instead of being independent conventionally, the $ReV(z)$ and $ImV(z, E)$ are tied together through the surface charge distribution $\rho(z)$. Except for the inner potential constant V_0, the parameters λ, α and z_1 are functional dependents of z_0. The number of variables for the SPB is hence reduced from four to one. Besides, the layout is more physically meaningful than the independent treatment of the correlated parameters.

This single-variable parameterization allows the calculation code to optimize z_0 automatically to give intensity matching between calculation and measurement at each value of energy. If the character z_0 remains constant, the $ReV(z)$ and the $ImV(z, E)$ will degrade to the conventional form, namely, one-dimensional uniform and monotonic energy dependence. Besides, as will be shown, the $z_0(E)$ profile varies with atomic arrangement. The connection of the SPB to the crystal geometry represents that the crystal structure and the SPB are interdependent as they are consequences of surface bond formation and charge localization. It becomes now possible to determine crystal and electronic properties simultaneously with VLEED by analyzing the structure dependent $z_0(E)$ profile. Therefore, the difficulty of VLEED in conventional wisdom has been overcome completely. On the other hand, the variation of the energy states represented by the $z_0(E)$ is an important aspect of chemisorption studies, can also be probed because of this approach.

However, processes such as band-transition and single-excitation that add humps to the monotonic damping (due to uniform DOS or constant work function) are not clear in the present expression. Fortunately, this can be compensated for by a z_0-optimizing method based on the premise that the work function depends on the DOS that is whatsoever not a constant.

This set of approaches correlate the chemical bond, valence DOS, and the SPB with all the observations in terms of microscopy, crystallography and spectroscopy. The significance of the approach is that it reflects the essential link of the quantities in terms of bond forming and its consequences on the behavior of atoms and valence electrons at the top layer of a surface by proper parameterization.

14.5   Summary

We have rationalized and formulated the energy dependent 3D-SPB and the M_2O tetrahedron models with expectation of the derivatives on the bond geometry, valence DOS and morphology of the chemisorbed surface. Parameterization of the SPB and crystal structures is necessary to correlate the features resolved using microscopy, crystallography, and the electron spectroscopy. Calculation would be simplified by reducing the number of independent parameters, which also reflects the real process of reaction on the performance of atoms and electrons in the surface top layer.

References

1. R. Jones, P.J. Jennings, O. Jepsen, Surface barrier in metals: a new model with application to W (001). Phys. Rev. B **29**(12), 6474 (1984)
2. G. Hitchen, S. Thurgate, P. Jennings, Determination of the surface-potential barrier of Cu(001) from low-energy-electron-diffraction fine structure. Phys. Rev. B **44**(8), 3939 (1991)
3. S.M. Thurgate, C. Sun, Very-low-energy electron-diffraction analysis of oxygen on Cu(001). Phys. Rev. B **51**(4), 2410–2417 (1995)
4. P. Jennings, S. Thurgate, G. Price, The analysis of LEED fine structure. Appl. Surf. Sci. **13**(1), 180–189 (1982)
5. A. Ermakov, E. Ciftlikli, S. Syssoev, I. Shuttleworth, B. Hinch, A surface work function measurement technique utilizing constant deflected grazing electron trajectories: Oxygen uptake on Cu(001). Rev. Sci. Instrum. **81**(10), 105109 (2010)
6. N.V. Smith, Phase analysis of image states and surface states associated with nearly-free-electron band gaps. Phys. Rev. B **32**(6), 3549 (1985)
7. J. Inkson, The effective exchange and correlation potential for metal surfaces. J. Phys. F: Met. Phys. **3**(12), 2143 (1973)
8. G. Malmström, J. Rundgren, A program for calculation of the reflection and transmission of electrons through a surface potential barrier. Comput. Phys. Commun. **19**(2), 263–270 (1980)
9. M. Lindroos, H. Pfnür, D. Menzel, Theoretical and experimental study of the unoccupied electronic band structure of Ru (001) by electron reflection. Phys. Rev. B **33**(10), 6684 (1986)
10. C. Bai, *Scanning tunneling microscopy and its application*. vol. 32 (Springer, 2000)
11. H. Rotermund, Investigation of dynamic processes in adsorbed layers by photoemission electron microscopy (PEEM). Surf. Sci. **283**(1), 87–100 (1993)
12. U. Döbler, K. Baberschke, J. Stöhr, D. Outka, Structure of c (2×2) oxygen on Cu(100): A surface extended X-ray absorption fine-structure study. Phys. Rev. B **31**(4), 2532 (1985)
13. G. Ertl, Reactions at well-defined surfaces. Surf. Sci. **299**, 742–754 (1994)
14. W. Jacob, V. Dose, A. Goldmann, Atomic adsorption of oxygen on Cu(111) and Cu(110). Appl. Phys. A **41**(2), 145–150 (1986)
15. Y. Kuk, F. Chua, P. Silverman, J. Meyer, O chemisorption on Cu(110) by scanning tunneling microscopy. Phys. Rev. B **41**(18), 12393 (1990)
16. J. Nørskov, Theory of adsorption and adsorbate-induced reconstruction. Surf. Sci. **299**, 690–705 (1994)
17. K.W. Jacobsen, Theory of the oxygen-induced restructuring of Cu(110) and Cu(100) surfaces. Phys. Rev. Lett. **65**(14), 1788 (1990)
18. H. Zeng, K. Mitchell, Further LEED investigations of missing row models for the Cu(100) − (22×2) R45°-O surface structure. Surf. Sci. **239**(3), L571–L578 (1990)
19. N. Lang, Vacuum tunneling current from an adsorbed atom. Phys. Rev. Lett. **55**(2), 230 (1985)
20. T.N. Rhodin, G. Ertl, *The nature of the surface chemical bond* (North-Holland Publishing Company: Sole Distributor for the USA and Canada Elsevier, North-Holland, 1979)
21. F. Besenbacher, J.K. Nørskov, Oxygen chemisorption on metal surfaces: general trends for Cu, Ni and Ag. Prog. Surf. Sci. **44**(1), 5–66 (1993)
22. M. Van Hove, G. Somorjai, Adsorption and adsorbate-induced restructuring: a LEED perspective. Surf. Sci. **299**, 487–501 (1994)
23. C.Q. Sun, O–Cu(001): II. VLEED quantification of the four-stage Cu_3O_2 bonding kinetics. Surf. Rev. Lett. **8**(6): 703–734 (2001)
24. C.Q. Sun, O–Cu(001): I. Binding the signatures of LEED, STM and PES in a bond-forming way. Surf. Rev. Lett. **8**(3–4): 367-402 (2001)
25. L. Pauling, The Nature of the Chemical Bond. 3rd edn. (Cornell University Press, Ithaca, NY, 1960)
26. C.Q. Sun, C.L. Bai, A model of bonding between oxygen and metal surfaces. J. Phys. Chem. Solids **58**(6), 903–912 (1997)
27. X. Zhang, Y. Huang, Z. Ma, Y. Zhou, W. Zheng, J. Zhou, C.Q. Sun, A common supersolid skin covering both water and ice. Phys. Chem. Chem. Phys. **16**(42), 22987–22994 (2014)

28. C.Q. Sun, *Relaxation of the Chemical Bond*. Springer Series in Chemical Physics, vol. 108 (Springer, Heidelberg, 2014), 807p

29. C.Q. Sun, Y. Sun, *The Attribute of Water: Single Notion, Multiple Myths*. Springer Series in Chemical Physics, vol. 113 (Springer, Heidelberg, 2016). 494 pp

30. C.Q. Sun, Oxidation electronics: bond-band-barrier correlation and its applications. Prog. Mater Sci. **48**(6), 521–685 (2003)

31. C.Q. Sun, Size dependence of nanostructures: impact of bond order deficiency. Prog. Solid State Chem. **35**(1), 1–159 (2007)

32. J.D. Jorgensen, Defects and superconductivity in the copper oxides. Phys. Today **44**, 34–40 (1991)

33. H. Kamimura, Y. Suwa, New theoretical view for high temperature superconductivity. J. Phys. Soc. Jpn. **62**(10), 3368–3371 (1993)

34. V.M. Goldschmidt, Crystal structure and chemical correlation. Berichte Der Deutschen Chemischen Gesellschaft **60**, 1263–1296 (1927)

35. D. Adams, H. Nielsen, J. Andersen, I. Stensgaard, R. Feidenhans, J. Sørensen, Oscillatory relaxation of the Cu(110) surface. Phys. Rev. Lett. **49**(9), 669 (1982)

36. P. Jennings, C.Q. Sun, Low-energy electron diffraction, in *Smart Surface Analysis Methods in Materials Science*, vol. 23, ed. by J. O'Connor, B. Sexton, R.S. Smart (Springer, 2013)

37. N. Lang, Theory of single-atom imaging in the scanning tunneling microscope, in *Scanning Tunneling Microscopy* (Springer, 1986), pp. 75–78

38. C.Q. Sun, A model of bonding and band-forming for oxides and nitrides. Appl. Phys. Lett. **72**(14), 1706–1708 (1998)

39. A.P. Cole, D.E. Root, P. Mukherjee, E.I. Solomon, T. Stack, A trinuclear intermediate in the copper-mediated reduction of O_2: four electrons from three coppers. Science **273**(5283), 1848 (1996)

40. F.M. Chua, Y. Kuk, P.J. Silverman, Oxygen chemisorption on Cu(110): An atomic view by scanning tunneling microscopy. Phys. Rev. Lett. **63**(4), 386–389 (1989)

41. F. Jensen, F. Besenbacher, E. Laegsgaard, I. Stensgaard, Dynamics of oxygen-induced reconstruction on Cu(100) studied by scanning tunneling microscopy. Phys. Rev. B **42**(14), 9206–9209 (1990)

42. C.Q. Sun, C.L. Bai, Modelling of non-uniform electrical potential barriers for metal surfaces with chemisorbed oxygen. J. Phys. Condens. Matter **9**(27), 5823–5836 (1997)

43. H. Rotermund, J. Lauterbach, G. Haas, The formation of subsurface oxygen on Pt(100). Appl. Phys. A **57**(6), 507–511 (1993)

44. J. Lauterbach, K. Asakura, H. Rotermund, Subsurface oxygen on Pt(100): kinetics of the transition from chemisorbed to subsurface state and its reaction with CO, H_2 and O_2. Surf. Sci. **313**(1–2), 52–63 (1994)

45. J. Lauterbach, H. Rotermund, Spatio-temporal pattern formation during the catalytic CO-oxidation on Pt(100). Surf. Sci. **311**(1), 231–246 (1994)

46. J. Boulliard, M. Sotto, On the relations between surface structures and morphology of crystals. J. Cryst. Growth **110**(4), 878–888 (1991)

47. R. Dietz, E. McRae, R. Campbell, Saturation of the image potential observed in low-energy electron reflection at Cu(001) surface. Phys. Rev. Lett. **45**(15), 1280 (1980)

48. M. Read, A. Christopoulos, Resonant electron surface-barrier scattering on W(001). Phys. Rev. B **37**(17), 10407 (1988)

49. A. Adnot, J. Carette, High-resolution study of low-energy-electron-diffraction threshold effects on W(001) surface. Phys. Rev. Lett. **38**(19), 1084 (1977)

50. H. Pfnür, M. Lindroos, D. Menzel, Investigation of adsorbates with low energy electron diffraction at very low energies (VLEED). Surf. Sci. **248**(1–2), 1–10 (1991)

51. J. Demuth, D. Jepsen, P. Marcus, Comments regarding the determination of the structure of c (2×2) sulfur overlayers on Ni(001). Surf. Sci. **45**(2), 733–739 (1974)

52. T. Fujita, Y. Okawa, Y. Matsumoto, K.-I. Tanaka, Phase boundaries of nanometer scale c (2×2)-O domains on the Cu(100) surface. Phys. Rev. B **54**(3), 2167 (1996)

53. E. McRae, Electron diffraction at crystal surfaces: I. Generalization of Darwin's dynamical theory. Surf. Sci. **11**(3): 479–491 (1968)
54. R.O. Jones, P.J. Jennings, LEED fine structure: origins and applications. Surf. Sci. Rep. **9**(4), 165–196 (1988)
55. J. Pendry, G.P. Alldredge, Low energy electron diffraction: the theory and its application to determination of surface structure. Phys. Today **30**, 57 (1977)
56. M. Nishijima, M. Jo, Y. Kuwahara, M. Onchi, Electron energy loss spectra of a Pd(110) clean surface. Solid State Commun. **58**(1), 75–77 (1986)
57. E. McRae, C. Caldwell, Absorptive potential in nickel from very low energy electron reflection at Ni(001) surface. Surf. Sci. **57**(2), 766–770 (1976)
58. C.Q. Sun, Spectral sensitivity of the VLEED to the bonding geometry and the potential barrier of the O–Cu(001) surface. Vacuum **48**(5), 491–498 (1997)
59. C. Hitchen, S. Thurgate, P. Jennings, A LEED fine structure study of oxygen adsorption on Cu(001) and Cu(111). Aust. J. Phys. **43**(5), 519–534 (1990)

Chapter 15
Methodology: Parameterization

Abstract Translating atomic dislocations into the tetrahedron bond geometry and correlating the ReV(z) and ImV(z, E) through Poisson's equation not only reduce largely the number of SPB freely adjustable parameters but also properly describes the true processes of VLEED. The integral of the elastic ReV(z) determines the phase shift and the integral of the inelasstic ImV(z, E) dictates the amplitude loss of the diffracted electron beams. Preliminary calculations justified consistently the validity of the parameterization, decoding method, calculation code and the bond and the SPB models.

Highlights

- The oxide tetrahedron and the 3D-SPB frame the STM/S observations.
- The SPB parameterization describes the spatial and energetic distribution of electrons.
- Relaxation of bond length and geometry sources the valence states and SPB.
- Parameterization is justified valid and meaningful.

15.1 Decoding Methodology

15.1.1 Data Calibration

Numerical processing the measured VLEED *I–E* spectra is essential to the understanding of the physics behind the experimental observations and hence to justify the models proposed. Digitization and calibration of the experimental data are the essential stage before computation. Inappropriate calibration of the measured data may mislead conclusions, and therefore, careful and reasonable calibration of the data is important prior to the numerical processing.

C. Q. Sun, *Electron and Phonon Spectrometrics*,
https://doi.org/10.1007/978-981-15-3176-7_15

15.1.1.1 Conventional Approach

Conventionally, one normalizes the individual I–E curve independently. Both the measured I_{exp}-E and the computed I_{cal}-E spectra are normalized by dividing the entire spectrum with the maximal peak intensities, I_{eM} and I_{cM}, of the specific spectrum, respectively; so that the maxima of all the normalized curves are equal to unit. Comparison between calculation and experimental results is then performed with the "least-square" R-factor method:

$$R = 1 - \sqrt{\frac{\sum_{i=1}^{N} [I_c(E_i) - I_e(E_i)]^2}{N(N-1)}}.$$

$I_c(E_i)$ and $I_e(E_i)$ are normalized intensities at selected energies E_i (step size of $E_i - E_{i-1} = 0.1$ eV was used). The R-value determines the degree of satisfaction. For an ideal agreement, $R = 1$. This convention is normally used and acceptable for qualitative simulation of a certain single VLEED spectrum. This treatment gives information of shape-similarity between the calculated and experimental spectral curves.

However, this method is inadequate for the functionalized SPB, because it is unable to differentiate with this method the intensity discrepancy among a series of I–E curves such as those produced by varying oxygen-exposure. This treatment will surely miss important information such as relative variation of the spectral intensity during the reaction.

15.1.1.2 Common-Standard Normalization

The relative intensity of one curve to another in a complete set of dynamic I–E spectra gives important information. Data collection is done under stable instrumental conditions. Therefore, it is necessary to calibrate all the experimental curves with a common scale to all the I–E profiles. The reasonable way is to choose one maximum $I_{eM}(E_i)$ from among all of the experimental spectra to calibrate all the I–E curves across. This ensures correct information on the relative change of intensities during the data processing, because chemical reaction changes not only the shape but also the intensity of the spectral peaks.

On the other hand, deviations of the absolute spectral intensities from the real status can be modulated by the damping potential $(1 \pm 0.2) \times \text{ImV}(E)$ that determines the spectral intensity. Modifying damping constant will offset the computed curves. Therefore, intensity change corresponds to the physical process that contributes to the inelastic damping.

Digitization of VLEED spectra is subject to the following considerations:

(1) The incident current I_0 between 6.0 and 16.0 eV was assumed to be constant within the instrumentation error. The inner potential constant V_0 was kept constant in calculation though it varies slightly with energy for the pure Cu(001)

surface in the normal LEED energies [1]. The VLEED spectra are only active in the upper valence band region, that is, 6.0–12.0 eV.

(2) Calculations revealed that the (00) beam reflectance (I_{00}/I_0) greater than 12.5% causes serious convergence problems. Measurements also showed that the reflectivity is about 10%. Hence, the maximal value of $I_{00}/I_0 = 10\%$ was assumed as the maximal reflectivity to calibrate all of the I–E curves. In each curve, the absolute intensity of the first and the last energy is vital to determine the constants of γ and δ in the inelastic SPB. For instance, once the SPB functions are defined, we can solve the damping equation with the initial conditions obtained by orthogonal-analysis technique.

(3) In the set of angular-resolved VLEED spectra [2], it was assumed that no structural change occurs to the bond geometry during the VLEED data acquisition. This assumption is also acceptable because increasing oxygen exposure promotes more the reaction than sample aging.

(4) Earlier data simulation for Cu(001) surface [3] provided values of $V_0 = 11.56$ eV, $z_0 = -2.5$ a.u. ($\cong 1.32$ Å approaches Cu atomic radius 1.276 Å) and $\lambda = 0.9$. These data were used to calibrate the local work function and as references for setting the SPB parameters.

(5) The z-scale difference of 0.45 Å in STM image offers reference for Δz_0 because both STM image and $z_0(x, y)$ are the convolution of the surface electron distribution. Taking the lateral convolution for the finite size of STM tip and modification by multiple diffraction and high-order diffraction in VLEED into consideration, it is reasonable to assume that the Δz_0 and Δz_{STM} are comparable and Δz_0 should be as small as possible.

15.1.2 Parameter Initialization

The second task before calculation is to initialize the SPB parameters. Orthogonal optimization technique was used first at several energies E_{0i} with parameters of z_{0i}, λ_i for the ReV(z) and ImV(E_{0i}). Least-square optimization was used to fit the $\lambda(z_0)$ plot for the constants. Both the STM image and barrier distribution imply that the SPB increases its saturation degree with the outward shift of the image plane, z_0. This match serves as the physical basis for the relation between z_0 and λ. The optimized z_{0M} and z_{0m} are -3.425 and -1.750 a.u., and the corresponding λ is 1.275 and 0.650.

There are four constants in the ImV(z, E). Matching the spectral intensities at the VLEED energy terminals of 6.0 and 16.0 eV gives the γ and δ values. Given the damping values at the energy terminals, one could solve the damping equation to obtain the γ and δ. The FWHM of the Gaussian functions for $z_1(z_0)$ and $\alpha(z_0)$ are determined by obtaining the smallest Δz_0 (near the <11> direction presents the maximal value).

The following summarizes the numerical processing of calculation:

- The SPB parameters were fixed first by using the known identities of the missing-row structure model [4] and the conventional SPB.
- To set the constants in the single-variable SPB functions, we calculated the VLEED data at a few energies by adjusting the most sensitive parameters.
- Uncertainty of the solutions was overcome by introducing the correlation among the SPB parameters as well as the local work function. The infinite number of solutions due to independently varying λ and z_0 were reduced to finite ones by the function of $\lambda(z_0)$. Introducing the local work function then ascertain the unique solution.

15.1.3   Calculation Methods

Conventionally, one can only change one parameter at a time in calculations. Although the computer can automatically do the numerical processing, this method is tedious and fruitless. The orthogonal optimizing method is widely used in optimization for experimental design, not only for reducing the time span of experiments but also for predicting the trends of results and correlation among the variables.

Complementing the orthogonal optimization, z_0-scanning and z_0-optimizing was carried out. After the SPB and geometrical parameters have been fixed, calculations were performed based on the single-variable parameterization by varying z_0 over -2.5 [refer to pure Cu(001)] ± 1.25 with 0.25 (a.u.) steps. Contour plot of z_0 versus E is then drawn with results satisfying $I_c(z_{0i}, E)/I_e(E) = 1.0 \pm 0.05$. The $z_0(E)$ contour plot is the unique yield of the z_0-scanning method, which gives the best fit of the measurement and shows all the possible solutions as well. This z_0-scanning and z_0-optimizing method is also convenient to compare different models and to refine the SPB shapes and the bond geometries.

Once the refinement of the $z_0(E)$ plot is completed, the program automatically fits the value of the $z_0(E_i)$ to give a desired level of agreement (normally $I_c(E_i)/I_e(E_i) = 1.00 \pm 0.01 \sim 0.03$). In contrast to the z_0 scanning method, the step of Δz_0 automatically varies from 0.25 to 0.0005 depending on the result of $\kappa = I_c(E_i)/I_e(E_i)$. If κ reaches the required precision, calculation will automatically turn to the next step E_{i+1}. This method yields simply the geometry-dependent $z_0(E)$ profile and a replication of the measured VLEED spectrum. Quantities such as the bond geometry, work function, barrier shapes and energy band structure are automatically given as the product of data processing. Therefore, careful calibration of the experimental data is essential.

15.1.4 Criteria for Model Justification

Criteria for the optimal $z_0(E)$ profiles, and correspondingly, the optimal bond geometry are essential. In convention, it is expected that the $z_0(E)$ approaches a constant. This is acceptable for pure metal surfaces with homogeneous energy states and small ion core corrugations but it is no longer true for a system with chemisorbed oxygen. The $z_0(E)$ profile should be in any form showing joint features of topography and valence DOS other than traditional constant or monotonically energy dependence. With respect to both the conventional wisdom and the features of the new SPB showing humps due to electron excitation, the criteria for selecting the $z_0(E)$ curves were set as follows:

- The number of solutions of $z_0(E)$ is finite;
- Δz_0 is as small as possible; and
- The fewest extra features appear on the $z_0(E)$ curve.

The z_0-scanning and z_0-optimizing methods are much more revealing than calculations by adjusting independently all parameters. These methods have enabled the best models and the match between the STM/S images and the VLEED refined contour plots, which reveals information on the electronic spatial distribution and variation of the valence DOS in the VLEED energy window.

15.2 Code Validation

15.2.1 Clean Cu(001)

Two identifies determine the VLEED fine-structure features. One is the surface lattice geometry subjecting to reconstruction and relaxation; the other is the SPB describing the behavior of surface electrons in both real and energy domains. Usually, presumption is made that the lattice geometry is already known and then one can explore the SPB parameters. In order to verify the validity of the multiple diffraction code, scattering from a Cu(001) clean surface was first calculated by using the one-dimensional SPB approximation and the monotonic damping functions with the known lattice structure. The conventional wisdom for clean metal surface is adequate to obtain qualitative agreement between theory and measurement [5].

VLEED fine-structure features are very sensitive to the elastic $ReV(z, z_0, \lambda)$ potential. A minimal difference of the parameters from earlier work is due to the multiple diffraction effects that are fully included in the multi-atom and multiple diffraction code [6].

A total of nine SPB parameters were divided into two groups. The orthogonal-optimization of the $L_{25}(5^6)$-$(\gamma, \delta, \eta, \beta, z_1, V_0)$ group and the $L_9(3^4)$-(z_0, λ, α) group were conducted to gain the best fit. The following describes the calculation procedures [10]:

(1)   Searching the dominant parameters and locking the value ranges of the parameters for possible best fit. Statistical analyses of the calculation results confirmed that the spectrum intensity is very insensitive to parameter α and β, thus, they can hence be taken as constants. The V_0 was kept constant at 11.56 eV as optimized for pure Cu(001) surface [2]. Doing so, reduces the number of independent variables from nine to six for the subsequent optimizations.

(2)   Utilizing correlations between the parameters. The combinations of (z_0, λ) and $(\gamma, \delta; z_1, \alpha; \beta, \eta)$ are correlated through the integration of $\mathrm{Re}V(z; z_0, \lambda)$ and $\mathrm{Im}V(z, E; \gamma, \delta; z_1, \alpha; \beta, \eta)$. Then we could fix one parameter and adjust the other one in each of the couples. Thus, the number of variables is further reduced from three pairs to three independent ones.

(3)   Repeating (i) and (ii) to refine the parameters. The parameter values for the given crystal structure could then be optimized.

Figure 15.1 compares the theoretical fitting (solid line) to the Cu(001) spectrum (broken line) measured at 70.0° incidence and 42.0° azimuth angles, near the <11> direction (45.0°). I_{00}/I_0 is the ratio of intensities of the measured (00) beam to the incident beam. The simulation result obtained with the multiple diffraction schemes and the one-dimensional SPB model.

Within the tolerance, the simulation is acceptable except at energies below 9.0 eV. Another similar fit was also obtained by subtracting the inner potential constant of the top layer by an amount of ~3.1 eV. Simulating the VLEED spectrum of a

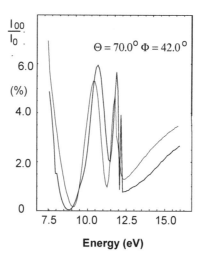

Fig. 15.1 Theoretical fitting (solid line) to the experimental (broken line) VLEED spectrum of a Cu(001) clean surface. Incident and azimuth angles are 70.0° and 42.0°, respectively. Result implies that the multiple-diffraction approach is sufficient and complete to include all necessary diffraction events and that the uniform SPB is adequate for clean metal surface. Reprinted with copyright permission from [7]

clean Cu(001) surface suggests that the multiple-diffraction approach is sufficiently accurate and complete to allow for the inclusion of all necessary diffraction events. Importantly, the conventional SPB model for pure metal (ideal Fermi system) is adequate to produce the correct intensity for the high-order diffraction events in the SPB region.

15.2.2 Oxygen on Cu(001): Model Comparison

To examine the code validity for the O-Cu(001) surface, calculations were performed with the same convention for clean Cu(001) surface by using geometries of several missing-row structural models to optimise the SPB parameters. The best model is then assigned to the one that provides the closest fit to the measured data.

Simulation was conducted to two typical VLEED I–E curves for 300-L oxygen chemisorbed Cu(001) surface. The incident angle (relative to the normal of the surface) is 69.0° and the azimuth angles of the compared curves are 23.5° (near <21> direction) and 43.5° (near <11> direction), respectively. Figure 15.2 compares the atomic positions in the compared models. The side view above shows atomic positions viewing along the -Cu-O-Cu- chain; the bottom is a top view of the Cu(001)- ($\sqrt{2} \times 2\sqrt{2}$) R45°-2O complex unit cell denoting the azimuth angle of incident beams. Table 15.1 lists the structural parameters used in calculations. Model A is derived from the M_2O tetrahedron structure. The five calculation parameters (D_{12}, DCu_x, DCu_z, DO_x, DO_z) for model B and C were optimized in earlier LEED calculations [8]. Parameters for structure D are derived from the effective-medium theory predictions [9].

By using the one-dimensional SPB with nine parameters, qualitative agreement between measurement and calculation is achieved for the spectrum collected only at 23.5° azimuth angle, far away from the <11> direction for all the compared structural models, as given in Fig. 15.3. The broken lines are the measured VLEED spectra. For the 23.5° azimuth, violent features on the curves at about 6.8 and 13.5 eV come from band-gap Bragg diffraction. The main Rydberg peak on the calculations (solid lines) at 13.0 eV is very sensitive to the SPB.

All the structures provide the match of the peak position at 13.0 eV but not ideally the intensities or the FWHM of this peak. The minor inconsistency below 12.0 eV and the FWHMs of all the simulations suggest the essentiality of SPB modification by involving the 3D effect and the energy dependent damping processes. The sharp peak at 15.5 eV and the intensity-difference between measured and calculated results from 14.0 to 16.0 eV of curves B, C and D could not be eliminated by adjusting SPB parameters. The main peak at 13.0 eV would disappear before a close match being reached, like curve A, in the range over 14.0–16.0 eV. The VLEED calculations at 23.5° azimuth suggest that the atomic positions in model A are closer to the true situation than those assigned in others. Unfortunately, none of the models can provide acceptable simulation at 43.5°, near the <11> direction. This indicates the

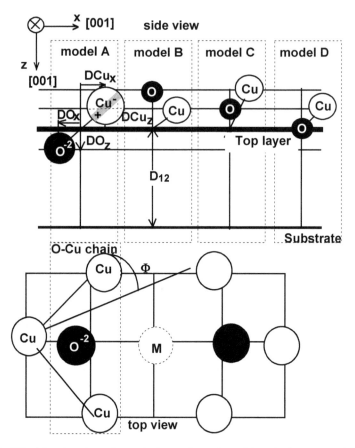

Fig. 15.2 Side (above) and top (bottom) views of atomic positions in the compared Cu(001)-($\sqrt{2} \times 2\sqrt{2}$)R45°-2O missing-row models along the O-Cu-O chain. Atomic positions in model A are converted from the M$_2$O bond geometry, in which alternations of atomic sizes and valence states are considered. These models represent almost all the documented Cu(001)-($\sqrt{2} \times 2\sqrt{2}$)R45°-2O atomic structures. Reprinted with copyright permission from [10, 11]

essentiality of the 3D feature of the SPB and the a nonuniformity and anisotropy of electron distribution.

Figure 15.4 compares the damping curves for Cu(001) and O-Cu(001) surface yielded in calculations. The damping of O-added (1.2, 9.0 eV) is much higher than that of the Cu(001) surface (0.78, 0.81 eV). Moreover, for O-added surface, the inelastic damping varies significantly with azimuth angles, a strong tendency of anisotropy. The features of nonuniformity and anisotropy of the SPB due to oxygen adsorption indeed go beyond the one-dimensional SPB description. Thus, information obtained in convention wise limits the comprehension of the chemisorbed surface.

Briefly, the one-dimensional SPB is unable to discriminate a structure model for oxygen chemisorbed system, though it works sufficiently well for the clean surface

Table 15.1 Structural information for compared models [10, 11]

Model		A	B	C	D
Variables	BA12	102.0			
Atomic Shifts (Å)	DCu$_x$	0.250	0.3	0.1	0.3
	DCu$_z$	−0.1495	−0.1	−0.2	−0.1
	DO$_x$	−0.1876	0.0	0.0	0.0
	DO$_z$	0.1682	−0.2	−0.1	0.0
	D$_{12}$	1.9343	1.94	2.06	1.94
Bond Length (Å)	BL1	1.628			
	BL2	1.776			
	BL3	1.926			
Bond Angle (°)	BA13	105.3			
	BA23	99.5			
	BA33	139.4			

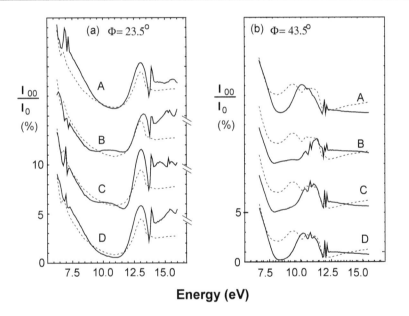

Fig. 15.3 Comparison of the calculation yields (solid curves) with four structural models at 23.5° and 43.5° azimuth angles for the O-Cu(001) surface. Broken curves are spectra measured at 69°-incidence angle. Agreement is unfeasible at the <11> direction with any structure models under the one-dimensional SPB assumption. Reprinted with copyright permission from [12]

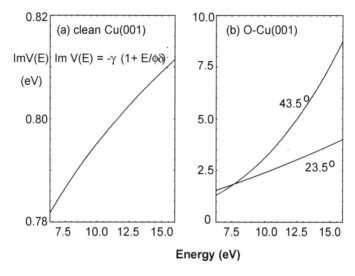

Fig. 15.4 The damping of Cu(001) is much lower than that of O-Cu(001) surface that also varies with azimuth angles, which indicates the essentiality of the anisotropy and nonuniformity of the SPB. Reprinted with copyright permission from [12]

of copper. Although the current model A is slightly favored at azimuth far away from the <11> direction, the convention itself cannot explain the outcome of calculations.

In the simulations, unexpected features present at the higher energies of the 23.5° curves (14.0–16.0 eV) for models B, C and D. In fact, these features do not come from numerical artifacts in the calculation code. These features arise from the crystal geometries. The sharp peak at ~15.0 eV and the relative intensity in the range of 14.0–16.0 eV suggests that the atomic positions in the corresponding models may not proper. On the other hand, the SPB was optimized based on structure B. The calculation results do not discriminate one structure is to be preferred over another except model A that gives an acceptable fit to the spectrum collected at 23.5° azimuth angle.

Furthermore, agreement of the 43.5° $I– E$ spectrum was never achieved for any structure models and for any variation of the one-dimensional SPB parameters. These uncertainties strongly against the monotonic damping and uniform SPB approximation for system with chemisorbed oxygen. The O-Cu(001) VLEED spectra with multiple features cannot be fitted simultaneously by the simple treatment as used in dealing with pure metals. Therefore, the 3D-SPB is necessary to consider that the SPB variables have an explicit energy and coordinate dependence to mimic the STM/S observations.

15.3 Cu$_3$O$_2$ Model Reality

15.3.1 Numerical Quantification

The z_0-optimizing method and the single-variable parameterized 3D-SPB functions allow for further evaluation of the structure models in Fig. 15.2. The z_0-optimizing method duplicates the measured data and yields the corresponding structure- and energy-dependent $z_0(E)$.

The program automatically fits the spectra by matching the computed $I_{cal}(E_i)$ to the experimental $I_{exp}(E_i)$ such that $I_{cal}(E_i)/I_{exp}(E_i) = 1.00 \pm 0.03$ at each step. Figure 15.5 shows the calculation results for 43.5° and 48.5° azimuth angles. The azimuths are selected such as these angles are far from affirmative insofar with conventional approach.

Figure 15.5 shows that the z_0-optimizing method makes all models work but different yields. Aside from the considerations and arguments insofar advanced for the current M$_2$O bond model A that is now further supported by the VLEED optimization. The structural models are selected by analyzing the $z_0(E)$ profiles. As given in Table 15.2, the minimal value of $\Delta z_{VLEED}/\Delta z_{STM}$ (43.5° curves provide maximal Δz) supports model A. The anisotropic $z_0(E)$ profiles suggest the essentiality of introducing the non-uniform SPB. To this end, the improved method of single-variable parameterization of the non-uniform SPB has been proven reliable and more revealing than the conventional one-dimensional-SPB wise.

15.3.2 Physical Indication

15.3.2.1 The $z_0(E)$ Plots

A comparison of the $z_0(E)$ profiles can differentiate the structural sets of atomic positions (Fig. 15.5 and Table 15.2). All the considered crystal structures offer spectral match near the <11> direction but the corresponding $z_0(E)$ curves are quitter different. This is right what one pursues. That the minor difference in atomic positions results in the observable variation of the $z_0(E)$ profile evidences that the VLEED is very sensitive to the crystal geometry. The bond geometrical dependence of the $z_0(E)$ curve allows one to judge a model by simply comparing the shape of the $z_0(E)$ profile with those of others.

One may assign the atomic-position (Table 15.2) as the one approaching to the true situation by carefully analyzing the shape of the $z_0(E)$ curve against criteria given in Sect. 15.1.4. From the perspective of energy, the $z_0(E)$ features below 7.5 eV coincide with the STS and PES profiles showing the occupied DOS below E_F on the O-Cu(110) and O-Cu(001) surfaces, which has been attributed to the nonbonding states, a characteristic of the sp^3-orbital hybridization of oxygen. The sharp features between 11.5 and 12.5 eV correspond to the Bragg reflections at the boundaries

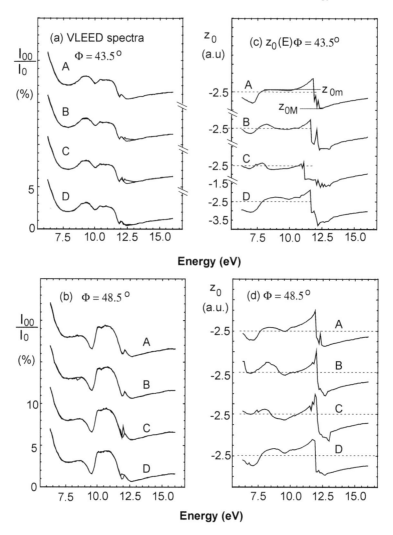

Fig. 15.5 Duplications of the VLEED spectra (a and b) and the corresponding $z_0(E)$ profiles (c and d) for models A-D at azimuth angles of 43.5° and 48.5°. Maximal Δz_{VLEED} (= $z_{0M} - z_{0m}$) that refers to the region giving a unique solution is obtained near the <11> direction. The spectral fit at the <11> direction is realized with the tetrahedral M_2O structure, the 3D-SPB, and the decoding method. As consequences of the crystal geometry, the $z_0(E)$ profiles vary with minor change of atomic geometry. This is exactly what the modeling approach expected. Reprinted with copyright permission from [10, 11]

Table 15.2 Quantitative analysis of VLEED spectrum at 43.5° azimuth for $Cu(001)$-$(2\sqrt{2}\times\sqrt{2})R45°$-2O surface. Reprinted with copyright permission from [13]

Models in Fig. 15.5	E (eV)	$-z_0$ (a.u.)	Intensity (theory)	Intensity experiment	Δz_{VLEED} (Å)	$\frac{\Delta z_{VLEED}}{\Delta z_{STM}}$
A	8.3	2.3861	0.02261	0.02216		
	12.6	3.3725	0.00454	0.00453	0.5218	1.16
B	8.2	2.2610	0.02214	0.02174		
	13.0	3.6154	0.00615	0.00553	0.7165	1.59
C	7.9	2.3852	0.02096	0.02070		
	12.6	3.6760	0.00771	0.00453	0.6828	1.52
D	8.9	2.2230	0.02984	0.02968		
	12.7	-3.6228	0.00565	0.00440	0.7406	1.65

$^*\Delta z_{STM} = 0.45$ Å. $\Delta z_{VLEED} = |z_{0M} - z_{0m}| \times 0.529$ (Å)
Energies between 11.0 and 12.0 eV are excluded for the multiple z_0 solutions provided. Potential barrier parameters are: $V_0 = 10.50$ eV, $\gamma = 0.9703$, $\delta = 4.4478$

of Brillouin zones. Consequently, the surrounding features relate to the electron-excitation at edges of different energy bands. Curves B and C present extra features around 8.0 eV with respect to A and D. It is difficult to specify this feature on the premise of valence DOS modification.

From the spatial point of view, the difference between z_{0m} and z_{0M} varies with the crystal structures. Structure A gives the smallest Δz_0 and it is closer to the STM scale-difference of 0.45 Å. Thus, the calculations favor again the atomic positions assigned in model A and then in structure D. Therefore, VLEED optimization supports the conclusion advanced by Besenbacher and Nørskov [9] that the oxygen atoms go underneath the first Cu layer for bonding. At the same time there is a pairing of Cu dipoles bridging over the missing row. The paring dipoles and the missing-row vacancies originate from the Cu_3O_2 surface bonding. Unfortunately, normal LEED optimization could hardly discriminate the difference of model B and C based on the minimization of the R-factor.

15.3.2.2 Surface Electronic Dynamics

The underlying physics of the observations verifies the modeling considerations. These results certainly deepen our insight into the behavior of surface electrons and quantify the localized features of O-added surfaces as observed with STM. At the dipole site, $z_1 \cong z_0$, $\alpha \cong \lambda^{-1}$. This describes that the metal dipoles enhance the SPB through the outward-shift of the wave function, giving a high degree of saturation. For the O-Cu(001) surface, the z_{0M} ($z_{0M}/z_0(Cu) = 3.37/2.50 \approx 1.35$) and the λ_M ($\lambda_M/\lambda(Cu) = 1.27/0.9 \approx 1.41$) are $\sqrt{2}$ times that of the clean Cu(001) surface. The conductive electrons colonize and form electron islands (metal dipoles). The values of z_{0M} and λ_M also quantify the protrusions in the STM image as that the higher the islands are, the denser the electron cloud is.

In the missing-row site, $z_1 \ll z_0$, $\alpha \gg \lambda^{-1}$, i.e., the missing-row vacancy is not occupied by "free electrons", which depresses the STM image with least SPB saturation. On the O-Cu(001) surface, the lowest saturation degree of the SPB is $(\lambda_m/\lambda(Cu) = 0.65/0.9\approx)$ $1/\sqrt{2}$ times that of the Cu(001) surface. Therefore, the electrons of the O-chemisorbed surfaces are very local. Therefore, the O-metal surface is a non-Fermi system absenting freely moving electrons. The O-Cu(001) surface consists of a dipole layer that lowers the work function. The electron localization is also responsible for the non-Ohmic rectifying of an oxide surface.

The current non-uniform SPB approach is able to account for the behavior of electrons on the Cu surfaces with chemisorbed oxygen. The formulations can well quantify the localized features as revealed by STM and STS. As an important factor influencing the interaction of incident electrons with the surface, the variation of the valence DOS can be obtained from the $z_0(E)$ profile. Because the $z_0(E)$ profiles vary with crystal structure, the difficulty of simultaneously quantifying crystal structure and electronic distribution has thus been overcome with simplified optimizations. The consistency among the VLEED profiles, $z_0(E)$ profile and saturation degree, and STM and STS observations evidences that the current SPB and the tetrahedrally-structured M_2O models are essentially appropriate.

15.4 Summary

The fitting of the VLEED spectra of O-Cu(001) surface verifies the essentiality of the energy dependent 3D-SPB, the tetrahedral M_2O bond geometry, and the spectrometric decoding skills. The single-variable parameterization provides agreement of the VLEED spectra at all the azimuth angles with all the compared models. The geometrical-dependent $z_0(E)$ profiles offer rich and profound information about the behavior of atoms and electrons at the surfaces in terms of bond geometry, valence DOS, and the 3D-SPB. Progress demonstrates the essentiality and efficiency of the premise of *Coordination Bonding and Electronic Dynamics* in dealong with chemical reaction.

References

1. P. Jennings, S. Thurgate, The inner potential in LEED. Surf. Sci. **104**(2), L210–L212 (1981)
2. C. Hitchen, S. Thurgate, P. Jennings, A LEED fine structure study of oxygen adsorption on Cu (001) and Cu (111). Aust. J. Phys. **43**(5), 519–534 (1990)
3. G. Hitchen, S. Thurgate, P. Jennings, Determination of the surface-potential barrier of Cu (001) from low-energy-electron-diffraction fine structure. Phys. Rev. B **44**(8), 3939 (1991)
4. H. Zeng, K. Mitchell, Further LEED investigations of missing row models for the Cu (100)-(22 × 2) R45°-O surface structure. Surf. Sci. **239**(3), L571–L578 (1990)

5. G. Hitchen, S. Thurgate, Determination of azimuth angle, incidence angle, and contact-potential difference for low-energy electron-diffraction fine-structure measurements. Phys. Rev. B **38**(13), 8668 (1988)

6. S.M. Thurgate, C. Sun, Very-low-energy electron-diffraction analysis of oxygen on Cu(001). Phys. Rev. B **51**(4), 2410–2417 (1995)

7. I. Bartoš, M. Van Hove, M. Altman, Cu (111) electron band structure and channeling by VLEED. Surf. Sci. **352**, 660–664 (1996)

8. A. Atrei, U. Bardi, G. Rovida, E. Zanazzi, G. Casalone, Test of structural models for Cu (001)-($\sqrt{2} \times 2\sqrt{2}$) R45°-O by LEED intensity analysis. Vacuum **41**(1), 333–336 (1990)

9. F. Besenbacher, J.K. Nørskov, Oxygen chemisorption on metal surfaces: general trends for Cu, Ni and Ag. Prog. Surf. Sci. **44**(1), 5–66 (1993)

10. C.Q. Sun, O-Cu(001): VLEED quantification of the four-stage Cu_3O_2 bonding kinetics. Surf. Rev. Lett. **8**(6), 703–734 (2001)

11. C.Q. Sun, O-Cu(001): I. binding the signatures of LEED, STM and PES in a bond-forming way. Surf. Rev. Lett. **8**(3–4), 367–402 (2001)

12. C.Q. Sun, C.L. Bai, A model of bonding between oxygen and metal surfaces. J. Phys. Chem. Solids **58**(6), 903–912 (1997)

13. D. Adams, H. Nielsen, J. Andersen, I. Stensgaard, R. Feidenhans, J. Sørensen, Oscillatory relaxation of the Cu (110) surface. Phys. Rev. Lett. **49**(9), 669 (1982)

Chapter 16
VLEED Capability and Sensitivity

Abstract VLEED offers such a unique tool that collects information of bond geometry, valence density-of states (DOS), and SPB outside the second atomic layer with high sensitivity and reliability. Based on the principle of Fourier transformation, VLEED calculation derives consistent information obtained with electron spectroscopy, crystallography, and morphology on an atomic scale.

Highlights

- The $ReV(z)$ diffracts elastically the electron beam and determines its phase shift.
- The $ImV(z, E)$ diffracts inelastically the electron beam and determines is amplitude loss.
- Covering the valence band energy, VLEED beam is subject to energy loss by absorption.
- VLEED signal is only sensitive to relaxation of the outermost two atomic layers.

16.1 Uniqueness of Solution

Independent treatment of the geometric and the SPB parameters in simulation of VLEED spectrum could lead to solution uncertainty [1], because the diffraction intensity depends on the arrangement of scattering centers with different cross-sections and the interference of the diffracted beams by the SPB. The geometrical arrangement determines the phase change of diffracted beams. The cross section of diffraction varies with the effective charges of the scattering centers. The interference determines the amplitude of the diffracted waves. Therefore, reaction with charge transportation and polarization modifies the atomic positions and the SPB. Proper combination of the geometry, SPB and the parametrization wise ensures the solution certainty in decoding the VLEED spectrum to represent the true situation happened during reaction.

C. Q. Sun, *Electron and Phonon Spectrometrics*,
https://doi.org/10.1007/978-981-15-3176-7_16

In the practice of verifying the solution uniqueness previously, all the SPB parameters were treated as independent, which has clarified the roles of the structural and SPB parameters in determining the features in a VLEED spectrum. Here describes the inspection of the uniqueness of solutions at selected energy values. Except for a selected pair of SPB parameters, for example, z_0 and λ, the structural and other SPB parameters were kept unchanged. Optimization parameters for the VLEED spectrum collected at 70° incidence and 43.5° azimuth were used as next iteration of calculation. The optimal $z_0(E)$ is used as reference to vary its parameters independently at a time of calculation. All other parameters were automatically generated with the SPB functions. The computer was assigned to do loops on the selected pair of variables with new values.

16.1.1   ReV(z; z₀, λ) Sensitivity

In fact, the integration of the $ReV(z, z_0, \lambda)$ combines the origin of the image plane z_0 and the saturation degree λ of the elastic SPB. The two parameters are correlated to determine the shape and integration of the $ReV(z; z_0,\lambda)$. VLEED concerns the integration area of the $ReV(z)$ function instead of the exact value of each in determining the phase shift and the beam reflectivity. Hence, it is essential and realistic to set the $\lambda(z_0)$ as a function of z_0 to reduce the number of solutions. When the z_0 moves away of the lattice, the SPB tends to be more saturated because of polarization, as observed from STM image. If the z_0 moves inward, the SPB will be less saturated because of ion or atomic vacancy formation.

Figure 16.1 shows the correlation between z_0 and λ at various energies. There are two groups of correlation curves in each panel corresponding to a $2n\pi$ phase change of one another. The infinite couples of (z_{0i}, λ_i) along each curve in a group provide matching between calculated and experimentally detected intensities $I_{cal}/I_{exp} = 1.00 \pm 0.05$. In order to establish the correlation between the couple of z_0 and λ parameters, one can collect data $((z_{0i}, \lambda_i), i = 1, 2, \ldots, 30)$ from each z_0-λ curve and then integrate the $ReV(z; z_{0i}, \lambda_i)$. The integration ranges from the second atomic plane (D_{12}) to infinitely far away (in practice the upper limit is chosen as -100 a.u.) from the surface:

$$I(Z_{0i}, \lambda_i) = \int_{D12}^{-\infty} ReV(z; z_{0i}, \lambda_i)dz \quad (i = 1, 2, \ldots, N = 30)$$

$N = 30$ is sufficient for statistical analysis. The average of the integration values of each curve in Fig. 16.1 is:

$$I = \frac{1}{N} \sum_{i=1}^{N} I(Z_{0i}, \lambda_i) \quad (N = 30),$$

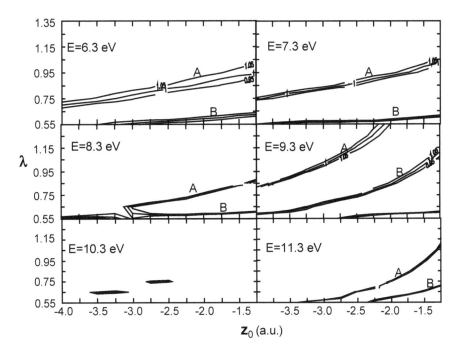

Fig. 16.1 Counter plots correlating the image plane z_0 to the saturation degree $1/\lambda$ at different energies for the VLEED spectrum from O-Cu(001) surface. Presence of groups A and B shows the 2π phase shift and each curve gives infinite number of solutions at a certain energy. The lacking of a universal constant for all the energies implies the solution uncertainty due to the independent treatment of the correlated SPB parameters (Reprinted with permission from [2])

and the standard deviation is:

$$D = \sqrt{\frac{\sum_{i=1}^{N}(I_i - I)^2}{N(N-1)}}$$

Table 16.1 summarizes the integration of ReV(z; z_0, λ) along the z_0-λ curves. Except for E = 10.3 eV at the boundary of the first Brillouin zone, all energies present at least two groups of curves that give a solution of $I_{cal}/I_{exp} = 1.00 \pm 0.05$. However, there is not a common value suitable for the integrations at all the considered energies. The absence of such an identical integration for all the energies means that no constant z_0 or λ is available for the entire specific spectrum. Therefore, the z_0 and λ varies from site to site at the surface. It is further justified that it is essentially reasonable to define the $z_0(E)$ parametrization.

Table 16.1 Integration of $\mathrm{Re}V(z, z_0, \lambda)$ along the correlation curves as typically shown in Fig. 16.1

E	Integration of $\mathrm{Re}V(z)$	Deviation	Deviation/Intensity
6.3-A	7.8032	0.0648	0.0083
7.3-A	8.0070	0.0631	0.0079
8.3-A	7.3052	0.0243	0.0033
9.3-A	8.5842	0.0131	0.0015
11.3-A	7.1091	0.0016	0.0002
6.3-B	6.6424	0.0453	0.0068
7.3-B	6.6531	0.0660	0.0099
8.3-B	6.7318	0.0742	0.0110
9.3-B	7.6051	0.0273	0.0036

Integration ranges from D_{12} (~3.5 a.u.) to -100 (a.u.)

16.1.2 $ImV(z; z_1, \alpha)$ Correlation

The spatial integration of the inelastic potential, $ImV(z, E)$, correlates the z_1 and α, which determines the amplitude change of the electron beams. The z_1 and the α in the Fermi-z function describe the spatial distribution of charge being equivalent to the E_F and the kT in the Fermi function. Figure 16.2 shows the z_1-α contour plots at different energies. Unlike the couple of z_0 and λ, the z_1 correlates with α uniquely through one curve at each energy. The z_1-α trends change differently from that of the z_0-λ plots in different energy ranges. These trends indicate that the spatial decay of the inelastic damping is rather local at the surface—varies from site to site and from energy to energy. If one parameter such as z_1 is fixed and then the α will be certain. The correlation further provides the experimental basis for the functionalization of the non-uniform SPB approximation. The diffracted beam intensity is sensitivity to the damping by charge quantity that is a combination of the individual parameters of the $ImV(z, E)$.

16.1.3 Solution Certainty

Correlation between any pair of the SPB parameters, or even the atomic positions, can be obtained by treating them independently, which leads to the uncertainty of solutions at any energy. The z_0-λ contour plot, for example, at 9.3 eV shows the $\mathrm{Re}V(z)$ is less saturated. The $1/\lambda$ decreases with the outward shift of the image plane—z_0 from surface, giving an infinite number of solutions. If one defines a function of $\lambda(z_0)$ that is orthogonal to the three z_0-λ correlation curves, the three groups of infinite solutions will then be reduced to three finite ones. If one treats all the SPB variables as functions of z_0, the solution uniqueness will be realized.

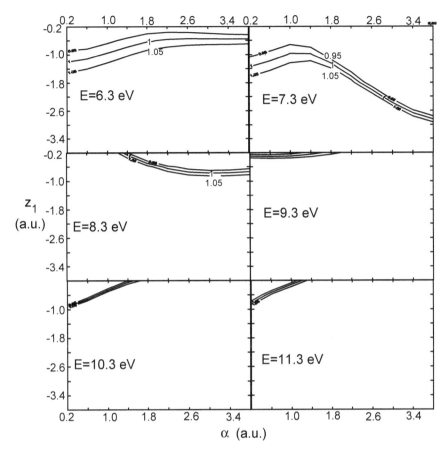

Fig. 16.2 Contour plots for the z_1—α correlation at different energies. Different trends of the single correlation curve ($I_{cal}/I_{exp} \approx 1$) at each energy value indicate the essentiality of the non-uniform spatial-decay of the inelastic damping. At energies of 10.3 and 11.3 eV, $ImV(z)$ saturates (α increase) with the inward shift of z_1; at 9.3 eV, α is smaller than 1.8 and z_1 is limited to -0.4 a.u. (Reprinted with permission from [3])

Hence, it is justified to limit all the SPB parameters as functions of one variable—the characteristic position of the electron distribution, $z_0(E)$.

Figure 16.3 shows the z_0-scanning solutions for different SPB and structural models. Plots A and B are results of identical structural parameters of the present Cu_3O_2 model with different SPB functions. As denoted, B is the standard case while A has $z_1 = z_0$ and $\alpha\lambda = 1$. This means that the spatial decay of damping is identical to the Fermi-part of the $ReV(z)$. C is the result of the same SPB functions as B but includes a structure of $c(2 \times 2)$ with oxygen being situated 0.85 Å above the unreconstructed lattice plane. The existence of structure C has already been excluded for the O-Cu(001) system [4, 5].

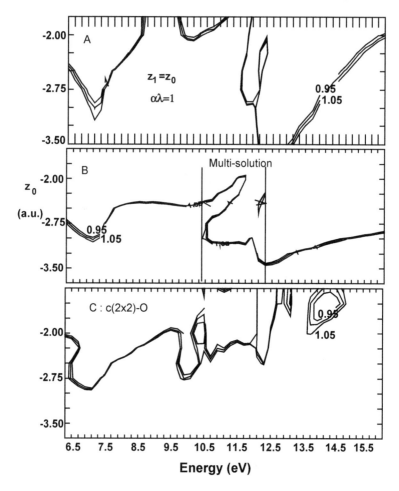

Fig. 16.3 The uniqueness of solution obtained with the combination of the Cu_3O_2 geometry, the parameterized SPB, and the z_0-optimizing strategy for the spectrum collected at 43.5° azimuth. Panel A compared with B using the same Cu_3O_2 structure but different SPB; panel B compared with C by the same SPB but different structure as indicated. B is preferred due to the unique solution, which justifies the validity of the current SPB and structural model and the solution certainty. (Reprinted with permission from [3])

Except for the region between 10.5 and 12.3 eV, a unique solution is assured with the single-variable parameterization of SPB. Obviously, plot A presents steeply varied dips with too large Δz_0 while plot C has no unique solution above 9.5 eV. Plot B, a yield of the combination of the current bond and non-uniform SPB models and the decoding skills, is preferable over others. Therefore, the approaches developed in so far are realistic and correct for the particular O-Cu(001) surface. As will be demonstrated, approaches have enabled the VLEED to provide simultaneously integrated information about the bond geometry, surface topography and the DOS in the valence band.

16.2 VLEED Capacity and Reliability

16.2.1 Decoding Procedures

Calculations were performed with the Cu_3O_2 pairing tetrahedron structure [6] and the optimized SPB parameters. Geometric variables required in calculations are converted from the Cu_3O_2 bond geometry. Table 16.2 shows that any variation of the bond geometry resulted in a collective dislocation of the surface atoms and the interface spacing.

As references for the sensitivity examination, two I-E curves were simulated with the z_0-optimizing method [7]. Table 16.2 lists the optimal structure parameters and Fig. 16.4 shows the optimized $z_0(E)$ profiles. Figures 16.5, 16.6, 16.7 duplicated the measured spectra in broken curves.

VLEED spectral sensitivity was examined by letting the code read in the initial optimal data in Table 16.2 (denoted *a and *b) and the $z_0(E)$ profiles in Fig. 16.4, and then repeat the calculation by adjusting the individual parameters to be examined.

16.2.2 Sensitivity to the Bond Geometry

Figure 16.5 shows the calculation results obtained by varying the Cu_3O_2 bond geometry. Besides the two bond parameters, BA12 and DCu_x, Q_2 was treated as an additional variable. For comparison purpose the effect of individual atomic displacement, DCu_z (denoted as D) was also examined. The structural sensitivity examination results in Fig. 16.5 revealed the following [10]:

(1) Adjusting the bond angle BA12 and the bond contraction coefficient Q_2 modulated the D_{12} spacing, which governs the fine-structure features between 7.0 and 11.0 eV, matching the spectral features collected at oxygen exposures of 25–200 L [10].

(2) Increasing DCu_x attenuates the general intensity at energies below the second band-gap (<12.0 eV). This trend agrees *remarkably* well with the spectral features for oxygen exposures >200 L and long term aging (>30 min) [10].

(3) Changing DCu_z causes nothing more than a slight change in intensity between 9.5 and 11.5 eV, without any matching to spectral measurements. This observation supports that the single atomic-shift is less realistic than the bond-geometry in describing the reaction dynamics.

Table 16.2 Correlation between the bond-geometry and atomic-shift in the pairing Cu_3O_2 tetrahedron on the Cu(001)-O surface [3]

Labels Fig. 16.4	Bond-geometry (in Å and °)				Atomic-dislocation (Å)				
	BA_{12}	DCu_x	D	Q_2	D_{12}	DO_x	DO_z	DCu_x	DCu_z
(a)-C	101.0	0.25	D_{av}	0.04	1.9086	−0.1852	0.1442	0.25	−0.1635
*a	102.0	0.25			1.9343	−0.1877	0.1682		−0.1495
*b	101.5	0.225			1.9251	−0.1863	0.1553	0.225	−0.1375
(a)-B	104.5	0.25			1.9968	−0.1956	0.2316		−0.1156
(a)-A	107.0	0.25			2.0567	−0.2055	0.2926		−0.0834
(b)-A	102.0	0.25	D_{av}	0.03	1.9545	−0.1879	0.1699	0.25	−0.1385
(b)-B				0.05	1.9141	−0.1876	0.1666		−0.1605
(b)-C				0.08	1.8531	−0.1870	0.1614		−0.1939
(c)-A	102.0	0.15	D_{av}	0.04	1.9343	−0.1877	0.1682	0.15	−0.0709
(c)-B		0.30						0.30	−0.1859
(c)-C		0.40						0.40	−0.2535
(d)-A	102.0	0.25	D_1	0.04	1.9343	−0.1877	0.1682	0.25	−0.0920
(d)-B			D_2						−0.2046

*a and *b indicate optimal parameters for the reference (broken lines) spectra in Figs. 16.5 and 16.6

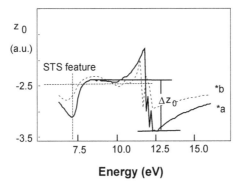

Fig. 16.4 The optimal $z_0(E)$ profiles were read as input by the calculation codes to conduct the sensitivity examination. Spectrum *a corresponds to the surface that is more fully developed than spectrum *b. The valley near 7.5 eV correspond to the nonbonding lone pair states as detected using STS. The violent features below 12.5 eV arise from Bragg diffraction. (Reprinted with permission from [8, 9])

(4) Geometry-sensitivity examination indicates that the reaction processes are dominated by the formation of the Cu_3O_2 bond while the SPB, as a subsequence of bonding, varies insignificantly with the tiny change in bond geometry.

16.2.3 Sensitivity to the SPB

Instead of the three constants (V_0, δ, γ) and one variable (z_0) in current SPB parametrization, eight independent SPB parameters were used as did in convention to examine the spectral sensitivity to the SPB. The δ (in $ImV(E)$) was not considered as it determines the slope of the $ImV(E)$ and its value depends on the calibration of the spectral intensity. The corresponding variations of the SPB parameters in Figs. 16.6 and 16.7 are varied by 10% rather than simply by an absolute amount. This is because most of the parameters, such as λ, z_1 and α, are no constants and thus it would be meaningful to change them in the equal percentage for comparison purpose. β and η are introduced as such to modulate the spatial decay of damping at different regions:

$$ImV(z, E) = \begin{cases} \beta \times ImV(E, z), z \leq z_1 \\ \eta \times ImV(E, z), z_1 < z < z_{SL} \\ ImV(E, z), z_{SL} < z \end{cases}$$

where z_{SL} is the position of the second atomic plane. The z-axis directs into the crystal.

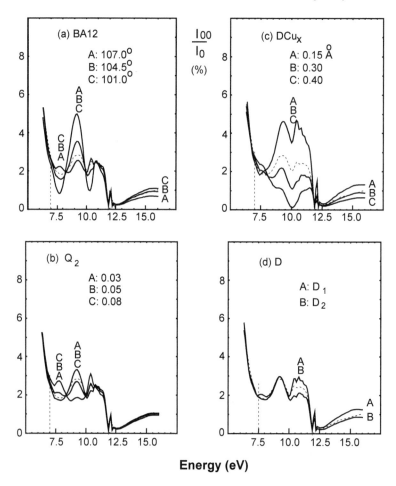

Fig. 16.5 VLEED spectral sensitivity to the bond geometry [11, 12]. Table 16.2 lists the parameters for examination. Broken lines are the reference spectrum. BA12 and Q_2 modulate the spectral features from 7.0 to 11.0 eV, featuring the change of 25-200 L oxygen coverage. DCu_x modulates the intensity between 7.5 and 12.0 eV, matching the features at higher exposure (>200 L). DCu_z (denoted D) has very little effect on the spectrum between 10.0 and 12.0 eV, producing no match to the measured profiles. The bond geometry or crystallography has little effect on features out of 7.0–12.5 eV or the valence band of copper. (Reprinted with permission from [11, 12])

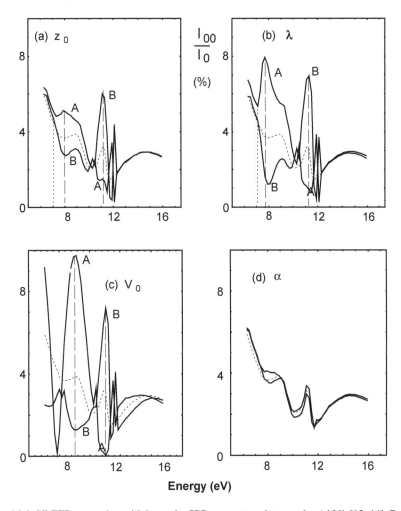

Fig. 16.6 VLEED spectral sensitivity to the SPB parameters that vary by ±10% [13, 14]. Panels (**a–c**) show the effect of the elastic potential on the phase change of the beams. A 20% change of the V_0, z_0 and λ in $ReV(z)$ has the similar effect causing a π phase change. Obviously, the elastic potential dominates the shape of the spectrum at energy covering the valence band ≤ 12.0 eV. Broken line is the reference spectrum [3]. (Reprinted with permission from [13, 14])

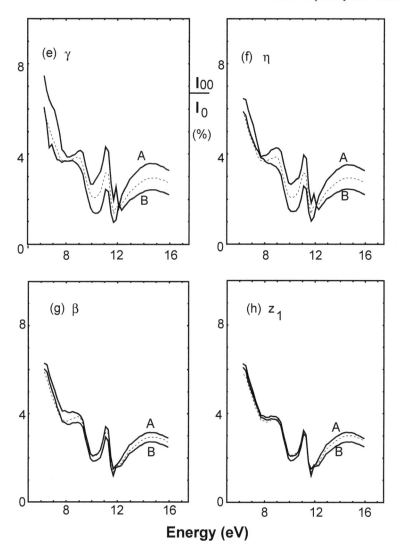

Fig. 16.7 (continues) Panels (**d–h**) are results from varying the ImV(z) parameters that determines the absolute intensity (wave function amplitude). The similar quantitative effect of γ and η on the intensity indicates that the topmost layer ($z \leq z_{SL}$) dominates the process of damping. Electrons from below the substrate second plane give no contribution, evidencing that the VLEED collects information from the top single atomic layer. (Reprinted with permission from [13, 14])

Results in Figs. 16.6 and 16.7 indicate the following [3]:

(1)   The elastic potential $ReV(z; z_0, \lambda, V_0)$ dominates the shape of the fine structures. The integration of the $ReV(z; z_0, \lambda, V_0)$ makes more sense than the individual limit in determining the phase change of the diffracted beam. Only a 20% change in these parameters leads to a phase change of π, resulting in an alteration from spectral maximum to spectral minimum at given energies, see Fig. 16.6a–c.

(2)   The shape of the fine structures is insensitive to the inelastic damping, $ImV(z)$. All the parameters in the inelastic damping, however, affect the absolute reflectivity or the amplitude of the wave function of the spectrum. Therefore, all the $ImV(z)$ parameters can be functionally dependent on the characteristic variable of the image plane z_0 ($\rho(z_0) = 0$; $\rho(z_1) = 0.5\rho_M$). The inelastic potential parameters could be used conveniently to compensate for the accuracy in calibrating the measured data.

(3)   By comparing the spectral intensity modulated by γ (for the whole range of z-dimension in $ImV(z, E)$) with the resultant of η and β, one can find easily that electrons out of the second atomic layer ($z > z_{SL}$) dominates the damping. The resultant of $\beta(z \leq z_1)$ and $\eta(z_1 < z \leq z_{SL})$ has the same quantitative effect as γ on the reflectivity. So electrons below the second atomic plane never come into play in the VLEED spectra. The $ImV(z, E)$ describes the valence DOS distribution in real and energy domains.

(4)   The spectrum is, however, insensitive to the variations of α, β and z_1. Thus, approaches in the functional dependence of the insensitive parameters on the characteristic one are substantially necessary. Although the effect of changing parameter δ (slope) in $ImV(E)$ was not considered, one may use it to compensate for the assumption in data acquisition that the incident current be constant during increasing the incident beam energy.

(5)   It is noticed that only the inelastic damping modulates the spectral intensity at energies being deeper than the valence band (>12.0 eV). This fact suggests that the electrons in deeper bands are less affected by the surface bonding, and as a result, they affect insignificantly the elastic potential at the surface.

(6)   Another important fact is that the spectral intensity above 7.0 eV (E_F) is modulated neither by the bond geometry nor the shape of the elastic potential (z_0 and λ) as indicated in Figs. 16.5 and 16.6. The intensity at this region is however sensitive to the inner potential V_0 constant and parameters in the inelastic damping. Therefore, spectral features above 7.0 eV varies with the surface electron density rather than the bond geometry or the barrier shape.

16.3   Summary

An examination of the VLEED spectral sensitivity to the bond-geometry and to the SPB parameters clarified the following:

(1)   The VLEED spectra are sensitive to the bond geometry instead of the individual atomic dislocations. Individually varying atomic position is not practical in simulating the process of surface bond forming. Agreement of the trends between the measurements and calculations realized by adjusting the bond variables challenges further efforts towards quantifying the O-Cu(001) bonding kinetics.

(2)   The shape of the *I-E* curve is sensitive to the integral of the elastic potential ReV(z), which determines the phase change of the diffracted electron beams.

(3)   The absolute intensity of the *I-E* curve is dominated by the integral of the inelastic damping ImV(Z, E). Thus the parameters in the ImV(z, E) can compensate for the inaccuracy of the calibration of experimental data.

(4)   It is uncovered that electrons out of the second layer predominate the inelastic damping. Therefore, LEED at very low energies is the unique technique that collects nondestructive information from a single atomic layer of a surface and covers the valence band energy.

The self-consistency between the outcomes and the constraints, on the sense of SPB integrations, the SPB is premised evidences sufficiently that the decoding technique and the modeling approaches are essentially correct, and therefore the capacity and the reliability of the VLEED in turn are fully uncovered. The present approach reveals the performance of bonds and electrons in terms of bond geometry, Bragg diffraction, intensity loss by the valence DOS, ImV(z, E), absorption, and the elastic SPB diffraction featured by the ReV(z).

References

1.   V. Pouthier, C. Ramseyer, C. Girardet, P. Zeppenfeld, V. Diercks, R. Halmer, Characterization of the Cu (110) − (2×1) O reconstruction by means of molecular adsorption. Phys. Rev. B **58**(15), 9998 (1998)

2.   C.Q. Sun, C.L. Bai, Modelling of non-uniform electrical potential barriers for metal surfaces with chemisorbed oxygen. J. Phys. Condensed Matter **9**(27), 5823–5836 (1997)

3.   C.Q. Sun, Spectral sensitivity of the VLEED to the bonding geometry and the potential barrier of the O-Cu(001) surface. Vacuum **48**(5), 491–498 (1997)

4.   T. Lederer, D. Arvanitis, G. Comelli, L. Tröger, K. Baberschke, Adsorption of oxygen on Cu (100). I. local structure and dynamics for two atomic chemisorption states. Phys. Rev. B **48**(20), 15390 (1993)

5.   R. Mayer, C.-S. Zhang, K. Lynn, Evidence for the absence of a c (2×2) superstructure for oxygen on Cu (100). Phys. Rev. B **33**(12), 8899 (1986)

6.   C.Q. Sun, Oxidation electronics: bond-band-barrier correlation and its applications. Prog. Mater Sci. **48**(6), 521–685 (2003)

7.   C.Q. Sun, Angular-resolved VLEED from O-Cu(001): Valence bands, chemical bonds, potential barrier, and energy states. Int. J. Mod. Phys. B **11**(25), 3073–3091 (1997)

8. C.Q. Sun, O-Cu(001): II. VLEED quantification of the four-stage Cu_3O_2 bonding kinetics. Surf. Rev. Lett. **8**(6), 703–734 (2001)
9. C.Q. Sun, O-Cu(001): I. Binding the signatures of LEED, STM and PES in a bond-forming way. Surf. Rev. Lett. **8**(3–4), 367–402 (2001)
10. C. Hitchen, S. Thurgate, P. Jennings, A LEED fine structure study of oxygen adsorption on Cu (001) and Cu (111). Aust. J. Phys. **43**(5), 519–534 (1990)
11. Y.L. Huang, X. Zhang, Z.S. Ma, Y.C. Zhou, W.T. Zheng, J. Zhou, C.Q. Sun, Hydrogen-bond relaxation dynamics: resolving mysteries of water ice. Coord. Chem. Rev. **285**, 109–165 (2015)
12. X.J. Liu, M.L. Bo, X. Zhang, L. Li, Y.G. Nie, H. Tian, Y. Sun, S. Xu, Y. Wang, W. Zheng, C.Q. Sun, Coordination-resolved electron spectrometrics. Chem. Rev. **115**(14), 6746–6810 (2015)
13. R. Koch, E. Schwarz, K. Schmidt, B. Burg, K. Christmann, K. Rieder, Oxygen adsorption on Co (101⁻0): Different reconstruction behavior of hcp (101⁻0) and fcc (110). Phys. Rev. Lett. **71**(7), 1047 (1993)
14. R. Koch, B. Burg, K. Schmidt, K. Rieder, E. Schwarz, K. Christmann, Oxygen adsorption on Co (1010). The structure of p (2×1) 2O. Chem. Phys. Lett. **220**(3–5), 172–176 (1994)

Chapter 17
Brillouin Zones, Effective Mass, Muffin-tin Potential, and Work Function

Abstract O-Cu(001) chemisorption derives the Cu_3O_2 pairing-tetrahedronand the $Cu^P{:}O^{2-}{:}Cu^P$ chains lined along the missing-raw edges, which roughen the SPB and the surface morphology. At the dipole site, The origin of the SPB moves $\sqrt{2}z_0$ outwardly with $\sqrt{2}\lambda_0$ saturation degree, at the atomic vacancy site, the SPB is characterized by $z_0/\sqrt{2}$ and $\lambda_0/\sqrt{2}$, with z_0 and λ_0 being the references for clean Cu(001) surface. Connecting the Bragg diffraction peaks results in the first and the second two-dimensional Brillouin zones with the $\langle 11 \rangle$ direction deformation corresponding to a 0.22 Å dislocation of the Cu^P towards the MR. Matching the theoretical Brillouin zones derived the electronic effective mass of $m_1^* = 1.10$, and $m_2^* = 1.14$ at the boundaries.

Highlights

- VLEED probes nondestructively the bond geometry and interatomic charge transfer.
- Cu_3O_2 bond formation resolves the missing-row vacancies and the O–Cu–O chains.
- Charge transition from Cu to O reduces the atomic muffin-tin inner potential constant.
- Dipole formation pushes the SPB origin outwardly with increasing degree of saturation.

17.1 Angular-Resolved VLEED Profiles

Hitchen, Thurgate and Jennings measured the VLEED reflectance from an O–Cu(001) surface under various experimental conditions [1, 2]. These results formed the unique kinetic-VLEED database available to date. The VLEED (00) beam I–E profiles to be decoded were collected from a Cu(001) surface exposed to 300 L oxygen. For the symmetry consideration, angles from 18.5° to 63.5° with 5° increment

C. Q. Sun, *Electron and Phonon Spectrometrics*,
https://doi.org/10.1007/978-981-15-3176-7_17

are sufficient to represent the full azimuth of the surface. The features of the measured VLEED profiles in Fig. 17.1 can be summarized as follows [3]:

(1) Two sharp, solitary troughs or violent peaks (denoted by dotted lines) appear on each curve. These two troughs move away from each other as the azimuth moves away from the ⟨11⟩ direction. The angular-resolved sharp features divide the VLEED energies (6.0–16.0 eV) into three regions with variation in boundary energies. Calibration of these critical positions gave the reduction in work function of ~1.1 eV [2, 3].

(2) The fine-structure feature (denoted by the dashed lines) splits into its separate components as the azimuth moves away from the symmetric point 45.0°. The first component vanishes outside the region of 33.5°–58.5°. The second one becomes narrower while moving away from the symmetric point. Further, the intensity of the second peak tends to be weaker when the azimuth is greater than 45.0°.

(3) The peak positions and intensities of the curves show a reduction in symmetry relative to the symmetric center. The reduction of the intensities at higher azimuth angles may come from the development of the reaction, as the long-term aging leads to a general attenuation of the spectral intensity. The deviation of the symmetry means that the reaction breaks the original C_{4v} group symmetry of the surface lattice.

17.2 Sharp Features: Brilliouin Zones and Energy Bands

17.2.1 Brillouin Zones and Effective Electron Masses

The Cu_3O_2 formation results in the primary unit cell of $Cu(001)-(\sqrt{2} \times 2\sqrt{2})R45°-2O^{-2}$. The presence of defects and impurities, such as the missing-row vacancy and the oxygen adsorbates, has no effect on constructing either the real or the reciprocal lattice [5, 6]. Displacements of lattice atoms, such as the DCu_x for the atoms closing to the missing-row, or at the ⟨11⟩ direction, however, deform both the real and the reciprocal lattice.

As the emergence of new diffraction beam is independent of the inner potential and the barrier shapes but depends on the incident and diffraction conditions, and the two-dimensional geometry of the surface lattice [7]. Emergence happens when the lateral components of the diffracted wave, with vector $k'_{//}$, and the incident wave $k_{//}$ satisfy the Bragg diffraction condition,

$$k'_{//} - k_{//} = g,$$

where g is the vector of a reciprocal lattice. This Bragg diffraction happens at the band-gap or at the boundary of a Brillouin zone. Therefore, the sharp features closing to the emergence of new beam on the VLEED spectra arise from the band-gap reflections at the boundaries of the first two Brillouin zones.

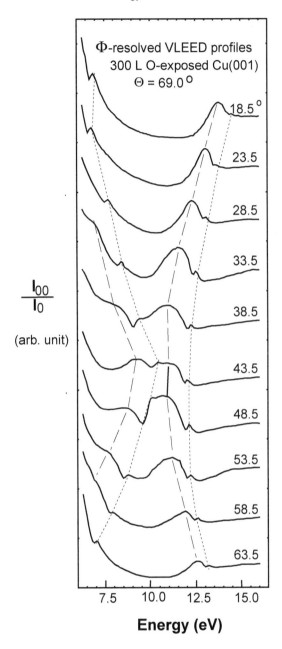

Fig. 17.1 Angular-resolved VLEED profiles collected from a Cu(001) surface exposed to 300 L oxygen at 69.0° incidence [4]. The azimuth angles are labeled for each curve. There is a "cross-point" at about 45°. Two small sharp Bragg peaks (denoted by dotted lines) appear on each curve. The broaden fine-structure feature (dashed lines) splits when the azimuth moves away from the ⟨11⟩ direction. Reprinted with copyright permission from [3]

Table 17.1 Positions of the sharp peaks, E_{p1} and E_{p2}, in the angular-resolved VLEED profiles of the O–Cu(001) surface [4]. The incident angle is kept constant at 69.0° [3]

Azimuth (°)	E_{p1} (eV)	E_{p2} (eV)
18.5	5.00	14.50
23.5	5.20	14.00
28.5	5.60	13.30
33.5	8.30	12.50
38.5	9.50	12.20
43.5	10.40	12.00
48.5	10.00	12.20
53.5	8.80	12.30
58.5	5.80	12.60
63.5	5.00	13.20

Accordingly, the sharp-peak positions in the angular-resolved VLEED data offers the two-dimensional Brillouin zones, as summarized in Table 17.1. The location of a peak E_p can be decomposed in k-space as (in atomic units: $m = e = \hbar = 1$, 1 a.u. $= 0.529$ Å and 27.21 eV):

$$E_p = \frac{\left[k_{\langle 10 \rangle}^2 + k_{\langle 10 \rangle}^2 + k_z^2 \right]}{2m^*}$$

with $k_{\langle 10 \rangle} = \sqrt{2m^* E_p} \sin \theta \cos \phi$, and $k_{\langle 01 \rangle} = \sqrt{2m^* E_p} \sin \theta \sin \phi$,   (17.1)

where m*, the effective mass of electron populated near the boundary of Brillouin zone, is introduced such that it compensates for the reduction of the diffracted k' due to its energy loss. θ and φ represent the incident and the azimuth angles of the incident beam. The wave vector is decomposed into the $k_{\langle 10 \rangle}$ and $k_{\langle 01 \rangle}$, which extend from the center to the boundary of the first Brillouin zone.

One can then build the first two Brillouin zones by adjusting the m* values to match the theoretical ideally situation, ($k_{\langle 10 \rangle} = k_{\langle 01 \rangle} = n\pi/a$, n = 1, 2) as represented by solid lines in Fig. 17.1. The effective masses of electrons surrounding the Brillouin zone boundaries are optimized with least-square method. The optimal values are: $m_1^* = 1.10$, and $m_2^* = 1.14$.

The increase of m* with energy coincides with the trend that the energy-loss of the diffracted beam ($k' = k/\sqrt{m^*}$) increases with the kinetic energy [8].

On the other hand, the first Brillouin zone contracts at point Y but expands near X. The contraction at Y correlates to the atomic shift of Cu^p along the $\langle 11 \rangle$ direction in the real lattice, DCu_x, while the expansion near X needs to be identified.

17.2.2 *Lattice Reconstruction*

The mutual-reciprocal relationship between the k-space with basic vector k_i and the r-space with basic vector a_i:

$$a_i \cdot k_i = 2\pi \delta_{ij}; \quad \delta_{ij} = \begin{cases} 1, i = j \\ 0, i \neq j \end{cases} \tag{17.2}$$

implies that the deviation of the experimental Brillouin zone from the theoretical form originates from the deformation of the primary unit cell. Therefore, one can trace inversely the DCu_x from the contraction of the Brillouin zone at Y.

The distance in k space, $\overline{\Gamma Y}$ $(=\frac{1}{2} |k_i + k_j| = 2\pi/a)$, correlates to a quantity $a_{(11)}$ through the mutual reciprocal relation (17.2),

$$a_{(11)} \cdot \left(k_i + k_j\right) = a_{(11)} \cdot 2\overline{\Gamma Y} = 2\pi, \tag{17.3}$$

which yields,

$a_{(11)} = \pi/\overline{\Gamma Y} = a/\sqrt{2}$,

corresponding to the shortest row spacing on the (001) surface. Taking the logarithm and deriving from both sides of Eq (17.3), we have

$$\frac{da_{(11)}}{a_{(11)}} + \frac{d\overline{\Gamma Y}}{\overline{\Gamma Y}} = 0,$$

yielding the lateral shift of the dipole along the $\langle 11 \rangle$ direction:

$$DCu_x = da_{(11)} = -\frac{a_{(11)}}{\overline{\Gamma Y}}d\overline{\Gamma Y} = -\frac{a^2}{2\pi}d\overline{\Gamma Y}. \tag{17.4}$$

Substituting $m_1^* = 1.10$, $\theta = 69.0°$ and $E_{p1} = 10.4$ eV $= 0.3804$ a.u. (at $43.5°$ close to the $\langle 11 \rangle$ direction) into Eq (17.1) yields $\overline{\Gamma Y}' = \sqrt{2m_1^* E_{p1}} = 0.8104$ (1/a.u.). On the other hand, by adopting $a = 2.555/0.529 = 4.8308$ (a.u.), one can have the theory value $\overline{\Gamma Y}' = \sqrt{2}\pi/a = 0.9217$ (1/a.u.).

Therefore,

$$DCu_x = -\frac{a^2}{2\pi}\left(\overline{\Gamma Y}' - \overline{\Gamma Y}\right)$$

$= 0.4133$ (a.u.)

$\cong 0.22$ (Å),

which coincides with the values (0.2–0.3 Å) optimized with LEED [10], XRD [11], and the current VLEED (~0.25 Å).

17.2.3 Valence Bands

The critical positions correspond to the boundaries of Brillouin zones, which are edges of the energy bands. Therefore, the VLEED energies are divided into three regions, as shown in Fig. 17.2. These regions correspond to the valence bands of copper. Variations of the bandwidth with azimuth are known as dispersion in the field of Solid State Physics.

The first sharp-peak positions varying from 7.0 to 10.4 eV can be ascribed as the bottom of Cu-3d band, which agrees with the known 3d band structure, 2.0–5.0 eV below E_F (5.0 eV), as detected by ARUPS [12, 13]. Accordingly, the region between the two sharp peaks corresponds to the Cu-3p band. The 3p band overlaps partially the 3d band due to the azimuth effect. The more delocalized 4s band between Fermi energy E_F and vacuum level E_0 (12.04 eV) fully covers both 3p and 3d band. Energies deeper than the second Brillouin zone correspond to deeper bands that are not of immediate concern in valence-electron transportation between oxygen and copper and out of the scope of VLEED.

17.3 Bond Geometry, Valence DOS, and 3D-SPB

Calculations were conducted by assuming that the crystal structure was unchanged during VLEED data collection. Figure 17.3a shows the z_0-optimizing that matches calculations to measurements and Panel b and c show the energy dependence of the

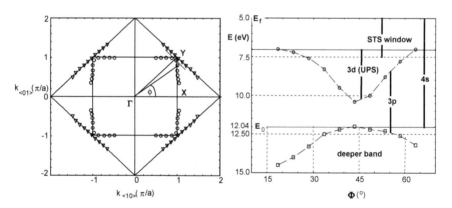

Fig. 17.2 a The first two Brillouin zones, derived from the critical positions on the angular-resolved VLEED profiles, are denoted by open circles and triangles, respectively. Compared in solid lines are the theoretical Brillouin zones. Deformation of the first Brillouin zone near the Y point corresponds to the DCu$_x$ dislocation in real lattice. **b** Band structure extracted from the angular-resolved VLEED profiles [9]. The boundary lines divide the VLEED energies into various bands as indicated. The current VLEED covers 4s, 3d, and 2p of copper, integrating the windows of STS and UPS. Reprinted with copyright permission from [4]

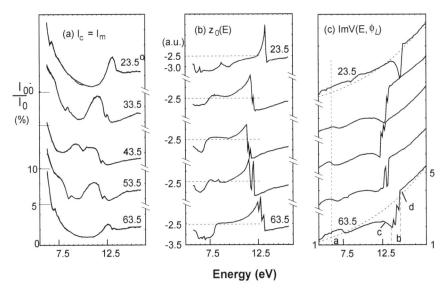

Fig. 17.3 Reproduction of the angular-resolved VLEED spectra from O–Cu(001) surface (3% precision) [4]. Besides the bond geometry and the inner potential constant, outcomes of the z_0-optimizing calculations contain **a** replications of the measured spectra and **b** the corresponding $z_0(E)$ profiles. Panel **c** shows damping curves with four significant features. Region a features the hybridized nonbonding states; region b represents band-gap reflection; region c and d are excitations occurring at edges of different bands. Compared in broken lines are the constant z_0 and monotonic ImV(E) for Cu(001) surface. Reprinted with copyright permission from [4]

optimized $z_0(E)$ and the inelastic damping $ImV(E, \phi_L)$. As references, the constant z_0 (−2.5 a.u.) and the monotonic $ImV(E, 5.0)$ for Cu(001) clean surface are compared in dotted lines.

Table 17.2 summarizes information available from the calculations [4].

Atomic dislocations, layer-spacing relaxation and in-plane lattice reconstruction as well as the adsorbate position are defined uniquely by the Cu_3O_2 bond geometry. The derived atomic arrangement agrees with the conclusion drawn from the effective-medium theory calculations [14]. Oxygen ions go underneath the first layer for bonding instead of sitting on top of the surface. Meanwhile, there is a pairing of the Cu dipoles bridging over the missing row vacancy.

As mentioned before, the $z_0(E)$ curves in Fig. 17.3b exhibits the joint features of topography and DOS spectroscopy. The shapes and the intensities of $z_0(E)$ curves, as well as the damping in Fig. 17.3c, varies apparently with azimuth. At 43.5°, closing to the ⟨11⟩ direction, the $z_0(E)$ curve provides the maximal $\Delta z_0 \sim 0.52$ Å closing to the scale difference of the STM image. This clearly indicates the nonuniformity and anisotropy of the SPB, as noted by Baribeau et al. [15] in their VLEED simulations. They indicated that the damping is no more isotropic due to oxygen adsorption. It is clear now why one was unable to fit the VLEED spectra near the ⟨11⟩ direction without inclusion of the nonuniform and anisotropic SPB into consideration.

Table 17.2 Information from the angular-resolved VLEED spectra of Cu(001)–$(2\sqrt{2} \times \sqrt{2})$R45°–$2O^{-2}$ surface [4]

Controlling	Q_1, Q_2	0.12, 0.04
	BA12 (°)	102.0
Variables	DCu_x (Å)	0.250
Atomic	DCu_z	−0.1495
Shifts	DO_x	−0.1876
(Å)	DO_z	0.1682
Relaxation	D_{12}	1.9343
Bond	BL1	1.628
Length	BL2	1.776
(Å)	BL3(2)	1.926
Bond	BA13	105.3
Angle	BA23	99.5
(°)	BA33	139.4
Inner potential	V_0 (eV)	10.50
ImV(E)	γ	0.9703
	δ	6.4478
Reduction of work function	$\Delta\phi$ (eV)	1.20
Microscopy features	Δz_{VLEED} (Å)	0.52
z_0(E)	<7.5 eV	Hybridization
Spectroscopy features	2nd BZ boundary	Band-gap reflection
(Energy-states)	Around BZ boundary	Band-edge excitation
BZs	Figure 17.2a	
Energy bands	Figure 17.2b	

The calculations yield the change in energy states that contribute to the damping. As indicated in Fig. 17.3b, c, there are several notable regions exhibiting DOS features:

(a)   Features above 7.5 eV are independent of azimuth angles. These humps coincide with the STS and PES signatures O–Cu(110) signatures around 7.1 eV, which have been specified as the contribution of nonbonding states of O^{-2}. The anti-resonance of this feature (intensity independence of azimuth and incident energy) on O–Cu(001) surface has been confirmed with PES by Warren et al., [16] indicating a strong one-dimensional localization of the $[O^{-2}{:}Cu^{-2}{:}O^{-2}]$ chain. Coincidence of the VLEED (reduction in work function) with the STS (polarization states) in the above-E_F (antibonding) and the below-E_F (nonbonding) features can be convincing evidence that the oxygen adsorbate hybridizes its *sp* orbits and polarizes electrons of its neighboring metal atoms.

(b)   Band-gap reflection along the boundary of the second Brillouin zone is appar-
        ent, while such reflection at the boundary of the first Brillouin zone is invisible
        near the symmetry point (45°). The invisible features are obviously due to the
        strong overlap of the Cu $3d3p4s$ bands near the $\langle 11 \rangle$ direction, as illustrated
        in Fig. 17.2.
(c, d) The damping surrounding the boundary of the second Brillouin zone features
        electron excitation at the edges of different bands. The excitation of electrons
        from the bottom of a band seems to be harder than that from the top edge of
        a band where the electrons might be denser.

17.4   Inner Potential Constant and Work Function

17.4.1   Beam Energy Reduced V_0

The muffin-tin inner-potential constant V_0 is an important parameter required to
achieve best fit between theoretical calculations and the experimental measurements
of LEED spectra. Earlier investigations [17] revealed that the V_0 decreases with the
energy increases of the incident beams (so-called E-reduced V_0), and it was suggested
that the V_0 be affected by the formation of the surface dipole-layer. Investigating the
energy dependence of the V_0 of Ni and Cu surfaces, Jennings and Thurgate [17]
found that the V_0 decreases exponentially with the increase of the incident beam
energy in the normal LEED regime. This reduction was explained with free-electron
like scheme that the exchange-interaction between the ions of metal surface and the
incident electrons will be weakened as the electron velocity increases.

Measurements showed, however, the V_0 changes insignificantly at energies lower
than 40 eV for the Cu surfaces, indicating the insignificance of energy effect on
the V_0 at very-low energies. Rundgren and Malmstrom [18] suggested that the V_0
decreases as energetic incident beams bombard the surface and drag out the surface
ion cores to neutralize the surface negative charges. The sputtering of the surface by
the incident beams reduces the quantity of net charges. If one electron was sputtered
away from the surface, the residual ion core will also reduce the net negative charge.
Therefore, the quantity of surface charges dominates the V_0.

17.4.2   Oxygen Reduced V_0

For system with chemisorbed oxygen, the variation of V_0 appeared to be much
more complicated. The amount of the reduction varies with crystal structure used
in calculation. The V_0 reduced by 1.21 eV for the Cu(001)–c(2 × 2)–2O and by
2.15 eV for the $(2\sqrt{2} \times \sqrt{2})R45°$–2O structure [19]. Best fit of the VLEED I–E
curves in this series using the Cu_3O_2 structure requires a 1.06 eV (9.2%) reduction

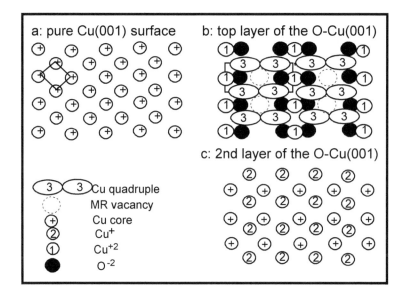

Fig. 17.4 Surface-charge distribution in both clean and O-chemisorbed Cu(001) surfaces. **a** Pure Cu(001) surface with Cu ion cores regularly arranged in the Fermi sea; **b** and **c** are the top and the second layer of the Cu(001)–$(2\sqrt{2} \times \sqrt{2})R45°$–$2O^{-2}$ phase. Panels **a** and **b** match closely to the STM images. Reprinted with permission from [30, 31]

in the V_0 from the bulk value of Cu(001), 11.56 eV. Similar conclusion has also been drawn by Pfnür et al. [20] who found in their VLEED calculations it necessary to use a step function to describe the reduction in the V_0 of the O–Ru top layer. It is certain that oxygen chemisorption results in a pronounced reduction in the V_0. The O-reduced V_0 in VLEED is suggested herewith to arise from the screening by the lone-pair induced surface dipoles.

In examining the VLEED spectral sensitivity to the potential parameters, we compared in Fig. 16.6c the effects of varying the V_0 on the calculated VLEED spectrum by using the Cu_3O_2 model. The solid curve duplicates the measurement with the optimized $V_0 = 10.50$ eV. The rest two curves result from varying the V_0 from 10.50 by $\pm 10\%$. The variation of the inner potential from -10 to 10% leads to a phase-change of $\Delta\Phi \cong \pi$. The optimal value of $V_0 = 10.5$ eV for the best fit is appropriate.

17.4.3 Oxygen Reduced Local $\phi_L(E)$

Another striking feature of oxygen-chemisorption is the reduction in the work function, $\phi_L(E)$. Characterization of the $\Delta\phi$ has hence been developed as one standard means used to determine the surface electronic properties. Hofmann et al. [21] and Benndorf et al. [22, 23] found in their experiments that the ϕ decreases when oxygen incorporates into the Cu(001) surface. From a HREELS study Dubois [24] concluded that oxygen chemisorbs three-fold hollow sites on the Cu(111) surface, either in or below the outmost plane of Cu atoms, resulting in a work function change. Ertl and Rhodin [25] suggested that the $\Delta\phi$ be due to the formation of the dipole layer. Lauterbach and Rotermund [26] [27][28] attributed the $\Delta\phi$ to an inversion of the oxygen dipole moment when it goes beneath the surface.

17.5   Mechanism Clarification

During the reaction, the $c(2 \times 2)$–$2O^{-1}$ domain, or the off-centered CuO_2 pairing pyramid evolves into the Cu_3O_2 pairing tetrahedron that gives rise to the Cu(001)–$(2\sqrt{2} \times \sqrt{2})R45°$–$2O^{-2}$ phase. STM images revealed creation of dipoles resulting in, respectively, the tensile and the compressive stress [29].

Figure 17.4 illustrates the variation of the surface-charge caused by the reaction. Compared with the clean-Cu(001) surface with ion cores arranged regularly in the Fermi sea, as shown in panel (a), panels (b) and (c) correspond to the outmost two atomic layers of the Cu(001)–$(2\sqrt{2} \times \sqrt{2})R45°$–$2O^{-2}$ phase. Cu_3O_2 bonding results in two hybridized-O^{-2} ions, one Cu^{+2}, two Cu^p, and a missing-row vacancy in a complex unit cell at the top layer. In the second layer, Cu atoms in the every other row along the [010] direction contributes one electron to the oxygen atom for tetrahedron formation. Two Cu^+ ions are left behind in a unit cell. Panels (a) and (b) illustrate satisfactorily the corresponding STM signatures of Jensen et al. [30]. In particular, panel (b) can account for the depressions and the "dumb-bell" protrusions that bridge over the missing rows.

17.5.1   Oxygen and Beam Energy Reduced V_0

The Cu_3O_2 structure accounts well for the V_0 reduction caused by the Cu_3O_2 bond formation. Initially, there are four Cu atoms in a Cu(001)–$(2\sqrt{2} \times \sqrt{2})R45°$–$2O^{-2}$ unit cell at the top layer. The chemisorption of O_2 removes one Cu atom out. The oxygen atom catches one electron from each of the top and second substrate layer. The total number of electrons at each layer contributing to the V_0 is reduced by the adsorption of oxygen (with 8 electrons for each O atom) and the removal of one Cu (with each Cu atom there are 29 electrons). The transportation of the two electrons from the second layer to the oxygen adsorbate also varies the V_0 of both the top and the second layers. Hence, the relativistic charge quantity in the top layer is:

$$([29e \times 2(Cu^p) + (29 - 2)e\ (Cu^{+2}) + 29e \times 0(MR)]$$
$$+ [(8 + 2)e \times 2(O^{-2})])/[29e \times 4(Cu)]$$
$$= 105/116 = 90.5\%$$
$$\cong 10.50/11.56 = 90.8\% \text{ (VLEED optimization)}$$

In the second layer, each Cu atom in every other row along the [100] direction donates one electron to the oxygen adsorbate. In the case of a Cu atom that has lost one electron, the residual ion core will also reduce the V_0 due to its residual positive charge. Every loss of an electron equals taking two electrons away from the sum of the negative charge, so that the relativistic variation of the V_0 of the second layer is:

$[29e \times 2(Cu) + (29 - 2)e \times 2(Cu^+)]/[29e \times 4(Cu)]$
$= 112/116 = 96.5\%$

Thus, the net charge of the top and the second layer is reduced by 9.5% and 3.5%, respectively. The net charge finally approaches to the bulk value in the third metallic layer and below, the V_0 of which is less affected by the reaction according to the current model. The V_0 gradually approaches to the bulk value of the clean Cu(001) when getting inside into the crystal, but the deeper layers are beyond the scope of VLEED. Agreement between the analysis and the numerical optimization of the V_0 for the top layer further evidences the reality and integrity of the Cu_3O_2 bond model for the Cu(001)–($\sqrt{2} \times 2\sqrt{2}$)R45°–$2O^{-2}$ surface reaction.

Although the analysis is simply based on a number count of electrons for the particular Cu(001)–($2\sqrt{2} \times \sqrt{2}$)R45°–$2O^{-2}$ phase, result reveals the correlation between the bond forming and the reduction of V_0 and work function. The V_0 relates to the charge quantity of the corresponding layer, and that at very-low energies the exchange-interaction between the incident beams and the surface ion is insignificant. The plasma excitation energy is often ~15 eV below the E_F and the incident energy is insufficient for ionization. This non-exchange-interaction adds another advantage to the VLEED for nondestructive detection.

17.5.2 Oxygen Reduced Local $\phi_L(E)$

As justified, the z-directional integration of the $\rho(x, y, z)$, from the second layer to infinitely far away of the surface, yields the local DOS $n(x, y)$ that contributes to the SPB and the work function in the form of $[n(x, y)]^{2/3}$ [29]. Since the VLEED integrates over large surface areas, all the quantities depending on coordinate (x, y) become E dependent. Therefore, at certain energy the VLEED integration results the $n(x, y)$ into the $n(E)$ that relates to the occupied DOS and hence the local work function.

The work function is dimensionless and it depends uniquely on the electron density at surface. However, the concept ϕ_L holds for large surface areas over which the VLEED integrates for the DOS, and the ϕ_L is also extended to being energy dependent. The work function depends uniquely on the $n(E)$, a z-dimensional integration of the $\rho(x, y, z)$ at energy E. The z_0 is the boundary of the $\rho(x, y, z)$ ($\rho(z_0) = 0$). The work function is determined by the z_0. An outward shift of the $\Delta z_0 \cong 1.0$ a.u. (from -2.3 to -3.3 a.u., closing to the grey scale of the STM image) corresponds to a $\Delta\phi \cong -1.2$ eV reduction. This quantity is consistent with the PEEM results of the O–Pt system due to oxygenation.

The Cu_3O_2 structure explains that the $\Delta\phi$ originates from the formation of the antibonding dipoles. Induction of electrons by either the O^{-1} or the nonbonding lone pairs of the O^{-2} fill up the antibonding sub-band. In real space, the dipoles buckle up with expansion of sizes and elevation of energy states. The buckling dipole changes nothing about the net charge of the surface layer but shifts the critical position z_0 outward of the surface, which increases the saturation degree of the SPB. Therefore, the dipoles have no apparent influence on the V_0 but occupy the empty DOS above E_F. Because VLEED integrates over large area of surface, it is impractical to try to discriminate the V_0 from site to site on the surface but one can recognize the local work function due to dipole formation.

17.6 Summary

The muffin-tin inner potential constant V_0 is related to the quantity of net charges while the local work function $\phi(x, y, E)$ depends on the z-dimensional distribution and energy dependence of the polarized electrons. The Cu_3O_2 bonding scheme explains the oxygen-reduced V_0 as arisen from charge transportation during reaction. The V_0 drops by 9.5% and 3.5% for the outermost two atomic layers but the VLEED is only sensitive to the first with 9.8% reduction. The buckling dipoles and the extent of polarization reduce the $\phi(x, y, E)$ by increasing the occupied DOS at surface while the dipoles affect little the V_0. It is also justified that at very-low energies the exchange-interaction between the incident beams and the surface ions is too weak to affect either the quantity or the distribution of the surface charges. Therefore, VLEED is an ideal means to collect non-destructive information about the behavior of surface electrons in terms of inner potential constant and work function.

Decoding the angle-resolved VLEED spectra has enabled determination of the first two Brillouin zones, the effective electron mass, lattice reconstruction, Cu_3O_2 bond geometry, energy bands and the energy states due to the nonbonding lone pairs of O^{-2}. Consistency between calculations and measurements further verifies the adequacy of the VLEED spectrometrics for exploring information of the bond geometry, SPB and the valence states for chemisorbed surfaces.

References

1. K.W. Jacobsen, Theory of the oxygen-induced restructuring of Cu (110) and Cu (100) surfaces. Phys. Rev. Lett. **65**(14), 1788 (1990)
2. G. Hitchen, S. Thurgate, Determination of azimuth angle, incidence angle, and contact-potential difference for low-energy electron-diffraction fine-structure measurements. Phys. Rev. B **38**(13), 8668 (1988)
3. G. Hitchen, S. Thurgate, Azimuthal angular dependence of LEED fine structure from Cu (001). Surf. Sci. **197**(1–2), 24–34 (1988)

4. C.Q. Sun, Angular-resolved VLEED from O-Cu(001): valence bands, chemical bonds, potential barrier, and energy states. Int. J. Mod. Phys. B **11**(25), 3073–3091 (1997)

5. M.A. Omar, *Elementary Solid State Physics: Principles and Applications* (Addison-Wesley, New York, 1993)

6. C. Kittel, *Intrduction to Solid State Physics*, 8th edn. (Willey, New York, 2005)

7. R.O. Jones, P.J. Jennings, LEED fine structure: origins and applications. Surf. Sci. Rep. **9**(4), 165–196 (1988)

8. E. McRae, C. Caldwell, Absorptive potential in nickel from very low energy electron reflection at Ni (001) surface. Surf. Sci. **57**(2), 766–770 (1976)

9. M. Read, A. Christopoulos, Resonant electron surface-barrier scattering on W (001). Phys. Rev. B **37**(17), 10407 (1988)

10. H. Zeng, R. McFarlane, R. Sodhi, K. Mitchell, LEED crystallographic studies for the chemisorption of oxygen on the (100) surface of copper. Can. J. Chem. **66**(8), 2054–2062 (1988)

11. T. Yokoyama, D. Arvanitis, T. Lederer, M. Tischer, L. Tröger, K. Baberschke, G. Comelli, Adsorption of oxygen on Cu (100). II. Molecular adsorption and dissociation by means of O K-edge x-ray-absorption fine structure. Phys. Rev. B **48**(20), 15405 (1993)

12. U. Döbler, K. Baberschke, J. Stöhr, D. Outka, Structure of c (2×2) oxygen on Cu (100): a surface extended x-ray absorption fine-structure study. Phys. Rev. B **31**(4), 2532 (1985)

13. R. DiDio, D. Zehner, E. Plummer, An angle-resolved UPS study of the oxygen-induced reconstruction of Cu (110). J. Vac. Sci. Technol., A **2**(2), 852–855 (1984)

14. F. Besenbacher, J.K. Nørskov, Oxygen chemisorption on metal surfaces: general trends for Cu, Ni and Ag. Progr. Surf. Sci. **44**(1), 5–66 (1993)

15. J. Baribeau, J. Carette, Observation and angular behavior of Rydberg surface resonances on W (110). Phys. Rev. B **23**(12), 6201 (1981)

16. S. Warren, W. Flavell, A. Thomas, J. Hollingworth, P. Wincott, A. Prime, S. Downes, C. Chen, Photoemission studies of single crystal CuO (100). J. Phys. Condens. Matter **11**(26), 5021 (1999)

17. P. Jennings, S. Thurgate, The inner potential in LEED. Surf. Sci. **104**(2), L210–L212 (1981)

18. J. Rundgren, G. Malmström, Surface-resonance fine structure in low-energy electron diffraction. Phys. Rev. Lett. **38**(15), 836 (1977)

19. S.M. Thurgate, C. Sun, Very-low-energy electron-diffraction analysis of oxygen on Cu(001). Phys. Rev. B **51**(4), 2410–2417 (1995)

20. H. Pfnür, M. Lindroos, D. Menzel, Investigation of adsorbates with low energy electron diffraction at very low energies (VLEED). Surf. Sci. **248**(1–2), 1–10 (1991)

21. P. Hofmann, R. Unwin, W. Wyrobisch, A. Bradshaw, The adsorption and incorporation of oxygen on Cu(100) at T $\geq$ 300 K. Surf. Sci. **72**(4), 635–644 (1978)

22. C. Benndorf, B. Egert, G. Keller, H. Seidel, F. Thieme, Oxygen interaction with Cu (100) studied by AES, ELS, LEED and work function changes. J. Phys. Chem. Solids **40**(12), 877–886 (1979)

23. C. Benndorf, B. Egert, G. Keller, F. Thieme, The initial oxidation of Cu (100) single crystal surfaces: an electron spectroscopic investigation. Surf. Sci. **74**(1), 216–228 (1978)

24. L. Dubois, Oxygen chemisorption and cuprous oxide formation on Cu (111): A high resolution EELS study. Surf. Sci. **119**(2–3), 399–410 (1982)

25. T.N. Rhodin, G. Ertl, The Nature of the Surface Chemical Bond. North-Holland Publishing Company, sole distributor for the USA and Canada Elsevier North-Holland (1979)

26. J. Lauterbach, H. Rotermund, Spatio-temporal pattern formation during the catalytic CO-oxidation on Pt (100). Surf. Sci. **311**(1), 231–246 (1994)

27. J. Lauterbach, K. Asakura, H. Rotermund, Subsurface oxygen on Pt (100): kinetics of the transition from chemisorbed to subsurface state and its reaction with CO, H_2 and O_2. Surf. Sci. **313**(1–2), 52–63 (1994)

28. H. Rotermund, J. Lauterbach, G. Haas, The formation of subsurface oxygen on Pt (100). Appl. Phys. A **57**(6), 507–511 (1993)

29. C.Q. Sun, Relaxation of the chemical bond. Spr. Ser. Chem. Phys. **108**, 807 (Springer-Verlag, Heidelberg, 2014)

30. F. Jensen, F. Besenbacher, E. Laegsgaard, I. Stensgaard, Dynamics of oxygen-induced reconstruction on Cu(100) studied by scanning tunneling microscopy. Phys. Rev. B **42**(14), 9206–9209 (1990)
31. C.Q. Sun, Oxygen-reduced inner potential and work function in VLEED. Vacuum **48**(10), 865–869 (1997)

Chapter 18
Four-Stage Cu₃O₂ Bonding Dynamics

Abstract VLLED has enabled quantification of the four-stage Cu_3O_2 pairing-tetrahedra formation in the Cu(001) surface transiting from O^- to O^{2-} with production of the missing rows, Cu-O-Cu chains, oppositely paired Cu^p crossing the missing rows. The surface stress turns from tensile in the O^- derived first phase to the O^{2-} derived second phase. The phase transition dynamics is beyond the scope of computations from the perspective of total energy minimization or structural optimization. Annealing relaxes the Cu_3O_2 bond geometry, the SPB, and the valence states accordingly while heating at a dull-red color de-hybridizes the *sp* orbits of oxygen, desorption will occur at higher temperatures.

Highlights

- Transition from O^{-1} to O^{-2} proceeds in four discrete stages with involvement of top two atomic layers.
- O^{-1} prefers the apical site of an off-centred pyramid and polarizes its nearest neighbours.
- Second O-Cu bond forms to a Cu underneath with missing-row formation and geometry relaxation.
- The lone pairs of Cu_3O_2 couple oppositely the Cu^p dipoles cross over the missing row.

18.1 Exposure-Resolved VLEED: Four-Stage Reaction Kinetics

The most striking challenge in studying oxygen chemisorption is the kinetics of bond formation. Numerous scientists have devoted to resolving this issue in the past century. The invention of the STM/S has propelled the progress with visulization

© The Editor(s) (if applicable) and The Author(s), under exclusive license
to Springer Nature Singapore Pte Ltd. 2020
C. Q. Sun, *Electron and Phonon Spectrometrics*,
https://doi.org/10.1007/978-981-15-3176-7_18

[1, 2]. A combination of the H$_2$O like tetrahedron and the paramerized VLEED decoding strategies and the excellent data collected by Hitchen, Thurgate, and Jennings [3] made the quantification of the bonding kinetics possible [4].

Figure 18.1a–c shows the exposure-resolved VLEED(00) beam reflectance I$_{00}$/I$_0$ versus incident beam energy measured at 70° incidence and 42° azimuth angles. The fine-structure features are very sensitive to the oxygen-exposure. Examination of the typical spectral peak intensities at 7.1, 9.1 and 10.3 eV revealed that the reaction progresses in four discrete stages as a function of oxygen exposure (Θ_O in Langmuir= 10^{-6} s torr) [4]:

(i) $\Theta_O \leq 30$ L: The peak at 7.1 eV decreases in magnitude until oxygen exposure reaches 30 L, while other peaks have little change.
(ii) 30 L > $\Theta_O \leq 35$ L: The decreased peak intensity at 7.1 eV recovers a bit.
(iii) 35 L > $\Theta_O \leq 200$ L: The first peak attenuates while one new peak at 9.1 eV emerges; and then both the peak at 9.1 eV and the peak at 10.3 eV have increasing maximum up to 200 L.
(iv) $\Theta_O > 200$ L: A general attenuation of the entire spectrum occurs.

Besides, the peak at 10.3 eV moves towards lower energy with increasing oxygen exposure. The spectral sensitivity examination suggested that the BA12 expansion and second O-Cu bond contraction (Q$_2$) modulate the intensities of the first two peaks while the diopole dislocation (DCu$_x$) depresses the intensity of the entire spectral features above 12.5 eV.

18.2 Geometrical Examination

Calculations used the same procedures described in previous section. The bond parameters and the z$_0$(E) curve for the 400 L oxygen exposure data are optimized first. The optimal parameters are Q$_2$ = 0.04, BA12 = 102.0°, DCu$_x$ = 0.25 Å as listed in Table 18.1. The optimal z$_0$(E) curve for 400 L is indicated in Figure 18.1d–f. Regardless of the accuracy of the single-variable parameterization and the extent to which the SPB varies with oxygen exposure, we then examined the structural sensitivity of the 400 L spectrum. Geometrical examination was conducted by adjusting one geometrical parameter at a time and maintaining the others and the z$_0$(E) curve undisturbed.

Figure 18.1d–f compare the calculation results from individually varying the three bond variables BA12, Q$_2$ or BL2, and the displacement DCu$_x$. Results show the following trends:

(1)   Simulation of the entire set of the exposure-resolved VLEED data can be made by keeping the oxygen-coverage constant at 0.5 monolayer. This implies that the extra oxygen atoms due to increasing exposure promote the reaction. The additional oxygen does not participate directly in the reaction after the saturation of adsorption. The promotion is treated as post-saturation effect. No quantitative correspondence between exposure (measured in Langmuir) and coverage (measured in monolayer—ML) can be established, as justified in many earlier studies

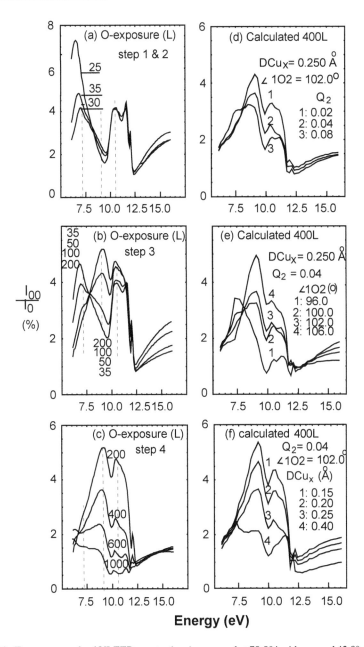

Fig. 18.1 Exposure-resolved VLEED spectra (**a–c**) measured at 70.0° incidence and 42.0° azimuth from the O-Cu(001) surface, and the calculated results (**d–f**) from varying individual bond variables. Variations of intensities in panel (**a–c**) at 7.1, 9.1 and 10.3 eV show four reaction stages. Calculation results (**d–f**) from varying bond variables Q_2, ∠1O2 and DCu_x for 400 L oxygen-exposure provide the resemblance of the measurements at different stages. (Reprinted with permission from [4])

Table 18.1 Four-stage O-Cu(001) surface bonding kinetics (Reprinted with permission from [4])

Reaction Stages		1 (<30 L: BL1 formation); 2 (30–35 L: BL2 & ∠1O2 change); 3 (35–200 L: ∠1O2 expansion); 4 (>200 L: DCu_x increase)								
Exposure	(L)	25	30	35	50	100	200	400	600	≥800
Bond	Q_2	0	0	0.04	0.04					
Geometry $Q_1 = 0.12$	BA12	92.5	94.0	98.0	100.0	101.0	102.0			
	DCu_x	0.125	0.150	0.150	0.150	0.150	0.150	0.250	0.355	0.450
Atomic shift (Å)	$- DCu_z$	0.1460	0.1440	0.1268	0.0938	0.0844	0.0709	0.1495	0.2239	0.2849
	$- DO_x$	0.1814	0.1796	0.1802	0.1831	0.1852	0.1877			
	DO_z	−0.0889	−0.0447	0.0618	0.1158	0.1422	0.1682			
	D_{12}	1.7522	1.7966	1.8287	1.8824	1.9086	1.9343			
Bond length (Å)	BL1	1.628								
	BL2	1.850		1.776						
	BL3	1.8172	1.8326	1.8833	1.8983	1.9053	1.9121	1.9262	1.9396	1.9505
Bond angle (°)	BA13	95.70	98.82	104.24	105.12	105.64	105.33	105.30	104.01	103.83
	BA23	91.80	93.11	95.75	96.46	96.83	95.18	99.52	101.67	103.43
	BA33	165.87	160.83	145.26	144.32	143.02	141.82	139.43	135.38	135.71

Empty space is identical in value to that in the previous cell. All information is provided by the controlling variables (BA12, Q_2, DCu_x). Error bars for bond variables are 0.010 Å and 0.2°

The SPB constants: $V_0 = 10.50$ eV, $\gamma = -0.9703$, $\delta = 6.4478$

[5, 6]. Further evidence for the post-saturation effect is that the relaxation of the layer spacing increases with oxygen exposure.

(2) The results of the geometrical sensitivity examination, especially panels (e) and (f), *incredibly* coincide with the trends in the measurements of panels (b) and (c), respectively. This observation indicates that the process of the Cu_2O bond forming dominates the spectral features while the SPB change is insignificantly sensitive to the oxygen exposures. It is therefore further evidenced that the bond variables are realistic and that the corresponding parameterization of the SPB is reasonably correct.

(3) The four discrete reaction stages can be simply expressed in terms of a corresponding variation of the bond variables. For instances, features appeared in the range from 35 to 200 L are dominated by the increase of $\angle 1O2$, while features for the exposure greater than 200 L are dominated by the DCu_x alone. From 30 to 35 L, the recovery of the peak at 7.1 eV can be realized by increasing the Q_2 with smaller $\angle 1O2$ and smaller DCu_x.

(4) Calculation by varying the DCu_z was also carried out to examine the effect of individual atomic shift. Variation of DCu_z gives a little change of the spectral intensity between 9.5 and 11.5 eV, indicating again that the individual atomic shift describes an untrue reaction kinetics.

It is feasible to find optimal bond parameters from the 400 L result for different oxygen exposures. By assuming that the $z_0(E)$ is insensitive to the exposures, we repeated calculations to match the intensities of the three peaks of all the measurements in Fig. 18.1a–c. The calculations were performed with a careful search over large range and smaller steps of the variables. Table 18.1 lists the optimal structural parameters for structures at various exposures.

18.3 Four-Stage Cu_3O_2 Bonding and Band Forming Kinetics

VLEED calculations at other exposures with the optimal geometry values in Table 18.1 confirmed that the SPB is relatively insensitive to the oxygen-exposure. Figure 18.2 shows the offset $z_0(E)$ profiles that produce the $ImV(E, \phi_L)$ curves and duplicate the measured spectra in Fig. 18.1a–c. It is seen that the $z_0(E)$ curves, in general, are insignificantly sensitive to the oxygen exposure. They are similar in shape except for minimal variations above 7.5 eV at 25 L exposure. This further evidences for the assumption that the VLEED reflectance is less sensitive to the SPB than to the bond geometry during exposure increasing. The slight outward-shift (-z direction, relative to −2.5 a.u. as indicated by broken lines) of the $z_0(E)$ curves at higher exposures increases the n(E); and as a result, the increase of the n(E) reduces the work function and attenuates the amplitude of the reflected beams.

The $z_0(E)$ is the contribution of occupied DOS, n(E), that is convoluted by real space (local spatial DOS n(x, y)) due to multiple-diffraction but the convolution is

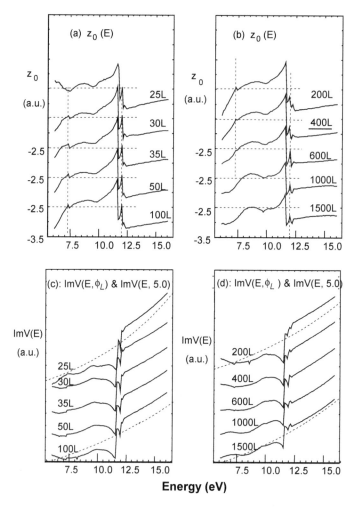

Fig. 18.2 Exposure-resolved $z_0(E)$ (**a, b**) and ImV(E, ϕ_L) (**c, d**) profiles exhibiting joint features of topography and spectroscopy reproduced the spectra in Fig. 18.1a–c. The z-axis directs into the bulk. The features on the $z_0(E)$ curves ($\leq$600 L) at 7.1 eV coincide with those appeared in STS [7]. Absence of the feature below 7.1 eV for 25 L indicates the O^{-1} dominates at this coverage. Violent features at 11.8–12.5 eV are attributed to the band-gap reflection. Broken lines represent the z_0 value (-2.5 a.u.) and the monotonic damping ImV(E, 5.0) for pure Cu(001) surface. The general outward-shift of the $z_0(E)$ profiles corresponds to the work function reduction [8]. (Reprinted with permission from [4])

minimally important in the VLEED energies. The non-constant form of the $z_0(E)$ arises from the fact the O-induced "rather local" attributes, which match the STM/S observations of the nonbonding (7.1 eV) and the bonding states below. The localization originates from the vacating, ionizing and polarizing of the surface atoms.

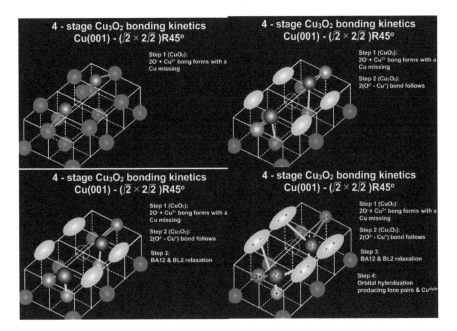

Fig. 18.3 VLEED derived four-stage Cu_3O_2 bond forming dynamics [9]

The $\Delta z_0(E)$ in the z-direction is about $(-2.3-(-3.3)$ atomic unit) 0.53 Å closing to ~0.45 Å as probed with STM [2].

The small features at 7.1 eV appearing in the curves of 30–600 L coincide with the peak at 2.1 eV below-E_F probed with STS from the O-Cu chain region and PES from O-Cu surfaces. The creation of new occupied states above 7.5 eV is identified as the contribution of nonbonding lone pair states, because of orbital hybridization of oxygen. Hence the absence of the features above 7.5 eV at 25 L implies that the sp-hybridization has not occurred yet at this stage in which O^{-1} dominates without orbital hybridization.

The presence of the bonding features around -5 eV below E_F, as resolved in PES, may go beyond the resolution of VLEED and convoluted by the spatial effect. The absence of the 7.1 eV features above 600 L oxygen-exposure is the annihilation of the hybridization state by the spatial effect on the reflect intensity, i.e., the saturation and outward-shift of the SPB for metal dipoles.

The violent features at 11.8–12.5 eV (~vacuum level) come from the band-gap reflection, and the surrounding features from electron-excitation near band edges. The shape similarity of all the $z_0(E)$ profiles at energies deeper than 7.5 eV indicates that energy states at the bottom of the valence bands and even the deeper $2p$-band are less affected by the chemisorption. Therefore, focus on the variation of valence states and its derivatives on the spectral features are on desirably correct track.

It should be noted that the VLEED data at 25 L could be simulated using the single $(\sqrt{2} \times 2\sqrt{2})R45°\text{-}2O^{-2}$ phase. The calculation result seems to conflict with the STM observations [1] that confirmed the co-existence of clean Cu(001), c(2 × 2)-2O^{-1} and the ($\sqrt{2}$ × 2$\sqrt{2}$)R45°-2O^{-2} phases at low oxygen exposures (<25 L). The extremely low damping of the clean Cu(001) surface [ImV(E = 6.0 eV, 16.0 eV) $\cong$ (0.78, 0.81 eV)], and the c(2 × 2)-2O^{-1} (1.0, 3.0 eV) [4] indicates that information from these two phases has been filtered out by the high damping of the ($\sqrt{2}$ × 2$\sqrt{2}$)R45°-2O^{-2} phase (1.3, 6.5 eV).

The high damping for the Cu$_3$O$_2$ phase can be gained by substituting $\gamma = -0.9703$, $\delta = 6.4478$ and $\phi_L \cong 4.0$ eV into the ImV(z, E). In fact, the relative number and the saturation degree of the dipoles dominate the intensity of the damping. O^{-1} induced dipoles composing the c(2 × 2)-2O^{-1} domain boundaries should be less saturated than those induced by the lone pairs of O^{-2} in the ($\sqrt{2}$ × 2$\sqrt{2}$)R45°-2O^{-2} phase. Fig. 18.3 illustrates the VLEED resolved four-stage Cu$_3$O$_2$ bonding dynamics. O$_2$ dissociates and bonds to a surface Cu atom to form the CuO$_2$ pairing-pyramid and then the second O-Cu bond follows with a Cu atom underneath associated with sp^3 orbital hybridization and lone pair production. The lone pair polarizes the neighboring Cu atom into dipoles. During the process of reaction, bond length and angle relaxatiom continues with production of the missing-row vacancy by isolating the vacated atom from other neighbors.

18.4   Aging and Annealing Effects on VLEED Profiles

VLEED *I-E* scans after aging and annealing of the specimen provide useful information about the bond formation, relaxation, and dissociation under such conditions. Figure 18.4a shows the effect of annealing and aging on the VLEED *I-E* curves of a 300 L oxygen-exposed Cu(001) surface. The time-resolved spectra to be decoded were collected at 72.0° incidence and 42.0° azimuth [3]. The experimental conditions and the fine-structure features of these spectra are summarized as follows:

(1)   Scan A was taken immediately after the clean Cu(001) surface exposed to 300 L oxygen. There are two broad peaks at 9.0, 11.0 eV due to Rydberg resonance and two sharp peaks at 10.5 and 12.0 eV due to Bragg diffraction.
(2)   Scan B was taken after 25 min aging and produced the same result, apart from a change in slope (goes up) below 9.5 eV.
(3)   Scan C was taken after 5 min of mild heating (~550 K), to a dull red color, and showed a change in structure. Besides the slope below 9.5 eV, the whole spectrum increases in intensity.
(4)   Scan D was taken after a further three hours aging. No change in structure from scan C is noted apart from a general attenuation in intensity of the spectrum and a significant intensity decrease below 9.5 eV. The changes between scan C and D are similar in effect to the result of oxygen exposure greater than 200 L.

In general, the spectral shape and intensity below 9.5 eV are more sensitive to the aging and annealing, which indicates that reaction modifies energy states of the lone pairs—orbital hybridization and dehybridization.

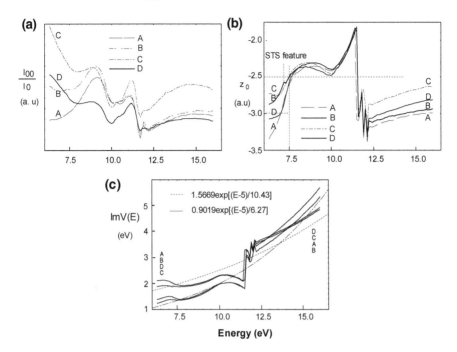

Fig. 18.4 Annealing and aging effect on the (**a**) VLEED spectra [3] and the corresponding (**b**) energy dependence of $z_0(E)$ profiles and (**c**) the damping $ImV(E)$. Profiles in (**b**) and (**c**) provide duplication of the VLEED spectra. Annealing increases the slope of the inelastic damping. Except for the $z_0(E)$ of plot B the hybridization feature at 7.1 eV is invisible for all the $z_0(E)$ curves. The lowering of the $z_0(E)$ above 7.5 eV (lone pair features) for scan C indicates the tendency of de-hybridization of oxygen due to annealing [10]. (Reprinted with permission from [3, 10])

18.5   De-Hybridisation of Oxygen

As a reference, the geometrical and SPB parameters were fixed first for curve D in Fig. 18.4a. The optimized $z_0(E)$ profiles and the bulk damping $ImV(E)$ for scans A–D are shown in Fig. 18.4b and c, respectively. Table 18.2 lists the best-fit structural parameters. The calculation procedures are the same as that used in the earlier Sections. A structure-sensitivity examination was performed based on the fixed parameters for curve D.

Figure 18.5 shows the results from varying individual bond variables of BL2 (Q_2), BA12, and DCu_x. Comparing with the bond variable, the DO_z is also examined to show the effect of individual atomic displacement. Calculated spectra indicate that the aging and annealing effect is not readily, as did the exposure effect, quantified by varying individual bond-parameter except for the long duration aging (from scan C to D).

Spectra under long duration aging can be simulated simply by changing the DCu_x alone. The results in Fig. 18.5 showed, however, features below 9.5 eV in Fig. 18.4a

Table 18.2 Cu$_3$O$_2$ structure varies with aging and annealing of the O-Cu(001) surface (Reprinted with permission from [10])

VLEED scans		A(g)	B(h)	C	D
Bond length (Å)	BL1	1.628	1.628	1.628	1.628
	BL2	1.776	1.776	1.776	1.776
	BL3	1.9053	1.9112	1.9202	1.9297
Bond angle (°)	BA12	101.00	101.25	101.50	102.00
	BA31	105.42	105.48	105.40	105.17
	BA32	96.83	95.52	98.77	100.03
	BA33	143.02	141.98	140.46	138.92
Atomic shift (Å)	DCu$_x$	0.150	0.175	0.225	0.275
	DCu$_z$	−0.0844	−0.1015	−0.1375	−0.1679
	DO$_x$	−0.1852	−0.1858	−0.1864	−0.1877
	DO$_z$	0.1442	0.1488	0.1553	0.1682
Layer spacing	D$_{12}$	1.9086	1.9150	1.9215	1.9343
Damping	γ	1.5669		0.9019	
Potential	δ	10.427		6.2736	

Controlling variables are BA12 and DCu$_x$. V$_0$ = 10.56 eV

can be modulated by the combination of BA12 and Q$_2$. Features in scan C, after five minutes annealing, can be produced by reducing DCu$_x$. The adjustment of DCu$_z$ produces no clear variation of the spectrum that could match observed trends and thus it is evidenced again that the displacements of individual atoms are untrue reaction.

The structure sensitivity examination suggested that the time-resolved VLEED data can be simulated but the bonding kinetics appears not as explicit as that happened when increasing oxygen exposure. Therefore, annealing and aging only modulate the production of lone pairs and the lone pair induced dipoles, and the bond length andgle relaxation.

The z$_0$(E) and ImV(E) profiles in Fig. 18.4b and c vary their shapes apparently at energies outside 7.5–12.5 eV. Features above 7.5 eV correspond to the nonbonding lone pair states of O^{-2}. The z$_0$(E) features above 12.5 eV are dominated by inelastic damping. The small feature at 7.1 eV of the z$_0$(E) profiles, an indicator of the *sp*-orbital hybridization of oxygen, varies with the aging and annealing conditions.

This observation suggests that the *sp*-hybridization be not fully-developed yet immediately after the specimen being exposed (scan A) to oxygen and that, the de-hybridization occurs due to annealing (scan C). The absence of the small feature from curve D is similar to that happened to the O-exposure greater than 600 L, due to the annihilation of the lone-pair features by the fully developed metal dipoles. The intensity of the hybridization states (<7.5 eV) in curve C is obviously weakened by annealing.

On the other hand, annealing treatment (for scan C and D) changes the slope of damping as indicated in Fig. 18.4c. At lower energies the ImV(E) becomes relatively

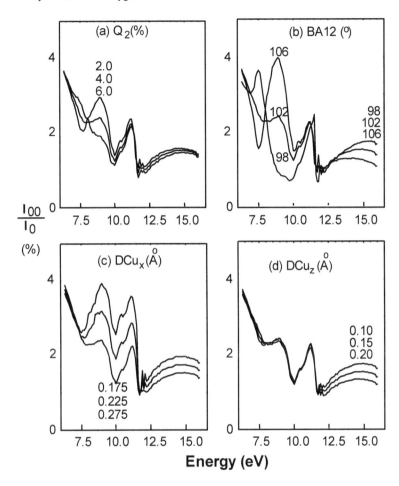

Fig. 18.5 Sensitivity of VLEED *I-E* spectrum (D) to the bond geometry [10]. Results indicate that the aging and annealing effect is not readily quantified by individual bond parameters except for the long duration aging (from Fig. 18.4a scan C to D) that can be described by increasing DCu_x (panel c) alone. Features below 9.5 eV of scans A–D in Fig. 18.4a can be modulated by the joint contribution of Q_2 (panel a) and BA12 (panel b) with the extent of lone pair development. (Reprinted with permission from [10])

lower while at higher energies the ImV(E) are higher. Features above 7.5 eV imply that the energy states in the upper of valence bands are more readily to be affected by annealing than states in the bottom of the valence band. The change of upper states corresponds to the formation of nonbonding lone pairs, namely, the hybridization of O^{-2}. The lower states correspond to the O-Cu bonding. Weakening the $z_0(E)$ and ImV(E) features at lower energy due to annealing indicates that oxygen de-hybridization takes place. Therefore, annealing supplies energy for oxygen to be de-hybridized, which forms also the basis of thermal desorption and bond switching of oxide tetrahedron [11].

18.6 Summary

The $CuO_2 \Rightarrow Cu_3O_2$ transition bonding kinetics on Cu(001) surface is quantified as
follows (refer to Fig. 18.3 and Table 18.1) [4]:

(1) $\Theta_O \leq 30$ L: The dissociated oxygen atoms form one contracting ($Q_1 = 12\%$)
    ionic bond with a Cu atom (labeled 1) on the surface. The DCu_x reaches 0.15
    Å and $\angle 1O2$ reaches 94.0°. O^{-1} is located at a position above the surface and
    forms an off-centered single or pairing-pyramid polarizing its rest neighbors.
(2) 30 L $> \Theta_O \leq 35$ L: O^{-1} forms the second contracting (increase Q_2 from 0 to
    4%) ionic bond with a Cu atom (labeled 2) in the substrate second layer, and as
    a result, the O^{-2} penetrates the bulk and evolves its position from the apical site
    of the pyramid to the center of the tetrahedron upon the second bond formation.
    Meanwhile, $\angle 1O2$ increases from 94.0 to 98.0°, associated with the isolation
    and evaporation of the missing-row atom.
(3) 35 L $> \Theta_O \leq 200$ L: The angle $\angle 1O2$ increases resulting in the relaxation D_{12},
    and simultaneously, angle $\angle 1O2$ increases from 98.0° to a saturation value of
    102.0° while other parameters have little change.
(4) $\Theta_O > 200$ L: The sp-orbital hybridization takes place with creation of the non-
    bonding lone pairs. The interaction between the O^{-2} and the lone-pair-induced
    Cu^P develops, which dominates the reaction at higher exposure and long-term
    aging. Lone pairs push the dipoles outwardly, and as a result, pairing dipoles
    form and bridge over the missing row. The DCu_x increases from 0.15 up to 0.45
    Å at about 800 L.

The variations of the structural-dependent $z_0(E)$ profiles (Fig. 18.2) agree with the
bonding kinetics. The features are summarized as follows:

(1) Features above 7.5 eV, particularly the small sharp peak at 7.1 eV, appearing in
    the 30–600 L curves are derivatives of the nonbonding lone-pair states of the
    O^{-2}-hybrid [12].
(2) The insignificant outward-shift at higher exposures corresponds to the reduc-
    tion of work function that is caused by the development of the oppositely
    coupled metal dipoles.
(3) The absence of the nonbonding states at 25 L because of absenting the
    O^{-2} sp-orbital hybridization, while at higher exposures the 7.1 eV feature is
    overtoned by the development of metal dipoles to be undetectable.
(4) Similarity of the exposure-resolved $z_0(E)$ profiles at energies greater than 7.5 eV
    implies that chemisorption has little effect on the electrons in the bottom of
    valence band or even deeper. The weak bonding states (~ -5.0 eV) may go
    beyond the resolution of VLEED due to the spatial convolution by saturated
    Cu^P. Charge transferring occurs only to electrons in the upper valence band.

The effect of aging and annealing on the O-Cu(001) system is slightly complex and
therefore it is less explicit than the effect of increasing oxygen exposure. However,
it is certain that long-term aging effects the same as higher oxygen exposure on the

spectral features that can be understood as the development of interaction between the nonbonding lone pairs and the lone pair induced Cu^p. The annealing provides a force to de-hybridize the oxygen, reducing the DCu_x and the lone pair DOS features rather than enhancing the bond formation.

A consistent understanding of the bonding kinetics and the corresponding variation of valence DOS has thus been obtained through both the bond-geometry and the $z_0(E)$ profiles, which further evidence that the modeling approaches are complete and realistic. The VLEED technique is unique and powerful in revealing dynamic and quantitative information on the valence DOS, SPB, and the bond formation at the surface.

References

1. T. Fujita, Y. Okawa, Y. Matsumoto, K.-I. Tanaka, Phase boundaries of nanometer scale c (2 × 2)-O domains on the Cu (100) surface. Phys. Rev. B **54**(3), 2167 (1996)
2. F. Jensen, F. Besenbacher, E. Laegsgaard, I. Stensgaard, Dynamics of oxygen-induced reconstruction on Cu(100) studied by scanning tunneling microscopy. Phys. Rev. B **42**(14), 9206–9209 (1990)
3. C. Hitchen, S. Thurgate, P. Jennings, A LEED fine structure study of oxygen adsorption on Cu(001) and Cu(111). Aust. J. Phys. **43**(5), 519–534 (1990)
4. C.Q. Sun, Exposure-resolved VLEED from the O-Cu(001): bonding dynamics. Vacuum **48**(6), 535–541 (1997)
5. M. Wuttig, R. Franchy, H. Ibach, Structural models for the Cu(100)(2 × 22) R45°-O phase. Surf. Sci. **224**(1), L979–L982 (1989)
6. H. Zeng, K. Mitchell, Further LEED investigations of missing row models for the Cu(100)-(22 × 2) R45°-O surface structure. Surf. Sci. **239**(3), L571–L578 (1990)
7. F.M. Chua, Y. Kuk, P.J. Silverman, Oxygen chemisorption on Cu(110): an atomic view by scanning tunneling microscopy. Phys. Rev. Lett. **63**(4), 386–389 (1989)
8. C.Q. Sun, C.L. Bai, Modelling of non-uniform electrical potential barriers for metal surfaces with chemisorbed oxygen. J. Phys. Condensed Matter **9**(27), 5823–5836 (1997)
9. C.Q. Sun, Oxidation electronics: bond-band-barrier correlation and its applications. Prog. Mater Sci. **48**(6), 521–685 (2003)
10. C.Q. Sun, Time-resolved VLEED from the O-Cu(001): atomic processes of oxidation. Vacuum **48**(6), 525–530 (1997)
11. C.Q. Sun, H. Xie, W. Zhang, H. Ye, P. Hing, Preferential oxidation of diamond {111}. J. Phys. D Appl. Phys. **33**(17), 2196–2199 (2000)
12. C.Q. Sun, S. Li, Oxygen-derived DOS features in the valence band of metals. Surf. Rev. Lett. **7**(3), 213–217 (2000)

Chapter 19
Perspectives

Abstract The dynamic VLEED, in junction with STM and PES, has enabled the hitherto comprehensive insight into the nature and dynamics of four-stage CuO_2 and subsequent Cu_3O_2 bond forming on the Cu(001) surface and its consequences on the behavior of atoms and valence electrons in terms of the tetrahedral bond geometry, four valence DOS features and the 3D SPB evolution. Derived quantititive information also includes the 2D Brillouin zones, effective masses of electrons at the boundaries, energy band dispersion, muffin-tin potential, work function and forces driving the missing rows, lattice reconstruction and relaxation, etc. VLEED forms such a promising means that reconciles crystallography, surface morphology and electron enenrgy spectroscopy overing the valence band and above, though sophisticated logic thinking and patience are necessary. Progeress proved that the impact of *Coordination Bonding and Electronic Dynamics* is immesely profound and essential to the understanding of the surface processes.

Highlights

- VLEED reconciles capabilities of electronic spectroscopy, crystallography and morphology.
- Orbital hybridization and surface bond contraction are key drivers for chemisorption.
- It is forbidden that one O forms more than one bond with any specific atom.
- The two bonds forms one after the other followed by sp^3-orbital hybridization.

The dynamic VLEED, in junction with STM and PES, has enabled the hitherto comprehensive insight into the nature and dynamics of oxide bond forming on the Cu (001) surface and its consequences on the behavior of atoms and valence electrons from the perspective of the bond geometry, valence DOS and the 3D SPB. Strategies of extracting information from the combination of STM, PES and the dynamic VLEED are proven essential, effective and correct. Major progress made is summarized as follows:

(1) VLEED is such a unique technique that it collects quantitative and nondestructive information from a single atomic layer and covers the valence band of a

C. Q. Sun, *Electron and Phonon Spectrometrics*,
https://doi.org/10.1007/978-981-15-3176-7_19

surface regarding the bond geometry, bond forming dynamics and the corresponding change of atomic locations, valence DOS and the surface potential. VLEED enables the current modeling approach to be verified and the modeling approach in turn uncovers the full capacity and reliability of the VLEED technique.

(2) VLEED spectral sensitivity examination revealed that a 20% change of the $ReV(z)$ parameters cause an antiphase shift, altering the intensity of spectral features oppositely; the spectral features are less sensitive to the $ImV(z)$ parameters; the bond geometry parameters are more practical in responsible for the spectral features than the individual atomic dislocations.

(3) The tetrahedron bond formation with a small number of parameters and physical constraints reflects the real process of reaction in which atoms move their positions collectively. The model indicates that oxidation is a dynamic process of electron transportation in which O^{-1} forms first and then O^{-2} follows with sp-orbital hybridization and lone pair production. O^{-2} prefers in the nearly central position of a tetrahedron. Importantly, the O^{-1} or nonbonding lone pair induces metal dipoles at the surface. Besides the electron transport between oxygen and metals, metal dipoles as well as the nonbonding states of the hybridized O^{-2} play key roles in the electronic processes and physical properties of oxidation. For example, the strongly localized electrons with low mobility reduce the work function and increase the contact electrical resistance.

(4) As a consequence of bond forming, the valence DOS of a metal is modified with four additional features of O-M bonding far below E_F, nonbonding lone pairs of oxygen just below E_F, antibonding metal dipoles above and the holes below the E_F. Therefore, oxygen possesses the special capacity of creating a gap or widens the existing band gap, and adding a sub-band above E_F reducing the work function, which agrees with the PES detection of a number of metal surfaces with chemisorbed oxygen.

(5) The single-variable parameterized SPB functions uncover the full capacity of VLEED in simultaneously determining the bond geometry, the shape of the SPB and the variation of the valence DOS and their interdependence. One can judge models by simply comparing the shape of the geometrical-dependent $z_0(E)$ curves.

(6) The decoding skills are proven more revealing than the traditional wisdom. The overall approaches not only reflect properly the real process of reaction but also reduce the numerical efforts and ensure the certainty of solution.

(7) O-Cu(001) reaction is a dynamic process that progresses in four discrete stages. The Cu_3O_2 configuration indicates that one Cu may donate more than one electron to different oxygen adsorbates but one oxygen atom can never get more than one electron from a specific Cu. This provides the basis of preferential oxidation of a certain orientation of crystalline such as diamond (111) faces [1].

(8) The O-Cu(001) surface bond forming is responsible for all the observations, whether static or dynamic. Atomic dislocation and surface relaxation are determined by the bond geometry. The strong localization of surface electrons and

variation of the energy states, as well as the reduction in both the work function and the inner potential constant result from the electron transportation from one energy level to the other of different atoms such as polarization and ionization.

(9) The various observations are consequences of bond formation that depends on the electronegativity, atomic size and lattice geometry of the host surface differing from case to case, and the controlling parameters seem not surprisingly to be the temperature, oxygen exposure and aging time.

(10) Although the surface morphologies and atomic geometries vary from case to case, the basic oxide tetrahedron, the DOS features, and the electronic process of oxidation are substantially the same in all the considered samples. The electronic processes of oxidation and its extension to technical applications has formed the subject of review, referring to [2].

(11) Most strikingly, this work reveals two essential events occurred at an oxygen chemisorbed surface. One is the *surface bond contraction* and the other is the *sp-orbital hybridization with lone pair and dipole formation,* which laid the foundations of the sequential work in this practationer's career life. The spontaneous bond contraction leads to a change of the binding energy and their derivatives. Surface bond contraction and the rise in surface-to-volume ratio have been found to be responsible for the size-dependency of nanometric solid, such as lattice contraction [21, 22], enhancement of surface stress and Young's modulus [22, 23], band gap expansion [24, 25] and dielectric suppression [26–28] nanometric semiconductors and diamond melting point rise or depression [29], enhancement of magnetization [30], modification of phase transition temperature [31–33]. Surface bond contraction has led to a consistent understanding of the size-dependent properties of nanometric solid, which has reconciled the performance of undercoordinated atoms associated with adatoms, defects, terrace edges, skins of solid and liquid, and nanostructures of various shapes.

(12) An extension of sp-orbital hybridization to the interaction of C, N and O to the fcc (001) surface of Rh and Ni, has enabled the quantification surface bond stress, and as a consequence, improves the adhesion between diamond and metals [34], and as a consequence, improves the adhesion between diamond and metals [34]. The band-gap generation mechanism has led to a finding that the PbZrTi oxide emits intensive blue light [35].

(13) Applying the knowledge to the interaction of O with the fcc ((001), (110), (111)) surfaces of Cu [36] and Rh, [37] as well as the hcp (($10\bar{1}0$), (0001)) of Ru [38] and Co [39, 40] has led to a consistent understanding of the multiphase reaction kinetics observed by means of STM [41, 42], LEED [43], PES and TDS [44], which leads to a conclusion [2] that though the surface morphologies, atomic geometries and phase ordering may vary from case to case, the basic oxide tetrahedron, the oxygen-derived DOS features, and the kinetics of bond forming are substantially the same in all the considered cases.

References

1. C.Q. Sun, H. Xie, W. Zhang, H. Ye, P. Hing, Preferential oxidation of diamond {111}. J. Phys. D-Appl. Phys. **33**(17), 2196–2199 (2000)
2. C.Q. Sun, Oxidation electronics: bond-band-barrier correlation and its applications. Prog. Mater Sci. **48**(6), 521–685 (2003)
3. C.Q. Sun, Y. Sun, The attribute of water: single notion, multiple myths. Springer Ser. Chem. Phys. **113**, 494 (2016). (Heidelberg, Springer)
4. X. Zhang, P. Sun, Y. Huang, T. Yan, Z. Ma, X. Liu, B. Zou, J. Zhou, W. Zheng, C.Q. Sun, Water's phase diagram: from the notion of thermodynamics to hydrogen-bond cooperativity. Prog. Solid State Chem. **43**, 71–81 (2015)
5. X. Zhang, Y. Huang, Z. Ma, L. Niu, C.Q. Sun, From ice supperlubricity to quantum friction: electronic repulsivity and phononic elasticity. Friction **3**(4), 294–319 (2015)
6. X.J. Liu, M.L. Bo, X. Zhang, L.T. Li, Y.G. Nie, H. TIan, Y. Sun, S. Xu, Y. Wang, W. Zheng, C.Q. Sun, Coordination-resolved electron spectrometrics. Chem. Rev. **115**(14), 6746–6810 (2015)
7. Y.L. Huang, X. Zhang, Z.S. Ma, Y.C. Zhou, W.T. Zheng, J. Zhou, C.Q. Sun, Hydrogen-bond relaxation dynamics: Resolving mysteries of water ice. Coord. Chem. Rev. **285**, 109–165 (2015)
8. C.Q. Sun, Relaxation of the chemical bond. Springer Ser. Chem. Phys. **108**, 807 (2014)
9. L. Pan, S. Xu, X. Liu, W. Qin, Z. Sun, W. Zheng, C.Q. Sun, Skin dominance of the dielectric electronic-phononic-photonic attribute of nanoscaled silicon. Surf. Sci. Rep. **68**(3–4), 418–445 (2013)
10. Z. Ma, Z. Zhou, Y. Huang, Y. Zhou, C. Sun, Mesoscopic superelasticity, superplasticity, and superrigidity. Sci. China Phys. Mech. Astron. **55**(6), 963–979 (2012)
11. W.-T. Zheng, C.Q. Sun, Underneath the fascinations of carbon nanotubes and graphene nanoribbons. Energy Environ. Sci. **4**(3), 627–655 (2011)
12. C.Q. Sun, Dominance of broken bonds and nonbonding electrons at the nanoscale. Nanoscale **2**(10), 1930–1961 (2010)
13. C.Q. Sun, Y. Sun, Y.G. Ni, X. Zhang, J.S. Pan, X.H. Wang, J. Zhou, L.T. Li, W.T. Zheng, S.S. Yu, L.K. Pan, Z. Sun, Coulomb repulsion at the nanometer-sized contact: a force driving superhydrophobicity, superfluidity, superlubricity, and supersolidity. J. Phys. Chem. C **113**(46), 20009–20019 (2009)
14. C.Q. Sun, Thermo-mechanical behavior of low-dimensional systems: The local bond average approach. Prog. Mater Sci. **54**(2), 179–307 (2009)
15. C.Q. Sun, Size dependence of nanostructures: impact of bond order deficiency. Prog. Solid State Chem. **35**(1), 1–159 (2007)
16. M.X. Gu, L.K. Pan, B.K. Tay, C.Q. Sun, Atomistic origin and temperature dependence of Raman optical redshift in nanostructures: a broken bond rule. J. Raman Spectrosc. **38**(6), 780–788 (2007)
17. W.T. Zheng, C.Q. Sun, Electronic process of nitriding: mechanism and applications. Prog. Solid State Chem. **34**(1), 1–20 (2006)
18. C.Q. Sun, *O-Cu(001): I. Binding the signatures of LEED, STM and PES in a bond-forming way.* Surface Review and Letters, **8**(3–4): 367-402, (2001)
19. C.Q. Sun, O–Cu (001) II VLEED quantification of the four-stage Cu_3O_2 bonding kinetics. Surf. Rev. Lett. **8**(6), 703–734 (2001)
20. C.Q. Sun, The sp hybrid bonding of C, N and O to the fcc(001) surface of nickel and rhodium. Surf. Rev. Lett. **7**(3), 347–363 (2000)
21. C.Q. Sun, The lattice contraction of nanometre-sized Sn and Bi particles produced by an electrohydrodynamic technique. J. Phys. Condensed Matter **11**(24), 4801–4803 (1999)
22. C.Q. Sun, B.K. Tay, S.P. Lau, X.W. Sun, X.T. Zeng, S. Li, H.L. Bai, H. Liu, Z.H. Liu, E.Y. Jiang, Bond contraction and lone pair interaction at nitride surfaces. J. Appl. Phys. **90**(5), 2615–2617 (2001)
23. X.J. Liu, J.W. Li, Z.F. Zhou, L.W. Yang, Z.S. Ma, G.F. Xie, Y. Pan, C.Q. Sun, Size-induced elastic stiffening of ZnO nanostructures: Skin-depth energy pinning. Appl. Phys. Lett. **94**, 131902 (2009)

24. C.Q. Sun, H.Q. Gong, P. Hing, H.T. Ye, Behind the quantum confinement and surface passivation of nanoclusters. Surf. Rev. Lett. **6**(2), 171–176 (1999)
25. C.Q. Sun, X.W. Sun, H.Q. Gong, H. Huang, H. Ye, D. Jin, P. Hing, Frequency shift in the photo-luminescence of nanometric SiOx: surface bond contraction and oxidation. J. Phys. Condensed Matter **11**(48), L547–L550 (1999)
26. C.Q. Sun, X.W. Sun, B.K. Tay, S.P. Lau, H.T. Huang, S. Li, Dielectric suppression and its effect on photoabsorption of nanometric semiconductors. J. Phys. D-Appl. Phys. **34**(15), 2359–2362 (2001)
27. H.T. Ye, C.Q. Sun, H.T. Huang, P. Hing, Dielectric transition of nanostructured diamond films. Appl. Phys. Lett. **78**(13), 1826–1828 (2001)
28. H.T. Ye, C.Q. Sun, H.T. Huang, P. Hing, Single semicircular response of dielectric properties of diamond films. Thin Solid Films **381**(1), 52–56 (2001)
29. C.Q. Sun, Y. Wang, B. Tay, S. Li, H. Huang, Y. Zhang, Correlation between the melting point of a nanosolid and the cohesive energy of a surface atom. J. Phys. Chem. B **106**(41), 10701–10705 (2002)
30. W.H. Zhong, C.Q. Sun, S. Li, H.L. Bai, E.Y. Jiang, Impact of bond-order loss on surface and nanosolid magnetism. Acta Mater. **53**(11), 3207–3214 (2005)
31. H.T. Huang, C.Q. Sun, T.S. Zhang, P. Hing, Grain-size effect on ferroelectric Pb(Zr1-xTix)O-3 solid solutions induced by surface bond contraction. Phys. Rev. B **63**(18), 184112 (2001)
32. H.T. Huang, C.Q. Sun, P. Hing, Surface bond contraction and its effect on the nanometric sized lead zirconate titanate. J. Phys. Condensed Matter **12**(6), L127–L132 (2000)
33. C.Q. Sun, W.H. Zhong, S. Li, B.K. Tay, H.L. Bai, E.Y. Jiang, Coordination imperfection suppressed phase stability of ferromagnetic, ferroelectric, and superconductive nanosolids. J. Phys. Chem. B **108**(3), 1080–1084 (2004)
34. C.Q. Sun, Y.Q. Fu, B.B. Yan, J.H. Hsieh, S.P. Lau, X.W. Sun, B.K. Tay, Improving diamond-metal adhesion with graded TiCN interlayers. J. Appl. Phys. **91**(4), 2051–2054 (2002)
35. C.Q. Sun, D. Jin, J. Zhou, S. Li, B. Tay, S. Lau, X. Sun, H. Huang, P. Hing, Intense and stable blue-light emission of Pb (ZrxTi1-x) O3. Appl. Phys. Lett. **79**(8), 1082–1084 (2001)
36. C.Q. Sun, Nature of the O-fcc(110) surface-bond networking. Mod. Phys. Lett. B **11**(25), 1115–1122 (1997)
37. C.Q. Sun, On the nature of the O Rh (110) multiphase ordering. Surf. Sci. **398**(3), L320–L326 (1998)
38. K. Meinel, H. Wolter, C. Ammer, A. Beckmann, H. Neddermeyer, Adsorption stages of O on Ru (0001) studied by means of scanning tunnelling microscopy. J. Phys. Condens. Matter **9**(22), 4611 (1997)
39. C.Q. Sun, On the nature of the triphase ordering. Surf. Rev. Lett. **5**(05), 1023–1028 (1998)
40. R. Koch, E. Schwarz, K. Schmidt, B. Burg, K. Christmann, K. Rieder, Oxygen adsorption on Co (101⁻ 0): Different reconstruction behavior of hcp (101⁻ 0) and fcc (110). Phys. Rev. Lett. **71**(7), 1047 (1993)
41. R. Koch, B. Burg, K. Schmidt, K. Rieder, E. Schwarz, and K. Christmann, Oxygen adsorption on Co (1010). The structure of p (2 × 1) 2O. Chem. Phys. Lett. **220**(3–5), 172–176 (1994)
42. S. Schwegmann, A.P. Seitsonen, H. Dietrich, H. Bludau, H. Over, K. Jacobi, G. Ertl, The adsorption of atomic nitrogen on Ru (0001): geometry and energetics. Chem. Phys. Lett. **264**(6), 680–686 (1997)
43. S. Schwegmann, H. Over, V. De Renzi, G. Ertl, The atomic geometry of the O and CO + O phases on Rh (111). Surf. Sci. **375**(1), 91–106 (1997)
44. K. Yagi, K. Higashiyama, H. Fukutani, Angle-resolved photoemission study of oxygen-induced c (2 × 4) structure on Pd (110). Surf. Sci. **295**(1), 230–240 (1993)

Part III
Multifield Phonon Dynamics

Phonon spectrometrics probes the bonding dynamics pertained to multifield perturbation or chemical reaction and offers information of the local bond length and energy, binding energy density, atomic cohesive energy, single bond force constant, elastic modulus, Debye temperature and the abundance-stiffness-fluctuation transition from one equilibrium to another.

Chapter 20
Wonders of Multifield Lattice Oscillation

Abstract Physical perturbation mediates material's properties by relaxing the inter-atomic bonding and electron configuration in various bands. Phonon spectroscopy probes bond relaxation or bond tranformation from one equilibrium to another under perturbation and electron spectroscopy fingerprints bond-relaxation induced electronic configuration, which have important impact to chemistry, physics, and material engineer and science. However, extracting information from the phonon spectroscopic measurements and gaining consistent insight into the physics behind observations are still infancy compared to crystallography and surface morphology because of lacking basic regulations. Conventional decomposition of spectral peaks or empirical simulation of spectral evolution under perturbation provide limited information with freely adjustable parameters and debating mechanisms. The featured multifield phonon spectrometrics aims to extracting atomistic, local, and quantitative information on the bonding and nonbonding dynamics and their correlation to performance of a variety of substances under external perturbations.

Highlights

- Substance evolves its structures and properties when subjecting to perturbation or reaction.
- Bond relaxation and the associated electronic energetics dictate material's performance.
- Correlating and resolving the bond-phonon-property of a substance is a high challenge.
- Phonon spectrometrics resolves unprecedented information and deepens the physical insight.

C. Q. Sun, *Electron and Phonon Spectrometrics*,
https://doi.org/10.1007/978-981-15-3176-7_20

20.1 Scope

This part starts with an overview in this chapter on the significance of bond relaxation, lattice oscillation, available experimental database, and theoretical approaches to naming challenges and potential opportunities. Focus is given on generalizing the trends of phonon frequency shift due to multifield activation in terms of atomic undercoordination, mechanical activation, thermal excitation and available theoretical descriptions. Thus, one can discriminate the intrinsic effects from those extrinsic artifacts and specify rooms for theoretical unification to gain consistent insight and information on bond relaxation by perturbations. It is exciting to find out that crystal size reduction creates three types of phonon frequency shifts unseen in the bulk—size-reduction induced E_{2g} blueshift, A_{1g} redshift, and the emerging of the low-frequency Raman (LFR) mode in the THz frequencies that undergo blueshift with the inverse of size. The Raman shift, bandgap, and elastic modulus follows a Debye thermal decay—drops nonlinearly and then linearly as temperature rises in a way of $1 - U(T)/E_{con}$ with $U(T)$ and E_{con} being the integral of the Debye specific heat and the atomic cohesive energy, respectively. Compression stiffens the phonon frequency nonlinearly and the mechanical compression enhances the effect of atomic undercoordination on the phonon relaxation. However, it is inspiring to note the large gap to be filled for the bond-phonon-property correlation. Overwhelming debating approaches exist from various, intrinsic and extrinsic, perspectives for a specific phenomenon. The conventional spectral peak decomposition and evolution simulation with freely-adjustable parameters with limited information hindered the progress in understanding the nature behind observations. It is urgent to develop a theory to reconcile the perturbation-bond-property correlation and to derive information on bonding dynamics, which drove the presented dedication made in past decade.

Chapter 21 is dedicated to developing the average-bond oscillating dynamics to incorporate the quantum approaches and Fourier transformation. Conventionally, the phonon relaxation is described from the perspective of Gibbs free energy, Grüneisen parameters, $\partial\omega/\partial x_i$, or their similarities, with x_i being the specific degree of freedom. In contrast, one may consider the Hamiltonian, Schrödinger equation, Lagrangian oscillation mechanics, Fourier transformation, and Taylor series of potentials with focus on the bond relaxation by external perturbation. The function dependence of a detectable quantity $Q(x_i)$ on the bond length and energy is necessary. To seeks for the relative change of a detectable quantity, $\Delta Q(x_i)/Q(x_{i0}) = f(x_i, d(x_i), E(x_i))$ with x_i being any stimulus taken as a hidden factor driving bond relaxation. The $Q(x_i)$ can be phonon frequency, bandgap, elastic modulus, etc. This way of approach reconciles the perturbations of atomic and molecular undercoordination, temperature, strain, pressure, electric polarization, etc., into one equation.

Chapter 22 examined the two-dimensional layered graphene, black phosphorus, and $(W, Mo)(S, Se)_2$ structures by theoretically reproducing their Raman shifts due the number-of-layers, orientational strain, compression, and thermal activation with derivative of new kind of information. The information includes the bond length and energy, band nature index, the dimer vibration frequency of reference, Debye temperature, atomic cohesive energy, thermal expansion coefficient, compressibility, elastic modulus, binding energy density, which should be the right capability of the spectrometrics. It is uncovered that the single dimer-vibration and the collective dimer-vibration govern, respectively, the number-of-layer reduction induced blue and red phonon frequency shift. The slope of the Debye thermal decay of the $\omega(T)$, $\partial\omega/\partial T$, at higher temperatures is the inverse of atomic cohesive energy and the pressure slope of the $\omega(P)$ profile, or the $\partial\omega/\partial P$, approaches the inverse of binding energy density.

Chapter 23 proved first the core-shell structures of water droplets and nanocrystals with derive of the skin thickness of 0.09 nm for the water droplets and two atomic diameters, 0.5 nm for CeO_2 nanocrystals. The skins follow the universal bond order-length-strength (BOLS) notion that specify the shorter and stiffer bonds between undercoordinated atoms. Examination of the sized crystals from nanoscale to the bulk of group IV, III-nitride, II-oxide under the perturbations of size reduction, pressure and temperature derived the same kind information described in Sect. 20.3.

Chapter 24 briefly introduces the recent progress in the spectrometrics of water ice and aqueous solutions. The effect of pressure, temperature, molecular undercoordination, and charge injection by acid, base, and salt solvation on the performance of water ice was systematically examined. Aqueous charge injection by solvation in the forms of anions, cations, electrons, lone pairs, molecular dipoles, and protons modulates the hydrogen bonding (O:H–O or HB) and electronic dynamics and properties of a solution. The modulation is through O:H formation, O:H–O bond relaxation, H $\leftrightarrow$ H anti-HB and O:$\Leftrightarrow$:O super-HB repulsions, electrostatically screened polarization, solute-solute interaction, and the solute H–O bond contraction due to bond order deficiency. Polarization by charge injection and molecular undercoordination modify the critical pressures and temperatures for the confined ice-quasisolid and the salted water-ice phase transition under heating and compression. Consistency between theoretical predictions and measurements confirms the ever-unaware issues such as quasisolid phase of negative thermal expansion (NTE) due to O:H–O bond segmental specific heat disparity. Molecular coordination deficiency and electrostatic polarization result in the supersolid phase. Excessive protons and lone pairs form the H_3O^+ hydronium and HO^- hydroxide, which turns an O:H–O bond into the H $\leftrightarrow$ H anti–HB upon acid solvation and O:$\Leftrightarrow$:O super–HB on base solvation. The aqueous molecular nonbonding can extend to other molecular crystals and to the negative thermal expansion of other solid substance.

Chapter 25 summarizes the attainment, limitation, and forward-looking directions. The phonon spectrometrics detects directly the bond responding to perturbation but the electron spectrometrics probes the behavior of electrons in various bands/levels due to bond relaxation, which are totally different but complement each other. This set of spectrometrics provides a powerful technique applicable to situations when electrons and phonons are involved and under any external perturbation—for atomistic, dynamic, local, and quantitative information on bond and electron performance and consistent insight into the nature of observations, an essential theoretical and experimental strategies for functional materials devising.

Appendix compares the advantages and limitations of the spectrometrics of electron emission, electron diffraction and multifield liquid and solid phonon spectrometrics.

20.2 Significance of Multifield Lattice Oscillation

Much attention has been paid to bond formation and dissociation by chemical reaction that revolves the material's properties in an abrupt way [1]. For instance, nitrogenation turns the metallic Gallium into the semiconductive GaN for intense blue light emission [2]. Oxidation transits Zn and Al into the wide bandgap ZnO semiconductors for electronic optical devices and into Al_2O_3 insulator for fast thermal energy dissipation [3]. However, attention is necessary to be paid to the bond gradual relaxation from one equilibrium to another or from formation to dissociation under continued external perturbation such as compressing, heating, stretching, atomic undercoordination by defect, nanostructure and surface formation, contamination by charge injection, doping and impurities. Bond relaxation and the associated energetics, localization, entrapment, and polarization of electrons in various bands mediate continually the performance of substance [4].

Variation of the size and shape of a crystal has created tremendous fascinations, which has formed the foundations for nanoscience and nanotechnology being recognized as a thrust to the science and technology of the concurrent century and future generations [5, 6]. Nanostructured materials perform differently from their bulk counterparts as the quantities like elastic modulus, dielectric constant, work function, band gap, critical temperatures for phase transition, remain no longer constant but change with the shape and size of the nanostructures. Atomic undercoordination shortens and stiffens the bonds between undercoordinated atoms (called confinement in occasions) [7–9]. Atomic undercoordination strengthens the nanocrystals but lowers its thermal stability, a competition of both strength and thermal stability results in the inverse Hall-Patch effect—hardest at the 10 s nanometer scale [7, 10]. Size reduction generally depresses the critical temperatures for nanocrystal phase transition [11]. Hetero-coordination may harden the twin grain boundaries [12] by energy densification or soften some other materials at the interfaces by polarization [13].

Multifield lattice oscillation of the sized crystals have received extensive attention [14–16] because the phonon behavior influences directly on the electrical and optical transport dynamics in semiconductors [17, 18], such as electron-phonon coupling, photoabsorption, photoemission and waveguide devices for light transportation. The Raman-active modes of Bi_2Se_3 nano-pallets shift a few wavenumbers lower as the thickness is decreased in the vicinity of ~15 nm [19], similar to that of the D and 2D modes in the number-of-layer resolved graphene [20, 21]. The LO mode of the CdS film thinner than 80 nm also showed the size-induced phonon frequency softening [22]. The frequency of the LO mode for a 9.6 nm-sized CdSe dot is slightly lower than that of the corresponding CdSe bulk at room temperature. As the CdSe crystal size is reduced to 3.8 nm, the peak frequency shifts to a lower frequency by about $3\,\mathrm{cm}^{-1}$ [23].

Considerable effort has also been paid to the study of bulk compounds due to their intriguing thermal and mechanical properties and to the promise that they offer potential applications in optoelectronics, waveguides, laser frequency doubling devices, high capacity computer memory cells, sensors, actuators, etc. [24, 25]. Materials under mechanical and thermal perturbation vary their structures and properties such as phase transition or mechanical hardness [26]. Compression hardens globally a substance and raises the vibration frequency and the critical pressure for phase transition of regular substance. Heating and stretching have a contrasting effect of compression to narrow the band gap and lower the work function. The volume concentration of nanopores below a certain value can harden the substance but above the critical value causes detrimental to the yield strength of the porous materials [7].

In conjunction to the Raman shift, the bulk modulus B or the inverse expansibility is related to the performance of a material such as acoustic transmission, Debye temperature, specific heat capacity, and thermal conductivity of the specimen, which keep constant at the ambient atmospheres. However, the B becomes tunable with the applied T and P stimuli [27–30]. Atomistic simulations have revealed that the B of a substance is softened under elevated temperature and stiffened under increased pressure [28].

The macroscopic properties of a substance depend functionally on the bond length, bond energy, and valence electron configuration. For instance, the band gap and dielectrics varies with the interatomic bond energy and the electronic occupancy in the conduction and in the valence bands [31]. Likewise, the local binding energy density determines material's elastic modulus and yield strength [32] and the atomic cohesive energy [26] determines the critical temperatures of phase transitions, catalytic activation, interatomic diffusion, etc. The competition of binding energy density and atomic cohesive energy determines the inverse Hall-Petch relationship and the maximal hardness of a crystal at the nanometer scale [33].

The curvature of the bond potentials at equilibrium $r = d$ determines the frequency of vibration ω in the form of $\mu\omega^2 x^2 = [U''(d) + U'''(d)x/3]x^2$ where the nonlinear term can be omitted in the harmonic approximation that is proper at the equilibrium. According to the dimensional analysis, the ω is proportional to the bond length d and energy E in the form of $(\Delta\omega)^2 \propto E/(\mu d^2)$ [34] with μ being the reduced mass of the vibrating dimer. Any perturbation will relax the bond and shift the phonon frequency. Therefore, external stimulus mediates the properties of a substance by relaxing the bond length and energy and the crystal potentials which provides one with opportunities to calibrate and control the property change of a substance.

20.3   Outline of Experimental Observations

20.3.1   Size Matter—Atomic Undercoordination

The Raman-active modes and the IR-active modes for the sized and the layered two-dimensional (2-D) structures show different trends of phonon frequency shift [35–39]:

(1) The transverse and the longitudinal optical (TO/LO) Raman modes shift toward either lower or higher frequency.

(2) The E_{2g} mode for WX_2 and TiO_2 and the G mode for graphene undergo blueshift as the feature size is reduced.

(3) The A_{1g} mode and the D/2D mode shift to lower frequency when the features size decreases.

(4) Low-frequency Raman (LFR) acoustic modes emerge at wave numbers of a few or a few tens cm^{-1}, or THz range (~33 cm^{-1}), and this mode undergoes a blueshift when the feature size is reduced. The LFR disappears at infinitely large crystal size.

The Raman spectra for (a) CeO_2 [40] and (b) Si [41] nanoparticles in Fig. 20.1 show consistently that crystal size reduction softens the phonons, broadens the linewidth, and asymmetrizes the line-shape of the characteristic phonons. The presence of oxygen vacancies or other impurities also affects the spectral line shapes and peak frequencies. The peak shape features the intrinsic population and the peak maximum is the highest probability. The peak area integration of the population function within a certain frequency range represents for the number of phonons contributing

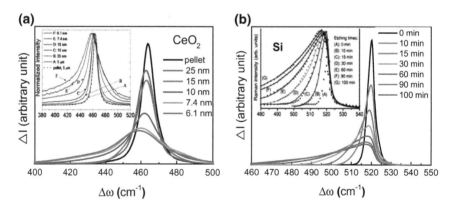

Fig. 20.1 Raman shifts for the size-resolved **a** CeO_2 (featured at 465 cm^{-1}) [40] and **b** Si (521 cm^{-1}) [41] nanoparticles being spectral peak area normalized. Insects compare the widely-used peak maximal intensity normalized spectra. Size reduction softens the phonon stiffness, broadens the linewidth, and asymmetrizes the line shape, being attributed to the phonon quantum confinement, nonlinear effect, or inhomogeneous strains. Reprinted with permission from [40, 41]

to the spectral features [6]. In place of the peak maximum normalization crossing all
the spectra for the same specimen of different sizes, as depicted in Fig. 20.1 insets,
the peak area normalization is physically meaningful as this process minimizes the
experimental artifacts such as scattering due to surface roughness.

The Raman shifts are very sensitive to the feature size of the examined substance
at the nanometer scale because of the raised number ratio of the undercoordinated
atoms in the skin shells [36]. For the layered graphene instance [42], the D and 2D
bands undergo a redshift but the G band shifted from 1582 to 1587 cm^{-1} when its
number-of-layer (n) is reduced from 20 to one [21, 43]. When the n is increased
from a few to multiple, the Raman peaks turn from the dominance of the monolayer
component to the dominance of the bulk graphite component [21]. The opposite
number-of-layer trends suggest that different yet unclear mechanisms govern the G
mode and the D/2D modes. One can also estimate the exact number of layers of a
graphene [44, 45] and the diameter of a single-walled CNT [46] as the frequency
of the radial breathing mode ω_{RBM} is inversely and empirically proportional to the
thickness of the graphene and the CNT diameter.

The nature of the G and the D modes in carbon allotropes is conventionally interpreted as the resonant excitation of the π states and the long-range polarizability
of the unpaired π bonding electrons [47, 48]. Visible Raman spectral data on amorphous, disordered, and diamond-like carbon are classified in three stages showing the
factors that control the peak positions, peak intensities, and peak widths of the G and
the D modes. The Raman spectra depend on the configuration of the sp sites in the
sp^2-bonded clusters. In cases where a fraction of sp^3 bonding is involved in the sp^2
clustering, such as in the as-deposited tetrahedral amorphous carbon (ta-C) or hydrogenated amorphous carbon (a-C:H) films, the visible Raman spectral parameters can
be used to derive the sp^3 fraction. However, the Raman modes and their shifts are
directly related to the oscillation of lattices in different geometries by perturbation
to the Hamiltonian.

From the dispersion of the Raman phonon frequencies and peak intensities with
excitation wavelength, Ferrari et al. [47, 48] derived the local bonding and structural
disorder of graphene. They found three basic features. Under visible light excitation,
graphene shows the D mode around 1350 cm^{-1} (1332 in diamond [49]) and the G
mode around 1600 cm^{-1} (1580 in graphite [50] and show thermal stiffening that
was attributed to the weakening of the electron-phonon coupling effect). Under UV
excitation, an extra T peak appears, which lies around 1060 cm^{-1} for the H-free
carbons and around 980 cm^{-1} for the hydrogenated carbons. The G peak shows
structural disorder being attributed to the stretching motion of sp^2 pairs. This G peak
disperses only in amorphous networks, with a dispersion rate proportional to the
degree of disorder. The dispersion of the D peak is strongest in the ordered carbon,
but it is weak for the amorphous carbon.

Figure 20.2 shows the D/2D and the G mode frequency evolution with the GNRs
thickness compared with bulk highly oriented pyrolytic graphite (HOPG) reference
[20, 21]. When the bulk graphite evolves into a monolayer GNR, the 2D peak
shifts downwardly from 2714 to 2678 cm^{-1} and the D peak changes from 1368

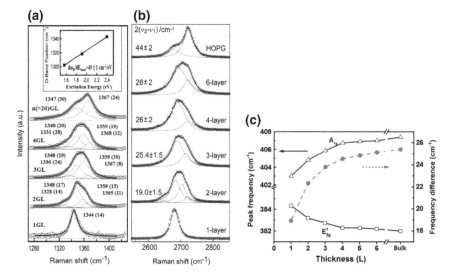

Fig. 20.2 Number-of-layer resolved Raman **a** D mode and **b** 2D band for the few-layered graphene nanoribbon (GNR) compared with bulk highly-oriented pyrolytic graphite (HOPG) reference [21], compared with **c** the E_{2g}^1 and A_{1g} peak frequency shift of the layered MoS$_2$ films [39]. The single-peak for the monolayer GNR 2D is centered at 2678 cm^{-1}. Inset **a** shows the D-band dω_D/dE_{ex} dispersion as a function of excitation energy of the incident light [20] and the red line in **c** shows the net frequency shift between the two modes. Reprinted with copyright permission from [20, 21, 39]

to 1344 cm^{-1}. In contrast, the G band undergoes a blueshift when the number-of-layer is reduced [43]. The G-mode blueshift follows the empirical relations [51]: $\omega_G(n) = 1581.6 + 5.5/n$, or $\omega_G(n) = 1581.6 + 11/(1 + n^{1.6})$.

Likewise, the layered MX$_2$ (M = W, Mo; X = S, Se) semiconductors show the same trends of phonon frequency relaxation of graphene [52–61]. The E_{2g}^1 phonon mode undergoes a blueshift and the A_{1g} mode a redshift as the MoS$_2$ number-of-layer is decreased [39]. Along with the phonon frequency shift, the number-of-layer reduction deepens the surface-potential-well of the MoS$_2$ [62], which evidences the BOLS prediction of the surface bond contraction and local bond potential depression [63].

20.3.2   Compression and Directional Uniaxial-Stain

Mechanical compression stiffens Raman phonons in general, as shown in Fig. 20.3 [64], while the uniaxial stretching softens and splits the Raman phonons of graphene and WX$_2$, see Fig. 20.4 [65]. The velocity of phonon stiffening varies with not only the bond nature of the substance but also the specific mode of the same material.

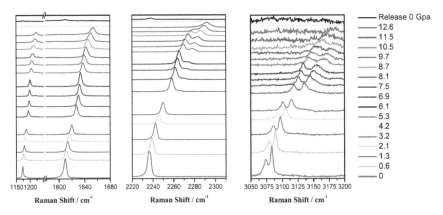

Fig. 20.3 Pressure dependence of the Raman phonon shift for a TPN (Terephthalonitrile) crystal. Reprinted with permission from [64]

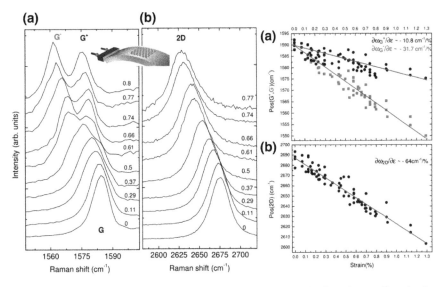

Fig. 20.4 Uniaxial strain softened Raman **a** G mode and **b** 2D mode of graphene collected using polarized light. The G peak splits into the G^+ and the G^- subbands. The strain-induced phonon softening is approximated using the Grüneisen constants, $\partial\omega/\partial\varepsilon$. Reprinted with permission from [65, 66]

Mohiuddin and co-workers [65] suggested that a uniaxial stretching strain can degenerate the E_{2g} or the equivalent G optical mode into two components: one is polarized along the strain and the other is perpendicular to the strain. Increasing the strain further softens and separates the G^+ and the G^- peaks, in agreement with first-principles calculations. Small strains also soften the 2D bands but do not split them.

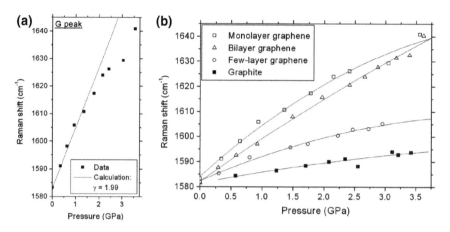

Fig. 20.5 Pressure-resolved **a** G mode frequency shift whose lopes deviate from the $\gamma = \partial\omega/\partial P$ Grüneisen constant at low pressures **b** for the monolayer, bilayer, few-layer and bulk graphite. Reprinted with permission from [67]

Figure 20.5 shows compression induced phonon shift described with a linear $\gamma = \partial\omega/\partial P$ Grüneisen parameter but at higher pressure, greater than 1.5 GPa, the measurement curve deviates much from the γ constant. Meanwhile, atomic undercoordination and mechanical compression enhance each other on the Raman blueshift of graphene [67]. The monolayer graphene shifts most, and the graphite shifts least their G mode compared with other layered graphene samples under the same pressure. The $\gamma = \partial\omega/\partial P$ deviates from its linear form to the measured nonlinear form as the graphite turns into the monolayer graphene.

20.3.3   Debye Thermal Decay

Figure 20.6 shows the thermal evolution of the frequency shift ($\Delta\omega$) and the full-width-at-half-maximum (FWHM or Γ) of the Raman characteristic phonons for GeSe [68]. The Γ describes the structure thermal fluctuation and the $\Delta\omega$ is the stiffness standing for the bond stretching vibration. The thermal fluctuation does not contribute to the systems energy on average, but the fluctuation is associated with the phonon thermal softening—Debye thermal decay.

The Raman phonon frequency thermal decay follows the general trend displayed by GeSe, diamond [69–72], and GaN [73] with or without involvement of interfaces or impurities, and even the AlN and InN vibration modes in the InAlN alloy [74] albeit different high-temperature Grüneisen slopes. Within the temperature range of −190 and 100 °C, the G peak position and its temperature coefficient of graphene shows the number-of-layer resolved manner. The $\Delta\omega$ shifts from 1582 to 1580 cm^{-1} and the linear slope χ changes from −0.016 to −0.015 cm^{-1}/°C when the graphene

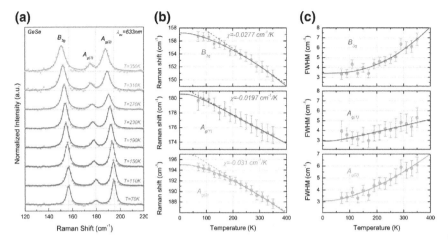

Fig. 20.6 Debye thermal decay of the mode-resolved **a** Raman spectra, **b** peak frequency, and **c** Γ for GeSe thin flakes measured using $\lambda = 633$ nm light excitation. The peak shift is approximated with the Grüneisen constant, $\chi = \partial\omega/\partial T$, at high temperatures. Reprinted with permission from [68]

turns from single- to bilayer under 488 nm laser excitation [45]. However, the graphite G mode showed thermal stiffening effect that was attributed to the weakening of the electron-phonon coupling effect [50].

These experimental observations show the following common features of the Debye thermal decay:

(1) Heating softens the $\Delta\omega$ and broadens the Γ of the phonon band.
(2) The $\Delta\omega$ and Γ change slowly at low temperatures and then transit into a linear form at higher temperatures toward melting.
(3) The slope χ of the linearly thermal decay and the transition temperature vary from mode to mode for a specific specimen.
(4) The overall temperature trends of the $\omega(T)$ follows the $\omega(T)/\omega(T_0) = 1 - U(T)$ with $U(T)$ being thermal integration of the Debye specific heat curve.

20.4 Overview on Theoretical Progress

20.4.1 Quantum Size Trends

20.4.1.1 Empirical Formulation

Compared with the huge volume of Raman database, the physical insight into the size dependent Raman shift is limited. Interpretations are often based on the continuum dielectric mechanism [75, 76], microscopic lattice dynamics [38, 77], multi-phonon resonant scattering [36], strain or phonon confinement [40–42]. Information could not be possible on the bond formation, dissociation, relaxation and vibration under given conditions. Particularly, for the size-reduction induced phonon frequency blueshift, redshift, and LFR creation are beyond capacity of available theories.

One often describes the size-dependent Raman shifts in the following hypothetic relation with multiple freely adjustable parameters [35, 38],

$$\omega(K) = \omega(\infty) + A_f \left(\frac{d_0}{K} \right)^\kappa,$$

where A_f and κ are adjustable parameters used to fit the measurement. The d_0 is the lattice constant that should contract with the solid dimension [78]. K is the dimensionless size of the crystal. For the LFR in the THz frequencies, $A_f > 0$, $\kappa = 1$. The LFR mode disappears when the particle size approaches infinity, $\omega(\infty) = 0$. For the redshift, $A_f < 0$. For Si example, $\omega(\infty) = 520$ cm^{-1}, κ varies from 1.08 to $= 1.44$, depending on experimental conditions.

The variation of atomic number-of-layer from monolayer to bulk leads to a redshift of photon energy or band gap E_G and the phonon frequency ω [79, 80]. The photon energy E_G and phonon frequency shift ω under perturbation of size N, pressure P, and temperature T follow the hypothetic empirical relationships with numerous adjustable parameters [25, 35, 81, 82]:

$$\left. \begin{array}{l} \omega(N) - \omega_0 \\ E_G(N) - E_0 \\ \omega(T) - \omega_0 \\ \omega(P) - \omega_0 \end{array} \right\} = \left\{ \begin{array}{l} -d(a/N)^q \\ AN^{-2} - BN^{-1} - C \\ \Delta\omega_e(T) + \Delta\omega_d(T) \\ kP + lP^2 \end{array} \right. \tag{20.1}$$

where ω_0 and E_0 are the reference phonon frequency and photon energy of the bulk; N denotes the number of atomic layers; a is the lattice constant; A, B, C, and d, q, k, and l are adjustable parameters to match the N- and P-dependent Raman shifts. The N^{-1} and N^{-2} terms stand for the potential and kinetic energies of the electron − hole pairs in the quantum confinement scheme. These hypothetic models can reproduce observations albeit the unclear physical meanings.

20.4.1.2 Hwang's Scheme

Hwang and co-workers [83] presented a sophisticated theory describing the size effect on the Raman shift of the optical and the LFR mode. The LFR is attributed to the quadruple vibration, lattice contraction, and the size-induced optical softening due to quantum confinement. The LFR mode at the THz regime is in acoustic category, which is specifically associated with the vibration of the nanoparticles. The phonon energies are size dependent and vary with materials of the host matrix. The LFR scattering from silver nanoclusters embedded in porous Al_2O_3 [84] and SiO_2

[85] was attributed to being arisen from the quadrupolar vibration modes that are enhanced by the excitation of the surface plasmas of the encapsulated Ag particles. The mode choice by LFR scattering is due to the stronger plasmon-phonon coupling. For an Ag particle smaller than 4 nm, the size dependence of the LFR peak frequency approximates Lamb's theory [86], which gives vibrational frequencies of a homogeneous elastic body in a spherical form.

The mechanism for the LFR mode enhancement is analogous to the case of surface-plasma enhanced Raman scattering from molecules adsorbed on rough metal surfaces. The surface acoustic phonons are eigen frequencies of a homogeneous elastic sphere under stress-free boundary conditions, which gives rise to a low-frequency ω that is in the range of THz regime. The LFR mode corresponds to the spheroidal and the torsional mode of vibrations of a spherical or an ellipsoidal particle. Spheroidal motions are associated with dilation and it depends strongly on the cluster material through the transverse and the longitudinal sound velocities, v_t and v_l, respectively. The sound velocity in a medium depends functionally on the Young's modulus and the mass density, i.e., $v \sim (Y/\rho)^{0.5} \sim \sqrt{E_b}$, where E_b is the bond energy [87].

One can also ascribe the polarized LFR peaks as the confined LA-like and the depolarized LFR as the TA-like acoustic phonons [88]. Interface between the nanoparticle and the support matrix leads to a redshift for both the polarized and the depolarized LFR peaks. This approach improves the fit to the measurement compared with Lamb's model. Atomistic simulations [89] suggested that the morphology of nanoscopic Ag grains (twined icosahedra, Mark's decahedra, and irregular nanograins) introduces a high degree of complexity into the phonon spectra with the total and the partial vibrational density of states and phonon localization.

Hwang et al. also suggested that the size-reduction induced lattice contraction stems the LFR blueshift. For example, CdS_xSe_{1-x} nanocrystals embedded in a borosilicate (B_2O_3–SiO_2) glass matrix undergoes the size-dependent compressive strain [90]. The lattice strain raises the surface tension when the crystal size is reduced. It is suggested that the compressive stress overcomes the redshift of the confined phonon due to negative dispersion and thus drives the LFR blueshift. The LFR blueshift is also related to the bond length and energy that are functionally dependent on the entropy, latent heat of fusion, and the melting point in a classical thermodynamic manner [91].

The high-frequency optical modes shift in opposite directions. The Raman blueshift is usually suggested to be activated by surface disorder [92], surface stress [93, 94], phonon quantum confinement, and surface chemical effect [95, 96]. The Raman shifts of TiO_2 particles are attributed to the effects of decreasing particle size on the force constants and vibrational amplitudes of the nearest neighbor bonds [97]. However, the effect of stress can usually be ignored for hydrogenated silicon, in which hydrogen atoms passivate the surface dangling bonds to reduce the bond strains and the residual stress [98, 99].

The phonon confinement model [95] ascribes the asymmetric Raman redshift to relaxation of the q-vector selection-rule for the excitation of the Raman active phonons due to their localization. The relaxation of the momentum conservation rule arises from the finite crystalline size and the diameter distribution of the nanosolid in the films. When the size is decreased, the rule of momentum conservation will be relaxed and the Raman active modes will not be limited to the center of the Brillouin zone [93]. The large surface-to-volume ratio of a nanodot strongly affects the optical properties mainly due to introducing surface polarization states [100].

Particle size reduction softens the phonons for all glasses including CdSe nanograins. Models based on assumptions that the materials are homogeneous and isotropic are valid only in the long-wavelength limit. When the size of the nanosolid is in the range of a few nanometers, the continuum dielectric models are faced limitations.

Hwang et al. [101] consider the effect of lattice contraction in explaining the versatile phonon redshifts of nanosolid CdSe embedded in different glass matrices. The following expresses the K dependent phonon shift with inclusion of lattice contraction,

$$\omega(K) = \omega_L + \Delta\omega_D(K) + \Delta\omega_C(K) \tag{20.1}$$

where ω_L is the LO phonon frequency of the bulk. $\Delta\omega_D(K)$ is the peak shift due to phonon dispersion and $\Delta\omega_C(K)$ is the peak shift due to lattice contraction. The dispersion term $\Delta\omega_D(K)$ follows,

$$\Delta\omega_D(K) = \left[\omega_L^2 - \beta_L^2\left(\frac{\mu_{np}}{Kd_0}\right)^2\right]^{\frac{1}{2}} - \omega_L \cong -\left(\frac{\beta_L^2}{2\omega_L}\right)\left(\frac{\mu_{np}}{Kd_0}\right)^2 \tag{20.2}$$

where the parameter β_L describes the dispersion assumed to be parabolic and μ_{np} is the nonzero n_p the root of the equation of $\tan(\mu_{np}) = \mu_{np}$. The bond contraction term $\Delta\omega_C(K)$ is given as [90]:

$$\Delta\omega_C(K) = \omega_L\left[\left(1 + \frac{3\Delta d(K)}{d}\right)^{-\gamma} - 1\right] \cong 3\gamma\omega_L\frac{\Delta d(K)}{d}$$

where,

$$\frac{\Delta d(K)}{d} = (\alpha' - \alpha)(T - T_g) - \frac{2\beta_c}{3}\left(\frac{\delta_\infty}{Kd_0} + \frac{b}{2(Kd_0)^2}\right)$$

$$\cong (\alpha' - \alpha)(T - T_g) - \frac{2\beta_c b}{3(Kd_0)^2} \tag{20.3}$$

γ is the Grüneisen parameter, α' and α are the linear thermal-expansion coefficients of the matrix host glass and the nanocrystal, respectively. T and T_g are the testing and the heat-treatment temperature, respectively. β_c and σ_∞ are the compressibility and the surface tension of the bulk, respectively, and b is the parameter describing the size-dependent surface tension of the crystal. The contribution of surface tension to the frequency for a bulk is small. The first term in Eq. (20.3) describes lattice contraction by thermal-expansion mismatching the matrix to the crystal, and the second term arises from the increase of surface tension with the decrease of crystal size. Substituting Eq. (20.3) into (20.2), yields,

$$\frac{\Delta\omega(K)}{\omega_L} = -3\gamma\left(\alpha' - \alpha\right)\left(T - T_g\right) - \left[\frac{1}{2}\left(\frac{\beta_L\mu_{np}}{\omega_L}\right)^2 - \gamma\beta_C b\right](Kd_0)^{-2}$$

$$= A - BK^{-2} \tag{20.4}$$

For a free surface, $\alpha' = \alpha$, and $b = 0$. There are some difficulties, however, to use this equation, as remarked on by Hwang et al. [101], since the thermal-expansion coefficient difference within the temperature range $T - T_g$ is hardly detectable. The value of B in Eq. (20.4) is given by the difference of the phonon negative dispersion and the size-dependent surface tension. Thus, a positive value of B suggests that the phonon negative dispersion exceeds the size-dependent surface tension and so causes a redshift of the phonon frequency. On the contrary, if the size-dependent surface tension is stronger than the phonon negative dispersion, blueshift occurs. In case of balance of the two effects, i.e., $B = 0$, the size dependence disappears. Furthermore, the parameter b introduced by the size-dependent surface tension is unknown. At the lower end of the size limit, the $\omega(K) \to -\infty$ diverges in a K^{-2} manner.

20.4.2 Grüneisen Notion for Compression and Thermal Excitation

Besides the hypothetically polynomial expressions [102, 103], one often uses the Grüneisen parameter, $\gamma = -\partial\omega/\partial\varepsilon$ or $\gamma_E = -\partial L_N\omega/\partial L_N\varepsilon$ to describe the effect of strain or compression on the phonon frequency shift. The Grüneisen parameter is the slope of the experimental ω–ε curve. The following addresses the E mode shift of graphene in terms of the Grüneisen parameter and the shear deformation potential β_{E2g} [65, 67]:

$$\begin{pmatrix} \gamma_{E2g} \\ \beta_{E2g} \end{pmatrix} = \frac{1}{\omega_{E2g}^0}\begin{pmatrix} -\frac{\partial\omega_{E2g}^h}{\partial\varepsilon_h} \\ \frac{\partial\omega_{E2g}^s}{\partial\varepsilon_s} \end{pmatrix} \tag{20.5}$$

where $\varepsilon_h = \varepsilon_{ll} + \varepsilon_{tt}$ is the hydrostatic component of the applied uniaxial strain and $\varepsilon_s = \varepsilon_{ll} - \varepsilon_{tt}$ is the shear component of the strain, l is along the strain direction, and t is the transverse direction; ω_{E2g}^0 is the referential G peak position under zero strain.

On the other hand, the thermal expansion coefficient $\alpha(t)$ varies non-linearly in the low temperatures and then increases with temperature toward a constant. The $\alpha(t)$ also varies with the feature size of nanostructures [104, 105]. Near-edge X-ray absorption fine structure spectroscopy (NEXAFS) investigations [106] suggested that in the 20–300 K range the first Au–Au neighbor distance in the 2.4–50.0 nm sized gold nanoparticles is different from that of the bulk counterpart. Atomic undercoordination has an opposite effect to compensate for thermal expansion on the lattice constant [9].

According to Cardona [107], the lattice thermal expansion of a cubic crystal could be expressed in terms of the Grüneisen parameter, γ_q, and the lattice vibration frequency ω_q,

$$\frac{\Delta d}{d_0} = \alpha T = \frac{\hbar}{3BV} \sum_q \gamma \omega_q \left[n_B(\omega_q) + \frac{1}{2} \right]$$

$$\propto \begin{cases} \frac{2kT}{BV_C} < \gamma_q > & (T > \theta_D) \\ \int_0^{\omega_D} < \gamma_q > \omega^3 \left\{ \left[\exp(\hbar\omega/kT) - 1 \right]^{-1} + 1/2 \right\} d\omega & (else) \end{cases}$$

$$\gamma_q \qquad = -\frac{\partial Ln(\omega_q)}{\partial Ln(V)}$$

$$(20.6)$$

where B is the bulk modulus and V the volume. V_C is the volume of the primary unit cell and $\langle \gamma_q \rangle$ the average of γ_q over all branches of the Brillouin zone. The $n_B(\omega_q)$ is the Bose-Einstein population function. From the phonon nonlinearities, Grüneisen [108] derived the volumetric thermal expansion coefficient that is proportional to the product of the specific heat and Grüneisen parameter,

$$\alpha = \frac{\gamma C_v}{V B_T} \qquad (20.7)$$

One may note that the bandgap, elastic modulus, and the phonon frequency follow the same trend of Debye thermal decay. The following equation empirically describes the thermal evolution of the photonic bandgap E_g [82, 109],

$$E_g(T) = E_{g0} - \frac{\beta T^2}{T + \theta_D} \qquad (20.8)$$

where β is the fitting parameter related to the T-dependent dilatation of the lattice and θ_D is the Debye temperature.

Typically, the thermal evolution of the elastic modulus Y or the bulk modulus B follows the empirical relationships [110, 111],

$$Y = Y_0 - b_1 T \exp\left(\frac{-T_0}{T} \right)$$

$$Y = Y_0 - \frac{3R\gamma\delta T}{V_0} H\left(\frac{T}{\theta_D} \right)$$

$$B_T^0 = B_{T=0}^0 \times \exp\left[\int_{T=0}^T \alpha_V^0(T)\delta^0(T)dT \right]$$

where,

$$H\left(\frac{T}{\theta_D}\right) = 3\left(\frac{T}{\theta_D}\right)^3 \int_0^{\frac{\theta_D}{T}} \frac{x^3 dx}{e^x - 1} \tag{20.9}$$

where Y_0 is the Young's modulus and $B_{T=0}^0$ the bulk modulus at $T = 0$ K. The bulk modulus B correlates to Y by $Y/B = 3 \times (1 - 2v)$, where v denoting the Poisson ratio that is negligibly small, and therefore, $Y \approx 3B$. The parameters b_1 and T_0 are arbitrary constants for data fitting. γ is Grüneisen parameter and δ is Anderson constant. The superscript 0 denotes quantities gained at one bar pressure. The α_V^0 is the coefficient of volume thermal expansion, and $\delta^0(T)$ is the Anderson-Grüneisen parameter.

Numerically, the first expression could fit the linear part and the last two could reproduce measurements over the entire temperature range despite the freely adjustable parameters such as the γ and δ that are hardly experimentally available. The physical origin for the thermally driven elastic softening is still open for exploration.

20.4.3 Phonon Optical-Acoustic Thermal Degeneration

The thermal evolution of the Γ and $\Delta\omega$ are often attributed to the anharmonic phonon–phonon interactions, lattice-mismatch, volume thermal expansion, and the optical phonon degeneration into multiple acoustic phonons [68, 112]. On the base of the extended Klemens–Hart–Aggarwal–Lax premise [113, 114], Balkanski et al. described the $\Delta\Gamma$(T) and $\Delta\omega$(T) in the following forms [115]:

$$\begin{pmatrix} \Delta\omega(T) \\ \Delta\Gamma(T) \end{pmatrix} = \begin{pmatrix} A & B \\ C & D \end{pmatrix} \begin{pmatrix} 1 + \frac{2}{e^{x/2}-1} \\ 1 + \frac{3}{e^{x/3}-1} + \frac{3}{(e^{x/3}-1)^2} \end{pmatrix} \tag{20.10}$$

where $x = \hbar\omega_0/k_B T$ with $\hbar$ being the Plank constant and k_B the Boltzmann constant. The $\hbar\omega_0$ is the phonon energy at $T = 0$ K from which the $\Delta\Gamma$(T) and $\Delta\omega$(T) shift. A, B, C and D are adjustable parameters. This notion attributes the $\Delta\Gamma$(T) and $\Delta\omega$(T) to the cubic and quartic anharmonicity of lattice potential, which makes the optical phonon decay into two (three phonon process, A) or three (four phonon process, B) components of acoustic phonons.

Comparatively, Kolesov considered an alternative on the process of phonon excitation in a $\Delta\omega$(T) function of anharmonic vibration of the chemical bonds [116]:

$$\Delta\omega \equiv (\chi_T + \chi_V)\Delta T = \left(\frac{d\omega}{dT}\right)_V \Delta T + \left(\frac{d\omega}{dV}\right)_T \Delta T$$

$$= \left(\frac{d\omega}{dT}\right)_V \Delta T + \left(\frac{d\omega}{dV}\right)_T \left(\frac{d\omega}{dT}\right)_V \Delta T. \tag{20.11}$$

where $\omega_i < \omega$. The $\langle n_i \rangle$ is the population function of ω_i selected from the thermal bath. The c and c_i are adjustable weighting factors.

Although Eqs. (20.10)–(20.11) reproduce equally well the $\Delta\omega(T)$ for diamond and silicon, the physical indication of the adjustable weighting factors is unclear.

20.5   Motivation and Objectives

Phonon spectroscopy is a widely-used tool of detection but the conventional approaches of spectral peak Gaussian decomposition or the empirical simulation of the spectral feature evolution under perturbation limited its advantage in revealing the bonding dynamics. One often simply decomposes a spectral peak into multiple Gaussian components with limited constraints albeit physical indications or simulates the spectral peak evolution in an empirical manner with multiple freely adjustable variables. Conventional phonon spectroscopy analysis has delivered information far less than it is supposed to be because of lacking the phonon-bond-stimulus correlation functions. The Grüneisen constant is an experimental derivative showing the linear dependence of the frequency shift on the stimulus. The optical phonon degeneration and the multiple phonon resonant scatting are extrinsic artifacts exist throughout the processes of experiments.

The key challenge is a theory to reproduce observations because of bond relaxation in length and energy and the bond-phonon-property correlation of the examined substance to reconcile as many as perturbations to the phonon frequency shift in terms of excited bond relaxation. One needs to correlate the $\Delta\omega$, $\Delta\Gamma$, ΔA (abundance) intrinsically to the bond length an energy relaxed by external stimuli. Therefore, bond relaxation is profoundly and ubiquitously important to the engineering of materials and thus should receive deserved attention.

This part is devoted to modeling the multifield lattice oscillation in the past decade to meet the following targets:

(1) Theoretical reproduction of the phonon spectroscopy measurements with physically meaningful parameters.
(2) Provision of atomistic, local, quantitative, elemental information on bonding dynamics from measurements.
(3) Comprehension of the physical mechanism and basic rules governing observations.
(4) Correlation of the detectable quantities, $\Delta Q(x_i)/Q(x_{i0}) = f(x_i, d(x_i), E(x_i))$, to the bond length and energy that relax upon perturbation by the degree of freedom x_i.

This part deals with the multifield lattice oscillation dynamics to amplify the *Coordination-resolved Electron Spectrometrics* [6, 118] and its analytical strategies to the current *Multifield Phonon Spectrometrics* [119–121] for consistent insight into phonon relaxation with derivative of conventionally-unexpected quantitative information. From the perspective of the local bond averaging (LBA) approach [7], we will be focused on formulating the Raman shifts in terms of the order, length and strength of the representative bond of the entire specimen and their response to the applied stimuli.

Agreement between the theory predictions and the measurements has led to consistent insight into the mechanism behind the fascinations with revelation of the information. The information includes the referential frequency $\omega(1)$ and its bulk shift, binding energy E_b, atomic cohesive energy E_{coh}, binding energy density E_{den}, Debye temperature θ_D, elastic modulus B, force constant k, compressibility β, and the effective atomic CN for the few-layer graphene and WX_2 and their bond lengths and energies.

References

1. C.Q. Sun, Oxidation electronics: bond-band-barrier correlation and its applications. Prog. Mater Sci. **48**(6), 521–685 (2003)
2. C. Liu, F. Yun, H. Morkoc, Ferromagnetism of ZnO and GaN: a review. J Mater. Sci. Mater. Electron. **16**(9), 555–597 (2005)
3. J.W. Li, X.J. Liu, L.W. Yang, Z.F. Zhou, G.F. Xie, Y. Pan, X.H. Wang, J. Zhou, L.T. Li, L.K. Pan, Z. Sun, C.Q. Sun, Photoluminescence and photoabsorption blueshift of nanostructured ZnO: skin-depth quantum trapping and electron-phonon coupling. Appl. Phys. Lett. **95**(3), 3184566 (2009)
4. G.W. Shim, K. Yoo, S.-B. Seo, J. Shin, D.Y. Jung, I.-S. Kang, C.W. Ahn, B.J. Cho, S.-Y. Choi, Large-area single-layer $MoSe_2$ and its van der Waals heterostructures. ACS Nano **8**(7), 6655–6662 (2014)
5. C. Tan, X. Cao, X.-J. Wu, Q. He, J. Yang, X. Zhang, J. Chen, W. Zhao, S. Han, G.-H. Nam, Recent advances in ultrathin two-dimensional nanomaterials. Chem. Rev. **117**(9), 6225–6331 (2017)
6. X.J. Liu, M.L. Bo, X. Zhang, L. Li, Y.G. Nie, H. Tian, Y. Sun, S. Xu, Y. Wang, W. Zheng, C.Q. Sun, Coordination-resolved electron spectrometrics. Chem. Rev. **115**(14), 6746–6810 (2015)
7. C.Q. Sun, Y. Sun, Y.G. Ni, X. Zhang, J.S. Pan, X.H. Wang, J. Zhou, L.T. Li, W.T. Zheng, S.S. Yu, L.K. Pan, Z. Sun, coulomb repulsion at the nanometer-sized contact: a force driving superhydrophobicity, superfluidity, superlubricity, and supersolidity. J. Phys. Chem. C **113**(46), 20009–20019 (2009)
8. M.C. Daniel, D. Astruc, Gold nanoparticles: assembly, supramolecular chemistry, quantum-size-related properties, and applications toward biology, catalysis, and nanotechnology. Chem. Rev. **104**(1), 293–346 (2004)
9. C.Q. Sun, Size dependence of nanostructures: impact of bond order deficiency. Prog. Solid State Chem. **35**(1), 1–159 (2007)
10. J.R. Trelewicz, C.A. Schuh, The Hall-Petch breakdown at high strain rates: optimizing nanocrystalline grain size for impact applications. Appl. Phys. Lett. **93**(17), 171916 (2008)
11. C.Q. Sun, W.H. Zhong, S. Li, B.K. Tay, H.L. Bai, E.Y. Jiang, Coordination imperfection suppressed phase stability of ferromagnetic, ferroelectric, and superconductive nanosolids. J. Phys. Chem. B **108**(3), 1080–1084 (2004)
12. Y.J. Tian, B. Xu, D.L. Yu, Y.M. Ma, Y.B. Wang, Y.B. Jiang, W.T. Hu, C.C. Tang, Y.F. Gao, K. Luo, Z.S. Zhao, L.M. Wang, B. Wen, J.L. He, Z.Y. Liu, Ultrahard nanotwinned cubic boron nitride. Nature **493**(7432), 385–388 (2013)
13. Y. Wang, Y. Pu, Z. Ma, Y. Pan, C.Q. Sun, Interfacial adhesion energy of lithium-ion battery electrodes. Extreme Mech. Lett. **9**, 226–236 (2016)

14. J. Linder, T. Yokoyama, A. Sudbø, Anomalous finite size effects on surface states in the topological insulator Bi_2Se_3. Phys. Rev. B **80**(20), 205401–205406 (2009)

15. B.F. Variano, N.E. Schlotter, D.M. Hwang, C.J. Sandroff, Investigation of finite size effects in a first order phase transition: high pressure Raman study of CdS microcrystallites. J. Chem. Phys. **88**(4), 2848–2851 (1988)

16. Y. Wang, N. Herron, Quantum size effects on the exciton energy of CdS clusters. Phys. Rev. B **42**(11), 7253–7255 (1990)

17. A.W. Schill, M.A. El-Sayed, Wavelength-dependent hot electron relaxation in PVP capped CdS/HgS/CdS quantum dot quantum well nanocrystals. J. Phys. Chem. B **108**(36), 13619–13625 (2004)

18. H. Borchert, D. Dorfs, C. McGinley, S. Adam, T. Möller, H. Weller, A. Eychmüller, Photoemission study of onion like quantum dot quantum well and double quantum well nanocrystals of CdS and HgS. J. Phys. Chem. B **107**(30), 7486–7491 (2003)

19. J. Zhang, Z. Peng, A. Soni, Y. Zhao, Y. Xiong, B. Peng, J. Wang, M.S. Dresselhaus, Q. Xiong, Raman spectroscopy of few-quintuple layer topological insulator Bi_2Se_3 nanoplatelets. Nano Lett. **11**(6), 2407–2414 (2011)

20. A.K. Gupta, T.J. Russin, H.R. Gutierrez, P.C. Eklund, Probing graphene edges via Raman scattering. ACS Nano **3**(1), 45–52 (2008)

21. D. Graf, F. Molitor, K. Ensslin, C. Stampfer, A. Jungen, C. Hierold, L. Wirtz, Spatially resolved raman spectroscopy of single- and few-layer graphene. Nano Lett. **7**(2), 238–242 (2007)

22. D.S. Chuu, C.M. Dai, Quantum size effects in CdS thin films. Phys. Rev. B **45**(20), 11805–11810 (1992)

23. A. Tanaka, S. Onari, T. Arai, Raman-scattering from CdSe microcrystals embedded in a germanate glass matrix. Phys. Rev. B **45**(12), 6587–6592 (1992)

24. O. Auciello, J.F. Scott, R. Ramesh, The physics of ferroelectric memories. Phys. Today **51**, 22 (1998)

25. J. Liu, Y.K. Vohra, Raman modes of 6H polytype of silicon carbide to ultrahigh pressures: a comparison with silicon and diamond. Phys. Rev. Lett. **72**(26), 4105–4108 (1994)

26. Z.W. Chen, C.Q. Sun, Y.C. Zhou, O.Y. Gang, Size dependence of the pressure-induced phase transition in nanocrystals. J. Phys. Chem. C **112**(7), 2423–2427 (2008)

27. J. Zhu, J.X. Yu, Y.J. Wang, X.R. Chen, F.Q. Jing, First-principles calculations for elastic properties of rutile TiO_2 under pressure. Chin. Phys. B **17**(6), 2216 (2008)

28. N. Iles, A. Kellou, K.D. Khodja, B. Amrani, F. Lemoigno, D. Bourbie, H. Aourag, Atomistic study of structural, elastic, electronic and thermal properties of perovskites Ba(Ti, Zr, Nb)O-3. Comput. Mater. Sci. **39**(4), 896–902 (2007)

29. B.B. Karki, L. Stixrude, S.J. Clark, M.C. Warren, G.J. Ackland, J. Crain, Structure and elasticity of MgO at high pressure. Am. Mineral. **82**, 51–60 (1997)

30. A. Bouhemadou, K. Haddadi, Structural, elastic, electronic and thermal properties of the cubic perovskite-type $BaSnO_3$. Solid State Sci. **12**, 630–636 (2010)

31. L.K. Pan, C.Q. Sun, T.P. Chen, S. Li, C.M. Li, B.K. Tay, Dielectric suppression of nanosolid silicon. Nanotechnology **15**(12), 1802–1806 (2004)

32. X.J. Liu, Z.F. Zhou, L.W. Yang, J.W. Li, G.F. Xie, S.Y. Fu, C.Q. Sun, Correlation and size dependence of the lattice strain, binding energy, elastic modulus, and thermal stability for Au and Ag nanostructures. J. Appl. Phys. **109**(7), 074319 (2011)

33. X.J. Liu, L.W. Yang, Z.F. Zhou, P.K. Chu, C.Q. Sun, Inverse Hall-Petch relationship of nanostructured TiO_2: skin-depth energy pinning versus surface preferential melting. J. Appl. Phys. **108**, 073503 (2010)

34. L.K. Pan, C.Q. Sun, C.M. Li, Elucidating Si-Si dimmer vibration from the size-dependent Raman shift of nanosolid Si. J. Phys. Chem. B **108**(11), 3404–3406 (2004)

35. J. Zi, H. Büscher, C. Falter, W. Ludwig, K. Zhang, X. Xie, Raman shifts in Si nanocrystals. Appl. Phys. Lett. **69**(2), 200–202 (1996)

36. X.X. Yang, J.W. Li, Z.F. Zhou, Y. Wang, L.W. Yang, W.T. Zheng, C.Q. Sun, Raman spectroscopic determination of the length, strength, compressibility, Debye temperature, elasticity, and force constant of the C–C bond in graphene. Nanoscale **4**(2), 502–510 (2012)

37. M. Fujii, Y. Kanzawa, S. Hayashi, K. Yamamoto, Raman scattering from acoustic phonons confined in Si nanocrystals. Phys. Rev. B **54**(12), R8373–R8376 (1996)
38. W. Cheng, S.F. Ren, Calculations on the size effects of Raman intensities of silicon quantum dots. Physical Review B **65**(20), 205305 (2002)
39. C. Lee, H. Yan, L.E. Brus, T.F. Heinz, J. Hone, S. Ryu, Anomalous lattice vibrations of single-and few-layer MoS2. ACS Nano **4**(5), 2695–2700 (2010)
40. J.E. Spanier, R.D. Robinson, F. Zheng, S.W. Chan, I.P. Herman, Size-dependent properties of CeO2-y nanoparticles as studied by Raman scattering. Phys. Rev. B **64**(24), 245407 (2001)
41. H. Mavi, A. Shukla, R. Kumar, S. Rath, B. Joshi, S. Islam, Quantum confinement effects in silicon nanocrystals produced by laser-induced etching and cw laser annealing. Semicond. Sci. Technol. **21**(12), 1627 (2006)
42. S. Kim, D. Hee Shin, C. Oh Kim, S. Seok Kang, S. Sin Joo, S.-H. Choi, S. Won Hwang, C. Sone, Size-dependence of Raman scattering from graphene quantum dots: interplay between shape and thickness. Appl. Phys. Lett. **102**(5), 053108 (2013)
43. A. Gupta, G. Chen, P. Joshi, S. Tadigadapa, P.C. Eklund, Raman scattering from high-frequency phonons in supported n-graphene layer films. Nano Lett. **6**(12), 2667–2673 (2006)
44. A.C. Ferrari, J.C. Meyer, V. Scardaci, C. Casiraghi, M. Lazzeri, F. Mauri, S. Piscanec, D. Jiang, K.S. Novoselov, S. Roth, A.K. Geim, Raman spectrum of graphene and graphene layers. Phys. Rev. Lett. **97**, 187401 (2006)
45. I. Calizo, A.A. Balandin, W. Bao, F. Miao, C.N. Lau, Temperature dependence of the Raman spectra of graphene and graphene multilayers. Nano Lett. **7**(9), 2645–2649 (2007)
46. A. Jorio, R. Saito, J.H. Hafner, C.M. Lieber, M. Hunter, T. McClure, G. Dresselhaus, M.S. Dresselhaus, Structural (n, m) determination of isolated single-wall Carbon nanotubes by resonant Raman scattering. Phys. Rev. Lett. **86**(6), 1118 (2001)
47. A.C. Ferrari, J. Robertson, Interpretation of Raman spectra of disordered and amorphous carbon. Phys. Rev. B **61**(20), 14095 (2000)
48. A. Ferrari, J. Robertson, Resonant Raman spectroscopy of disordered, amorphous, and diamondlike carbon. Phys. Rev. B **64**(7), 075414 (2001)
49. R.S. Krishnan, Raman spectrum of diamond. Nature **155**(3928), 171 (1945)
50. H. Yan, D. Song, K.F. Mak, I. Chatzakis, J. Maultzsch, T.F. Heinz, Time-resolved Raman spectroscopy of optical phonons in graphite: phonon anharmonic coupling and anomalous stiffening. Phys. Rev. B **80**(12), 121403 (2009)
51. H. Wang, Y. Wang, X. Cao, M. Feng, G. Lan, Vibrational properties of graphene and graphene layers. J. Raman Spectrosc. **40**(12), 1791–1796 (2009)
52. A. Ramasubramaniam, Large excitonic effects in monolayers of molybdenum and tungsten dichalcogenides. Phys. Rev. B **86**(11), 115409 (2012)
53. H.S. Lee, S.-W. Min, Y.-G. Chang, M.K. Park, T. Nam, H. Kim, J.H. Kim, S. Ryu, S. Im, MoS$_2$ nanosheet phototransistors with thickness-modulated optical energy gap. Nano Lett. **12**(7), 3695–3700 (2012)
54. M. Dragoman, A. Cismaru, M. Aldrigo, A. Radoi, A. Dinescu, D. Dragoman, MoS$_2$ thin films as electrically tunable materials for microwave applications. Appl. Phys. Lett. **107**(24), 243109 (2015)
55. S. Das, M. Demarteau, A. Roelofs, Nb-doped single crystalline MoS$_2$ field effect transistor. Appl. Phys. Lett. **106**(17), 173506 (2015)
56. X. Zhang, Y. Huang, Z. Ma, L. Niu, C.Q. Sun, From ice supperlubricity to quantum friction: electronic repulsivity and phononic elasticity. Friction **3**(4), 294–319 (2015)
57. M.R. Laskar, D.N. Nath, L. Ma, E.W. Lee, C.H. Lee, T. Kent, Z. Yang, R. Mishra, M.A. Roldan, J.-C. Idrobo, S.T. Pantelides, S.J. Pennycook, R.C. Myers, Y. Wu, S. Rajan, p-type doping of MoS$_2$ thin films using Nb. Appl. Phys. Lett. **104**(9), 092104 (2014)
58. B. Radisavljevic, A. Radenovic, J. Brivio, V. Giacometti, A. Kis, Single-layer MoS$_2$ transistors. Nat. Nanotechnol. **6**(3), 147–150 (2011)
59. A. Sarathy, J.-P. Leburton, Electronic conductance model in constricted MoS$_2$ with nanopores. Appl. Phys. Lett. **108**(5), 053701 (2016)

60. R. Thamankar, T.L. Yap, K.E.J. Goh, C. Troadec, C. Joachim, Low temperature nanoscale electronic transport on the MoS$_2$ surface. Appl. Phys. Lett. **103**(8), 083106 (2013)
61. Z. Huang, W. Han, H. Tang, L. Ren, D.S. Chander, X. Qi, H. Zhang, Photoelectrochemical-type sunlight photodetector based on MoS$_2$/graphene heterostructure. 2D Mater. **2**(3), 035011 (2015)
62. V. Kaushik, D. Varandani, B.R. Mehta, Nanoscale mapping of layer-dependent surface potential and junction properties of CVD-grown MoS$_2$ domains. J. Phys. Chem. C **119**(34), 20136–20142 (2015)
63. X. Zhang, Y.G. Nie, W.T. Zheng, J.L. Kuo, C.Q. Sun, Discriminative generation and hydrogen modulation of the Dirac-Fermi polarons at graphene edges and atomic vacancies. Carbon **49**(11), 3615–3621 (2011)
64. D. Li, K. Zhang, M. Song, N. Zhai, C. Sun, H. Li, High-pressure Raman study of Terephthalonitrile. Spectrochim. Acta Part A Mol. Biomol. Spectrosc. **173**, 376–382 (2017)
65. T.M.G. Mohiuddin, A. Lombardo, R.R. Nair, A. Bonetti, G. Savini, R. Jalil, N. Bonini, D.M. Basko, C. Galiotis, N. Marzari, K.S. Novoselov, A.K. Geim, A.C. Ferrari, Uniaxial strain in graphene by Raman spectroscopy: G peak splitting, Gruneisen parameters, and sample orientation. Phys. Rev. B **79**(20), 205433 (2009)
66. Z.H. Ni, T. Yu, Y.H. Lu, Y.Y. Wang, Y.P. Feng, Z.X. Shen, Uniaxial Strain on Graphene: Raman spectroscopy study and band-gap opening. ACS Nano **2**(11), 2301–2305 (2008)
67. J.E. Proctor, E. Gregoryanz, K.S. Novoselov, M. Lotya, J.N. Coleman, M.P. Halsall, High-pressure Raman spectroscopy of graphene. Phys. Rev. B **80**(7), 073408 (2009)
68. A. Taube, A. Łapińska, J. Judek, N. Wochtman, M. Zdrojek, Temperature induced phonon behaviour in germanium selenide thin films probed by Raman spectroscopy. J. Phys. D Appl. Phys. **49**(31), 315301 (2016)
69. M.S. Liu, L.A. Bursill, S. Prawer, R. Beserman, Temperature dependence of the first-order Raman phonon lime of diamond. Phys. Rev. B **61**(5), 3391–3395 (2000)
70. H. Herchen, M.A. Cappelli, First-order Raman spectrum of diamond at high temperatures. Phys. Rev. B **43**(14), 11740 (1991)
71. W. Borer, S. Mitra, K. Namjoshi, Line shape and temperature dependence of the first order Raman spectrum of diamond. Solid State Commun. **9**(16), 1377–1381 (1971)
72. J.B. Cui, K. Amtmann, J. Ristein, L. Ley, Noncontact temperature measurements of diamond by Raman scattering spectroscopy. J. Appl. Phys. **83**(12), 7929–7933 (1998)
73. M.S. Liu, L.A. Bursill, S. Prawer, K. Nugent, Y. Tong, G. Zhang, Temperature dependence of Raman scattering in single crystal GaN films. Appl. Phys. Lett. **74**(21), 3125–3127 (1999)
74. M. Tangi, P. Mishra, B. Janjua, T.K. Ng, D.H. Anjum, A. Prabaswara, Y. Yang, A.M. Albadri, A.Y. Alyamani, M.M. El-Desouki, Bandgap measurements and the peculiar splitting of E2H phonon modes of InxAl1-xN nanowires grown by plasma assisted molecular beam epitaxy. J. Appl. Phys. **120**(4), 045701 (2016)
75. M.C. Klein, F. Hache, D. Ricard, C. Flytzanis, Size dependence of electron-phonon coupling in semiconductor nanospheres—the case of CdSe. Phys. Rev. B **42**(17), 11123–11132 (1990)
76. C. Trallero-Giner, A. Debernardi, M. Cardona, E. Menendez-Proupin, A.I. Ekimov, Optical vibrons in CdSe dots and dispersion relation of the bulk material. Phys. Rev. B **57**(8), 4664–4669 (1998)
77. C.J. Zhang, P. Hu, The possibility of single C-H bond activation in CH$_4$ on a MoO$_3$-supported Pt catalyst: a density functional theory study. J. Chem. Phys. **116**(10), 4281–4285 (2002)
78. C.Q. Sun, S. Li, B.K. Tay, Laser-like mechanoluminescence in ZnMnTe-diluted magnetic semiconductor. Appl. Phys. Lett. **82**(20), 3568–3569 (2003)
79. W. Lu, H. Nan, J. Hong, Y. Chen, C. Zhu, Z. Liang, X. Ma, Z. Ni, C. Jin, Z. Zhang, Plasma-assisted fabrication of monolayer phosphorene and its Raman characterization. Nano Res **7**(6), 853–859 (2014)
80. V. Tran, R. Soklaski, Y. Liang, L. Yang, Layer-controlled band gap and anisotropic excitons in few-layer black phosphorus. Phys. Rev. B **89**(23), 235319 (2014)
81. G. Viera, S. Huet, L. Boufendi, Crystal size and temperature measurements in nanostructured silicon using Raman spectroscopy. J. Appl. Phys. **90**(8), 4175–4183 (2001)

82. R. Cuscó, E. Alarcón Lladó, J. Ibáñez, L. Artús, J. Jiménez, B. Wang, M.J. Callahan, Temperature dependence of Raman scattering in ZnO. Phys. Rev. B **75**(16), 165202–165213 (2007)

83. Y.N. Hwang, S.H. Park, D. Kim, Size-dependent surface phonon mode of CdSe quantum dots. Phys. Rev. B **59**(11), 7285 (1999)

84. B. Palpant, H. Portales, L. Saviot, J. Lerme, B. Prevel, M. Pellarin, E. Duval, A. Perez, M. Broyer, Quadrupolar vibrational mode of silver clusters from plasmon-assisted Raman scattering. Phys. Rev. B **60**(24), 17107–17111 (1999)

85. M. Fujii, T. Nagareda, S. Hayashi, K. Yamamoto, Low-frequency Raman-scattering from small silver particles embedded in SiO_2 thin-films. Phys. Rev. B **44**(12), 6243–6248 (1991)

86. H. Lamb, On the variations of an elastic sphere. Lond. Math. Soc. Proc. **13**, 233–256 (1882)

87. M.A. Omar, *Elementary Solid State Physics: Principles and Applications* (Addison-Wesley, New York, 1993)

88. J. Zi, K.M. Zhang, X.D. Xie, Microscopic calculations of Raman scattering from acoustic phonons confined in Si nanocrystals. Phys. Rev. B **58**(11), 6712–6715 (1998)

89. G.A. Narvaez, J. Kim, J.W. Wilkins, Effects of morphology on phonons in nanoscopic silver grains. Phys. Rev. B **72**(15), 155411 (2005)

90. G. Scamarcio, M. Lugara, D. Manno, Size-dependent lattice contraction in CdS1-xSex nanocrystals embedded in glass observed by Raman-scattering. Phys. Rev. B **45**(23), 13792–13795 (1992)

91. L.H. Liang, C.M. Shen, X.P. Chen, W.M. Liu, H.J. Gao, The size-dependent phonon frequency of semiconductor nanocrystals. J. Phys. Condens. Matter **16**(3), 267–272 (2004)

92. A. Dieguez, A. Romano-Rodriguez, A. Vila, J.R. Morante, The complete Raman spectrum of nanometric SnO_2 particles. J. Appl. Phys. **90**(3), 1550–1557 (2001)

93. Z. Iqbal, S. Veprek, Raman-scattering from hydrogenated microcrystalline and amorphous-silicon. J. Phys. C Solid State Phys. **15**(2), 377–392 (1982)

94. E. Anastassakis, E. Liarokapis, Polycrystalline Si under strain: Elastic and lattice-dynamical considerations. J. Appl. Phys. **62**, 3346–3352 (1987)

95. H. Richter, Z.P. Wang, L. Ley, The one phonon Raman spectrum in microcrystalline silicon. Solid State Commun. **39**(5), 625–629 (1981)

96. I.H. Campbell, P.M. Fauchet, The effects of microcrystal size and shape on the one phonon Raman spectra of crystalline semiconductors. Solid State Commun. **58**(10), 739–741 (1986)

97. H.C. Choi, Y.M. Jung, S.B. Kim, Size effects in the Raman spectra of TiO_2 nanoparticles. Vib. Spectrosc. **37**(1), 33–38 (2005)

98. X. Wang, D.M. Huang, L. Ye, M. Yang, P.H. Hao, H.X. Fu, X.Y. Hou, X.D. Xie, Pinning of photoluminescence peak positions for light-emitting porous silicon—an evidence of quantum-size effect. Phys. Rev. Lett. **71**(8), 1265–1267 (1993)

99. J.L. Andujar, E. Bertran, A. Canillas, C. Roch, J.L. Morenza, Influence of pressure and radio-frequency power on deposition rate and structural-properties of hydrogenated amorphous-silicon thin-films prepared by plasma deposition. J. Vacuum Sci. Technol. Vac. Surf. Films **9**(4), 2216–2221 (1991)

100. L. Banyai, S.W. Koch, *Semiconductor Quantum Dots* (World Scientific, Singapore, 1993)

101. Y.N. Hwang, S.H. Shin, H.L. Park, S.H. Park, U. Kim, H.S. Jeong, E.J. Shin, D. Kim, Effect of lattice contraction on the Raman shifts of CdSe quantum dots in glass matrices. Phys. Rev. B **54**(21), 15120–15124 (1996)

102. X. Yang, J. Li, Z. Zhou, Y. Wang, W. Zheng, C.Q. Sun, Frequency response of graphene phonons to heating and compression. Appl. Phys. Lett. **99**(13), 133108 (2011)

103. G. Ouyang, C.Q. Sun, W.G. Zhu, Pressure-stiffened Raman phonons in group III nitrides: a local bond average approach. J. Phys. Chem. B **112**(16), 5027–5031 (2008)

104. J.L. Hu, W.P. Cai, C.C. Li, Y.J. Gan, L. Chen, In situ x-ray diffraction study of the thermal expansion of silver nanoparticles in ambient air and vacuum. Appl. Phys. Lett. **86**(15), 151915 (2005)

105. L. Li, Y. Zhang, Y.W. Yang, X.H. Huang, G.H. Li, L.D. Zhang, Diameter-depended thermal expansion properties of Bi nanowire arrays. Appl. Phys. Lett. **87**(3), 031912 (2005)

106. T. Comaschi, A. Balerna, S. Mobilio, Temperature dependence of the structural parameters of gold nanoparticles investigated with EXAFS. Phys. Rev. B **77**(7), 075432 (2008)
107. M.T. Cardona, M. L. W., Isotope effects on the optical spectra of semiconductors. Rev. Mod. Phys. **77**(4), 1173–1224 (2005)
108. E. Grüneisen, The state of a body. Handb. Phys. **10**, 1–52. (NASA translation RE2-18-59W)
109. K. O'Donnell, X. Chen, Temperature dependence of semiconductor band gaps. Appl. Phys. Lett. **58**(25), 2924–2926 (1991)
110. J.B. Wachtman, W.E. Tefft, D.G. Lam, C.S. Apstein, Exponential temperature dependence of youngs modulus for several oxides. Phys. Rev. **122**(6), 1754 (1961)
111. J. Garai, A. Laugier, The temperature dependence of the isothermal bulk modulus at 1 bar pressure. J. Appl. Phys. **101**(2), 2424535 (2007)
112. X. Wang, Z. Chen, F. Zhang, K. Saito, T. Tanaka, M. Nishio, Q. Guo, Temperature dependence of Raman scattering in β-$(AlGa)_2O_3$ thin films. AIP Adv. **6**(1), 015111 (2016)
113. P.G. Klemens, Anharmonic decay of optical phonons. Phys. Rev. **148**(2), 845–848 (1966)
114. T.R. Hart, R.L. Aggarwal, B. Lax, Temperature dependence of Raman scattering in silicon. Phys. Rev. B **1**(2), 638–642 (1970)
115. M. Balkanski, R.F. Wallis, E. Haro, Anharmonic effects in light-scattering due to optical phonons in silicon. Phys. Rev. B **28**(4), 1928–1934 (1983)
116. B.A. Kolesov, How the vibrational frequency varies with temperature. J. Raman Spectrosc. **48**(2), 323–326 (2017)
117. I. Calizo, S. Ghosh, W. Bao, F. Miao, C. Ning Lau, A.A. Balandin, Raman nanometrology of graphene: temperature and substrate effects. Solid State Commun. **149**(27-28), 1132–1135 (2009)
118. C.Q. Sun, *Atomic Scale Purification of Electron Spectroscopic Information (US 2017 Patent No. 9,625,397B2)*. (United States, 2017)
119. X.X. Yang, C.Q. Sun, *Raman Detection of Temperature (CN 106908170A)* (2017)
120. Y.L. Huang, X.X. Yang, and C.Q. Sun, *Spectrometric Evaluation of the Force Constant, Elastic Modulues, and Debye Temperature of Sized Matter (Disclosure at Evaluation)* (2018)
121. Y. Gong, Y. Zhou, Y. Huang, C.Q. Sun, *Spectrometrics of the O:H–O bond Segmental Length and Energy Relaxation (CN 105403515A)* (China, 2018)

Chapter 21
Theory: Multifield Oscillation Dynamics

Abstract A physical perturbation mediates intrinsically the performance of a substance through relaxing the length and energy of the chemical bonds and associated electrons in various energy bands. From the perspective of Fourier transformation, one can formulate the bond oscillation frequency $\Delta\omega\,(z, d, E, \mu)$ for a variety of materials by perturbing the Hamiltonian. Reproduction of the excited $\Delta\omega$ by a perturbation x_i such as bond-order-imperfection, electric polarization, compression, tension, and thermal activation turns out information on the bond length $d(x_i)$, bond energy $E(x_i)$, single-bond force-constant, binding energy density, mode cohesive energy, Debye temperature, elastic modulus, etc., complementing the electron spectrometrics. Exercises proved the immense power of the phonon spectrometrics in revealing the nature behind the lattice vibration in terms of multifield single-bond oscillation dynamics in liquid and solid phases.

Highlights

- Bond order, length, and strength stem the phonon frequency, bandgap, and elasticity.
- Perturbation relaxes the bond and evolves the phonon frequency and related properties.
- Phonon relaxation fingerprints the intrinsic manner of bond response to stimulus.
- Spectrometrics enables the unprecedented information on bond-phonon-property cooperativity.

21.1 Lattice Oscillation Dynamics

21.1.1 Single-Body Hamiltonian

There are three major approaches to the oscillation dynamics of a bond that relaxes in length and energy by perturbing its crystal potential through applying an external force: resolution to the Schrödinger equation [1], monatomic and diatomic chain

© The Editor(s) (if applicable) and The Author(s), under exclusive license   393
to Springer Nature Singapore Pte Ltd. 2020
C. Q. Sun, *Electron and Phonon Spectrometrics*,
https://doi.org/10.1007/978-981-15-3176-7_21

dispersion [2], and the Lagrangian oscillation mechanics for the coupled oscillators [3, 4]. The oscillation can also be categorized as an isolated dimer oscillator, a coupled oscillator pair, and collective vibrations in which multiple dimers are simultaneously involved.

Bond relaxation and associated electronic redistribution in the real and energy spaces mediate the structure and properties of a substance [5]. An electron in a solid or in a liquid is subject to its intra-atomic potential of $v_{atom}(r)$ and the superposition of all interatomic potentials, time dependent $U(r, t)$, involved in the single-body Hamiltonian [1]:

$$i\hbar \frac{\partial}{\partial t} |v, r, t \ge \left[-\frac{\hbar^2 \nabla^2}{2m} + v_{atom}(r) + U(r, t) \right] |v, r, t >$$

$$where,$$

$$U(r) = \sum_i u_i(r) = \sum_{n=0} \left(\frac{d^n U(r)}{n! dr^n} \right)_{r=d_z} (r - d_z)^n$$

(21.1)

the $|v, r, t >$ is the Bloch wave function that describes the electronic spatial-temporal behavior at site r in the v_{th} energy level [6]. The single-body approach approximates the long-range interactions and the many-body effects as a background of mean field.

For an electron in a certain core level follows the dispersion defined by the tight-binding theory [2]. The first term in Eq. (21.1) is the electronic kinetic energy. The coupling of the intra-atomic potential $v_{atom}(r, t)$ and the respective wave Bloch functions, $E_v(0) = < v, r, t | v_{atom}(r, t) | v, r, t >$, defines the v_{th} energy level of an isolated atom. The $E_v(0)$ is the reference from which the specific core level shifts upon perturbation such as bond formation with different numbers or types of neighboring atoms [7]. The core level shifts when the electrons are subjecting to perturbation $U(R)(1 + \Delta)$ in terms of the exchange integral dominance and the overlap integral as a secondary [7]. The exchange integral and the overlap integral are both depend on the perturbed bond energy $E_b(1 + \Delta)$ with Δ being the perturbation.

The energy gap E_g between the conduction band and the valence band depends on the first Fourier coefficient of the crystal potential $U(r)$, according to the nearly-free electron approximation [2],

$$\begin{cases} E_g = 2|U_1| \propto \langle E_b \rangle \\ U_1 = \int U_{cry}(r) e^{ik \cdot r} dr \end{cases}$$

(21.2)

The bandgap is proportional to the bond energy $\langle E_b \rangle$ as well. The $e^{ik \cdot r}$ is the Bloch wave function approaching the nearly-free electrons. The crystal potential $U_{cry}(r)$ determines the intrinsic E_g that is different from the optical band gap with involvement of electron-phonon coupling [61, 62]. Therefore, a perturbation changes the band gap and the core level shift in the same way $\Delta E_g(x_i) \propto \Delta E_b(x_i)$ intrinsically.

For lattice oscillation, the wave function describes the vibrating oscillator. The V_{atom} is replaced with V_{dimer} for the intra-dimer interaction [1]. The crystal potential $U_{cry}(r)$ adds a perturbation to the V_{dimer}, which expands into a Taylor series at equilibrium ($r = d$). The general solution to the Schrödinger equation for the oscillation is a Fourier transformation function. Both Taylor and Fourier series are intrinsically correlated through their coefficients. A Taylor series can be considered a special case of the generalized Fourier series, with an orthonormal base of power functions and a properly defined inner product [8].

For the oscillating system, the $E_v(0)$ is replaced with the $\hbar\omega_0$ that is the reference dimer vibration energy and $\omega = \omega_0(1 + \Delta)$ is the shift under perturbation. The ω varies with the curvature of the resultant potential in the form of $\mu\omega^2 = [U(d)(1 + \Delta)]''$ at the relaxed equilibrium in which the nonlinear contribution is within the limit of instrumental detection. An examination [9] of the contribution of the anharmonic correction to the H–O vibration frequency (3200 cm^{-1}) revealed that the addition of the anharmonic vibrational potential only shifts the H–O phonon by about 100 cm^{-1} or less without deriving new vibrating signatures.

One can replace the integration of the coupled wave functions and interatomic potentials with the interatomic potential energy directly to simplify the discussion. At equilibrium, the coordinate (d, E) for the potential curve gives directly the bond length (d) and bond energy (E). Despite the possible rendering of precision of the solution, this approximation allows one to focus on the nature origin behind and the varying trends of phononic measurements. Consideration of the relative frequency shift due to perturbation further improves the precision due to simplification. Furthermore, the spectroscopic measurements fingerprint the perturbation to the collection of all possible potentials without needing to discriminate them one from another.

21.1.2 Atomic Chains

Raman scattering arises from the radiating dipole moment induced by the electric field of the incident electromagnetic radiation. The laws of momentum and energy conservation govern the interaction between the incident photons and the phonons being activated. When one considers a solid containing numerous Bravais unit cells and each cell contains n atoms, there will be 3n modes of vibrations. Among the 3n modes, there will be three acoustic modes, LA, TA$_1$ and TA$_2$ and 3(n − 1) optical modes. The acoustic mode stands for the in-phase motion of the mass center of the unit cell or the entire solid.

For an decoupled monatomic chain and a diatomic chain of the same spring force constant β, one can solve the lattice vibration equations to derive the phonon dispersion relations in the reciprocal $k = 2\pi/\lambda$ space with $\mu = m_1 m_2/(m_1 + m_2)$ being the reduced mass of the oscillator [2]:

$$\begin{cases} \omega^2(k) = \frac{2\beta}{\mu} \sin^2\left(\frac{ka}{2}\right) \propto \frac{\beta}{\mu} & (monatomic\ chain) \\ \omega_\pm^2(k) = \frac{\beta}{\mu}\left[1 \pm \sqrt{1 - \frac{4\mu^2 \sin^2\left(\frac{ka}{2}\right)}{m_1 m_2}}\right] \propto \frac{\beta}{\mu}[1 \pm \delta(k)] & (complex\ atomic\ chain) \end{cases}$$

$$(21.3)$$

Table 21.1 Correlation between the macroscopic detectable quantities and the Taylor series coefficients

Microscopic q	Bond identities	Macroscopically detectable Q	No
$E_z = U(d)$	Bond energy E_z	Core level shift ΔE_v, band gap E_G,	(7.1)
$\left.\frac{dU(r)}{dr}\right\|_{r=d} = 0 \propto \left[\frac{E}{d}\right]$	Bond length d	Mass density d^3, strain $\Delta d/d$	(7.2)
$f = -\frac{dU(r)}{dr}$	At non-equilibrium	Force	(7.3)
$p \propto -\frac{dU(r)}{r^2 dr} \propto \frac{U(r)}{r^3}$		Pressure $-\frac{\partial U(r)}{\partial r} / \frac{\partial V}{\partial r}$	(7.4)
$\kappa = \left.\frac{dU(r)}{dr^2}\right\|_{r=d} \propto \left[\frac{E}{d^2}\right]$	Force constant k	Bond stiffness Yd, dimer vibration frequency: $\omega = k/\mu = \left(E/\mu d^2\right)^{1/2} \propto (Yd)^{1/2}$	(7.5)
$B \propto \left.\frac{d^3 U(r)}{dr^3}\right\|_{r=d} \propto \left[\frac{E}{d^3}\right]$	Energy density E_{den}	Elastic modulus $B, Y \propto -V \frac{\partial P}{\partial V}$	(7.6)
$T_C \propto z E_z$	Atomic cohesive energy E_{coh}	Critical temperature for phase transition; energy band width $E_{v,w} \propto 2z E_z$	(7.7)

These solutions agree with the single-bond approach shown in Table 21.1, irrespective of the acoustic or the optical phonon. Since the wavelength of the IR and the visible light ($500 < \lambda < 1500$ nm) and is much greater than the lattice constant in a 10^{-1} nm order. The single-bond approximation is valid if one is focused on the frequency shift of a specific vibration mode within a certain range near the Brillouin zone center, see Fig. 21.1.

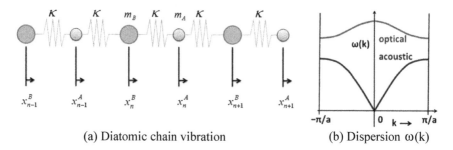

(a) Diatomic chain vibration   (b) Dispersion $\omega(k)$

Fig. 21.1 Illustration of **a** the diatom chain vibration (K is also the force constant β) and **b** the dispersion of the acoustic and optical phonon. The acoustic phonon dispersion holds for monatomic chain as well. The IR and visible light is within the tiny $k = 2\pi/\lambda$ ($\lambda \gg a$) value at the Brillouin zone center [2]

21.1.3 Lagrangian Mechanics of Coupled Oscillators

An alternative yet efficient way to deals with oscillating system is solving the Lagrangian oscillation equation. In particular for the segmented O:H–O bond that performs as an asymmetrical oscillator pair coupled by the O:⇔:O Coulomb repulsion and bridged by the H atom as the coordination origin [4]. The motion of the oscillator pair follows Lagrangian equation [10]:

$$\frac{d}{dt}\left(\frac{\partial L}{\partial\left(\frac{dq_i}{dt}\right)}\right) - \frac{\partial L}{\partial q_i} = Q_i \tag{21.4}$$

The Lagrangian $L = T - U$ consists of the total kinetic energy T and the total potential energy U of the oscillating system with involvement of the Q_i non-conservative forces due to perturbation. The external non-conservative forces include mechanical compression, molecular undercoordination, electrification, thermal excitation, and any radiation absorption [11]. The action of a Q_i force relaxes the bond from one equilibrium to another. The time dependent $q_i(t)$, represents the generalized variables, denoting the coordinates of an atom composed the oscillator. The kinetic energy T sums vibration energies of the individual atoms, in the form of $2T_i = m_i(dq_i(t)/dt)^2$. The potential energy U is composed of all interatomic interactions. The u_i is the coordinate of the ith atom.

Lagrangian resolution of the O:H–O coupled oscillator pair [10] resulted in the segmental $\omega_x(k_x)$ dispersion ($x = L$ for the O:H and H for the H–O segment), with k_x being the segmental force constant and k_C the curvature of the O:⇔:O repulsive coupling potential [3, 4]. The m_x is the reduced mass of the oscillator.

$$\omega_x = (2\pi c)^{-1}\sqrt{\frac{k_x + k_C}{m_x}} \tag{21.5}$$

Because of the segmental k_x disparity, this dispersion specifies that under any perturbation, the O ions dislocate in the same direction but by different amounts along the O:H–O with respect to the H as the coordination origin. The O:H relaxes always more than the H–O. Consequently, if one segment becomes longer, its phonon turns to be softer, and vice versa. Decoupling the k_C, the dispersion degenerates into the isolated oscillators, which is the non-segmented A-B type bond approximation, which is equivalent to Eq. (21.3) in the Brillouin zone center and $k_x = \beta$ being the force constant.

21.1.4   Collective Oscillation

The collective oscillation means a certain atom vibrates in concert with its z-coordinated neighbors. The vibration amplitude is the dislocation of an atom with its equilibrium position $x = r - d_0$. The high-order terms of the potentials contribute to the nonlinear behavior. For a dimer oscillator, the atomic coordination number is $z = 1$; otherwise, the short-range interaction on each atom results from its $z > 1$ neighboring coordinating atoms, the atomic vibrating dislocation is the contribution from all the surrounding coordinates, z. Since the vibration amplitude $x \ll d_0$, the mean contribution from each coordinate to the force constant and to the magnitude of the atomic dislocation as the first-order approximation [12],

$$k_1 = k_2 = \cdots = k_z = \mu_i(c\omega)^2$$

and,

$$x_1 = x_2 = \cdots = x_z = (r - d_0)^2/z$$

Therefore, the total energy of a certain atom with its z neighbors is the sum over all coordinates,

$$u(r) = -zE_b + \left.\frac{zd^2u(r)}{2!dr^2}\right|_{d_0} (r - d_0)^2 + \cdots \tag{21.6}$$

This relation leads to an expression for the phonon frequency as a function of bond order z, bond length d_z, and bond energy E_z, in terms of the curvature of the superposition of all (z) the neighboring potentials involved. For single dimer oscillation, $z = 1$.

21.2   Taylor Coefficients Versus Observables

The crystal potential $U_{cry}(r)$ and the associated electron distribution uniquely determine the performance of a substance that is represented by an average of the whole pack of bonds in the substance [7, 13–16]. As exemplified in Table 21.1, the Taylor coefficients correspond directly to the detectable quantities in terms of the dimensionality.

For instance, the interatomic potential determines the energy band structure including the core level shift and forbidden band gap between the conduction and the valence band [7]. One can take phonons as individual particles, so the integrals are related directly to the interatomic potential energies. This approximation avoids the combination of wave functions and simplifies the approximation, which may be subject to some precision with focus on the nature of origin and the trend of change.

Considering the proportional relations of the dimensionality would be adequate as one is focusing on the relative change of a known bulk property as the standard reference upon perturbation.

From the dimensionality analysis of Table 21.1 (line 5 and 6), one can correlate the atomic site resolved elastic modulus $B_i(z_i)$ and the vibration frequency shift $\Delta\omega(z_i)$ with $[\Delta\omega(z_i)]^2/[B_i d_i] \equiv 1$ if the z_i and remain unchanged. These relationships apply to any interatomic potential $U(r)$ as the B_i and at the ith atomic site are related only to the bond order, length, and energy at the equilibrium.

21.3 Single Bond Multifield Oscillations

21.3.1 Bond Length and Energy Relaxation

The interatomic bond is the basic unit of structure and energy-storage. The bond relaxes upon perturbation, which in turn mediates the electronic energetics and the detectable quantities of a substance when subjecting to perturbation. Any perturbation will transmit the initial $U(r, t)$ into another equilibrium $U(r, t)(1 + \Delta)$ by relaxing the length and energy of the interatomic bond. If the effects of atomic CN deficiency, strain, stress, and thermal excitation come into play simultaneously, the length $d(z, \varepsilon, T, P, ...)$ and energy, $E_i(z, \varepsilon, T, P, ...)$ of the representative bond will change in the following ways [16],

$$\begin{cases} d(z, \varepsilon, P, T, ...) = d_b \prod_J (1 + \varepsilon_J) = d_b \left[[1 + (C_z - 1)] \left(1 + \int_0^\varepsilon d\varepsilon \right) \frac{\left(1 + \int_{T_0}^T \alpha(t)dt\right)\left(1 - \int_{P_0}^P \beta(p)dp\right) \cdots}{d_b} \right] \\ E(z, \varepsilon, P, T, ...) = E_b \left(1 + \sum_J \Delta_J\right) = E_b \left[1 + \left(C_z^{-m} - 1\right) \frac{-d_z^2 \int_0^\varepsilon \kappa(\varepsilon)\varepsilon d\varepsilon - \int_{T_0}^T \eta(t)dt - \int_{V_0}^V p(v)dv \cdots}{E_b} \right] \end{cases}$$

where

$$\begin{cases} d_b = d(z_b, 0, P_0, T_0) \\ E_b = E(z_b, 0, P_0, T_0) \end{cases} \tag{21.8}$$

The ε_J is the strain and the Δ_J is the energy perturbation due to the applied stimulus. The summation and the production are proceeded over all the Jth stimulus of all the degrees of freedom $(z, \varepsilon, T, P, ...)$. The C_z is the bond contraction coefficient depending on the atomic coordination numbers $(z$ or CN$)$. $C_z - 1$ is the undercoordination induced strain. The m is the bond nature index that correlates the bond length to energy. The $\alpha(t)$ is the temperature-dependent thermal expansion coefficient. The $\eta(t) = C_v(t/_D)/z$ is the Debye specific heat of the representative bond for a z-coordinated atom. The $\beta = -\partial v/(v\partial p)$ is the compressibility $(p < 0$, compressive stress$)$ or extensibility $(p > 0$ tensile stress$)$ that is proportional to the inverse of elastic bulk modulus. The $k(\varepsilon)$ is the strain-dependent of the single bond force constant.

One can extend the Grüneisen parameter to inspect the bonding dynamics under compression and heating. Eq. (21.8) gives the relative change of bond length and energy,

$$
\begin{cases}
\frac{\Delta d}{d_b} = \frac{1}{d_b}\left(\int_{T_0}^{T} \alpha(t)dt\right)\left(\int_{P_0}^{P} \beta(p)dp\right) \\
\frac{\Delta E}{E_b} = \left\{-\frac{1}{E_b}\left[\int_{T_0}^{T} \eta(t)dt + \int_{V_0}^{V} p(v)dv\right]\right\}
\end{cases}
$$

Employing the single bond specific-heat $\eta(t) = C_v(t)/z_b$, and $d(VP) = VdP + PdV$, and $PdV = PVdP \times dV/(VdP) = -\beta PVdP$, and

$$
\frac{\Delta\omega}{\omega} = \frac{\Delta E}{2E} - \frac{\Delta d}{d};
$$

Yields the extended Grüneisen parameters,

$$
\left.\begin{array}{c}
\gamma_T = \frac{d\omega}{\omega_0 dt} \\
\gamma_P = \frac{d\omega}{\omega_0 dp}
\end{array}\right\} = -\left\{
\begin{array}{c}
\frac{C_v(t/\Theta_D)}{2z_b E_b} + \frac{\alpha(t)}{d_b} \\
\beta(p)\left[-\frac{p}{2E_b/V_b} + \frac{1}{d_b}\right]
\end{array}\right.
$$

The compressibility $(p < 0)$, $\beta = -\partial v/(V \partial p)$, is an inversion of its elastic modulus in dimension. The Grüneisen parameter integrates information on how the bond length and bond energy change under P and T perturbation and counts the intrinsic specific heat, Debye temperature, compressibility, thermal expansion coefficient. The $z_b E_b = E_{coh}$ and $E_b/V_b = E_{den}$. Likewise, one can amplify the Grüneisen parameter to cover more stimuli.

Figure 21.2 illustrates the bond relaxation of a regular dimer oscillator. At equilibrium, the (d, E_b) coordinate corresponds to the bond length and bond energy.

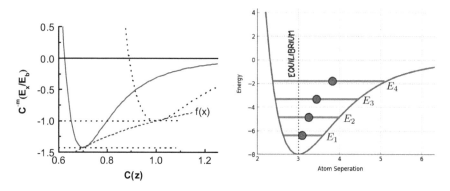

Fig. 21.2 The long-range, mono-well potential for a dimer oscillator in a regular substance [13]. The bond relaxes in length and energy under a stimulus $(x = z, P, T, \varepsilon, \text{etc.})$ along the $f(x) = [d_b \Pi(1 + \varepsilon_J), E_b(1 + \Sigma \Delta_J)]$ path transiting the potential curve from one equilibrium to the other [13, 15]

Perturbation acts on the potential and dislocates the (d, E_b) to a new equilibrium $[d_b \Pi(1 + \varepsilon_J), E_b(1 + \Sigma \Delta_J)]$. For instance, a compression or a tension perturbs the $U(R)$ in a manner given in Fig. 21.2a. Compression stores energy into a bond by shortening and stiffening it while tension does it contrastingly, along the $f(P)$ path. Thus, the bond is relaxed in length and energy and all the detectable properties of the substance vary. Thermal activation not only fluctuates the vibrating oscillator but also elongates and softens the bond. The perturbation may modify the shape of the potential function, but it is not that important or out of immediate concern when consider property at quasi-equilibrate state.

21.3.2 Phonon Frequency Shift

Instead of the extrinsic Raman scattering process, one shall emphasize that the solution to the Hamiltonian of a vibration system is a Fourier series with multiple terms whose frequencies are folds of that of the primary mode [1]. Therefore, the frequency of the 2D mode should be twofold that of the primary D mode of diamond. This fact may clarify the origin of the 2D mode that is often referred to the double resonant Raman scattering process. Any perturbation to the Hamiltonian such as the interlayer van der Waals force, the dipole-dipole interaction, or the nonlinear effect only relaxes the folded frequencies to deviate from the ideal values. The fact that the number-of-layer reduction induced D peak shifting from 1367 to 1344 cm^{-1} and the 2D peak shifting from 2720 to 2680 cm^{-1}, is right within this scheme.

The opposite trends of the Raman frequency shifts due to the change of the number-of-layer of graphene indicate that the origin of the G mode is different from that of the D/2D modes; therefore, one cannot expect to unify them simultaneously using single model. On the other hand, the applied strain, pressure, temperature or the atomic-CN variation can modulate the length and energy of the involved bonds, or their representative, and hence the phonon frequencies change with bond relaxation. Band splitting is expected to happen if the uniaxial strain is applied in a direction along or perpendicular to a C–C bond in the graphene of C_{6v} group symmetry. The extent of band splitting depends on the extent of the mismatch between a certain bond and the direction of the strain.

One can measure the Raman frequency of a particular x mode as, $\omega_x = \omega_{x0} + \Delta\omega_x$, where ω_{x0} is the referential dimer vibration frequency, from which the Raman shift $\Delta\omega_x$ proceeds. The ω_{x0} may vary with the frequency of the incident photon. Incorporating the variables of atomic coordination, strain, temperature, and pressure $(x_i = z, , T, P)$ into the expressions for bond length and bond energy, see Eq. (21.2), one can have the general form of the relative Raman shift,

$$\frac{\omega(z, \varepsilon, P, T) - \omega(1, \varepsilon, P_0, T_0)}{\omega(z_b, 0, P_0, T_0) - \omega(1, 0, P_0, T_0)} = \frac{z d_b}{d(z, \varepsilon, P, T)} \left(\frac{E(z, \varepsilon, P, T)}{E_b} \right)^{\frac{1}{2}} \quad (21.9)$$

As the first-order approximation, the vibration frequency shift $\Delta\omega_x(z, d_z, E_z, \mu)$ from the reference $\omega_x(1, d_b, E_b, \mu)$ depends functionally on the order z, length d_z, and energy E_z of the representative bond for the entire specimen and the reduced mass of the representative dimer,

$$\Delta\omega_x(z, d_z, E_z, \mu) = \omega_x(z, d_z, E_z, \mu) - \omega_x(1, d_b, E_b, \mu)$$

$$= \Delta\omega = \sqrt{\left.\frac{d^2\mu(r)}{\mu dr^2}\right|_{r=d_z}} \propto \frac{1}{d_z}\left(\frac{E_z}{\mu}\right)^{\frac{1}{2}} \times \begin{cases} 1(G, E_g) \\ z(D, A_g) \end{cases} \quad (21.10)$$

Considering the coordination-resolved mode of vibration, the z takes the values of $z = 1$ and $z > 1$. For instance [16, 17], for the D/2D modes of graphene and the A_g mode for 2-D structures, $z > 1$ is involved, which endows the phonon frequency redshift due to the vibrations of a collection of z oscillators. For the G mode of graphene, and the E_{2g} 141 cm^{-1} mode of TiO$_2$, $z \equiv 1$, which ensures the phonon frequency blueshift when the feature size is reduced because of the dominance of dimer oscillation.

21.4 Formulation of Multifield Perturbation

21.4.1 Atomic Undercoordination

21.4.1.1 BOLS-LBA Approach

The BOLS notion [15] suggests that bonds between undercoordinated atoms become shorter and stronger. The local density of charge and energy becomes higher and local potential well becomes deeper associated with quantum entrapment of local charge and energy. The locally densely entrapped charge will in turn polarize the valence electrons of atoms at the open end of the crystal. The polarization is subject to the edge atom whose outermost orbit is half-occupied [7]. Hence, the BOLS modulates the local atomic cohesive energy, the binding energy density, valence electron distribution by perturbing the Hamiltonian of the entire specimen and their relevant properties. The BOLS notion is expressed as follows:

$$\begin{cases} d_z = d_b C_z = 2\{1 + \exp[(12 - z)]/(8z)\}^{-1} \text{ (bond contraction)} \\ E_z = E_b C_z^{-m} \qquad\qquad\qquad\qquad \text{(bond strengthening)} \end{cases} \quad (21.11)$$

where z and b denote an atom with z neighbors and in the bulk specimen as a standard, respectively. The z spans from the outermost surface to the center of the solid up to three atomic layers, as there is no bond order loss that occurs when $z > 3$. The bond contraction coefficient C_z varies only with the effective CN (or z) of the atom of interest regardless of the nature of the bond or the solid dimensions, except for the hydrogen bond (O:H–O) in water ice. Molecular undercoordination shortens the H–O covalent bond but lengthens the O:H nonbond because of the coupling between electron pairs on adjacent O^{2-} [4].

The local bond average (LBA) represents the true Fourier transformation of the phonon relaxation dynamics, which sorts the constituent bonds according to their force constants or vibrational frequencies. In contrast to the volume partition approximation focusing on the value of a quantity in the partitioned volume, the LBA approach connects the deviation of the quantity from its known bulk value under an applied external stimulus. The volume partition approximation therefore describes only the local representative atomic bonds, disregarding the manner of distribution and the number of bonds. In the absence of phase transitions, the nature and total number of bonds keep unchanged. The LBA applies to all specimens of interest: crystalline, non-crystalline, solid and liquid, and those with or without defects or impurities. One can therefore focus on the performance of the representative bond, or the average of all bonds, toward the bond-phonon-property cooperativity.

A typical example is the graphene and carbon nanotube (CNT) that follows the BOLS prediction. Using the diamond C–C bond length of 0.154 and 0.142 nm for graphite, one can readily derive the effective CN for the bulk graphite as $z_g = 5.335$ from the bond contraction coefficient C_z. For the C atom in the bulk diamond, the effective CN is 12 instead of 4 because the diamond structure is an interlock of two fcc unit cells. Given the atom cohesive energy in diamond, 7.37 eV [18], and the bond nature index m = 2.56, one can derive, from the $E_z = C_z^{-m} E_b$ relation, the single C–C bond energy in the diamond is $E_b = 7.37/12 = 0.615$ eV and it is $E_3 = 1.039$ eV in the monolayer graphene of z = 3. The cohesive energy per atom in graphene is 3.11 eV/atom.

Theoretical reproduction of the elastic modulus enhancement [19–21], melting temperature depression of the SWCNT [19, 22], and the C $1s$ core level shift of the graphene edge, graphene interior, graphite, and diamond [23], confirmed consistently that the C–C bond at the graphene edge contracts by 30% from 0.154 to 0.107 nm with a 152% bond energy gain [19, 20]. The bond contraction and polarization dictate the ribbon width dependence of the band gap expansion of GNR [24], and the Dirac-Fermi polaritons generation and hydrogenation [25]. The C–C bond between the 3-coordinated atoms in GNR contracts by 18.5% to 0.125 nm with a 68% increase of bond energy [20]. The Young's modulus of the SWCNT was determined to be 2.595 TPa with respect to the bulk modulus of 865 GPa. The effective wall thickness of the SWCNT is determined to be 0.142 nm instead of the layer spacing 0.34 nm. It has been found [26] that breaking a C–C bond of the 2-coordinated carbon atom near the vacancy requires 7.50 eV per bond that is 32% higher than the energy (5.67 eV/bond) required for breaking one bond of a 3-coordinated carbon atom in a suspended graphene. This fact further evidences for the BOLS-LBA prediction of the shorter and stronger bonds between undercoordinated atoms.

21.4.1.2 Atomic-Site and Crystal-Size Resolved Shift

Any perturbation x mediates a detectable property Q at an atomic site or for a cluster of given size and shape in the following core-shell manners [15]:

$$\frac{\Delta Q(x)}{Q(x_0)} = \begin{cases} \frac{\Delta q(x)}{q(x_0)} \\ \sum_{j \leq 3} \gamma_j \frac{\Delta q(x)}{q(x_0)} \end{cases}$$

$$\gamma_j = \frac{V_j}{V} \cong \frac{N_j}{N} = \frac{\tau C_j}{K};$$

$$z_1 = 4(1 - 0.75/K), z_2 = z_1 + 2, z_3 = z_2 + 4(spherical\ dot) \tag{21.12}$$

The q being the density of the Q is a function of bond length d, bond energy E, and bond order z. γ_j is the volume ratio of the jth atomic layer over that of the entire specimen with subscript j counting from the outermost atomic layer inward up to three. $\tau = 1, 2, 3$ is the dimensionality for the thin plate, a cylindrical rod, and a spherical dot. The K is the feature size of the nanocrystal, which is the number of atoms lined along the feature size. For bonds between atoms at sites of point defects, monatomic chains, monolayer atomic sheets or monolayer skins, no weighted sum is considered.

The nanostructure core-shell configuration and the LBA approach yields the following form with $z_{ib} = z_i/z_b$ for the size-reduction induced Raman redshift ($z > 1$) and blueshift ($z = 1$) [27],

$$\omega(K) - \omega(1) = [\omega(\infty) - \omega(1)](1 + \Delta_R)$$

$$or$$

$$\frac{\omega(K) - \omega(\infty)}{\omega(\infty) - \omega(1)} = \Delta_R < 0$$

$$\Delta_R = \sum_{i \leq 3} \gamma_i \left(\frac{\omega_i}{\omega_b} - 1 \right) = \begin{cases} \sum_{i \leq 3} \gamma_i \left(z_{ib} C_i^{-(\frac{m}{2}+1)} - 1 \right)(z = z) \\ \sum_{i \leq 3} \gamma_i \left(C_i^{-(\frac{m}{2}+1)} - 1 \right)(z = 1) \end{cases} \tag{21.13}$$

where ω_{x0} and ω_{xi} correspond to the vibration frequency of an atom inside the bulk and in the ith surface atomic shell. $\omega_x(1)$ is the vibrational frequency of an isolated dimer, which is the reference point for the optical redshift upon nanosolid and bulk formation. The frequency decreases from the dimer value with the number of atomic CN and then reaches the bulk value ($z = 12$) that is experimentally detectable.

The LFR arises from the vibration of the entire nanosolid interacting with the host matrix or other grains. The optical mode is the relative motion of the individual atoms in a complex unit cell that contains more than one atom. For the elemental solids with a simple such as the fcc structure of Ag, there presents only acoustic modes. The structure for silicon or diamond is an interlock of two fcc structures that contains in each cell two atoms in nonequivalent positions, so there will be three acoustic modes and three optical modes. The complex structure of compound ensures multiple optical modes.

21.4.2 Thermal Excitation: Debye Thermal Decay

21.4.2.1 Lattice Debye Thermal Expansion

In place of the Grüneisen notion description, one may consider the thermal expansion coefficient $\alpha(t)$ for the LBA as follows [28]. From the definition, one can translate $L = L_0\left(1 + \int_0^T \alpha(t)dt\right)$ to

$$\alpha(t) \cong \frac{dL}{L_0 dt} = \frac{1}{L_0}\left(\frac{\partial L}{\partial u}\right)\frac{du}{dt} \propto -\frac{\eta_1(t)}{L_0 F(r)} = A(r)\eta_1(t) \qquad (21.14)$$

where $(\partial L/\partial u) = -F^{-1}$ is the inverse of the gradient of the interatomic potential $u(r \approx d)$ nearby equilibrium. The du/dt is the specific heat of Debye approximation. The dynamic $A(r) = (-L_0 F(r))^{-1}$ approaches the inverse of binding energy closes to but not equals equilibrium. Compared to the Grüneisen's volumetric thermal expansion coefficient (TEC) [29], the $\alpha(t) = (VB_T)^{-1}\gamma\eta_1(t)$, $\gamma/(VB_T) = [L_0 F(r)]^{-1}$ is almost a constant in the dimension of inverse energy.

Normally, the $\alpha(T > \theta_D)$ is in the order of $10^{-(6\sim7)}$ K^{-1}. The smaller expansion coefficient for nanoparticles [30–34] indicates an increase of the potential gradient, or stronger bond near the equilibrium—a narrowed shape of the interatomic potential for the contracted bond [13]. Thermal expansion is harder for the contracted bonds of a nanograin than those in the bulk standard. Approximating $A(r \approx d)$ to be a constant, the $\alpha(t)$ follows closely the single-bond Debye specific heat, $\eta(t)$. The current approach covers the full-range of T-dependent $\alpha(T)$ showing exceedingly good agreement with the measured data for AlN, Si$_3$N$_4$, and GaN as shown in Fig. 21.3 and Table 21.2 [28].

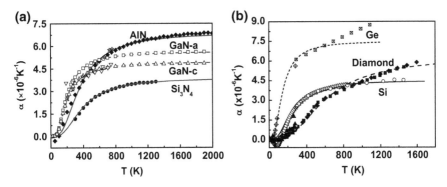

Fig. 21.3 LBA reproduction (solid lines [28]) of the measured (symbols) temperature dependence of the TECs for **a** AlN [41], Si$_3$N$_4$ [41], and GaN [34]), **b** Si [42, 43], Ge [44], and Diamond [33, 45–47]. Table 21.2 lists information derived from the fitting. (Reprinted with permission from [28])

Table 21.2 Parameters derived from fitting to the TECs and lattice parameters in Fig. 21.3 [28]

	Refs [18, 48]	$\alpha(t)$		$l(t)$			Mean	
	θ_D (K)	θ_D (K)	$A(r)$	θ_D (K)	$A(r)$	l_0 (Å)	θ_D (K)	$A(r)$
Si	647	1000	0.579	1100	0.579	5.429	1050	0.579
Ge	360	600	0.966	500	1.035	5.650	550	1.001
C	1860	2500	0.811	2150	0.792	3.566	2325	0.802
AlN	1150	1500	0.888	1500	0.946^b 0.881^c	3.110^b 4.977^c	1500	0.882
Si$_3$N$_4$	1150	1600	0.502	1400	0.888	7.734	1500	0.695
GaN	600	850	0.637	800	0.637^b 0.618^c	3.189^b 5.183^c	825	0.631

[a]The reference Debye temperature is documented in ref [49] for a-axis[b] and c-axis[c]

However, the negative TECs is beyond the scope of the present expression. Material having negative TECs often gives the negative Grüneisen parameters of the transverse acoustic phonons near the Brillouin-zone boundary. Generally, most materials expand upon being heated, although some exceptions undergo cooling expansion such as graphite [35], graphene oxide [36], ZrWO$_3$ [37, 38], and compounds composed of N, F, and O [39] because of the involvement of multiple-type interactions and their correlations like water ice [40]. Therefore, a superposition of the segmental specific heats and considers the coupling of the inter- and intra-molecular interactions would be necessary to reproduce the negative TECs [11].

21.4.2.2   Thermal Decay: Debye Temperature and Cohesive Energy

The specific heat $\eta_1(t)$ and its thermal integration is the conventionally termed internal energy, $U(T/\theta_D)$. The relation of $1 + x \sim \exp(x)$ is employed when the $x \ll 1$ for the term of bond expansion. According to Debye approximation, the T-induced bond weakening, ΔE_T, is the integration of the specific heat $\eta(t)$ from 0 K to T, which follows the relation,

$$\Delta E_T = \int_0^T \eta(t)dt = \frac{\int_0^T C_v(t)dt}{z} = \int_0^T \left[\int_0^{\frac{\theta_D}{T}} \frac{9R}{z} \left(\frac{T}{\theta_D} \right)^3 \frac{y^4 e^y}{(e^y - 1)^2} dy \right] dt$$

$$= \frac{9RT}{z} \left(\frac{T}{\theta_D} \right)^3 \int_0^{\frac{\theta_D}{T}} \frac{y^3 e^y}{e^y - 1} dy \qquad (21.15)$$

where R, θ_D, and C_v are the ideal gas constant, the Debye temperature, the specific heat, respectively. The $y = \theta_D/T$ is the reduced form of temperature. The $\eta(t)$ is the specific heat per bond, which closes to a constant value of $3R/z$ at high temperature.

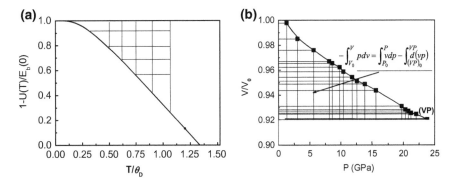

Fig. 21.4 The average-bond bond oscillating dynamics. **a** Thermal energy loss (shaded area) corresponds to the difference between unity and the specific heat integration and **b** the mechanical deformation energy gain (shaded area) to the classical form of free energy, $-\int_{V_0}^{V} pdv = \int_{P_0}^{P} vdp - \int_{(VP)_0}^{VP} d(vp)$ [50]

Figure 21.4 illustrates the Debye temperature dependence of the remnant energy of a bond subjecting to heating. The $U(T/\theta_D)$ is the loss of energy by thermalization toward bond dissociation and $1 - U(T/\theta_D)/E_b(0)$ is the remnant fraction of the bond energy.

The Raman shifts transform gradually from the nonlinear to the linear dependence of temperature—called Debye thermal decay. The slow decrease in Raman shift at remarkably low temperatures arises from the small $\int_0^T \eta(t)dt$ values when the specific heat $\eta(T)$ is proportional to T^3 at extremely low temperatures. Experimental observations show that Raman shift decreases linearly with increasing T at higher temperatures. With $T \gg \theta_D$, the specific heat C_v is considered a constant approaching $3R$.

21.4.3 Mechanical Compression: Elasticity and Energy Density

Likewise, the compressive distortion energy density gain, ΔE_P [13],

$$
\begin{cases}
\Delta E_P = -\int_{V_0}^{V} pdv = -V_0 \int_0^P p(x)\frac{dx}{dp}dp = V_0 P^2 \left[\frac{1}{2}\beta - \frac{2}{3}\beta P\right] \\
x(P) = \frac{V}{V_0} = 1 - \beta P + \beta' P^2; \ \frac{dx}{dp} = -\beta + 2\beta' P
\end{cases} \tag{21.16}
$$

or

$$
P(x) = 1.5B_0(x^{-7/3} - x^{-5/3})\left[1 + 0.75(B_0' - 4)(x^{-2/3} - 1)\right]
$$
$$
(Birch - M\ddot{u}rnaghan)
$$

V_0 denotes the volume of any size at zero temperature and the ambient pressure of reference. The $x(P)$ is a form of the equation of states. Matching the measured x-P curve with the $x(P)$ function and the Birch-Mürnaghan (BM) equation [51], one can obtain the nonlinear compressibility coefficients, β, β' and the bulk modulus B_0 and its first-order derivative B_0' of the specimen. The relation of $\beta B_0 \cong 1$ holds general for reference [18].

Figure 21.4b illustrates the P-V profile and its integration. In the P-V profile, only the gridded part $-\int_{V_0}^{V} pdV \cong -\int_{p_0}^{P} p \frac{dV}{dp} dp$ contributes to the energy density of the entire body [52]. For the single bond, the atomic volume, $V(z, t, p)$, is replaced by the bond length, $d(z, t, p)$, and the p is replaced by the force, f, acting along the bond. The $\beta(p)$ remains constant at fixed temperature within the regime of elastic deformation and then the integration $\int_0^P \beta(p)dp = \beta P$ can be simplified, unless phase transition occurs [53, 54].

For a given specimen of fixed size, we can obtain the analytical form for the T and P dependence of the B and $\Delta\omega$ by combining Eqs. (21.15) and (21.16),

$$
\left.\begin{array}{c} \frac{B(T, R)}{B(0, 0)} \\[2mm] \frac{\Delta\omega(T, P)}{\Delta\omega(0, 0)} \end{array}\right\} \cong \left\{\begin{array}{l} \left(1 + \frac{\Delta E_P - \Delta E_T}{E_0}\right) \exp\left[3\left(-\int_0^T \alpha(t)dt + \int_0^P [\beta - \beta' p]dp\right)\right] \\[4mm] \left(1 + \frac{\Delta E_P - \Delta E_T}{E_0}\right)^{1/2} \exp\left(-\int_0^T \alpha(t)dt + \int_0^P [\beta - \beta' p]dp\right) \end{array}\right.
\tag{21.17}
$$

21.4.4 Uniaxial Stretch: Single Bond Force Constant

The data of tensile strain for the monolayer graphene [55] and MoS_2 [56] enabled the derived information of the average-bond force constant and the relative direction between the strain and a specific bond in the specimen. The strain-effect is given as [57],

$$
\frac{\omega(z, \varepsilon) - \omega(1, 0)}{\omega(z, 0) - \omega(1, 0)} = \frac{d(0)}{d(\varepsilon)}\left(\frac{E(\varepsilon)}{E(0)}\right)^{1/2} = \frac{\left(1 - d^2 \int_0^\varepsilon \kappa\varepsilon d\varepsilon / E_{z1}\right)^{1/2}}{1 + \varepsilon}
$$

$$
\cong \frac{\left[1 - \kappa'\left(\lambda\varepsilon'\right)^2\right]^{1/2}}{1 + \lambda\varepsilon'}
\tag{21.18}
$$

With $\kappa' = \kappa d_z^2 / (2E_z) = $ constant.

To feature the orientation mismatch between a bond and the uni-axial strain, one can introduce a strain coefficient λ bounded by $0 \le \lambda \le 1$. The $\varepsilon = \lambda\varepsilon'$ is for a bond that is not along the applied strain. The $\kappa = 2E_z\kappa' d_z^{-2}$ being constant at limited strain is the effective force constant of all bonds to a given atom. One can only measure the average strain of the entire specimen other than that of a certain bond with force constant κ_0. The λ characterizes the actual strain of the orientated bonds in a basic unit from the complex of multiple bonds in different orientations. The strain can be along ($\lambda = 1$) or perpendicular ($\lambda = 0$) or random ($0 < \lambda < 1$) to a certain bond in the specimen. The bond along the strain is subject to a maximal extension and its phonon frequency shifts most; the bond perpendicular to the strain is subject to zero extension and keeps its phonon frequency. It is therefore not surprising that the mechanical strain splits the phonon band and the extent of splitting depends on the angle between the bond and the strain [16].

21.5 From Spectroscopy to Spectrometrics

A phonon spectral peak, probed either from Raman scattering or infrared absorption, called a vibration mode or a phonon band, features the Fourier transformation of all bonds vibrating in the similar frequency, irrespective of their locations or orientations in the real space of the solid, liquid, or the vapor phase of a substance in characteristic peaks. Fourier transformation gathers the vibration modes according to the bond stiffness and population in terms of the intrinsic vibrating dimers' force constants, or curvatures of the resultant potentials acting on the vibrating dimer at equilibrium. Multifield activation, electron entrapment and polarization, and light radiation contribute to the crystal potentials whose curvatures determine intrinsically the lattice oscillation dynamics.

Amplifying the phonon spectroscopy to spectrometrics aims to capturing the conventionally-unexpected information on the bonding dynamics by avoiding spectral peak decomposition or empirical simulation using hypothetic models. Compared with the electron spectrometrics that applies to conductive or semiconductive substance under high vacuum, the phonon spectrometrics extends to liquids and insulating species under multifield perturbation. The phonon spectrometrics directly captures bond relaxation information but the electron spectrometrics captures the effect of bond relaxation on the electronic dynamics in various energy levels or bands [7].

When an external perturbation is applied, the interatomic potential will change, and the characteristic phonon frequency will evolve or transit from its first state to a new one in terms of abundance, stiffness, and fluctuation order, which gives profound information on the bond relaxation that triggers the electronic structure and properties of the substance. External perturbations include atomic irregular-coordination, mechanical activation, thermal excitation, charge injection by doping and aqueous solvation, electrification and magnification, radiation, etc. Correlating the phonon relaxation to the applied perturbation yields the ever-unexpected information.

Besides direct calculation of the $\omega(x)$ to match the measured frequency shift of a substance under x perturbation, the other convenient method is to find the fraction of abundance transition and frequency shift using the differential phonon spectrometrics (DPS) strategy [58, 59]. The DPS purifies merely the transition in the phonon abundance and bond stiffness by conditioning, which monitors the phonon relaxation both statically and dynamically with high accuracy and high sensitivity without needing any approximation or assumption. The fraction coefficient, being the DPS peak integral, represents the fraction of bonds, or number of phonons transiting from the standard reference to the conditioned states under the change of the degree of freedom such as the solute concentration C in aqueous solutions and crystal size D for nanostructures,

$$f_x(C) = \int_{\omega_m}^{\omega_M} \left[\frac{I_{solution}(C, \omega)}{\int_{\omega_m}^{\omega_M} I_{solution}(C, \omega)d\omega} - \frac{I_{H_2O}(0, \omega)}{\int_{\omega_m}^{\omega_M} I_{H_2O}(0, \omega)d\omega} \right] d\omega.$$

$$\omega_{COG} = \frac{\int_{\omega_m}^{\omega_M} \omega I(\omega)d\omega}{\int_{\omega_m}^{\omega_M} I(\omega)d\omega} \tag{21.18}$$

The division of the fraction coefficient, $f(C)/C$, is proportional to the number of bonds per solute in the hydration shells, which characterizes the hydration shell size and its local electric field. The ω_{COG} is the center-of-gravity (COG) of the frequency, which obtained by integrating the population $I(\omega_x)$ from $\omega_{xm} = 0$ to $\omega_{xM} = 0$ [60].

Independent peak area normalization is necessary, which not only overcomes the limitation of intensity normalization but also removes artifacts of detection, which is of physically meaningful. In contrast, one often normalizes the peak by a maximal peak intensity across all peaks in experiments. The peak intensity normalization crossing a set of spectra of all samples deforms the peak shape and deviates the spectral fine-structure information from the true situation.

21.6 Summary

From the perspective of Fourier transformation and LBA consideration, one can resolve the multifield bond oscillation dynamics and direct measure the $\Delta\omega(z, d, E,)$ without involving the Grüneisen constants or the extrinsic multiple phonon resonant scattering or the optical phonon degeneration. Reproduction of the excited $\Delta\omega$ by perturbations such as bond-order-imperfection, compression, tension, thermal activation, and charge injection by solvation turns out the ever-unexpected information on the bond length, bond energy, single-bond force-constant, binding energy density, mode cohesive energy, Debye temperature, elastic modulus, etc., and thus one can reconcile the perturbation-relaxation-property cooperativity of a substance. Complementary to electronic spectrometrics, this set of theoretical and phonon spetrometrics provides not only efficient means for the local, dynamics, and quantitative information on the multifield bond oscillation dynamics but also comprehension of the physics behind observations.

References

1. W.G. Han, C.T. Zhang, A theory of nonlinear stretch vibrations of hydrogen-bonds. J. Phys.-Condens. Matter **3**(1), 27–35 (1991)
2. M.A. Omar, *Elementary Solid State Physics: Principles and Applications* (Addison-Wesley, New York, 1993)
3. Y.L. Huang, X. Zhang, Z.S. Ma, G.H. Zhou, Y.Y. Gong, C.Q. Sun, Potential paths for the hydrogen-bond relaxing with $(H_2O)(N)$ cluster size. J. Phys. Chem. C **119**(29), 16962–16971 (2015)
4. Y.L. Huang, X. Zhang, Z.S. Ma, Y.C. Zhou, W.T. Zheng, J. Zhou, C.Q. Sun, Hydrogen-bond relaxation dynamics: resolving mysteries of water ice. Coord. Chem. Rev. **285**, 109–165 (2015)
5. G.W. Shim, K. Yoo, S.-B. Seo, J. Shin, D.Y. Jung, I.-S. Kang, C.W. Ahn, B.J. Cho, S.-Y. Choi, Large-area single-layer $MoSe_2$ and its van der waals heterostructures. ACS Nano **8**(7), 6655–6662 (2014)
6. X. Zhang, J.L. Kuo, M.X. Gu, P. Bai, C.Q. Sun, Graphene nanoribbon band-gap expansion: broken-bond-induced edge strain and quantum entrapment. Nanoscale **2**(10), 2160–2163 (2010)
7. X.J. Liu, M.L. Bo, X. Zhang, L. Li, Y.G. Nie, H. Tian, Y. Sun, S. Xu, Y. Wang, W. Zheng, C.Q. Sun, Coordination-resolved electron spectrometrics. Chem. Rev. **115**(14), 6746–6810 (2015)
8. W. Kossek, Is a taylor series also a generalized fourier series? Coll. Math. J. **49**(1), 54–56 (2018)
9. Y. Shi, Z. Zhang, W. Jiang, Z. Wang, Theoretical study on electronic and vibrational properties of hydrogen bonds in glycine-water clusters. Chem. Phys. Lett. **684**, 53–59 (2017)
10. Y. Huang, X. Zhang, Z. Ma, Y. Zhou, G. Zhou, C.Q. Sun, Hydrogen-bond asymmetric local potentials in compressed ice. J. Phys. Chem. B **117**(43), 13639–13645 (2013)
11. C.Q. Sun, X. Zhang, J. Zhou, Y. Huang, Y. Zhou, W. Zheng, Density, elasticity, and stability anomalies of water molecules with fewer than four neighbors. J. Phys. Chem. Lett. **4**, 2565–2570 (2013)
12. L.K. Pan, C.Q. Sun, C.M. Li, Elucidating Si–Si dimmer vibration from the size-dependent Raman shift of nanosolid Si. J. Phys. Chem. B **108**(11), 3404–3406 (2004)
13. C.Q. Sun, Y. Sun, Y.G. Ni, X. Zhang, J.S. Pan, X.H. Wang, J. Zhou, L.T. Li, W.T. Zheng, S.S. Yu, L.K. Pan, Z. Sun, Coulomb repulsion at the nanometer-sized contact: a force driving Superhydrophobicity, superfluidity, superlubricity, and supersolidity. J. Phys. Chem. C **113**(46), 20009–20019 (2009)
14. L. Pan, S. Xu, X. Liu, W. Qin, Z. Sun, W. Zheng, C.Q. Sun, Skin dominance of the dielectric electronic-phononic-photonic attribute of nanoscaled silicon. Surf. Sci. Rep. **68**(3–4), 418–445 (2013)
15. C.Q. Sun, Size dependence of nanostructures: Impact of bond order deficiency. Prog. Solid State Chem. **35**(1), 1–159 (2007)
16. X.X. Yang, J.W. Li, Z.F. Zhou, Y. Wang, L.W. Yang, W.T. Zheng, C.Q. Sun, Raman spectroscopic determination of the length, strength, compressibility, Debye temperature, elasticity, and force constant of the C–C bond in graphene. Nanoscale **4**(2), 502–510 (2012)
17. X.J. Liu, L.K. Pan, Z. Sun, Y.M. Chen, X.X. Yang, L.W. Yang, Z.F. Zhou, C.Q. Sun, Strain engineering of the elasticity and the Raman shift of nanostructured TiO(2). J. Appl. Phys. **110**(4), 044322 (2011)
18. C. Kittel, *Intrduction to Solid State Physics*, 8th edn. (Willey, New York, 2005)
19. C.Q. Sun, H.L. Bai, B.K. Tay, S. Li, E.Y. Jiang, Dimension, strength, and chemical and thermal stability of a single C–C bond in carbon nanotubes. J. Phys. Chem. B **107**(31), 7544–7546 (2003)
20. E.W. Wong, P.E. Sheehan, C.M. Lieber, Nanobeam Mechanics: elasticity, strength, and toughness of nanorods and nanotubes. Science **277**(5334), 1971–1975 (1997)
21. M.R. Falvo, G.J. Clary, R.M. Taylor, V. Chi, F.P. Brooks Jr., S. Washburn, R. Superfine, Bending and buckling of carbon nanotubes under large strain. Nature **389**(6651), 582–584 (1997)

22. A. Bai, F. Seiji, Y. Kiyoshi, Y. Masamichi, Surface superstructure of carbon nanotubes on highly oriented pyrolytic graphite annealed at elevated temperatures. Jpn. J. Appl. Phys. **37**(6B), 3809–3811 (1998)

23. T. Balasubramanian, J.N. Andersen, L. Wallden, Surface-bulk core-level splitting in graphite. Phys. Rev. B **64**, 205420 (2001)

24. W.T. Zheng, C.Q. Sun, Underneath the fascinations of carbon nanotubes and graphene nanoribbons. Energy Environ. Sci. **4**(3), 627–655 (2011)

25. X. Zhang, Y.G. Nie, W.T. Zheng, J.L. Kuo, C.Q. Sun, Discriminative generation and hydrogen modulation of the Dirac-Fermi polarons at graphene edges and atomic vacancies. Carbon **49**(11), 3615–3621 (2011)

26. C.O. Girit, J.C. Meyer, R. Erni, M.D. Rossell, C. Kisielowski, L. Yang, C.H. Park, M.F. Crommie, M.L. Cohen, S.G. Louie, A. Zettl, Graphene at the edge: stability and dynamics. Science **323**(5922), 1705–1708 (2009)

27. J.W. Li, L.W. Yang, Z.F. Zhou, X.J. Liu, G.F. Xie, Y. Pan, C.Q. Sun, Mechanically stiffened and thermally softened Raman modes of ZnO crystal. J. Phys. Chem. B **114**(4), 1648–1651 (2010)

28. M.X. Gu, Y.C. Zhou, C.Q. Sun, Local bond average for the thermally induced lattice expansion. J. Phys. Chem. B **112**(27), 7992–7995 (2008)

29. E. Grüneisen, The state of a body. Handb. Phys. **10**, 1–52. NASA translation RE2-18-59 W

30. J.L. Hu, W.P. Cai, C.C. Li, Y.J. Gan, L. Chen, In situ x-ray diffraction study of the thermal expansion of silver nanoparticles in ambient air and vacuum. Appl. Phys. Lett. **86**(15), 151915 (2005)

31. L. Li, Y. Zhang, Y.W. Yang, X.H. Huang, G.H. Li, L.D. Zhang, Diameter-depended thermal expansion properties of Bi nanowire arrays. Appl. Phys. Lett. **87**(3), 031912 (2005)

32. T. Comaschi, A. Balerna, S. Mobilio, Temperature dependence of the structural parameters of gold nanoparticles investigated with EXAFS. Phys. Rev. B **77**(7), 075432 (2008)

33. G.A. Slack, S.F. Bartram, Thermal expansion of some diamondlike crystals. J. Appl. Phys. **46**(1), 89–98 (1975)

34. R.R. Reeber, K. Wang, Lattice parameters and thermal expansion of GaN. J. Mater. Res. **15**(1), 40–44 (2000)

35. Q.H. Tang, T.C. Wang, B.S. Shang, F. Liu, Thermodynamic properties and constitutive relations of crystals at finite temperature. Sci. China-Phys. Mech. Astron, G **55**, 933 (2012)

36. Y.J. Su, H. Wei, R.G. Gao, Z. Yang, J. Zhang, Z.H. Zhong, Y.F. Zhang, Exceptional negative thermal expansion and viscoelastic properties of graphene oxide paper. Carbon **50**(8), 2804–2809 (2012)

37. C. Martinek, F.A. Hummel, Linear thermal expansion of 3 tungstates. J. Am. Ceram. Soc. **51**(4), 227–228 (1968)

38. T.A. Mary, J.S.O. Evans, T. Vogt, A.W. Sleight, Negative thermal expansion from 0.3 to 1050 Kelvin in ZrW_2O_8. Science, **272**(5258), 90–92 (1996)

39. X. Zhang, J.L. Kuo, M.X. Gu, X.F. Fan, P. Bai, Q.G. Song, C.Q. Sun, Local structure relaxation, quantum trap depression, and valence charge polarization induced by the shorter-and-stronger bonds between under-coordinated atoms in gold nanostructures. Nanoscale **2**(3), 412–417 (2010)

40. C.Q. Sun, Y. Sun, The attribute of water: single notion, multiple myths. Springer Ser. Chem. Phys. **113**, 494pp (2016) (Heidelberg: Springer-Verlag)

41. R.J. Bruls, H.T. Hintzen, G. de With, R. Metselaar, J.C. van Miltenburg, The temperature dependence of the Gruneisen parameters of MgSiN2, AlN and beta-Si_3N_4. J. Phys. Chem. Solids **62**(4), 783–792 (2001)

42. K.G. Lyon, G.L. Salinger, C.A. Swenson, G.K. White, Linear thermal-expansion measurements on silicon from 6 to 340 K. J. Appl. Phys. **48**(3), 865–868 (1977)

43. R.B. Roberts, Thermal-expansion reference data—silicon 300-850 K. J. Phys. D-Appl. Phys. **14**(10), L163–L166 (1981)

44. H.P. Singh, Determination of thermal expansion of germanium rhodium and iridium by x-rays. Acta Crystallogr. Sect. A, A **24**, 469–471 (1968)

45. T. Sato, K. Ohashi, T. Sudoh, K. Haruna, H. Maeta, Thermal expansion of a high purity synthetic diamond single crystal at low temperatures. Phys. Rev. B **65**(9), 092102 (2002)

46. C. Giles, C. Adriano, A.F. Lubambo, C. Cusatis, I. Mazzaro, M.G. Honnicke, Diamond thermal expansion measurement using transmitted X-ray back-diffraction. J. Synchrotron Radiat. **12**, 349–353 (2005)

47. K. Haruna, H. Maeta, K. Ohashi, T. Koike, Thermal-expansion coefficient of synthetic diamond single-crystal at low-temperatures. Jpn. J. Appl. Phys. Part 1-Regul. Pap. Short Notes & Rev. Pap. **31**(8), 2527–2529 (1992)

48. http://www.infoplease.com/periodictable.php

49. B. Johansson, N. Martensson, Core-level binding-energy shifts for the metallic elements. Phys. Rev. B **21**(10), 4427–4457 (1980)

50. Z.W. Chen, C.Q. Sun, Y.C. Zhou, O.Y. Gang, Size dependence of the pressure-induced phase transition in nanocrystals. J. Phys. Chem. C **112**(7), 2423–2427 (2008)

51. F. Birch, Finite elastic strain of cubic crystals. Phys. Rev. **71**(11), 809–824 (1947)

52. G. Ouyang, C.Q. Sun, W.G. Zhu, Pressure-stiffened Raman phonons in group III nitrides: a local bond average approach. J. Phys. Chem. B **112**(16), 5027–5031 (2008)

53. M. Pravica, Z. Quine, E. Romano, X-ray diffraction study of elemental thulium at pressures up to 86 GPa. Phys. Rev. B **74**(10), 104107 (2006)

54. B. Chen, D. Penwell, L.R. Benedetti, R. Jeanloz, M.B. Kruger, Particle-size effect on the compressibility of nanocrystalline alumina. Phys. Rev. B **66**(14), 144101 (2002)

55. F. Ding, H. Ji, Y. Chen, A. Herklotz, K. Dorr, Y. Mei, A. Rastelli, O.G. Schmidt, Stretchable graphene: a close look at fundamental parameters through biaxial straining. Nano Lett. **10**(9), 3453–3458 (2010)

56. C. Rice, R. Young, R. Zan, U. Bangert, D. Wolverson, T. Georgiou, R. Jalil, K. Novoselov, Raman-scattering measurements and first-principles calculations of strain-induced phonon shifts in monolayer MoS_2. Phys. Rev. B **87**(8), 081307 (2013)

57. X.X. Yang, Y. Wang, J.W. Li, W.H. Liao, Y.H. Liu, C.Q. Sun, Graphene phonon softening and splitting by directional straining. Appl. Phys. Lett. **107**(20), 203105 (2015)

58. C.Q. Sun, Atomic scale purification of electron spectroscopic information (US 2017 patent No. 9,625,397B2). 2017: United States

59. Y. Gong, Y. Zhou, C. Sun, Phonon spectrometrics of the hydrogen bond (O:H–O) segmental length and energy relaxation under excitation, B.O. Intelligence, Editor. 2018: China

60. A. Wong, L. Shi, R. Auchettl, D. McNaughton, D.R. Appadoo, E.G. Robertson, Heavy snow: IR spectroscopy of isotope mixed crystalline water ice. Phys. Chem. Chem. Phys. **18**(6), 4978–4993 (2016)

61. L. K. Pan, C. Q. Sun, Coordination imperfection enhanced electron-phonon interaction. J. Appl. Phys. **95**(7), 3819–3821 (2004)

62. L. Pan, S. Xu, X. Liu, W. Qin, Z. Sun, W. Zheng, C. Q. Sun, Skin dominance of the dielectric–electronic–phononic–photonic attribute of nanoscaled silicon. Surf. Sci. Rep. **68**(3–4), 418–445 (2013)

Chapter 22
Layered Structures

Abstract Multifield perturbation on the phonon-frequency shift of the two-dimensional black phosphorus (BP), graphene, and MX_2 (M=Mo, W; X=S, Se) semiconductors features directly the bond order-length-energy (z, d, E). Theoretical reproduction of the spectroscopic measurement has turned out quantitative information of the bond length, bond energy, bond nature index, binding energy density, atomic cohesive energy, bond force constant, Debye temperature, and elastic modulus for the layered structures, and discrimination of the bonding origins for the phonon frequency shift. Progress exemplifies not only the essentiality of the theoretical framework but also the immense power of the theory-driven multifield phonon spectrometrics in probing the atomistic, dynamics, and quantitative information on the local bonding dynamics under perturbation.

Highlights

- Atomic-undercoordination induced phonon relaxation imprints the bond order-length-energy.
- Compression stiffened phonons yield the binding energy density and elastic modulus.
- Phonon Debye thermal decay turns out Debye temperature and atomic cohesive energy.
- Stretching induced phonon band softening and splitting turns out single bond force constant.

22.1 Wonders of the 2D Structures

Two-dimensional (2-D) substance such as black phosphorus (BP), graphene nanorribbon (GNR), and transition metal dichalcogenides MX_2 (M=Mo, W; X=S, Se) have emerged as an amazing group of materials with high tunability of chemical and physical properties that do not demonstrate by their bulk parents [1–3]. For example, the few-layered BP shows high carrier mobility, up to 10^3 cm^2 V^{-1} s^{-1}, and high current on/off ratios, up to 10^5 Hz at room temperature. Unrolling the single-walled carbon nanotubes (SWCNTs) triggers many intriguing properties that cannot be seen from the CNT or a large graphene sheet [4, 5]. In addition to the observed edge Dirac fermion states [6, 7] with ultrahigh electrical and thermal conductivity [8], and unexpected magnetization [9, 10], the bandgap of a GNR expands monotonically with the inverse of ribbon width [5, 11, 12].

One can turn the indirect bandgap of the bulk MoS$_2$ from 1.2 eV to the direct bandgap of 1.8 eV by simply reducing the number of layers from infinitely large into single [13, 14]. The indirect bandgap also degenerates into a direct for the few-layered MoSe$_2$ [15]. Bandgap transition from indirect to direct depends not only on the number of layers but also on the material and the operation conditions [16]. Kelvin and conductive atomic force microscopy (AFM) revealed that the surface potential well goes deeper linearly with thickness from the bulk value of -7.2 to -427 mV when the bulk MoS$_2$ turns into monolayer [17]. These entities make the 2-D semiconductors an appealing type of materials for applications as flexible, miniaturized, and wearable electronic devices and energy management such as photovoltaic cells [18, 19], field effect transistors [20–22], light-emitting diodes [23, 24], and photodetectors [25–27].

22.2 Orbital Hybridization and Structure Configuration

Figure 22.1 illustrates the bond configuration for the monolayer BP, GNR, and MX$_2$. The C, P, S and Se atoms undergo the sp^3-orbital hybridization with three (for C $2s^2p^2$ and P $3s^2p^3$), and two (for S $3s^2p^4$ and Se $4s^2p^4$) bonds to their neighbors. Each of the sp^2-orbital hybridized C atom bonds covalently to its three neighbors in a plane and leaves one unpaired electron for the π bonding. Each P atom has three tetrahedrally directed bonds to its three neighbors and one dangling lone pair. In the same group of oxygen, S and Se share the same s^2p^4 electronic configuration in the 3rd and 4th outermost electronic shells. The S and Se atoms form each a tetrahedron

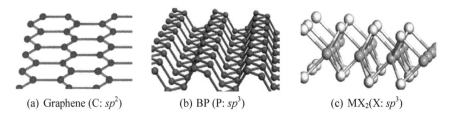

(a) Graphene (C: sp^2)   (b) BP (P: sp^3)   (c) MX$_2$(X: sp^3)

Fig. 22.1 Bond configuration for the **a** GNR, **b** BP, and **c** MX$_2$. The C, P, S and Se atoms undergo the sp orbital hybridization with three (C $2s^2p^2$; P $3s^2p^3$), and two (S $3s^2p^4$, Se $4s^2p^4$) bonds to their neighbors, which generate (**a**) one unpaired π electron and (**b**) one and (**c**) two electron lone pairs pointing to the open end of the sandwiched structures. The main layer consists of **a** one, **b** two and **c** three sublayers interacting with the next main layer through van der Waals forces

with two bonds to the sandwiched M atoms and two electron lone pairs exposing outwardly to the terminals of the sandwiched main layer. The extent of the orbital hybridization for the P, S and Se is less obvious than nitrogen and oxygen because of the orbital order occupancy of the valance electrons but their bond configuration does show the same trend [28].

The main layers of these materials are composed of (a) one, (b) two and (c) three sublayers interacting with their next main layers through van der Waals interaction, which makes the mechanical filtration of these layered structures possible. For the MX_2 honeycomb-like structure, they can be viewed as a positively charged plane of the transition-metal M^{4+} cations sandwiched between two planes of negatively charged chalcogen X^{2-} anions [29]. Each of the sandwiched M^{4+} cation bonds to its four neighboring X^{2-} anions. The MX_2 layer performs like a giant atom with the positive M^{4+} core and the negative X^{2-} shells. The MX_2 main layers interact one another through the short-range $X^{2-}:\Leftrightarrow:X^{2-}$ repulsion (the same as the $O^{2-}:\Leftrightarrow:O^{2-}$ super hydrogen-bond in basic solutions [30]) and the slightly longer $M^{4+} \sim X^{2-}$ Coulomb attraction. Such a set of interactions stabilizes the layered structures.

Figure 22.2 illustrates the atomic dislocations in the Raman-active and the IR-active vibration modes [31–35]:

(1) The transverse optical (TO) mode and the longitudinal optical (LO) Raman modes shift in frequency with the number-of-layer of the 2-D materials.
(2) The E_{2g} TO mode for the WX_2 and TiO_2 and the G mode for graphene undergo blueshift while the A_{1g} LO mode and the D mode shift to lower frequency as the feature size is reduced.

Furthermore, ribbons of these layered structures have zigzag- and armchair-edge. Most importantly, bonds between undercoordinated atoms in the layered structures are shorter and stronger and the bonds at edges are even shorter and stronger than they are in the bulk trunks. The bond contraction and the associated core electron quantum entrapment and valence electron polarization govern the performance of these undercoordinated systems [36, 37].

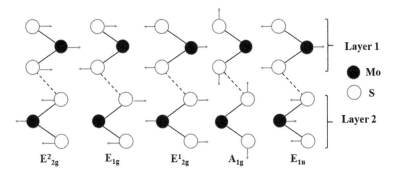

Fig. 22.2 Atomic displacements of the four Raman-active modes (subscripted with g) and one IR-active mode (E_{1u}) in the unit cell of the bulk MoS_2 crystal as viewed along the [1000] direction. Reprinted with copyright permission from [35]

22.2.1   *Phonon Frequency Tunability*

A huge database has been established toward the phonon spectroscopy/microscopy of the layered 2-D structures and their nanoribbons subjecting to various stimulation. The stimuli include the number-of-layer (n) [38–42], uniaxial strain (ε) [43–45], mechanical compression (P) [46], thermal excitation (T) [47, 48], defect density and location [49], substrate interaction [50, 51], hydrogenation or passivation [52, 53], impurity or doping [54], incident photon polarization [55], edge conditioning [56, 57], etc.

Graphene shows two Raman active modes as do its allotropes. As the first-order approximation, the G band (~1580 cm^{-1}) was suggested to arise from the in-plane vibration of the sp^2 bonding network [58]. The 2D band (2680 cm^{-1}) was thought to be a second-order process of double resonant scattering [59]. In the presence of undercoordinated defect or edge atoms, the defect-induced D band appears at frequencies around 1345 cm^{-1} with intensity being varied with edge conditions compared with the D band of bulk diamond featured at 1331 cm^{-1} [56, 57]. The monolayer BP shows the A_g^1, A_g^2 and B_{2g} and the MX$_2$ shows the A_{1g} and E_{2g}^1 Raman active modes.

The phonon behavior of the layered BP, GNR, and MX$_2$ [17, 60, 61] follows the same trend but by different amounts under the same excitation. The D/2D bands of graphene [40, 62] and the A_{1g} mode of the MoS$_2$ and the WS$_2$ [35] undergo a redshift when the number-of-layer is reduced. For instance, under 514.5 nm light radiation, the D mode shifts from 1367 to 1344 cm^{-1} and the 2D mode changes from 2720 to 2680 cm^{-1} when the bulk graphite evolves into the monolayer graphene [41]. In contrast, a blueshift happens to the G mode that shifts from 1582 to 1587 cm^{-1} when the n is reduced from twenty to one [40, 41]. The B_{2g} and A_{2g} modes for BP [63, 64] and the E_{2g}^1 mode for MoS$_2$ and WS$_2$ [35] undergo a blueshift as well when their number-of-layer is reduced. When the n is increased from a few to multiple, the Raman peaks turn from the dominance of the monolayer component to the dominance of the bulk component [40]. It is even amazing that the spectral intensity of the D band is one order higher at the armchair edge than it is at the zigzag edge of the GNR [56, 57].

Introducing the uniaxial strain to the 2-D substances by bending the layered structures on the elastic substrates can lower the work function to improve their carrier mobilities and raise their photoelectron emissivity [25, 26, 65]. The uniaxial tensile strain softens and splits the phonon bands of graphene [43] and MX$_2$ and narrows the bandgap of MX$_2$ [66]. The extent of the phonon band splitting depends on the strain and its relative direction between the strain and a specific bond of the layered structures [45, 67]. Under compressive strain, the Raman shift extrapolates the trend of tension in an opposite direction [68].

Mechanical compression depresses the V-P profile but raises all the phonon frequencies [46, 69–71]. Contrastingly, heating softens the Raman phonon to lower frequencies [72]. Heating from ultra-low to a temperature well-above room-temperature softens the A_g^1, A_g^2 and B_{2g} modes of BP [70, 71] and the A_{1g} and E_{2g}^1 phonons of

WS_2 and MoS_2 as well in the Debye thermal decay fashion [60, 73]. Heating from 0 to 300 K also reduces the excitonic transition energy of the WSe_2 in a similar manner [74]. Both the bandgap and phonon frequency initially drop at low temperature slowly and then linearly at high temperatures, being the form of the difference between a constant and the Debye specific heat integration [75]. The effects of mechanical compression and thermal excitation hold universal to both the layered and the bulk materials.

Current understanding of the multifield effect on the phonon frequency shift of the layered substance is focused on two perspectives. One is the continuum bond oscillating dynamics correlating the phonon frequency evolution to the stimuli in a plain function, like the Gibbs free energy. The other is Raman photon scattering and phonon decaying dynamics with involvement of the electronic contribution and the geometry of the Brillouin zone. The effect of T on the Raman shift is ascribed as optical phonon decaying into multiple acoustic components [76–80] and the pressure effect is described in terms of Grüneisen notion—the slope of the Raman shift with respect to pressure, $\partial\omega/\partial P$ or $\partial(Ln\omega)/\partial(LnP)$ [43]. A "two-phonon double-resonance scattering" is suggested to govern the strain effect on phonon band splitting and softening [67]. The number-of-layer effect on the Raman shift follows hypothetically the inverse number-of-layer $n^{-\gamma}$ with γ being a tunable parameter [31, 81].

The afore-mentioned empirical approaches could fit to the measurements independently despite multiple freely adjustable parameters involved. However, consistent insight into the "external stimulus—bond relaxation—phonon frequency–macroscopic property" correlation and finding means for detection remain yet to be explored. For instance, mechanisms govern the opposite trends of the phonon frequency shifts under the number-of-layer reduction and the edge-discriminated photon reflectivity is beyond the scope of available premises.

One may be concerned with what dictates the chemical and physical properties of the layered structures. According to Pauling [82], the nature of the chemical bond bridges the structure and property of a crystal and a molecule. Therefore, bond formation and relaxation and the associated energetics, localization, entrapment, and polarization of electrons mediate the macroscopic performance of substance accordingly [28]. Compared with the coordination-resolved electron spectrometrics [37] that probes the local bond length and energy and the electronic dynamics under high vacuum, the Raman spectrometry features are very sensitive to various stimuli in amiable manners, which offers rich information on the local bonding dynamics under external excitation.

Advancing the phonon spectroscopy to spectrometrics aims at quantitative information that the spectroscopy can hardly offer. The phonon and electron spectrometrics complement each other providing comprehensive information on the performance of bonds and electrons of a system under perturbation. The key challenge is how to correlate and formulate the spectral feature to the bonding identities and the way of responding to excitation.

The aim of this section is to show that a combination of the BOLS-LBA [36, 75] and the phonon spectrometrics could reconcile the (n, ε, T, P) effect on the Raman shifts with clarification, formulation, and quantification of the bond and phonon relaxation dynamics.

22.3 Numerical Reproduction of Phonon Relaxation

22.3.1 Number-of-Layer Dependence

22.3.1.1 Formulation

For the z dependence of the phonon frequency shift follows,

$$D_L(z) = \frac{\Delta\omega(z)}{\Delta\omega(z_b)} = \frac{\omega(z) - \omega(1)}{\omega(z_b) - \omega(1)} = \frac{d_b}{d_z}\left(\frac{E_z}{E_b}\right)^{1/2}\begin{cases} \frac{z}{z_b} \ (LO) \\ 1 \ (TO) \end{cases}$$

where,

$$\frac{d_b}{d_z}\left(\frac{E_z}{E_b}\right) = \left(\frac{C_z}{C_b}\right)^{-(1+m/2)} \tag{22.1}$$

This formulation clarifies that the direct z > 1 involvement derives the redshift and the other blueshift of phonons. Numerical matching to the experimental result yields the $\omega(1)$ and m with $\omega(z_b)$ and $\omega(z)$ as input for calculations, as demonstrated in the following subsections.

22.3.1.2 Graphene and Black Phosphorus

Bond-order variation and the pattern change of *sp*-orbit hybridization make carbon allotropes a group of materials varying from diamond, graphite, C_{60}, nanotube (CNT), nanobud (CNB), graphene, and GNRs with properties that are amazingly different. For instances, graphite is an opaque conductor, but diamond is an insulator transparent to light of all wavelengths; the former shares nonbonding unpaired (or π-bond) electrons due to sp^2-orbital hybridization compared with the latter of the ideally sp^3-orbital hybridization. CNT and graphene demonstrate anomalous properties including high tensile strength, electrical conductivity, ductility and thermal conductivity. GNR performs, however, quite differently from CNT or the infinitely large graphene because of the involvement of the two-coordinated edge atoms and the associated Dirac Fermions [83–86]. Raman phonons relax with the applied stimuli such as the temperature and pressure as well as the allotropic coordination environment [87, 88].

The database as a function of the number-of-layer [40, 62] of graphene and BP [64] enabled the verification of the framework for thickness dependence of the bond oscillating dynamics. In numerical calculations for graphene, the known effective bond length $d_g = 0.142$ nm and $z_g = 5.335$ for the bulk graphite, $z = 3$ for the single layer graphene, and $m = 2.56$ for carbon were taken as input. Errors in measurements render only the accuracy of the derivatives but not the nature and the trends of observations.

Taking $z_g = 5.335$ for the bulk graphite as a reference, one can obtain the reference $\omega(1)$ and the z-dependent vibration frequency $\omega(z)$ [88]. With the known 2D peak shifting from 2720 to 2680 cm^{-1} and the D peak from 1367 to 1344 cm^{-1} when the graphite (z_g) turns to be the monolayer ($z = 3$) graphene [40, 62, 89], and the G mode shifting from 1582 to 1587 cm^{-1} [40, 41], one can calibrate the z-dependent relative shift of these vibration modes. For the monolayer graphene ($z = 3$, $m = 2.56$),

$$C_x\left(z, z_g\right) = \frac{\omega_x(z) - \omega_x(1)}{\omega_x(z_g) - \omega_x(1)} = \left(\frac{C_z}{C_{z_g}}\right)^{-(m/2+1)} \begin{cases} \frac{z}{z_g} & (D, 2D) \\ 1 & (G) \end{cases}$$

$$= \left(\frac{0.8147}{0.9220}\right)^{-2.28} \times \begin{cases} \frac{3.0}{5.335} = 0.7456 & (D, 2D;\ z = 3) \\ 1 \quad\ \ = 1.3260 & (G;\ z = 1) \end{cases}$$

And the reference frequency,

$$\omega_x(1) = \frac{\omega_x(z) = \omega_x(z_g)C_x(z, z_g)}{1 - C_x(z, z_g)} = \frac{\omega_x(3) - \omega_x(z_g)C_x(3, z_g)}{1 - C_x(3, z_g)}$$

$$= \begin{cases} 1276.8\ (D) \\ 1566.7\ (G) \quad (\text{cm}^{-1}) \\ 2562.6\ (2D) \end{cases}$$

The z dependent phonon frequency for graphene,

$$\omega(z) = \omega(1) + \left[\omega\left(z_g\right) - \omega(1)\right] \times D_L(z)$$
$$= \begin{cases} 1276.8 + 90.2 \times D_D(z) & (D) \\ 2563.6 + 157.4 \times D_{2D}(z) & (2D) \\ 1566.7 + 16.0 \times D_G(z) & (G) \end{cases} \qquad (22.2)$$

Figure 22.3 shows the BOLS-LBA reproduction of the z-dependent Raman frequencies of the D/2D modes [40, 43, 68] and the G mode. Likewise, inset b is the z-dependence of the BP A_g^1 mode [64]. The DPS for the layered BP revealed that reduction of the number-of-layer from multiple to single transits the A_{2g} phonon from 466 to 470 cm^{-1} and the B_{2g} from 438 to 441 cm^{-1}. The n-resolved phonon positive shift indicates that both the A_{2g} and B_{2g} modes arise from oscillation of an invariant number of bonds [90]. The $\omega(1)$ for the BP was determined to be $A_g^1(360.20$ cm$^{-1})$, $B_{2g}(435.00$ cm$^{-1})$, and $A_g^2(462.30$ cm$^{-1})$.

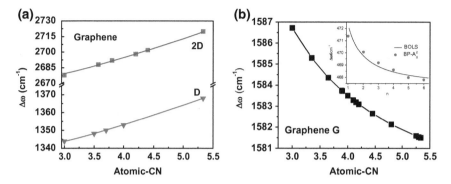

Fig. 22.3 BOLS-LBA reproduction of the z-resolved Raman shift for **a** the D/2D modes and **b** the G mode for the layered graphene [40, 41, 62] and A_g mode for the layered BP (inset f) [64] with derivative of information given in Table 22.1. The scattered data counts the number of layers. Reprinted with copyright permission from [90]

22.3.1.3 (W, Mo)(S, Se)$_2$

The DPS in Fig. 22.4 resolves the n-induced phonon frequency relaxation of (W, Mo)S$_2$ [91]. Phonons transit from the bulk component (DPS valleys) to the under-coordinated edge component (DPS peaks) of the layered structures. The frequency shifts of the two vibration modes suggests that the collective oscillation of bonds to the z-neighbors govern the A_{1g} redshift while the single bond vibration drives the

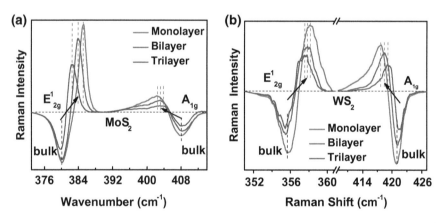

Fig. 22.4 Number-of-layer resolved Raman DPS for **a** MoS$_2$ [35, 92] and **b** WS$_2$ [93]. Phonons transit from the bulk component (DPS valleys at 380 and 408 cm^{-1} for the E$_{2g}$ and the A$_{1g}$ of MoS$_2$ and at 356 and 421 cm^{-1} for the WS$_2$) to the edge (n-resolved DPS upward-shift peak for the E$_{2g}$ and downward for the A$_{1g}$) of the layered structures. The collective oscillation of z-neighbors govern the A$_{1g}$ redshift while the single bond vibration drives the E$^1_{2g}$ blueshift. Reprinted with copyright permission from [91]

E_{2g}^1 blueshift, which are the same to the D/2D mode and the G mode of graphene, respectively.

With the known $\omega(n = \infty)$ for the bulk and $\omega(n = 1)$ as input parameters, one can derive information show in Table 22.1 for the layered MX_2 by repeating the same calculation iteration of graphene [91]. Figure 22.5 shows the BOLS-LBA matching to the phonon frequency shift of the layered MX_2 with the derive m and $\omega_L(1)$ tabulated in Table 22.1. Theoretical reproduction of the n-dependence of Raman shift verifies that the dimer bond interaction dictates the E_{2g}^1 mode blueshift but the collective interaction of an atom with its z-neighbors governs the A_{1g} mode redshift.

22.3.1.4   Atomic-CN Versus the Number-of-Layer

The scattered data in Figs. 22.3 and 22.5 correspond to different numbers of layers while the lateral axis is the effective atomic CN, which gives rise to the z − n correlation as shown in Fig. 22.6. When the n is greater than 6, the z saturates at the bulk value of 5.335 (z = 2.55 + 0.45n) for graphite and the bulk value of 12 (z = 2.4 + 1.6n) for the layered MX_2. Consistency between predictions and measurements of the z-dependent Raman shifts and the z-n transformation function evidence the essentiality and appropriateness of the BOLS framework for the lattice vibration in these 2D structures.

22.3.2   Strain-Induced Phonon Softening and Band Splitting

The data of tensile strain for the monolayer graphene [68] and MoS_2 [94] enabled the derived information of the average-bond force constant and the relative direction between the strain and a specific bond. The strain-effect is given as [95],

$$\frac{\omega(z, \varepsilon) - \omega(1, 0)}{\omega(z, 0) - \omega(1, 0)} = \frac{d(0)}{d(\varepsilon)} \left(\frac{E(\varepsilon)}{E(0)} \right)^{1/2}$$

$$= \frac{\left(1 - d^2 \int_0^\varepsilon \kappa \varepsilon d\varepsilon / E_{z1} \right)^{1/2}}{1 + \varepsilon}$$

$$\cong \frac{\left[1 - \kappa' \left(\lambda \varepsilon' \right)^2 \right]^{1/2}}{1 + \lambda \varepsilon'} \tag{22.3}$$

With $\kappa' = \kappa d_z^2/(2E_z) = $ constant

To feature the orientation mismatch between a bond and the uni-axial strain, one can introduce a strain coefficient λ bounded by $0 \leq \lambda \leq 1$. The $\varepsilon = \lambda \varepsilon'$ is for a bond that is not along the applied strain. The $\kappa = 2E_z \kappa' d_z^{-2}$ being constant at limited strain is the effective force constant of all bonds. One can only measure the average strain of the entire specimen other than that of a certain bond with force constant κ_0. The

Table 22.1 Quantitative information derived from reproducing the (n, ε, T, P) dependent Raman shift of the layered BP, GNR, and MX$_2$ [32, 90, 96, 100]

Stimuli	Quantity		Graphene	BP	MoS$_2$	WS$_2$	MoSe$_2$	WSe$_2$
n	ω(1) (cm^{-1})		1276.8 (D) 2562.6 (2D) 1566.7 (G)	360.20 (A$_{1g}$) 435.00 (B$_{2g}$) 462.30 (A$_{2g}$)	399.65 (A$_{1g}$) 377.03 (E$^1_{2g}$)	416.61 (A$_{1g}$) 352.26 (E$^1_{2g}$)	237.65 (A$_{1g}$) 265.12 (E$^1_{2g}$)	254.84 (A$_{1g}$) 246.42 (E$^1_{2g}$)
	Bond nature index m		2.56	4.60	4.68	2.42	2.11	1.72
	Bond length (nm)	Monolayer	0.125	0.224/0.223 [146]	0.249/0.241 [147]	0.243/0.242 [147]	0.250/0.254 [147]	0.264/0.255 [147]
		Bulk	0.154 (Diamond)	0.255	0.284	0.276	0.286	0.301
	Bond energy (eV)	Monolayer	1.04	0.527	0.338	0.538	0.575	0.613
		Bulk	0.615 [148] (Diamond)	0.286 [148]	0.181	0.390	0.435	0.487
ε	Force constant κ (Nm^{-1})		6.28	–	2.56	–	–	–
T	Debye θ$_D$ (K)		540	466/400 [149]	250	530	193	170
	Atomic E$_{coh}$ (eV/atom)		3.11	2.11	1.35	2.15	2.30	2.45
	Thermal expansion α(10^{-6}K^{-1})		9.00	22.0 [150]	1.90/8.65	10.10/-	7.24/12.9	6.80/10.6 (α$_a$/α$_c$)
P	Energy density E$_{den}$ (eV/nm^3)		320	9.46	21.9	29.6	–	–
	β/β′ (10^{-3}GPa^{-1}/GPa^{-2})		1.15/0.0763	0.32/2.20 [151]	13.30/0.96	25.60/0.88	–	–
	B$_0$/B$'_0$ (GPa/-)		690/5 (704/1) [152]	84.10/4.69 [153]	47.7/10.6	63.00/6.50 [110, 154]	–	–

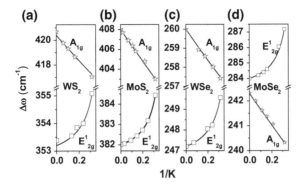

Fig. 22.5 BOLS-LBA reproduction of the z-resolved Raman shift for **a–d** Mo(S$_2$, Se$_2$) and W(S$_2$, Se$_2$) with derivative of information given in Table 22.1. The scattered data corresponds to the number of layers. Reprinted with copyright permission from [90]

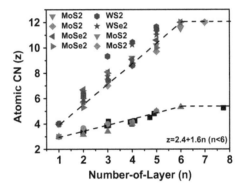

Fig. 22.6 The atomic CN correlates to the number-of-layer linearly up to n = 6. For n > 6, the z saturates to the bulk values for graphite (z = 5.335) and for the fcc standard (z = 12). Reprinted with copyright permission from [32, 91]

λ corresponds to the actual strains of the orientated bonds in a basic unit. The strain can be along ($\lambda = 1$) or perpendicular ($\lambda = 0$) or random ($0 < \lambda < 1$) to a certain bond in the specimen. The bond along the strain is subject to a maximal extension and its phonon frequency shifts most; the bond perpendicular to the strain is subject to zero extension and keeps its phonon frequency, as illustrated in Fig. 22.7a inset. It is therefore not surprising that the mechanical strain splits the phonon band and the extent of splitting depends on the angle between the bond and the strain [32].

Combining Eqs. (22.2) and (22.3) yields the joint z and ε effect on the Raman shifts for GNR:

$$\omega(z, \varepsilon) = \omega(1, 0) + [\omega(z_b, 0) - \omega(1, 0)]D_L(z) \times \frac{\left[1 - \kappa'\left(\lambda\varepsilon'\right)^2\right]^{1/2}}{1 + \lambda\varepsilon'}$$

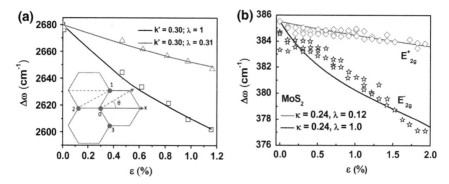

Fig. 22.7 BOLS-LBA reproduction of strain effect on the **a** 2D mode of graphene [68] and **b** the E_{2g} mode of MoS$_2$ [94]. Inset **a** illustrates the geometric relation between the C_{3v} bonds to the C_{2v} uni-axial strain. One extreme situation at $\theta = 0°$ (along a C–C bond), $\varepsilon_1 = \varepsilon_3 = \lambda\varepsilon_2 < \varepsilon_2$ and the other at $\theta = 30°$ (perpendicular to a C–C bond), $\varepsilon_1 = \varepsilon_2 > \varepsilon_3 \sim 0$. Reprinted with copyright permission from [96]

$$\left.\begin{array}{c} 1276.8 \\ = 2562.6 \\ 1566.7 \end{array}\right\} + D_L(z) \times \frac{\left[1 - \kappa'\left(\lambda\varepsilon'\right)^2\right]^{1/2}}{1 + \lambda\varepsilon'} \times \left\{\begin{array}{ll} 90.2 & (D) \\ 157.4 & (2D) \\ 16.0 & (G) \end{array}\right. \left(\text{cm}^{-1}\right)$$

$$(22.4)$$

Figure 22.7 shows reproduction of the tensile strain induced redshifting and band splitting for (a) the 2D mode of graphene [67] and the E_{2g} mode for MoS$_2$ [94]. Reproduction of the measurements turned out the reduced force constant, $\kappa' = \kappa d_z^2/(2E_z) = 0.30$, corresponding to $\kappa = 6.283$ N/m for a C–C bond in graphene ($z = 3$). The compressive [43, 68] strain only extrapolates the curve to opposite direction of the tensile strain of graphene [68]. Inset a illustrates the geometric relation between the C–C bonds denoted 1, 2, and 3, to the uni-axial tensile strain. The $\lambda = 1.0$ for the lower branch indicates that the strain is along a certain C–C bond of the GNR and the M–S bond of the MoS$_2$. The upper branches of $\lambda = 0.31$ and 0.12 are for other oriented bond in graphene and MoS$_2$, respectively.

For the GNR instance, the strain is along bond 2, $\theta = 0°$ $\lambda = 1$. The strain $\varepsilon_1 = \varepsilon_3 = \lambda\varepsilon_2 < \varepsilon_2$, the ε_2 is the maximum; at $\theta = 30°$, the strain is perpendicular to bond 3, $\varepsilon_1 = \varepsilon_2 > \varepsilon_3 \sim 0$. The hexagonal bond configuration allows one to focus on the angle ranging from 0° to 30° between a specific bond and the strain. There should be a branch remaining the original frequency as it is subject to $\varepsilon_3 \sim 0$ at $\theta = 30°$. The $\lambda = 0.31$ for GNR and 0.12 for MoS$_2$ and the $\lambda = 1.0$ for the lower branches in both cases mean that one C–C bond and one Mo–S bond projected to the surface are along the strain. The bonds labeled 1 and 3 shown are elongated by 31% of bond 2. The other Mo–S bond projection is by 12%. The Mo–S bonds projecting along the layer. As the λ changes with the relative direction between the strain and the bond, any extent of splitting and frequency variation with strain can be reproduced.

From the C_{3v} bond configuration shown as inset in Fig. 22.7a, and the derived effective force constant $\kappa = 6.283$ N/m, one can estimate the force constant of the single C–C bond in the monolayer graphene. One may define the C–C bond force constant k_0, the bonds labeled 1 and 3 are approximated as in parallel and the resultant $\kappa_{13} = 2\kappa_0$. This resultant bond connects to bond 2 in series and therefore the resultant force constant of the three bonds is $\kappa_{123} = 2\kappa_0/3$. Hence, the C–C bond force constant $\kappa_0 \approx 9.424$ N/m. Likewise, the force constant κ per M–X bond is derived as $\kappa_0 \approx 2.56$ N/m.

22.3.3 Mechanical Compression and Thermal Excitation

22.3.3.1 Formulation

Using the $1 + x \cong \exp(x)$ at $x \ll 1$ approximation for thermal expansion, one can formulate the thermal and pressure effects as follows ($y = P, T$) [87, 97],

$$
\frac{\omega(z, y) - \omega(1, y_0)}{\omega(z, y_0) - \omega(1, y_0)} = \frac{d(y_0)}{d(y)} \left(\frac{E(y)}{E(y_0)} \right)^{1/2} \cong
\begin{cases}
(1 - \Delta_T)^{1/2} \times \exp\left(-\int_{T_0}^{T} \alpha(t)dt \right) \\[4mm]
(1 + \Delta_P)^{1/2} \times \exp\left(+\int_{P_0}^{P} \beta(p)dp \right)
\end{cases}
$$

$$(22.5)$$

The thermally- and mechanically-induced energy perturbations Δ_T and Δ_P follow the relations [75],

$$
\begin{cases}
\Delta_T = \int_{T_0}^{T} \frac{\eta(t)dt}{E_z} = \int_{T_0}^{T} \frac{C_v(t/\theta_D)dt}{z E_z} \\[4mm]
C_v(\tau, T) = \tau R \left(\frac{T}{\theta_D} \right)^{\tau} \int_{0}^{\theta_D/T} \frac{x^{\tau+1} e^x}{(e^x - 1)^2} dx
\end{cases}
$$

and,

$$
\Delta_P = -\int_{V_0}^{V} \frac{p(v)dv}{E_z} = -\frac{V_0}{E_z} \int_{1}^{X} p(x)dx
$$

$$= \frac{\int_{P_0}^{P} v(p)dp - \int_{(VP)_0}^{VP} d(pv)}{E_{den}}$$

With,

$$\begin{cases} p(x) = \frac{3B_0}{2}\left(x^{-7/3} - x^{-5/3}\right) \times \left[1 + 3\frac{(B_0'-4)(x^{-2/3}-1)}{4}\right] & (B-M) \\ x(p) = 1 - \beta P + \beta' P^2 & (Polynominal) \end{cases} \quad (22.6)$$

The Δ_T is the integral of the specific heat reduced by the bond energy in a τ-dimensional Debye approximation. When the measuring temperature T is higher than θ_D, the two-dimensional specific heat C_v converges a constant of τR (R is the idea gas constant). The atomic cohesive energy $E_{coh} = zE_z$ and the θ_D are the uniquely adjustable parameters in calculating the Δ_T. The Δ_P is calculated based on the integral of the Birch-Mürnaghan equation [98, 99]. V_0 is the volume of a bond without compression.

The variables in Δ_P are the binding energy density $E_{den} = E_z/V_0$ and the compressibility β and its first derivative, β'. The x(P) is another form of the equation of states, which can be derived from the BM equation to obtain the nonlinear compressibility. Matching the BM equation to the measured x(P) curve, one can derive the nonlinear compressibility β and β', the bulk modulus B_0 ($\beta B_0 \cong 1$) and its first-order differentiation B_0'. Substituting the integrals (22.6) into (22.5), one can reproduce the P- and T-dependent Raman shift with derivatives of the θ_D, α, and E_{den} and the compressibility derived from the x(P) relation with the known E_{coh}.

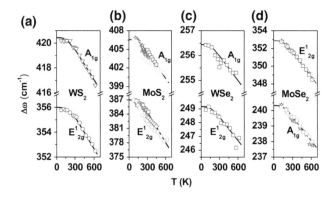

Fig. 22.8 BOLS-LBA reproduction (solid lines) of the measured (scattered) T-dependent A_{1g} and E_{2g}^1 modes for **a** WS$_2$ [102], **b** MoS$_2$ [103–105], **c** WSe$_2$ [60], and **d** MoSe$_2$ [60]. Reprinted with copyright permission from [100]

22.3.3.2 MX$_2$ Debye Thermal Decay

Figure 22.8 shows the theoretical reproduction of the thermally softened Raman shifts for the MX$_2$ with derived E$_{coh}$, and θ_D given in Table 22.1. The $\omega_x(1)$ and $\Delta\omega_x$ can also be obtained from matching the number-of-layer effect on the phonon frequency shift. Theoretical match is realized by adjusting the θ_D and the E$_{coh}$ without needing any hypothetic parameters [100], as the Raman shift features the $(E/d^2)^{1/2}$ that relaxes functionally with the integration of Debye specific heat. The slow decrease of the Raman shift at very low temperatures arises from the small $\int_0^T \eta dt$ values as the specific heat $\eta(t)$ is proportional to T^τ for the τ-dimensional system at very low temperatures. These results imply that the θ_D determines the width of the shoulder, the 1/E$_{coh}$ and the thermal expansion coefficient determine the slope of the curve at high temperatures.

Results derived from the matching to measurements agree with the BOLS-LBA expectation of the bond thermal relaxation premise. The atomic cohesive energy for the MX$_2$ follows the E$_{coh-MoS_2}$ (1.35 eV) $<$ E$_{coh-WS_2}$ (2.15 eV) $<$ E$_{coh-MoSe_2}$ (2.30 eV) $<$ E$_{coh-WSe_2}$ (3.21 eV) order. The MoS$_2$ shows higher rates of thermal expansion and bond energy attenuation than those of others, due to its intrinsically weaker atomic cohesive energy. In addition, the E$_{coh}$ also ranks the Debye temperature θ_D and the frequency offset $\Delta\omega_x$. Although the E$_{2g}^1$ mode of the WS$_2$ deviates slightly, the entire statistical trend holds. Theoretical production of the experimental observation suggests that the bond nature determines the thermal relaxation dynamics.

One may refer the atomic cohesive energy of the compounds to their electronegativity difference between the constituent M and X elements. The electronegativity values are W(1.7), Mo(1.8), S(2.5), and Se(2.4) [101]. Differences in electronegativity values of the constituent elements M and X of the compounds represent the polarity of chemical bonds, which are in the order: MoSe$_2$(0.6) $<$ WSe$_2$(0.7) $=$ MoS$_2$(0.7) $<$ WS$_2$(0.8).

22.3.3.3 BP and (W, Mo)S$_2$

Figure 22.9 shows the BOLS-LBA reproduction of the measured pressure and temperature dependent Raman shifts for the layered BP [70, 106, 107]. The reproduction of the P-stiffened and T-softened Raman shifts results in the compressibility β, energy density E$_{den}$, elastic modulus B, and the Debye temperature featured in Table 22.1 [90]. The compressibility $\beta = 0.32$ GPa^{-1} and energy density E$_{den} = 9.46$ eV nm^{-3}. The $\theta_D = 466$ K determines the width of the curve shoulder and the mode cohesive energy E$_{coh-m} = 2.11$ eV determines the slope of the curve at high temperature.

The database as a function of temperature [108] and pressure [46] for graphene and the data for the pressure dependence for (W, Mo)S$_2$ [109] phonon frequency shifts allow for information derived from the theoretical matching to measurements. Figure 22.10a shows the BOLS-LBA reproduction of the pressure-resolved phonon relaxation for (W, Mo)S$_2$ with derived information summarized in Table 22.1.

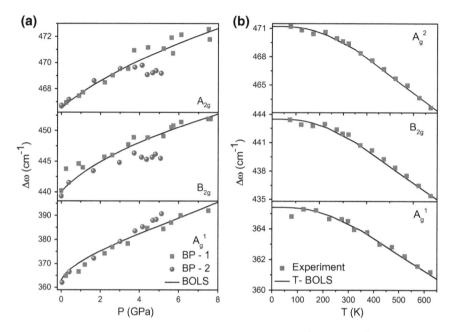

Fig. 22.9 Theoretical reproduction of the **a** P- and **b** T-dependent A_g^1, B_{2g}, and A_g^2 mode frequency shift for the few-layered BP [70, 106, 107]. Reprinted with copyright permission from [90]

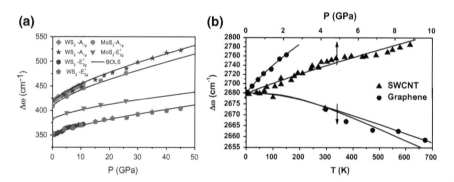

Fig. 22.10 BOLS-LBA reproduction of the **a** compression-resolved phonon frequency shift of WS$_2$ [110] and MoS$_2$ [109] and **b** P and T resolved phonon frequency shift of carbon allotropes [72, 111, 112]

22.3.3.4   Carbon Allotropes

With the derived $\omega_x(1)$ and the measured $\omega_x(z_g)$, one can reproduce the T and P dependent Raman shift with the quantified Debye temperature, mode cohesive energy, compressibility, and binding energy density of carbon allotropes as given in Table 22.2. Matching to the measured T-dependent Raman shift of the 2D mode for

Table 22.2 Quantitative information derived from reproducing the measured T- and P- resolved Raman shift for graphene, SWCNT, C_{60}, CNB, graphite and diamond [42]

Parameters		SWCNT	Graphite	C_{60}	CNB	Diamond
		G mode				D mode
P	CN	3.0	5.335	2.465	5.605	12
	A (ω-P slope)	0.024 [115]	0.028 [114]	0.0165 [112]	0.0273 [111]	0.033 [117]
	$\alpha(10^{-6}\ K^{-1})$	8	8 [155]	1	5	0.8 [156]
	$B_0/B_0'(GPa/-)$	865/5.0	39/10.0 [152]	–	–	446/3.6 [157]
	$\beta/\beta'(10^{-3}GPa^{-1}/GPa^{-2})$	1.156/0.0770	18.440/0.4427	–	–	2.120/0.0035
	E_{den} (eV/nm³)	456.32	347.68	–	–	249.59
T	d(nm)	0.125	0.142	0.118	0.143	0.154
	E_b (eV)	1.038	0.756	1.22	0.743	0.614 [158]
	$E_{mode-coh}$ (eV)	0.817	0.700	1.188	0.718	0.594
	$\theta_D(K)$	600	1000	650	550	2230 [159]

graphene [72] CNB [111], C_{60} [112] in Fig. 22.11b turns out that $\theta_D = 540$ K, with the given atomic cohesive energy of 3.11 eV/atom [32]. The θ_D is about 1/3 fold of the melting point 1605 K for the SWCNT [113]. At T ~ $\theta_D/3$, the Raman shift turns gradually from the nonlinear form to a linear at higher temperatures. The slow decrease of the Raman shift at very low temperatures arises from the small $\int_0^T \eta dt$ values as the specific heat $\eta(t)$ is proportional to T^τ for the τ-dimensional system at very low temperatures [108].

Figure 22.11a also shows the compression effect on the 2D mode of graphite [51, 114], SWCNT [115, 116], and diamond [117]. These results imply that the θ_D determines the width of the shoulder, the $1/E_{mod-coh}$ and the thermal expansion coefficient determine the slope of the linear form at high temperatures. The mode-cohesive energy E_{m-coh} (0) determines the slope of T-frequency curve at high temperature ($T \geq \theta_D/3$). By the known relation $E_{m-coh}(z) = C_z^{-m}E_{m-coh}(\text{bulk})$ and the known bulk mode-cohesive energy as well as the fixed m value, one can estimate the effective CN of the C_{60} and CNB phase, as given in Table 22.2.

The mode cohesive energy is lower than the atomic cohesive energy, which also varies from phase to phase for the same phonon mode. For instance, the G mode cohesive energy is 0.594 eV while the atomic cohesive energy is 7.37 eV for diamond. The former corresponds to energy activating the specific mode vibration while the latter to the energy of atomic evaporation of the specific crystal. From the derivatives, the C_{60} has lower effective CN than graphene because of the strain of the curvature. The bond energy, atomic cohesive energy, and mode cohesive energy of C_{60} are the highest among all the phases. Therefore, the C_{60} is strongest. The Debye temperatures of other allotropes are much lower than that of diamond, 2230 K.

Matching to the measured T-dependent Raman shift of the 2D mode for graphene in Fig. 22.10b turns out that $\theta_D = 540$ K, with the given atomic cohesive energy of

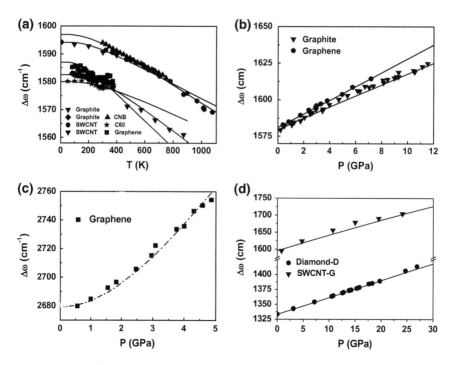

Fig. 22.11 Theoretical reproduction of the **a** *T*-dependent Raman shifts of diamond [117], CNB [111], C_{60} [112], graphene [72], graphite [114, 118], and SWCNT [115, 116], at the ambient pressure. **b** *P*-dependent Raman shifts of diamond [119], SWCNT [120], graphene [46], graphite [121], and graphene [46] and **c** graphene [46] at low pressures and **d** diamond [119], SWCNT [120], measured at room temperature with derived information summarized in Table 22.1 [32, 42]. Reprinted with permission from [32, 42]

3.11 eV/atom. The θ_D is about 1/3 fold of the melting point 1605 K for the SWCNT [113]. At T ~ θ_D/3, the Raman shift turns gradually from the nonlinear form to the linear when the temperature is increased. The match to the measured P-dependent Raman shift in Fig. 22.10b gives rise to the compressibility of $\beta = 1.145 \times 10^{-3}$ (GPa^{-1}) and $\beta' = 7.63 \times 10^{-5}$ (GPa^{-2}) and the energy density of 320 eV/nm^3.

Figure 22.11b-d show the BOLS-LBA reproduction of the P-dependent Raman shifts for carbon allotropes. Matching to the measurements yields the binding energy density (E_{den}) and the compressibility, as listed in Table 22.2. The energy densities of graphene and SWCNT's are higher than that of diamond, which is consistent with XPS measurement [75]. Since the elastic modulus is proportional to the binding energy density, the elastic modulus of graphene and SWCNT are higher, which explains why they are hardly compressible compared with diamond. The elastic modulus for C_{60} and CNB will be readily derived provided with the P-dependent Raman shift.

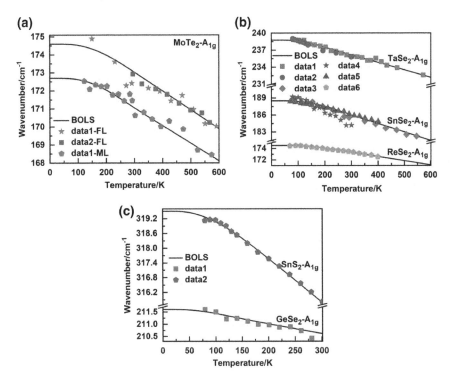

Fig. 22.12 BOLS reproduction of the measured A_{1g} modes for few layer [122, 123] (FL) and monolayer [122] (ML). **a** MoTe$_2$, **b** (Ta, Sn, Re)-Se$_2$ [124–128], and **c** (Sn, Ge)-(S$_2$, Se$_2$) [128, 129]

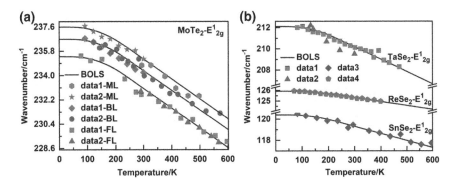

Fig. 22.13 BOLS reproduction of the measured E^1_{2g} modes for monolayer [123, 130] (ML), bilayer [122, 130] (BL) and few-layer [122, 123] (FL). **a** MoTe$_2$, **b** (Ta, Sn, Re)-Se$_2$ [124–127]

22.3.3.5 Debye Thermal Decay of Phonon Frequency

Figures 22.12 and 22.13 show the BOLS theory reproduction of the Debye thermal decay of the A_{1g} and the E_{2g}^1 vibration modes for MX_2. Results follow the universal trend of thermal relaxation, which gives rise to the Debye temperature θ_D, the referential frequency $\omega_L(1)$ and atomic cohesive energy E_z, as shown in Table 22.3.

Figure 22.14 shows the BOLS reproduction of the temperature dependent A_g^3 and B_{3g} modes of SnS, SnS and GeSe films. They are different from the hexagonal honeycomb structure of MDCs. The SnS, SnS, and GeSe nanomaterials exhibit wrinkled honeycomb structures. Because the Debye temperature for the SnS is greater than that for SnSe and GeSe, the low temperature nonlinear range of the former is distinctly larger than the latter.

The temperature dependent Raman frequency of the monolayer (ML), bilayer (BL), and few-layer (FL) of $MoTe_2$ follows the BOLS approximation, see Figs. 22.12a and 22.13a. Results show that atomic undercoordination shortens the bond length and strengthen the bond energy. The E_{2g}^1 mode undergoes a blueshift but

Table 22.3 Information derived from theoretical match to the Debye thermal decay of metal dichalcogenides and metal monochalcogenides

Stimuli	$\omega_A(1)$ (cm^{-1})	$\omega_E(1)$ (cm^{-1})	E_0 (eV)	E_z (eV)	θ_D (K)
ReSe$_2$	171.12	124.91	0.68	3.39	181 [136]
TaSe$_2$	232.12	206.73	–	5.23	150 [160]
MoTe$_2$	168.12[a]/169.80[c]	228.66[a]/230.01[b]/232.82[c]	0.58	4.22	169 [161]
SnSe$_2$	181.11	117.32	0.53	4.45	170 [137]
SnS$_2$	293.52	–	4.16	3.51	136 [162]
GeSe$_2$	198.52	–	1.21	2.29	180 [129]
SnSe	133.85[d]	88.46[e]	0.49	0.83	204 [163]
SnS	150.13[d]	84.96[e]	0.69	0.80	270 [164]
GeSe	175.62[d]	135.86[e]	0.65	0.69	214 [165]

[a]Monolayer of the nanomaterials
[b]Bilayer of the nanomaterials
[c]Fewlayer of the nanomaterials
[d]A_g^3 modes
[e]B_{3g} modes

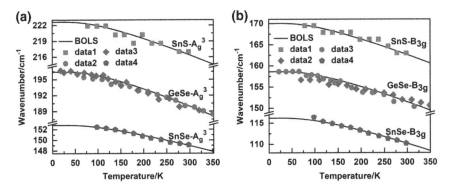

Fig. 22.14 BOLS reproduction of the measured for **a** A_g^3 and **b** B_{3g} modes of SnS [131], SnS [132] and GeSe [133, 134] films

the A_{1g} mode a redshift as the $MoTe_2$ number-of-layer decreases. This is similar to the situation of MoS_2 [91]—collective vibration dictates the redshift while the dimer vibration governs the blue shift.

Moreover, the atomic cohesive energy E_z determines the linear slope at high temperature of the temperature dependent Raman frequency for different nanomaterials in the same temperature range. When the temperature changes from 0 to 600 K, the atomic cohesive energy shows the following relationships: $E_{z\text{-}TaSe_2}$ (5.23 eV) > $E_{z\text{-}SnSe_2}$ (4.45 eV) > $E_{z\text{-}MoTe_2}$ (4.22 eV) > $E_{z\text{-}ReSe_2}$ (3.39 eV). So compared to the other three materials, $TaSe_2$ have the largest linear slope at high temperature, so the rate of decrease in Raman frequency is the fastest for continuous heating. In addition, as shown in Fig. 22.12c, the slope for SnS_2 at high temperatures is clearly larger than for $GeSe_2$ for temperatures between 0 and 300 K. So the order of their atomic cohesive energy calculated using bond relaxation theory is $E_{z\text{-}SnS_2}$ (3.51 eV) > $E_{z\text{-}GeSe_2}$ (2.29 eV). A similar situation occurs for the Raman frequency temperature dependence of SnS, SnSe and GeSe. The agreement between the BOLS calculations and experimental measurements confirms that the atomic cohesive energy E_z governs the slope at high temperature for thermal relaxation in these Raman spectroscopies.

22.3.4 Thermal Relaxation of Bandgap Energy

The bandgap energy of nanomaterials is important foundation for its applications in the field of semiconductors and optoelectronic devices. Because self-heating of the device is unavoidable, when electronic equipment is used, this directly affects the bandgap energy of nanomaterials. This means it is important to understand the physical mechanism of the temperature dependent the bandgap energy of nanomaterials. Thermal excitation weakens the bond, which causes thermal softening of the bandgap energy of these 2D film materials—see Fig. 22.15. The Debye temperature determines the nonlinear range at low temperature, while the atomic cohesive

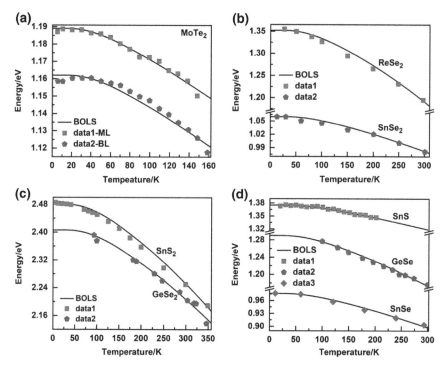

Fig. 22.15 BOLS reproduction of the measured bandgap energy for **a** monolayer [135] (ML) and bilayer [135] (BL) MoTe$_2$, **b** ReSe$_2$ [136] and SnSe$_2$ [137], **c** SnS$_2$ [138] and GeSe$_2$ [139] and **d** (Sn, Ge)-(S, Se) [140–142]

energy governs the linear slope at high temperature, which is similar to the temperature dependence of the Raman frequency of these 2D films. BOLS approximation derived both atomic cohesive energy E'_z and bulk bandgap energy E_0 were used to calculate the temperature dependence of the bandgap energy of these 2D films—see Table 22.3.

22.3.5 Edge Discriminative Raman Reflectivity

DFT computations [84] and scanning tunneling microscopy [143] confirmed that the Dirac-Fermi polaritons generate preferentially at the zigzag-edge of graphene [144] and at sites surrounding graphite surface atomic vacancies [145] because of the longer and uniform $\sqrt{3}d$ distance between the dangling bond electrons along the edge. The dangling bond electrons at the armchair-edge and the reconstructed zigzag-edge of graphene (5 and 7 atomic rings) tend to form quasi-triple-bond between the edge atoms of the alternative shorter d distance. The isolation and polarization of the unpaired dangling bond electrons at the zigzag-edge by the locally and deeply entrapped core and bonding electrons may scatter the incident radiation substantially and hence lowers the Raman D band reflectivity at the zigzag-edge compared with that at the arm-chaired edge [56, 57].

22.4 Summary

The notion of multifield bond oscillation has reconciled the following thrusts:

(1) One can join the continuum approaches to the quantum theory by focusing on the perturbed bond relaxation to the phonon frequency.
(2) Phonon frequency shift under various perturbations could be reproduced from the perspective of bond relaxation.
(3) Results improve the understanding about the multifield driven bond-phonon-property correlation of a substance.
(4) Theoretical reproduction results in quantitative information on bonding dynamics derived from the combination of the multifield bond oscillation framework and its driven phonon spectrometrics.

Being able to connect the continuum to quantum premises, the combination of the BOLS-LBA notion for the multifield oscillation dynamics and the enabled Raman phonon spectrometrics has enabled the quantitative information about the perturbation-relaxation-property that is beyond the scope of other available approaches. The practice of the average-bond bond oscillating dynamics has reconciled (z, ε, P, T) effects on the Raman shifts of BP, GNR, and MX_2 with not only clarification of the intrinsic phonon relaxation origin but also information on the bonding dynamics and their properties to the 2-D structures. Consistency between theory expectations and Raman measurements clarified and quantified the following:

(1) The undercoordination induced bond contraction and bond strength gain govern the number-of-layer resolved phonon frequency shift. The collective vibration of bonds between a certain atom and its z neighbors dictates phonon frequency redshift while the dimer interaction dictates the phonons undergoing blueshift occurred when the number-of-layer decreases. Matching the number-of-layer reduction induced phonon frequency shift offers the local bond length, bond energy, band nature index, referential wavenumber $\omega(1)$ and its bulk shift, and the effective coordination numbers for the layered 2-D structures.
(2) Bond stretching by the directional strain resolves the phonon red-shifting and phonon band splitting. Reproduction of the strain effect on phonon relaxation gives rise to the average-bond force constant and the relative direction between a certain bond and the strain.
(3) Bond thermal expansion softens all phonons—Debye thermal decay. One can derive the mode cohesive energy E_{coh} and Debye temperature θ_D from reproducing the Debye thermal decay of phonon frequencies.

(4) Bond mechanical compression stiffens all phonons. Theoretical matching to the effect of compression derives the binding energy density E_{den}, compressibility β, the bulk modulus B_0 and their first-order derivatives β' and B'_0.

Findings proved the essentiality of the BOLS-LBA notion that has empowered the Raman spectroscopy immensely in gaining quantitative information about the dynamics of the representative bond and the properties of a specimen. Most strikingly, freely hypothetic adjustable parameters are avoided in the entire process of data analysis. The attainments may inspire a new yet practical way of thinking about the stimulus-bond-property correlation and approaches dealing with substance from the perspective of average-bond approach with interdisciplinary nature, offering references for functional materials design and synthesis. Extending the exercises to covering more stimuli and more properties of the substance would be even more challenging, fascinating and rewarding.

References

1. S. Liu, N. Huo, S. Gan, Y. Li, Z. Wei, B. Huang, J. Liu, J. Li, H. Chen, Thickness-dependent Raman spectra, transport properties and infrared photoresponse of few-layer black phosphorus. J Mater Chem C **3**(42), 10974–10980 (2015)
2. W. Zhao, Z. Ghorannevis, L. Chu, M. Toh, C. Kloc, P.-H. Tan, G. Eda, Evolution of electronic structure in atomically thin sheets of WS_2 and WSe_2. ACS Nano **7**(1), 791–797 (2012)
3. X. Fan, C.-H. Chang, W. Zheng, J.-L. Kuo, D.J. Singh, The electronic properties of single-layer and multi-layer MoS_2 under High pressure. J. Phys. Chem. C **119**, 10189–10196 (2015)
4. C.Q. Sun, S.Y. Fu, Y.G. Nie, Dominance of broken bonds and unpaired nonbonding pi-electrons in the band gap expansion and edge states generation in graphene nanoribbons. J. Phys. Chem. C **112**(48), 18927–18934 (2008)
5. O. Hod, V. Barone, J.E. Peralta, G.E. Scuseria, Enhanced half-metallicity in edge-oxidized zigzag graphene nanoribbons. Nano Lett. **7**(8), 2295–2299 (2007)
6. L. Brey, H. Fertig, Electronic states of graphene nanoribbons studied with the Dirac equation. Phys. Rev. B **73**(23), 235411 (2006)
7. F.M.D. Pellegrino, G.G.N. Angilella, R. Pucci, Strain effect on the optical conductivity of graphene. Phys. Rev. B **81**(3), 035411 (2010)
8. S.H.M. Jafri, K. Carva, E. Widenkvist, T. Blom, B. Sanyal, J. Fransson, O. Eriksson, U. Jansson, H. Grennberg, O. Karis, R.A. Quinlan, B.C. Holloway, K. Leifer, Conductivity engineering of graphene by defect formation. J. Phys. D Appl. Phys. **43**(4), 045404 (2010)
9. Y.W. Son, M.L. Cohen, S.G. Louie, Energy gaps in graphene nanoribbons. Phys. Rev. Lett. **97**(21), 216803 (2006)
10. M. Fujita, K. Wakabayashi, K. Nakada, K. Kusakabe, Peculiar localized state at zigzag graphite edge. J. Phys. Soc. Jpn. **65**(7), 1920–1923 (1996)
11. L. Yang, C.-H. Park, Y.-W. Son, M.L. Cohen, S.G. Louie, Quasiparticle Energies and Band Gaps in Graphene Nanoribbons. Phys. Rev. Lett. **99**(18), 186801 (2007)
12. V. Barone, O. Hod, G.E. Scuseria, Electronic Structure and Stability of Semiconducting Graphene Nanoribbons. Nano Lett. **6**(12), 2748–2754 (2006)
13. K.F. Mak, C. Lee, J. Hone, J. Shan, T.F. Heinz, Atomically thin MoS_2: a new direct-gap semiconductor. Phys. Rev. Lett. **105**(13), 136805 (2010)
14. A. Kumar, P. Ahluwalia, Electronic structure of transition metal dichalcogenides monolayers 1H-MX_2 (M=Mo, W; X=S, Se, Te) from ab-initio theory: new direct band gap semiconductors. Eur. Phys. J. B **85**(6), 1–7 (2012)

15. S. Tongay, J. Zhou, C. Ataca, K. Lo, T.S. Matthews, J. Li, J.C. Grossman, J. Wu, Thermally driven crossover from indirect toward direct bandgap in 2D semiconductors: MoSe$_2$ versus MoS$_2$. Nano Lett. **12**(11), 5576–5580 (2012)

16. W. Zhao, R. Ribeiro, M. Toh, A. Carvalho, C. Kloc, A. Castro Neto, G. Eda, Origin of indirect optical transitions in few-layer MoS$_2$, WS$_2$, and WSe$_2$. Nano Lett. **13**(11), 5627–5634 (2013)

17. V. Kaushik, D. Varandani, B.R. Mehta, Nanoscale mapping of layer-dependent surface potential and junction properties of CVD-Grown MoS$_2$ domains. J. Phys. Chem. C **119**(34), 20136–20142 (2015)

18. A. Kuc, N. Zibouche, T. Heine, Influence of quantum confinement on the electronic structure of the transition metal sulfide TS$_2$. Phys. Rev. B **83**(24), 245213 (2011)

19. A. Ramasubramaniam, Large excitonic effects in monolayers of molybdenum and tungsten dichalcogenides. Phys. Rev. B **86**(11), 115409 (2012)

20. L. Li, Y. Yu, G.J. Ye, Q. Ge, X. Ou, H. Wu, D. Feng, X.H. Chen, Y. Zhang, Black phosphorus field-effect transistors. Nat. Nanotechnol. **9**(5), 372–377 (2014)

21. H. Liu, A.T. Neal, Z. Zhu, Z. Luo, X. Xu, D. Tománek, P.D. Ye, Phosphorene: an unexplored 2D semiconductor with a high hole mobility. ACS Nano **8**(4), 4033–4041 (2014)

22. B. Radisavljevic, A. Radenovic, J. Brivio, V. Giacometti, A. Kis, Single-layer MoS$_2$ transistors. Nat. Nanotechnol. **6**(3), 147–150 (2011)

23. H. Li, Z. Yin, Q. He, H. Li, X. Huang, G. Lu, D.W.H. Fam, A.I.Y. Tok, Q. Zhang, H. Zhang, Fabrication of single-and multilayer MoS$_2$ film-based field-effect transistors for sensing NO at room temperature. Small **8**(1), 63–67 (2012)

24. K.F. Mak, K. He, J. Shan, T.F. Heinz, Control of valley polarization in monolayer MoS$_2$ by optical helicity. Nat. Nanotechnol. **7**(8), 494–498 (2012)

25. H.S. Lee, S.-W. Min, Y.-G. Chang, M.K. Park, T. Nam, H. Kim, J.H. Kim, S. Ryu, S. Im, MoS$_2$ nanosheet phototransistors with thickness-modulated optical energy gap. Nano Lett. **12**(7), 3695–3700 (2012)

26. S. Kim, A. Konar, W.-S. Hwang, J.H. Lee, J. Lee, J. Yang, C. Jung, H. Kim, J.-B. Yoo, J.-Y. Choi, High-mobility and low-power thin-film transistors based on multilayer MoS$_2$ crystals. Nat Commun **3**, 1011 (2012)

27. C. Zhu, Z. Zeng, H. Li, F. Li, C. Fan, H. Zhang, Single-layer MoS$_2$-based nanoprobes for homogeneous detection of biomolecules. J. Am. Chem. Soc. **135**(16), 5998–6001 (2013)

28. G.W. Shim, K. Yoo, S.-B. Seo, J. Shin, D.Y. Jung, I.-S. Kang, C.W. Ahn, B.J. Cho, S.-Y. Choi, Large-area single-layer MoSe$_2$ and its van der Waals heterostructures. ACS Nano **8**(7), 6655–6662 (2014)

29. C. Ataca, H. Sahin, S. Ciraci, Stable, single-layer MX$_2$ transition-metal oxides and dichalcogenides in a honeycomb-like structure. J. Phys. Chem. C **116**(16), 8983–8999 (2012)

30. Y. Zhou, D. Wu, Y. Gong, Z. Ma, Y. Huang, X. Zhang, C.Q. Sun, Base-hydration-resolved hydrogen-bond networking dynamics: quantum point compression. J. Mol. Liq. **223**, 1277–1283 (2016)

31. J. Zi, H. Büscher, C. Falter, W. Ludwig, K. Zhang, X. Xie, Raman shifts in Si nanocrystals. Appl. Phys. Lett. **69**(2), 200–202 (1996)

32. X.X. Yang, J.W. Li, Z.F. Zhou, Y. Wang, L.W. Yang, W.T. Zheng, C.Q. Sun, Raman spectroscopic determination of the length, strength, compressibility, Debye temperature, elasticity, and force constant of the C–C bond in graphene. Nanoscale **4**(2), 502–10 (2012)

33. M. Fujii, Y. Kanzawa, S. Hayashi, K. Yamamoto, Raman scattering from acoustic phonons confined in Si nanocrystals. Phys. Rev. B **54**(12), R8373–R8376 (1996)

34. W. Cheng, S.F. Ren, Calculations on the size effects of Raman intensities of silicon quantum dots. Phys. Rev. B **65**(20), 205305 (2002)

35. C. Lee, H. Yan, L.E. Brus, T.F. Heinz, J. Hone, S. Ryu, Anomalous lattice vibrations of single-and few-layer MoS$_2$. ACS Nano **4**(5), 2695–2700 (2010)

36. C.Q. Sun, Size dependence of nanostructures: Impact of bond order deficiency. Prog. Solid State Chem. **35**(1), 1–159 (2007)

37. X.J. Liu, M.L. Bo, X. Zhang, L. Li, Y.G. Nie, H. Tian, Y. Sun, S. Xu, Y. Wang, W. Zheng, C.Q. Sun, Coordination-resolved electron spectrometrics. Chem. Rev. **115**(14), 6746–6810 (2015)

38. Y. Hao, Y. Wang, L. Wang, Z. Ni, Z. Wang, R. Wang, C.K. Koo, Z. Shen, J.T.L. Thong, Probing layer number and stacking order of few-layer graphene by Raman spectroscopy. Small **6**(2), 195–200 (2010)

39. H. Wang, Y. Wang, X. Cao, M. Feng, G. Lan, Vibrational properties of graphene and graphene layers. J. Raman Spectrosc. **40**(12), 1791–1796 (2009)

40. D. Graf, F. Molitor, K. Ensslin, C. Stampfer, A. Jungen, C. Hierold, L. Wirtz, Spatially resolved Raman spectroscopy of single- and few-layer graphene. Nano Lett. **7**(2), 238–242 (2007)

41. A. Gupta, G. Chen, P. Joshi, S. Tadigadapa, P.C. Eklund, Raman scattering from high-frequency phonons in supported n-graphene layer films. Nano Lett. **6**(12), 2667–2673 (2006)

42. X.X. Yang, Z.F. Zhou, Y. Wang, J.W. Li, N.G. Guo, W.T. Zheng, J.Z. Peng, C.Q. Sun, Raman spectroscopic determination of the length, energy, Debye temperature, and compressibility of the C–C bond in carbon allotropes. Chem. Phys. Lett. **575**, 86–90 (2013)

43. T.M.G. Mohiuddin, A. Lombardo, R.R. Nair, A. Bonetti, G. Savini, R. Jalil, N. Bonini, D.M. Basko, C. Galiotis, N. Marzari, K.S. Novoselov, A.K. Geim, A.C. Ferrari, Uniaxial strain in graphene by Raman spectroscopy: G peak splitting, Gruneisen parameters, and sample orientation. Phys. Rev. B **79**(20), 205433 (2009)

44. T. Yu, Z. Ni, C. Du, Y. You, Y. Wang, Z. Shen, Raman mapping investigation of graphene on transparent flexible substrate: the strain effect. J. Phys. Chem. C **112**(33), 12602–12605 (2008)

45. M. Huang, H. Yan, T.F. Heinz, J. Hone, Probing strain-induced electronic structure change in graphene by Raman spectroscopy. Nano Lett. **10**(10), 4074–4079 (2010)

46. J.E. Proctor, E. Gregoryanz, K.S. Novoselov, M. Lotya, J.N. Coleman, M.P. Halsall, High-pressure Raman spectroscopy of graphene. Phys. Rev. B **80**(7), 073408 (2009)

47. I. Calizo, S. Ghosh, W. Bao, F. Miao, C. Ning Lau, A.A. Balandin, Raman nanometrology of graphene: temperature and substrate effects. Solid State Commun. **149**(27–28), 1132-1135 (2009)

48. J.L. Dattatray, U. Maitra, L.S. Panchakarla, U.V. Waghmare, C.N.R. Rao, Temperature effects on the Raman spectra of graphenes: dependence on the number of layers and doping. J. Phys.: Condens. Matter **23**(5), 055303 (2011)

49. E.H. Martins Ferreira, M.V.O. Moutinho, F. Stavale, M.M. Lucchese, R.B. Capaz, C.A. Achete, A. Jorio, Evolution of the Raman spectra from single-, few-, and many-layer graphene with increasing disorder. Phys. Rev. B, **82**(12), 125429–125438 (2010)

50. S. Berciaud, S. Ryu, L.E. Brus, T.F. Heinz, Probing the Intrinsic properties of exfoliated graphene: Raman spectroscopy of free-standing monolayers. Nano Lett. **9**(1), 346–352 (2008)

51. I. Calizo, W. Bao, F. Miao, C.N. Lau, A.A. Balandin, The effect of substrates on the Raman spectrum of graphene. Appl. Phys. Lett. **91**, 201904 (2007)

52. Z. Luo, T. Yu, K.J. Kim, Z. Ni, Y. You, S. Lim, Z. Shen, S. Wang, J. Lin, Thickness-dependent reversible hydrogenation of graphene layers. ACS Nano **3**(7), 1781–1788 (2009)

53. L.M. Xie, L.Y. Jiao, H.J. Dai, Selective etching of graphene edges by hydrogen plasma. J. Am. Chem. Soc. **132**(42), 14751–14753 (2010)

54. H.J. Shin, W.M. Choi, D. Choi, G.H. Han, S.M. Yoon, H.K. Park, S.W. Kim, Y.W. Jin, S.Y. Lee, J.M. Kim, J.Y. Choi, Y.H. Lee, Control of electronic structure of graphene by various dopants and their effects on a nanogenerator. J. Am. Chem. Soc. **132**(44), 15603–15609 (2010)

55. L.G. Cancado, M.A. Pimenta, B.R.A. Neves, M.S.S. Dantas, A. Jorio, Influence of the atomic structure on the Raman spectra of graphite edges. Phys. Rev. Lett. **93**(24), 247401 (2004)

56. B. Krauss, P. Nemes Incze, V. Skakalova, L.P. Biro, K.V. Klitzing, J.H. Smet, Raman scattering at pure graphene zigzag edges. Nano Lett. **10**(11), 4544–4548 (2010)

57. Y. You, Z. Ni, T. Yu, Z. Shen, Edge chirality determination of graphene by Raman spectroscopy. Appl. Phys. Lett. **93**(16), 163112–163115 (2008)

58. M.A. Pimenta, G. Dresselhaus, M.S. Dresselhaus, L.G. Cancado, A. Jorio, R. Saito, Studying disorder in graphite-based systems by Raman spectroscopy. Phys. Chem. Chem. Phys. **9**(11), 1276–1290 (2007)

59. M.S. Dresselhaus, A. Jorio, R. Saito, Characterizing graphene, graphite, and carbon nanotubes by Raman spectroscopy. Annu. Rev. Condens. Matter Phys. **1**(1), 89–108 (2010)

60. D.J. Late, Temperature dependent phonon shifts in single-layer WS_2. ACS Appl. Mater. Inter. **6**(2), 1158–1163 (2014)

61. M. Staiger, P. Rafailov, K. Gartsman, H. Telg, M. Krause, G. Radovsky, A. Zak, C. Thomsen, Excitonic resonances in WS_2 nanotubes. Phys. Rev. B **86**(16), 165423 (2012)

62. A.K. Gupta, T.J. Russin, H.R. Gutierrez, P.C. Eklund, Probing graphene edges via Raman scattering. ACS Nano **3**(1), 45–52 (2008)

63. A. Favron, E. Gaufrès, F. Fossard, P. Lévesque, A. Phaneuf-L'Heureux, N. Tang, A. Loiseau, R. Leonelli, S. Francoeur, and R. Martel, Exfoliating pristine black phosphorus down to the monolayer: photo-oxidation and electronic confinement effects. arXiv:14080345 (2014)

64. X. Wang, A.M. Jones, K.L. Seyler, V. Tran, Y. Jia, H. Zhao, H. Wang, L. Yang, X. Xu, F. Xia, Highly anisotropic and robust excitons in monolayer black phosphorus. Nat. Nanotechnol. **10**(6), 517–521 (2015)

65. J. Jiang, S. Xiu, M. Zheng, T. Jia, H. Liu, Y. Zhang, G. Chen, Indirect–direct bandgap transition and gap width tuning in bilayer MoS_2 superlattices. Chem. Phys. Lett. **613**, 74–79 (2014)

66. H.J. Conley, B. Wang, J.I. Ziegler, R.F. Haglund Jr., S.T. Pantelides, K.I. Bolotin, Bandgap engineering of strained monolayer and bilayer MoS_2. Nano Lett. **13**(8), 3626–3630 (2013)

67. D. Yoon, Y.-W. Son, H. Cheong, Strain-dependent splitting of the double-resonance Raman scattering band in graphene. Phys. Rev. Lett. **106**(15), 155502 (2011)

68. F. Ding, H. Ji, Y. Chen, A. Herklotz, K. Dorr, Y. Mei, A. Rastelli, O.G. Schmidt, Stretchable fraphene: a close look at fundamental parameters through biaxial straining. Nano Lett. **10**(9), 3453–3458 (2010)

69. P. Johari, V.B. Shenoy, Tuning the electronic properties of semiconducting transition metal dichalcogenides by applying mechanical strains. ACS Nano **6**(6), 5449–5456 (2012)

70. D.J. Late, Temperature dependent phonon shifts in few-layer black phosphorus. ACS Appl. Mater Inter. **7**(10), 5857–5862 (2015)

71. S. Appalakondaiah, G. Vaitheeswaran, S. Lebegue, N.E. Christensen, A. Svane, Effect of van der Waals interactions on the structural and elastic properties of black phosphorus. Phys. Rev. B **86**(3), 035105 (2012)

72. I. Calizo, A.A. Balandin, W. Bao, F. Miao, C.N. Lau, Temperature dependence of the Raman spectra of graphene and graphene multilayers. Nano Lett. **7**(9), 2645–2649 (2007)

73. H. Li, Q. Zhang, C.C.R. Yap, B.K. Tay, T.H.T. Edwin, A. Olivier, D. Baillargeat, From bulk to monolayer MoS_2: evolution of Raman scattering. Adv. Func. Mater. **22**(7), 1385–1390 (2012)

74. A. Arora, M. Koperski, K. Nogajewski, J. Marcus, C. Faugeras, M. Potemski, Excitonic resonances in thin films of WSe_2: from monolayer to bulk material. Nanoscale 10421–10429 (2015)

75. C.Q. Sun, Y. Sun, Y.G. Ni, X. Zhang, J.S. Pan, X.H. Wang, J. Zhou, L.T. Li, W.T. Zheng, S.S. Yu, L.K. Pan, Z. Sun, Coulomb Repulsion at the nanometer-sized contact: a force driving superhydrophobicity, superfluidity, superlubricity, and supersolidity. J. Phys. Chem. C **113**(46), 20009–20019 (2009)

76. P.G. Klemens, Anharmonic decay of optical phonons. Phys. Rev. **148**(2), 845–848 (1966)

77. T.R. Hart, R.L. Aggarwal, B. Lax, Temperature dependence of Raman scattering in silicon. Phys. Rev. B **1**(2), 638–642 (1970)

78. M. Balkanski, R.F. Wallis, E. Haro, Anharmonic effects in light-scattering due to optical phonons in silicon. Phys. Rev. B **28**(4), 1928–1934 (1983)

79. J. Liu, Y.K. Vohra, Raman modes of 6H polytype of silicon carbide to ultrahigh pressures: a comparison with silicon and diamond. Phys. Rev. Lett. **72**(26), 4105–4108 (1994)

80. R. Cuscó, E. Alarcón Lladó, J. Ibáñez, L. Artús, J. Jiménez, B. Wang, M.J. Callahan, Temperature dependence of Raman scattering in ZnO. Phys. Rev. B, **75**(16), 165202–165213 (2007)

81. G. Viera, S. Huet, L. Boufendi, Crystal size and temperature measurements in nanostructured silicon using Raman spectroscopy. J. Appl. Phys. **90**(8), 4175–4183 (2001)

82. L. Pauling, The Nature of the Chemical Bond, 3rd edn. (Cornell University Press, Ithaca, NY, 1960)
83. W.T. Zheng, C.Q. Sun, Underneath the fascinations of carbon nanotubes and graphene nanoribbons. Energy Environ. Sci. **4**(3), 627–655 (2011)
84. X. Zhang, Y.G. Nie, W.T. Zheng, J.L. Kuo, C.Q. Sun, Discriminative generation and hydrogen modulation of the Dirac-Fermi polarons at graphene edges and atomic vacancies. Carbon **49**(11), 3615–3621 (2011)
85. C.O. Girit, J.C. Meyer, R. Erni, M.D. Rossell, C. Kisielowski, L. Yang, C.H. Park, M.F. Crommie, M.L. Cohen, S.G. Louie, A. Zettl, Graphene at the edge: stability and dynamics. Science **323**(5922), 1705–1708 (2009)
86. K.S. Novoselov, Z. Jiang, Y. Zhang, S.V. Morozov, H.L. Stormer, U. Zeitler, J.C. Maan, G.S. Boebinger, P. Kim, A.K. Geim, Room-temperature quantum hall effect in graphene. Science **315**(5817), 1379–1379 (2007)
87. X. Yang, J. Li, Z. Zhou, Y. Wang, W. Zheng, C.Q. Sun, Frequency response of graphene phonons to heating and compression. Appl. Phys. Lett. **99**(13), 133108 (2011)
88. Y. Wang, X. Yang, J. Li, Z. Zhou, W. Zheng, C.Q. Sun, Number-of-layer discriminated graphene phonon softening and stiffening. Appl. Phys. Lett. **99**(16), 163109 (2011)
89. C. Thomsen, S. Reich, Double resonant Raman scattering in graphite. Phys. Rev. Lett. **85**(24), 5214–5217 (2000)
90. Y. Liu, X. Yang, M. Bo, X. Zhang, X. Liu, C.Q. Sun, Y. Huang, Number-of-layer, pressure, and temperature resolved bond–phonon–photon cooperative relaxation of layered black phosphorus. J. Raman Spectrosc. **47**(11), 1304–1309 (2016)
91. Y. Liu, M. Bo, Y. Guo, X. Yang, X. Zhang, C.Q. Sun, Y. Huang, Number-of-layer resolved (Mo, W)-(S2, Se2) phonon relaxation. J. Raman Spectrosc. **48**(4), 592–595 (2017)
92. M. Boukhicha, M. Calandra, M.-A. Measson, O. Lancry, A. Shukla, Anharmonic phonons in few-layer MoS_2: Raman spectroscopy of ultralow energy compression and shear modes. Phys Rev B **87**(19), 195316 (2013)
93. W. Zhao, Z. Ghorannevis, K.K. Amara, J.R. Pang, M. Toh, X. Zhang, C. Kloc, P.H. Tan, G. Eda, Lattice dynamics in mono-and few-layer sheets of WS_2 and WSe_2. Nanoscale **5**(20), 9677–9683 (2013)
94. C. Rice, R. Young, R. Zan, U. Bangert, D. Wolverson, T. Georgiou, R. Jalil, K. Novoselov, Raman-scattering measurements and first-principles calculations of strain-induced phonon shifts in monolayer MoS_2. Phys. Rev. B **87**(8), 081307 (2013)
95. X.X. Yang, Y. Wang, J.W. Li, W.H. Liao, Y.H. Liu, C.Q. Sun, Graphene phonon softening and splitting by directional straining. Appl. Phys. Lett. **107**(20), 203105 (2015)
96. Y. Liu, X. Yang, M. Bo, C. Ni, X. Liu, C.Q. Sun, Y. Huang, Multifield-driven bond–phonon–photon performance of layered (Mo, W)–(S2, Se2). Chem. Phys. Lett. **660**, 256–260 (2016)
97. M.X. Gu, L.K. Pan, T.C. Au Yeung, B.K. Tay, C.Q. Sun, Atomistic origin of the thermally driven softening of Raman optical phonons in group III nitrides. J. Phys. Chem. C, **111**(36): 13606–13610 (2007)
98. F. Birch, Finite elastic strain of cubic crystals. Phys. Rev. **71**(11), 809–824 (1947)
99. F.D. Murnaghan, The compressibility of media under extreme pressures. Proc. Natl. Acad. Sci. U.S.A. **30**(9), 244–247 (1944)
100. Y. Liu, Y. Wang, M. Bo, X. Liu, X. Yang, Y. Huang, C.Q. Sun, Thermally driven (Mo, W)-(S2, Se2) phonon and photon energy relaxation dynamics. J. Phys. Chem. C **119**(44), 25071–25076 (2015)
101. S. Hind, P. Lee, KKR calculations of the energy bands in $NbSe_2$, MoS_2 and alpha $MoTe_2$. J. Phys. C: Solid State Phys. **13**(3), 349 (1980)
102. D.J. Late, S.N. Shirodkar, U.V. Waghmare, V.P. Dravid, C. Rao, Thermal expansion, anharmonicity and temperature-dependent raman spectra of single-and few-layer $MoSe_2$ and WSe_2. Chem. Phys. Chem. **15**(8), 1592–1598 (2014)
103. R. Yan, J.R. Simpson, S. Bertolazzi, J. Brivio, M. Watson, X. Wu, A. Kis, T. Luo, A.R. Hight Walker, H.G. Xing, Thermal conductivity of monolayer molybdenum disulfide obtained from temperature-dependent Raman spectroscopy. ACS Nano **8**(1): 986–993, (2014)

104. M. Thripuranthaka, R.V. Kashid, C.S. Rout, D.J. Late, Temperature dependent Raman spectroscopy of chemically derived few layer MoS_2 and WS_2 nanosheets. Appl. Phys. Lett. **104**(8), 081911 (2014)
105. N.A. Lanzillo, A. Glen Birdwell, M. Amani, F.J. Crowne, P.B. Shah, S. Najmaei, Z. Liu, P.M. Ajayan, J. Lou, M. Dubey, S.K. Nayak, apos, T.P. Regan, Temperature-dependent phonon shifts in monolayer MoS_2. Appl. Phys. Lett. **103**(9), 093102 (2013)
106. C. Vanderborgh, D. Schiferl, Raman studies of black phosphorus from 0.25 to 7.7 GPa at 15 K. Phys. Rev. B, **40**(14): 9595 (1989)
107. S. Sugai, T. Ueda, K. Murase, Pressure dependence of the lattice vibration in the orthorhombic and rhombohedral structures of black phosphorus. J. Phys. Soc. Jpn. **50**(10), 3356–3361 (1981)
108. L. Zhang, Z. Jia, L. Huang, S. O'Brien, Z. Yu, Low-temperature Raman spectroscopy of individual single-wall carbon nanotubes and single-layer graphene. J. Phys. Chem. C **112**(36), 13893–13900 (2008)
109. N. Bandaru, R.S. Kumar, D. Sneed, O. Tschauner, J. Baker, D. Antonio, S.-N. Luo, T. Hartmann, Y. Zhao, R. Venkat, Effect of pressure and temperature on structural stability of MoS_2. J. Phys. Chem. C **118**(6), 3230–3235 (2014)
110. N. Bandaru, R.S. Kumar, J. Baker, O. Tschauner, T. Hartmann, Y. Zhao, R. Venkat, Structural stability of WS_2 under high pressure. Int. J. Mod. Phys. B **28**(25), 1450168 (2014)
111. M. He, E. Rikkinen, Z. Zhu, Y. Tian, A.S. Anisimov, H. Jiang, A.G. Nasibulin, E.I. Kauppinen, M. Niemela, A.O.I. Krause, Temperature dependent Raman spectra of carbon nanobuds. J. Phys. Chem. C **114**(32), 13540–13545 (2010)
112. M. Matus, H. Kuzmany, Raman spectra of single-crystal C60. Appl. Phys. A Mater. Sci. Process. **56**(3), 241–248 (1993)
113. C.Q. Sun, H.L. Bai, B.K. Tay, S. Li, E.Y. Jiang, Dimension, strength, and chemical and thermal stability of a single C–C bond in carbon nanotubes. J. Phys. Chem. B **107**(31), 7544–7546 (2003)
114. N.J. Everall, J. Lumsdon, D.J. Christopher, The effect of laser-induced heating upon the vibrational raman spectra of graphites and carbon fibres. Carbon **29**(2), 133–137 (1991)
115. Z. Zhou, X. Dou, L. Ci, L. Song, D. Liu, Y. Gao, J. Wang, L. Liu, W. Zhou, S. Xie, D. Wan, Temperature dependence of the Raman spectra of individual carbon nanotubes. J. Phys. Chem. B **110**(3), 1206–1209 (2006)
116. Chiashi, S., Murakami, Y., Miyauchi, Y., S. Maruyama, Temperature dependence of Raman scattering from single-walled carbon nanotubes: undefined radial breathing mode peaks at high temperatures. Jpn. J. Appl. Phys. **47**(4), 2010–2015 (2008)
117. J.B. Cui, K. Amtmann, J. Ristein, L. Ley, Noncontact temperature measurements of diamond by Raman scattering spectroscopy. J. Appl. Phys. **83**(12), 7929–7933 (1998)
118. I. Calizo, F. Miao, W. Bao, C.N. Lau, A.A. Balandin, Variable temperature Raman microscopy as a nanometrology tool for graphene layers and graphene-based devices. Appl. Phys. Lett. **91**(7), 071913–071916 (2007)
119. H. Boppart, J. van Straaten, I.F. Silvera, Raman spectra of diamond at high pressures. Phys. Rev. B **32**(2), 1423–1425 (1985)
120. A. Merlen, N. Bendiab, P. Toulemonde, A. Aouizerat, A. San Miguel, J.L. Sauvajol, G. Montagnac, H. Cardon, P. Petit, Resonant Raman spectroscopy of single-wall carbon nanotubes under pressure. Phys. Rev. B **72**(3), 035409–035415 (2005)
121. M. Hanfland, H. Beister, K. Syassen, Graphite under pressure: equation of state and first-order Raman modes. Phys. Rev. B **39**(17), 12598–12603 (1989)
122. D.J. Late, Temperature-dependent phonon shifts in atomically thin $MoTe_2$ nanosheets. Appl. Mater. Today **5**, 98–102 (2016)
123. H. Zhang, W. Zhou, X. Li, J. Xu, Y. Shi, B. Wang, F. Miao, High temperature Raman investigation of few-layer $MoTe_2$. Appl. Phys. Lett. **108**(9), 091902 (2016)
124. J. Tsang, J. Smith Jr., M. Shafer, Effect of charge density wave fluctuations on the frequencies of optic phonons in $2H-TaSe_2$ and-$NbSe_2$. Solid State Commun. **27**(2), 145–149 (1978)
125. Z. Yan, C. Jiang, T. Pope, C. Tsang, J. Stickney, P. Goli, J. Renteria, T. Salguero, A. Balandin, Phonon and thermal properties of exfoliated $TaSe_2$ thin films. J. Appl. Phys. **114**(20), 204301 (2013)

126. A. Taube, A. Łapińska, J. Judek, M. Zdrojek, Temperature dependence of Raman shifts in layered ReSe₂ and SnSe₂ semiconductor nanosheets. Appl. Phys. Lett. **107**(1), 013105 (2015)

127. A.S. Pawbake, A. Date, S.R. Jadkar, D.J. Late, Temperature dependent raman spectroscopy and sensing behavior of few layer SnSe₂ nanosheets. Chem. Sel. **1**(16), 5380–5387 (2016)

128. S.V. Bhatt, M. Deshpande, V. Sathe, S. Chaki, Effect of pressure and temperature on Raman scattering and an anharmonicity study of tin dichalcogenide single crystals. Solid State Commun. **201**, 54–58 (2015)

129. X. Zhou, X. Hu, S. Zhou, Q. Zhang, H. Li, T. Zhai, Ultrathin 2D GeSe₂ rhombic flakes with high anisotropy realized by Van der Waals epitaxy. Adv. Funct. Mater. **27**(47), 1703858 (2017)

130. J. Park, Y. Kim, Y.I. Jhon, Y.M. Jhon, Temperature dependent Raman spectroscopic study of mono-, bi-, and tri-layer molybdenum ditelluride. Appl. Phys. Lett. **107**(15), 153106 (2015)

131. J. Xia, X.-Z. Li, X. Huang, N. Mao, D.-D. Zhu, L. Wang, H. Xu, X.-M. Meng, Physical vapor deposition synthesis of two-dimensional orthorhombic SnS flakes with strong angle/temperature-dependent Raman responses. Nanoscale **8**(4), 2063–2070 (2016)

132. S. Luo, X. Qi, H. Yao, X. Ren, Q. Chen, J. Zhong, Temperature-dependent Raman responses of the vapor-deposited tin selenide ultrathin flakes. J. Phys. Chem. C **121**(8), 4674–4679 (2017)

133. T. Fukunaga, S. Sugai, T. Kinosada, K. Murase, Observation of new Raman lines in GeSe and SnSe at low temperatures. Solid State Commun. **38**(11), 1049–1052 (1981)

134. A. Taube, A. Łapińska, J. Judek, N. Wochtman, M. Zdrojek, Temperature induced phonon behaviour in germanium selenide thin films probed by Raman spectroscopy. J. Phys. D Appl. Phys. **49**(31), 315301 (2016)

135. I.G. Lezama, A. Arora, A. Ubaldini, C. Barreteau, E. Giannini, M. Potemski, A.F. Morpurgo, Indirect-to-direct band gap crossover in few-layer MoTe₂. Nano Lett. **15**(4), 2336–2342 (2015)

136. C. Ho, P. Liao, Y. Huang, T.-R. Yang, K. Tiong, Optical absorption of ReS₂ and ReSe₂ single crystals. J. Appl. Phys. **81**(9), 6380–6383 (1997)

137. P. Manou, J. Kalomiros, A. Anagnostopoulos, K. Kambas, Optical properties of SnSe₂ single crystals. Mater. Res. Bull. **31**(11), 1407–1415 (1996)

138. L.A. Burton, T.J. Whittles, D. Hesp, W.M. Linhart, J.M. Skelton, B. Hou, R.F. Webster, G. O'Dowd, C. Reece, D. Cherns, Electronic and optical properties of single crystal SnS₂: an earth-abundant disulfide photocatalyst. J. Mater. Chem. A **4**(4), 1312–1318 (2016)

139. L. Tichý, H. Ticha, The temperature dependence of the optical gap of glassy GeSe₂. Mater. Lett. **15**(3), 198–201 (1992)

140. T. Raadik, M. Grossberg, J. Raudoja, R. Traksmaa, J. Krustok, Temperature-dependent photoreflectance of SnS crystals. J. Phys. Chem. Solids **74**(12), 1683–1685 (2013)

141. A. Elkorashy, Temperature dependence of the optical energy gap in tin monoselenide single crystals. J. Phys. Chem. Solids **50**(9), 893–898 (1989)

142. S. Vlachos, A. Lambros, A. Thanailakis, N. Economou, Anisotropic indirect absorption edge in GeSe. Phys. Status Solidi (b) **76**(2), 727–735 (1976)

143. M.M. Ugeda, D. Fernández-Torre, I. Brihuega, P. Pou, A.J. Martínez-Galera, R. Pérez, J.M. Gómez-Rodríguez, Point defects on graphene on metals. Phys. Rev. Lett. **107**(11), 116803 (2011)

144. T. Enoki, Y. Kobayashi, K.I. Fukui, Electronic structures of graphene edges and nanographene. Int. Rev. Phys. Chem. **26**(4), 609–645 (2007)

145. M.M. Ugeda, I. Brihuega, F. Guinea, J.M. Gómez-Rodríguez, Missing atom as a source of carbon magnetism. Phys. Rev. Lett. **104**, 096804 (2010)

146. Y. Du, C. Ouyang, S. Shi, M. Lei, Ab initio studies on atomic and electronic structures of black phosphorus. J. Appl. Phys. **107**(9), 093718 (2010)

147. J. Kang, S. Tongay, J. Zhou, J. Li, J. Wu, Band offsets and heterostructures of two-dimensional semiconductors. Appl. Phys. Lett. **102**(1), 012111 (2013)

148. C.Q. Sun, Y. Sun, Y.G. Nie, Y. Wang, J.S. Pan, G. Ouyang, L.K. Pan, Z. Sun, Coordination-resolved C–C Bond Length and the C 1s binding energy of carbon allotropes and the effective atomic coordination of the few-layer graphene. J. Phys. Chem. C **113**(37), 16464–16467 (2009)

149. D.T. Morelli, Thermal conductivity of high temperature superconductor substrate materials: Lanthanum aluminate and neodymium aluminate. J. Mater. Res. **7**(09), 2492–2494 (1992)
150. R.W. Keyes, The electrical properties of black phosphorus. Phys. Rev. **92**(3), 580 (1953)
151. T. Akai, S. Endo, Y. Akahama, K. Koto, Y. Marljyama, The crystal structure and oriented transformation of black phosphorus under high pressure. Int. J. High Press Res. **1**(2), 115–130 (1989)
152. S. Reich, C. Thomsen, P. Ordejon, Elastic properties of carbon nanotubes under hydrostatic pressure. Phys. Rev. B **65**(15), 153407–153411 (2002)
153. Y. Akahama, H. Kawamura, S. Carlson, T. Le Bihan, D. Häusermann, Structural stability and equation of state of simple-hexagonal phosphorus to 280 GPa: phase transition at 262 GPa. Phys. Rev. B **61**(5), 3139 (2000)
154. Z.-H. Chi, X.-M. Zhao, H. Zhang, A.F. Goncharov, S.S. Lobanov, T. Kagayama, M. Sakata, X.-J. Chen, Pressure-induced metallization of molybdenum disulfide. Phys. Rev. Lett. **113**(3), 036802 (2014)
155. F.C. Marques, R.G. Lacerda, A. Champi, V. Stolojan, D.C. Cox, S.R.P. Silva, Thermal expansion coefficient of hydrogenated amorphous carbon. Appl. Phys. Lett. **83**(15), 3099–3101 (2003)
156. R. Kalish, Ion-implantation in diamond and diamond films: doping, damage effects and their applications. Appl. Surf. Sci. **117**, 558–569 (1997)
157. D.L. Farber, J. Badro, C.M. Aracne, D.M. Teter, M. Hanfland, B. Canny, B. Couzinet, Experimental evidence for a high-pressure isostructural phase transition in osmium. Phys. Rev. Lett. **93**(9), 095502–095506 (2004)
158. C. Kittel, *Intrduction to Solid State Physics*, 8th edn. (Willey, New York, 2005)
159. W.R. Panero, R. Jeanloz, X-ray diffraction patterns from samples in the laser-heated diamond anvil cell. J. Appl. Phys. **91**(5), 2769–2778 (2002)
160. M. Naito, S. Tanaka, Carrier scattering mechanisms in 2H-TaSe$_2$. Phys. B **105**(1), 136–140 (1981)
161. S. Helmrich, R. Schneider, A.W. Achtstein, A. Arora, B. Herzog, S.M.D. Vasconcellos, M. Kolarczik, O. Schoeps, R. Bratschitsch, U. Woggon, Exciton-phonon coupling in mono- and bilayer MoTe$_2$. 2D Mater **5**, 045007 (2018)
162. A. Shafique, A. Samad, Y.-H. Shin, Ultra low lattice thermal conductivity and high carrier mobility of monolayer SnS$_2$ and SnSe$_2$: a first principles study. Phys. Chem. Chem. Phys. **19**(31), 20677–20683 (2017)
163. F. Liu, P. Parajuli, R. Rao, P. Wei, A. Karunarathne, S. Bhattacharya, R. Podila, J. He, B. Maruyama, G. Priyadarshan, Phonon anharmonicity in single-crystalline SnSe. Phys. Rev. B **98**(22), 224309 (2018)
164. Y.M. Han, J. Zhao, M. Zhou, X.X. Jiang, H.Q. Leng, L.F. Li, Thermoelectric performance of SnS and SnS–SnSe solid solution. J. Mater. Chem. A **3**(8), 4555–4559 (2015)
165. M. Sist, C. Gatti, P. Nørby, S. Cenedese, H. Kasai, K. Kato, B.B. Iversen, High-temperature crystal structure and chemical bonding in thermoelectric germanium selenide (GeSe). Chem.-Eur. J. **23**(28), 6888–6895 (2017)

Chapter 23
Sized Crystals

Abstract An incorporation of the BOLS-LBA notion and the phonon spectrometrics have reconciled the effects of crystal size, pressure, and temperature on the phonon frequency and elasticity relaxation of the group IV, ZnO, TiO$_2$, CeO$_2$, CdS, CdSe, and Bi$_2$Se$_3$ nanostructures and the group IV, III–V, II–IV bulk crystals with derivative of ever-unexpected information. Integral of the DPS peaks turns out the skin-shell thickness of nanostructured crystals and quasi-liquid water. Reproduction of the size trend of phonon frequency shift derives the bond nature index, bond length and energy, effective coordination number, intergrain coupling strength, dimer vibration frequency. Theoretical match to Debye thermal decay and compression of measurements has derived the cohesive energy, Debye temperature, binding energy density, and elasticity.

Highlights

- DPS resolves the skin thickness of the core-shell structured nanocrystals and liquid droplets.
- Intergrain coupling emits THz waves with frequency proportional to the inverse of grain size.
- Dimer vibration stems the TO and collective oscillation originates LO phonon frequency shift.
- Phonon frequency and elasticity hold the same trends of compression and Debye thermal decay.

23.1 Skin Thickness of the Core-Shelled Structures

Using the DPS, one can confirm the core-shell structure and determine the skin thickness of solid nanocrystals and water nanodroplets. The DPS was conducted to separate the skin bond vibration frequencies from their bulk mixture by differencing the bulk reference spectrum from those collected from the sized samples of the

C. Q. Sun, *Electron and Phonon Spectrometrics*,
https://doi.org/10.1007/978-981-15-3176-7_23

same substance. The DPS not only determines the skin-shell thickness but also distinguishes the performance of bonds in terms of length and stiffness and electrons in the skin shells in terms of binding energy shift.

Figure 23.1 insets show the peak-area normalized Raman spectra collected from the sized CeO_2 nanocrystals [1] and from the $H_2O(95\%) + D_2O(5\%)$ droplets [2] under the ambient conditions. At the ambient, water nanodroplet is in the supersolid state of higher melting point and lower freezing and evaporation temperature [3]. The measurements were focused on the D–O mode (centered at 2500 cm^{-1}) for the liquid water and the vibration mode centered at 464 cm^{-1} for the CeO_2. The spectral peak area normalization to one unit aims to minimizing the experimental artifacts. Insets c and d illustrate the core-shell structure and its potential well, which shows the BOLS effect on bond contraction and potential well depression in the undercoordinated skin region up to three atomic layers [4].

The DPS thus resolves the phonons transiting their population from the bulk (valley) to the skin (peak). The DPS blueshift from the valley to the peak represents

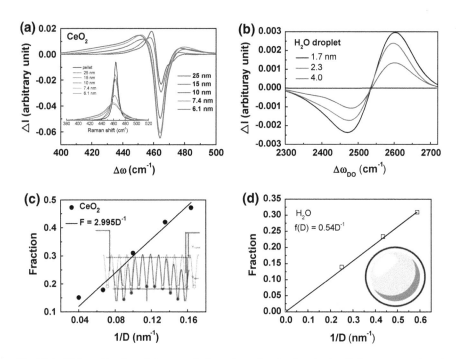

Fig. 23.1 DPS verification of the core-shell configuration for the sized **a** CeO_2 nanocrystals and **b** H_2O nanodroplets with insets a and b showing the spectral area normalized peaks (original spectral data are sourced from [1, 2]). Insets c and d illustrate the core-shell structure and its BOLS defined potential well that turns to be deeper in the skin covering shells. The fraction coefficient f(D) = $\Delta V/V = 3\Delta R/R = 6\Delta R/D$ defines the skin-shell thickness **b** $\Delta R = 0.5$ nm for CeO_2 nanocrystals and **d** $\Delta R = 0.09$ nm of the H–O dangling bond length for H_2O nanodroplets. Reprinted with permission from [7]

the skin bond stiffness gain and the redshift is associated with the undercoordination-induced subjective polarization of the surface electrons that screen and split the local potentials [5]. The CeO_2 skin covering sheets show both redshift and blueshift dominated by the bond contraction (blueshift) and the electron polarization (redshift). However, water droplets show only the D-O bond contraction featured at 2500 cm^{-1} as the polarization softens the O:H phonons in the THz regime, from 200 to 75 cm^{-1} [6]. The polarization on the high-frequency D–O bond is indistinguishable.

An integration of the DPS peak is the number/volume fraction of bonds transiting from the bulk to the skin covering shell of the nanostructures. For a spherical structure of D across, the $f(D) = \Delta V/V = 3\Delta R/R = 6\Delta R/D$, which gives rise to the shell thickness ΔR of 0.5 nm for the CeO_2 nanocrystals and 0.09 nm for the H_2O droplets. The 0.5 nm is two atomic diameters and 0.09 nm is the length of the H–O dangling bond that is 10% shorter than it is in the bulk water [3]. The direct measure of the skin-shell thickness is beyond the scope of any available methods, which proves for the universal core-shelled configured liquid and solid nanostructures. The unusual performance of bonds and electrons in the skin-shells and the varied skin/volume ratio dictate the size dependency of nanostructures in their performance [4].

23.2 Nanocrystals: Size and Thermal Effects

23.2.1 Raman Shift of the Core-Shelled Crystals

Based on the proven core-shell configuration of nanostructures, one can derive the size-induced phonon relaxation in a spherical nanostructure of the dimensionless form of size K that is the number of atoms of d across lined along the radius of a spherical dot. Measurements show that the Raman shift $\omega(K) - \omega(\infty)$ changes with the inverse of crystal size AK^{-1}. The A is the slope. A > 0 or <0 corresponds to the phonon frequency a blueshift or a redshift.

At the i-th atomic layer, the effective atomic CN approximates the relation: $z_1 = 4(1 - 0.75/K)$, $z_2 = z_1 + 2$, $z_3 = z_2 + 4$, and $z_4 = 12$. The i counts from the outermost layer inward up to a maximal of three. Incorporating the BOLS notion into the spectrometric measurements of the core-shelled nanostructures, yields,

$$\omega(K) - \omega(\infty) = \begin{cases} \frac{\pm A}{K} & (Measurement) \\ \Delta_R(\omega(\infty) - \omega(1)) & (Theory) \end{cases}$$

with

$$\Delta_R = \sum_{i \le 3} \gamma_i \left(\frac{\omega_i}{\omega_b} - 1 \right) = \begin{cases} \sum_{i \le 3} \gamma_i \left(z_{ib} C_i^{-(\frac{m}{2}+1)} - 1 \right)(z = z; \ LO \ mode) \\ \sum_{i \le 3} \gamma_i \left(C_i^{-(\frac{m}{2}+1)} - 1 \right)(z = 1; \ TO \ mode) \end{cases} \tag{5}$$

Hence, the frequency shift from the reference dimer vibration frequency $\omega(1)$ to the bulk value, $\omega(\infty) - \omega(1) = -A/(\Delta_R K)$, is a constant as $\Delta_R \propto K^{-1}$. The $\gamma_i = \Delta V/V = \tau C_i / K$ is the skin/bulk volume ratio and m is the bond nature index correlating bond energy to its length. One can obtain the reference frequency $\omega(1)$ given the slope A of the experimental profile. The sum over the skin subshells represents that the skin bond relaxation dictates the intrinsic while the size and geometry govern the trends of the size dependency of the phonon frequency shift for nanostructures.

23.2.2 Skin Dominated Size Dependency

23.2.2.1 Group IV Nanocrystals

With the core-shell configuration and the known bulk frequency shifts as references, one can reproduce the size dependence, or z-resolved Raman shift of nanostructures. Figure 23.2 shows the size K-resolved Raman shift for the group IV nanostructures. Except for the layered graphene G mode that undergoes blueshift with its inverse thickness $1/K$, the rest vibration modes for the group IV elemental nanocrystals follow the linear $\Delta\omega - 1/K$ dependence. The linear $1/K$ dependence can be expressed empirically in terms of the extended Grüneisen constant, $\partial\omega/\partial(1/K)$, though the physics behind the trends is tremendously propounding—undercoordination induced skin bond contraction originates and the skin/bulk volume ratio governs the magnitude of the phonon frequency shift.

The CN-reduction derived phonon frequency redshift occurs to the covalent system of Ge, Si, diamond, and the D and 2D modes of graphite, which evidence that the phonons of the covalently bonded systems are dominated by the collective oscillation of the associated oscillators. However, the G mode dimer oscillation exists

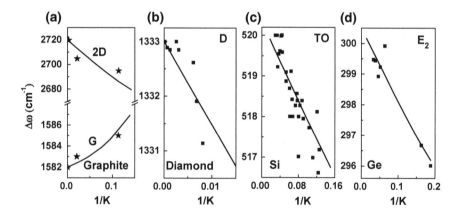

Fig. 23.2 BOLS reproduction of the K-resolved Raman shift for **a** Graphite [8] and Graphene [9–12], **b** Diamond [13, 14], **c** Si [15], and **d** Ge [16, 17] with derived information given in Table 23.1

Table 23.1 Vibration frequencies of isolated dimers of various nanosolids and their redshift upon bulk formation derived from simulating the size-dependent redshift of Raman optical modes

Material	m	Mode	$\omega(\infty)$ (cm^{-1})	$\omega(1)$ (cm^{-1})
Graphene/Graphite	2.56	2D	2565.1	2562.6
		G	1587.0	1566.7
Diamond		D	1333.3	1276.8
Si	4.88	TO	520.0	502.3
Ge	4.88	E_{2g}	302.0	290.6
TiO$_2$	5.34	A_{1g}	612.0	610.3
		E_{2g}	639.0	600.0
		E_{1g}	144.0	118.4
CeO$_2$	4.0	LO	464.5	415.1
SnO$_2$		A_{1g}	637.5	613.8
InP		LO	347.0	333.5
ZnO		E_{2g}	441.5	380.0
CdS$_{0.65}$Se$_{0.35}$		LO$_1$ CdSe-like	203.4	158.8
		LO$_2$ CdS-like	303.0	257.7
CdSe		LO	21.0	195.2
CdS		LO	106.2	106.6
Bi$_2$Se$_3$		A_{1g}^2	72.6	40.6

in the anisotropic graphite and graphene. Theoretical reproduction of the measured Raman shift derived the bond nature index, the reference $\omega(1)$ as Table 23.1 listed. The bond length and energy follow the BOLS regulation.

23.2.2.2 TiO$_2$

The anatase TiO$_2$ has six Raman active modes of 3E$_g$ (at 144, 196, and 639 cm^{-1}), 2B$_{1g}$ (397 and 519 cm^{-1}), and 1A$_{1g}$ (513 cm^{-1}). The rutile TiO$_2$ has four Raman active modes of A$_{1g}$ (612 cm^{-1}), B$_{1g}$ (143 cm^{-1}), E$_g$ (447 cm^{-1}), and B$_{2g}$ (826 cm^{-1}). The A$_{1g}$ (612 cm^{-1}) mode of rutile phase undergoes a redshift while the E$_g$ (144 cm^{-1}) mode of anatase phase undergoes a blueshift when the solid size is reduced [18–21]. The size trends of the phonon relaxation and the Yong's modulus $Y(K)$ follow the linear dependence of the inverse size of the crystals but at slightly different slopes, both of which are governed by the bond length and energy relaxation in the skin subshells.

One can determine the m value; the $\omega(1)$, and its bulk shift $\omega(\infty) - \omega(1)$ from matching the BOLS prediction to the measured/calculated size-reduction stiffened $Y(K)$ and the softened/stiffened Raman modes. The size-reduction induced Raman E$_g$ mode blueshift is governed by the same mechanism of the graphene G mode arising from dimer oscillator.

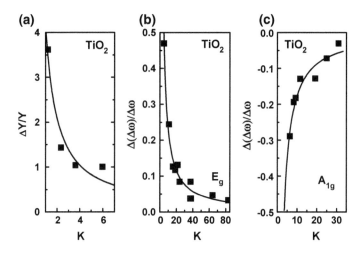

Fig. 23.3 Theoretical (solid curves) reproduction of the measured (scattered data) K-dependence of the **a** elastic modulus [22], **b** the E_g mode at 144 cm^{-1} and **c** the A_g mode at 612 cm^{-1} of nanostructured TiO$_2$ [18, 21]

Figure 23.3 shows the theoretical reproduction of the measured size dependence of the (a) Y(K) [22], and the $\Delta\omega(K)$ (b) of the E_g(144 cm^{-1}) mode of anatase phase [21], and (c) the A_{1g} (612 cm^{-1}) mode of rutile phase TiO$_2$ at the ambient [18]. The optimized bond nature index $m = 5.34$ [23]. The A_{1g} (612 cm^{-1}) mode of rutile phase undergoes a redshift while the E_g (144 cm^{-1}) mode of anatase phase goes a blueshift with size reduction. Decoding the measured size dependence of the Raman optical shift turned out the vibrational information on TiO$_2$ dimers $\omega(1) = 610.25$ cm^{-1} and their bulk shifts of 1.75 cm^{-1} for the A_{1g} mode of the rutile phase, and $\omega(1)$ = 118.35 cm^{-1} and their bulk shifts of 25.65 cm^{-1} for the E_g mode of the anatase phase.

23.2.2.3   CeO$_2$, SnO$_2$, ZnO, CdS, CdSe, and Bi$_2$Se$_3$

The LO phonons of CeO$_2$ [1, 24], SnO$_2$ [25], InP [26], ZnO [27, 28], CdS$_{0.65}$Se$_{0.35}$ [29], and CdSe [30, 31] show consistently the size-reduction induced redshift. The LO frequency goes lower when the CdS film is thinner than 80 nm and the CdSe nanodot turns to be smaller than 9.6 nm [32]. When the CdSe crystal size is reduced to 3.8 nm, the peak frequency shifts lower by about 3 cm^{-1} with respect to the bulk [31]. The Raman-active modes of Bi$_2$Se$_3$ nanoplatelets show a few wave numbers redshift as the thickness is decreased in the vicinity of ~15 nm [33], similar to that of the D and 2D modes in the number-of-layer resolved graphene [9, 10].

Repeating the same iteration of numerical reproduction of the measurements, one would be able to determine the $\omega(1)$ or the $\omega(\infty) - \omega(1)$ and the bond nature index. Figure 23.4 shows examples of the size softened optical LO phonons, which justifies

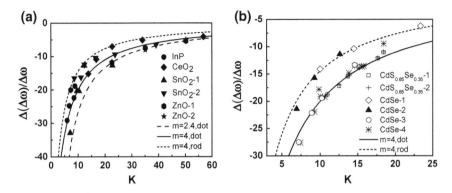

Fig. 23.4 BOLS reproduction of the measured (scattered datum) size-resolved LO phonons of **a** InP [26], CeO$_2$ [1, 24], SnO$_2$ [25], ZnO [27, 28], **b** CdS$_{0.65}$Se$_{0.35}$ [29], and CdSe [30, 31] nanoparticles with derived information given in Table 23.1

the validity of the BOLS with derived information about the corresponding dimer vibration frequency of reference.

23.2.3 Intergrain Interaction Derived THz Phonons

The LFR phonon frequency shifts linearly with the inverse size K, see Fig. 23.5. When K increases toward infinity, the LFR disappears, $\omega(\infty) = 0$, which implies that the LFR peaks not only undergo a blueshift but also are originated purely from intergrain interactions. The slope varies with the nature of the substance and the strengths of the nanoparticle-substrate interactions. Therefore, the regular Raman LO mode redshift arises from collective vibration of the lower coordinated atoms in the skin shell of the nanograin, whereas the LFR blueshift is predominated by intergrain interactions. Linearization of the size trend of the LFR gives information about the strength of the interparticle interactions, as summarized in Table 23.2. The mechanical coupling between nanoclusters is the key to the LFR [34].

23.2.4 Joint Effect of Size Reduction and Thermal Excitation

Figure 23.6 shows the theoretical reproduction of the joint effect of crystal size and thermal excitation on the phonon frequency shift of CdSe nanocrystals [49, 50]. The reproduction shows consistently that the crystal dimensionality reduction from bulk to a rod or to a dot at the nanometer scale shortens the length and enhances the energy of bonds involved in the collective oscillation, showing the redshift. However, the characteristic phonon frequency follows the Debye thermal decay that provides

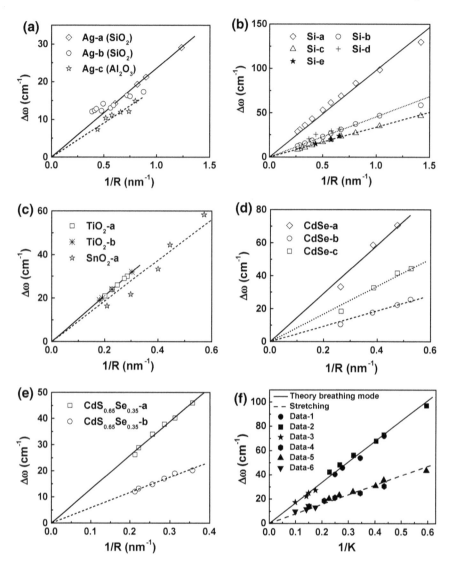

Fig. 23.5 Size-reduction derived LFR mode for **a** Si [15, 35–39], **b** Ag-a [40], Ag-b [41], Ag-c [42], **c** TiO$_2$ [43], SnO$_2$ [24], **d** CdSe-a, CdSe-b and CdSe-c [44], **e** CdS$_{0.65}$Se$_{0.35}$-a (in glass) [29], and **f** ZnO [45–47] nanoparticles. Reprinted with permission from [48]

information of the Debye temperature and the atomic cohesive energy. The inverse of the slope at higher temperatures is the inverse of E$_{coh}$, which is the nature behind the Grüneisen parameter, $\partial \omega / \partial T$.

Table 23.2 Linearization of the LFR acoustic modes of various nanosolids

Sample	A (slope)
Ag-(a, b)	23.6 ± 0.72
Ag-c	18.2 ± 0.56
TiO-(a, b)	105.5 ± 0.13
SnO-a	93.5 ± 5.43
CdSe-1-a	146.1 ± 6.27
CdSe-1-b	83.8 ± 2.84
CdSe-1-c	46.7 ± 1.39
CdS-a	129.4 ± 1.18
CdS-b	58.4 ± 0.76
Si	97.77; 45.57; 33.78

Table 23.3 Quantitative information derived from theoretical matching to the measured temperature and size dependence of the Raman shift for CdSe nanoparticles and nanorods [50]

	Input		Output					
	R (nm)	$A(d\omega/dT)$	$<z>$	d_z (nm)	E_z (eV)	E_{coh} (eV)	θ_D (K)	$\Delta\omega(z)$ (cm^{-1})
Bulk	∞	0.0142	12	0.2940	0.153	1.84	450	215.0
Dot	7.8	0.0162	9.14	0.2883	0.169	1.57	300	213.0
	4.8	0.0168	7.36	0.2824	0.204	1.47	–	211.6
	3.1	0.0170	4.81	0.2666	0.250	1.40	–	209.9
Rod	8.0		10.14	0.2906	0.160	1.65	–	213.1
	6.0	0.0157	9.52	0.2892	0.164	1.54	–	211.7
	3.5	0.0174	7.76	0.2840	0.176	1.42	–	210.1

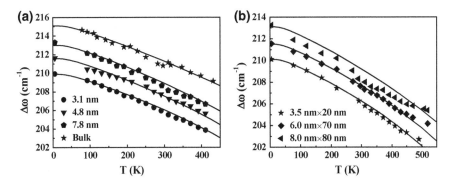

Fig. 23.6 Theoretical reproduction (solid curves) of the measured (scattered data) [49] temperature-dependent Raman shifts of **a** CdSe nanoparticles and **b** cylindrical nanorods of different radius and lengths derived with information given in Table 23.3. Reprinted with permission from [50]

23.2.5 Vibration Amplitude and Frequency of the Skin Atom

At a given temperature, the vibrational amplitude and frequency of a given atom follows Einstein's relation, $\mu(c\omega x)^2/2z = k_B T$, which yields the magnitude of vibration: $x \propto z^{1/2}\omega^{-1}$. The frequency and magnitude of vibration for a surface atom ($z = 4$) follows,

$$\frac{\omega_1}{\omega_b} = z_{ib}C_1^{-(\frac{m}{2}+1)} = \begin{cases} \dfrac{0.88^{-3.44}}{3} = 0.517\,(Si) \\[2mm] \dfrac{0.88^{-3/2}}{3} = 0.404\,(Metal) \end{cases}$$

$$\frac{x_i}{x_b} = \left(\frac{z_i}{z_b}\right)^{\frac{1}{2}}\frac{\omega_b}{\omega_1} = \left(\frac{z_b}{z_1}\right)^{\frac{1}{2}}C_1^{(\frac{m}{2}+1)}$$

$$= \begin{cases} \sqrt{3} \times 0.88^{3.44} = 1.09\,(Si) \\[2mm] \sqrt{3} \times 0.88^{3/2} = 1.43\,(Metal) \end{cases} \tag{1}$$

The vibrational amplitude of an atom at the surface is indeed greater [51, 52] than that of an atom in the bulk while the frequency becomes lower. The magnitude and frequency are sensitive to the bond nature index m value and vary with the curvature of a spherical dot when $K > 3$.

23.3 Bulk Crystals: Compression and Thermal Excitation

23.3.1 Group IV Semiconductors

Figure 23.7 shows the reproduction of the Debye thermal decay of the Raman shift and Young's modulus of group IV elemental solids with derived information of atomic cohesive energy $E_{coh}(0)$ and Debye temperature given in Table 23.4. Deviation between the derived $E_{coh}(0)$ for the $\omega(T)/\omega(0)$ and $Y(T)/Y(0)$ exists for diamond and Si but the $E_{coh}(0)$ for Ge is consistent [53].

23.3.2 Group III-Nitrides

Figure 23.8 presents the theoretical reproduction of the pressure and temperature dependent Raman shift of III–V semiconductors. Table 23.5 summarizes information of the reference frequencies $\omega(1)$, mode cohesive energy (E_{coh}) derived from the reproduction. The known Debye temperature (θ_D), melting point, and thermal expansion coefficient were used as input parameters.

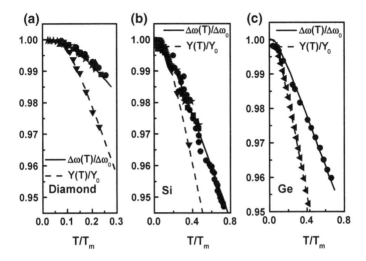

Fig. 23.7 Theoretical reproduction of the measured Debye thermal decay of the Raman shift and Young's modulus for **a** Diamond [54–58], **b** Si [59, 60], and **c** Ge [59, 61–63] with derived and input information given in Table 23.4. Reprinted with permission from [53]

Table 23.4 Information on the T_m, the Debye temperature, θ_D, T-dependent thermal expansion coefficient, $\alpha(T)$, and the atomic cohesive energy $E_{coh}(0)$ from theoretical reproduction the Debye thermal decay of the elastic modulus Y and the vibration frequency ω for Si, Ge, and Diamond [53]

	α $(10^{-6}$ K$^{-1})$ [64]	T_m (K)	θ_D (K)	$E_{coh}(0)$ (eV)			
				Raman	Y	Mean	Ref. [65]
Si	4.5	1647	647	2.83	4.33	3.58	4.03
Ge	7.5	1210	360	2.52	2.65	2.58	3.85
Diamond	55.6	3820	1860	6.64	5.71	6.18	7.37

23.3.3 TiO₂ and ZnO

TP-BOLS theoretical matching of the documented size [18, 22], temperature [80, 81], and pressure [80] dependence of the B and $\Delta\omega$ [82] for TiO$_2$ and ZnO allows for the verification of the developed solutions and extract information as given in Table 23.6. Generally, the frequency of the transverse optical (TO) phonon undergoes a redshift upon the radius R of nanosolid being decreased and almost all the modes are stiffened under high pressure and softened at elevated temperatures [20, 81, 83]. However, for the TiO$_2$, the A$_{1g}$ (612 cm^{-1}) mode of rutile phase undergoes a redshift while the E$_g$ (144 cm^{-1}) mode of anatase phase undergoes a blueshift when the solid size is reduced [18–21].

Theoretical matching to the measured and calculated size [18, 22], temperature [80, 81], and pressure [80] dependence of the B and $\Delta\omega$ [82] for TiO$_2$ at room

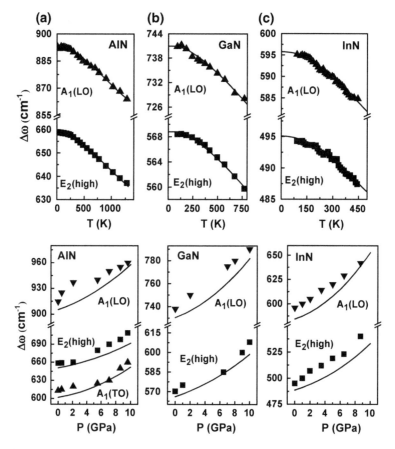

Fig. 23.8 Pressure and temperature dependent Raman shift for the bulk **a** AlN [66–68], **b** GaN [69–72], and **c** InN [73, 74]. Extracted intrinsic phonon frequency $\omega(1)$ and mode cohesive energies $E_{coh}(0)$ for various modes are tabulated in Table 23.5. Reprinted with permission from [75]

temperature allows one to verify the developed solutions and extract information as given in Table 23.6. Reproduction of TiO_2 phonons turns out m = 5.34 [23]. Reproduction of the E_g mode of TiO_2 revealed that the respective phonon frequency is contributed by only one neighbor (z = 1).

In the T-dependent curves, the shoulder is related to the θ_D and the slope at higher temperature depends on the atomic cohesive energy, $E_{coh} = zE_z$. Matching the two sets of B(T) and the $\Delta\omega(T)$ data will improve the reliability of the derivatives. The theoretical match of the measured T dependent B(T) [80] and the $\Delta\omega(T)$ [81] of the E_g (639 cm^{-1}) mode for anatase phase TiO_2 in Fig. 23.9 turns out the $\omega(1) =$ 600 cm^{-1} and the θ_D of 768 K, which is in good agreement with the reported value of 778 K [84]. The cohesive energy E_{coh} is derived as 1.56 eV. At $T \leq \theta_D/3$, the relative B and $\Delta\omega$ turns from nonlinear to linear when the temperature is increased.

Table 23.5 Melting temperature (T_m), Debye temperature (θ_D), and the thermal expansion coefficient $\alpha(T)$ for group III-N bulk crystals. The intrinsic Raman frequency at $\omega(1)$ is extrapolated from experimental data to 0 K. The fitting results, that is the mode cohesive energy at 0 K, $E_{coh}(0)$, of various Raman active modes [75, 76]

	T_m (K)	θ_D (K)	α [64] (Ref)	Raman mode	$\omega(1)$ (cm^{-1})	$E_{coh}(0)$ (eV)	σ (10^{-4})
AlN	3273	1150	[77]	E_2(high)	658.6	1.13	0.94
				A_1(LO)	892.6	1.21	1.54
				A_1(TO)	613	0.71	1.74
				E_1(LO)	914.7	1.31	1.52
				E_1(TO)	671.6	1.19	1.04
GaN	2773	600	[78]	E_2(high)	570.2	1.44	0.57
				A_1(LO)	738	0.97	1.20
				A_1(TO)	534	1.26	1.28
				E_1(LO)	745	0.95	0.68
				E_1(TO)	561.2	1.59	1.12
InN	1373	600	[79]	E_2(high)	495.1	0.76	1.15
				A_1(LO)	595.8	0.50	0.92

Table 23.6 Parameters derived from theoretical reproduction of the size, pressure, and temperature dependence of the bulk modulus and the Raman shift for TiO$_2$ [85] and ZnO [85, 90, 102]

Stimulus	Quantity	TiO$_2$ [80, 81, 96]	ZnO [90, 102]
T	Cohesive energy $E_c = z_b E_0$(eV/atom)	1.56	0.75
	Debye temperature, θ_D (K)	768 (778 [84])	310
P	Bulk modulus, B_0/B'_0(GPa/-)	143/8.86 (167/- [103])	160/4.4
	Compressibility, β/β' (GPa^{-1}/GPa^{-2})	6.84×10^{-3} $(1.21 \pm 0.05) \times 10^{-4}$	6.55×10^{-3} 1.25×10^{-4}
	Energy density, E_{den} (eV/Å^3)	0.182	0.097

Theoretical reproduction of the P-dependence leads to quantitative information of the compressibility and the binding energy density. Reproduction of the experimental x-P curve [82, 86, 87] derived the values of $\beta = 6.84 \times 10^{-3}$ GPa^{-1}, $\beta' = -1.21 \pm 0.05 \times 10^{-4}$ GPa^{-2}, $B_0 = 143$ GPa, and $B'_0 = 8.86$ [85]. Matching the prediction to the measurement derives the binding energy density 0.182 eV/Å^3. The theoretical $\Delta\omega(P)$ curve based on the relation $[\Delta\omega]^2/[Yd] \equiv 1$ showed consistency between theory prediction and measurements of the P-dependent B(P) [80] and $\Delta\omega(P)$ [20] for the E$_g$ (639 cm^{-1}) mode for anatase phase TiO$_2$.

ZnO with wurtzite structure belongs to the C_{6v}^4 symmetry group. At the Γ point of the Brillouin zone, optical phonons have the following irreducible representation: $\Gamma_{opt} = A_1 + 2B_1 + E_1 + 2E_2$, where the A_1 and the E_1 polar modes can be split

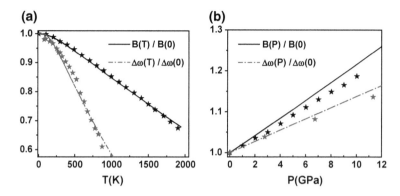

Fig. 23.9 **a** Temperature [81] and **b** pressure [20, 80] dependence of the bulk modulus and Raman shift A_{1g} mode [18, 85] of TiO_2. Theoretical matching gives rise to the mode cohesive energy and Debye temperature as listed in Table 23.6. Reprinted with permission from [85]

into TO and LO phonons, with all being the Raman and infrared active. The non-polar E_2 modes are the Raman active, while the B_1 modes are the Raman inactive [88–90]. These Raman modes contain important information. For example, the E_2 (high) mode represents the O-O bending vibrations [91].

With the known Debye temperature $\theta_D = 310$ K, bulk modulus B $= 160$ GPa and $B'_0 = 4.4$ [92], the frequency $\omega(1)$ [89, 90, 93], and the thermal expansion coefficient $\alpha(t)$ [94] as input, one can find the mode cohesive energy by matching the theory to the experimental results. Theoretical match to the measured T dependent $B(T)$ [80] and the $\Delta\omega(T)$ [81] of the E_g (639 cm^{-1}) mode for anatase phase TiO_2 in Fig. 23.9a leads to the $\omega(1) = 600$ cm^{-1} and the θ_D of 768 K, which is in good agreement with the reported value of 778 K [84]. The cohesive energy E_{coh} is derived as 1.56 eV. At $T \le \theta_D/3$, the relative B and $\Delta\omega$ turns from nonlinear to linear when the temperature is increased. The slow decrease of the B and $\Delta\omega$ at very low temperatures arises from the small $\int_0^T \eta(t)dt$ values as the specific heat $\eta(t)$ is proportional to T^3 at very low temperatures.

Figure 23.10a, b shows the theoretical match to the measured Debye thermal decay of the (a) $Y(T)$ and $\omega(T)$, (b) band gap $E_g(T)$ and (c) the pressure dependent $\omega(P)$ and $Y(P)$ for ZnO crystals [28]. Compared with the thermally softened ZnO Young's modulus, various modes of ZnO, the pressure-induced elastic stiffening results from bond compression and bond strengthening owing to mechanical work hardening.

Figure 23.10c and d present theoretical match with the measured pressure-dependent Raman shift of E_1(LO, 595 cm^{-1}), E_2(high, 441.5 cm^{-1}), E_1(TO, 410 cm^{-1}), A_1(TO, 379 cm^{-1}) and B_1(LO, 302 cm^{-1}) phonon modes for ZnO at room temperature [90, 95]. Agreement between predictions and experimental observations allows one to determine the $\omega(1)$ of E_1 (LO, 510 cm^{-1}), E_2 (high, 380 cm^{-1}), E_1 (TO, 355 cm^{-1}), A_1 (TO, 330 cm^{-1}), and B_1 (LO, 271 cm^{-1}) modes. The change of the bond energy is dependent on the ambient temperature and pressure. Therefore, the competition between the thermal expansion the pressure-induced compression determines the blueshift of Raman peaks.

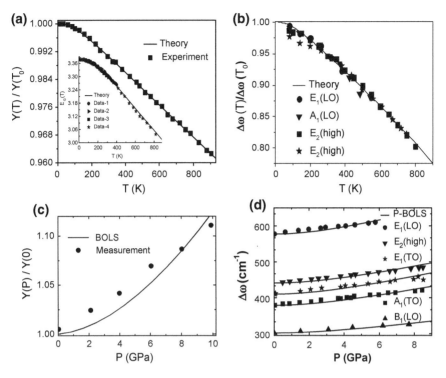

Fig. 23.10 Debye thermal decay of **a** the Young's modulus, Raman shifts [96], and band gap $E_g(T)$ [97–100] at the atmospheric pressure for ZnO with derivative of $\theta_D = 310$ K and $E_b = 0.75$ eV/bond. **c** Pressure dependence of elastic modulus [101] and **d** optical modes ($E_1(LO)$, $E_2(high)$, $E_1(TO)$, $A_1(TO)$ and, $B_1(LO)$) for ZnO [90, 95] with derivative of the binding energy density (E_d) of 0.097 eV/$Å^3$. Reprinted with permission from [28]

23.3.4 Other Compounds

Figure 23.11 presents the T-BOLS reproduction of the temperature dependent Raman shift of II–VI semiconductors. Table 23.7 summarizes information of the atomic cohesive energy (E_{coh}), Debye temperature (θ_D) with comparison of the documented θ_D.

Figure 23.12a, b show the theoretical match to the measured elasticity thermal decay of KCl, MgO, Al_2O_3, Ma_2SO_4 crystals. Figure 23.12c, d shows the theoretical match of the reported T dependence of $l(T)$ for $BaXO_3$ cubic perovskites [106]. The agreement between the theory and the reported temperature dependence of $l(T)$ of cubic perovskites justifies the validity of the Debye thermal expansion, as given in Table 23.7.

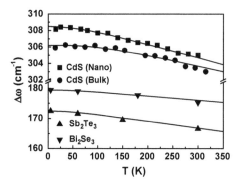

Fig. 23.11 Temperature dependence of Raman shift for CdS and CdSe [104] and Bi_2Se_3 [31–33, 105] and Sb_2Te_3 [31–33, 105] with derived information as given in Table 23.7

Table 23.7 Parameters derived from fitting to the T dependent of the Young's modulus for given materials. The T_m and θ_D for bulk are input and the atomic cohesive energy $E_b(0)$ are derived from the fitting [50, 53, 85, 103]

Sample	T_m (K)	θ_D (K)	$E_{coh}(0)$ (eV/atom)
KCl [107]	1044	214	0.57
$MgSO_4$ [107]	1397	711	2.80
Al_2O_3 [108]	2303	986	3.90
MgO [107]	3100	885	1.29
$BaTiO_3$ [106]	1862	580	2.50
$BaZrO_3$ [106]	2873	680	0.74
$BaNbO_3$ [106]	–	740	1.10
CdS (bulk)	1678 [109]	450/460 [110]	2.13
CdS (2 nm)	979.5 [111]	300/300 [110]	1.72
Sb_2Te_3	629 [112]	165/162 [113]	1.09
Bi_2Se_3	710 [114]	185/182 [115]	1.24

23.4   Summary

Incorporating the BOLS-LBA theory to the multifield phonon spectrometrics has enabled reconciliation of the crystal size, pressure, and temperature effect on the phonon and elasticity relaxation of the core-shelled ZnO, TiO_2, CeO_2, CdS, CdSe, and Bi_2Se_3 nanostructures and the group IV, III-N, II-O bulk crystals with derivative of the conventionally-unexpected information. Reproduction of the size trend of phonon frequency shift has derived the bond nature index, bond length and energy, effective atomic coordination number, intergrain interaction strength, dimer referential vibration frequency. Theoretical match to Debye thermal decay derives cohesive energy and Debye temperature. Reproduction of pressure effect leads to binding

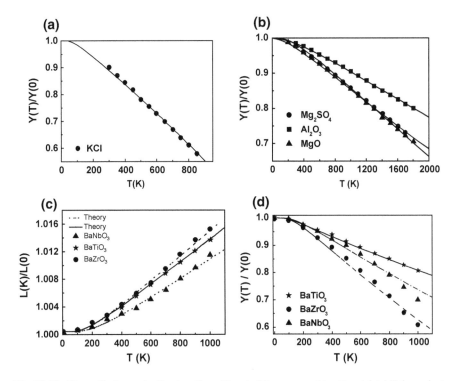

Fig. 23.12 Theoretical reproduction (continued lines) of the measured (scattered data) T-dependent Young's modulus and lattice constant for various specimens [80, 81, 106], with derived information of mean atomic cohesive energy as listed in Table 23.7

energy density and elasticity. The phonon and elasticity relaxation under perturbation results merely bond length and energy relaxation, which has little to do with the multiple phonon resonant scatter or the optical phonon degeneration into multiple acoustic phonons.

References

1. J.E. Spanier, R.D. Robinson, F. Zheng, S.W. Chan, I.P. Herman, Size-dependent properties of CeO$_2$-y nanoparticles as studied by Raman scattering. Phys. Rev. B **64**(24), 245407 (2001)
2. S. Park, D.E. Moilanen, M.D. Fayer, Water dynamics: the effects of ions and nanoconfinement. J. Phys. Chem. B **112**(17), 5279–5290 (2008)
3. Y.L. Huang, X. Zhang, Z.S. Ma, Y.C. Zhou, W.T. Zheng, J. Zhou, C.Q. Sun, Hydrogen-bond relaxation dynamics: resolving mysteries of water ice. Coord. Chem. Rev. **285**, 109–165 (2015)
4. C.Q. Sun, B.K. Tay, X.T. Zeng, S. Li, T.P. Chen, J. Zhou, H.L. Bai, E.Y. Jiang, Bond-order-bond-length-bond-strength (bond-OLS) correlation mechanism for the shape-and-size dependence of a nanosolid. J. Phys.-Condens. Matter **14**(34), 7781–7795 (2002)

5. X.J. Liu, M.L. Bo, X. Zhang, L. Li, Y.G. Nie, H. Tian, Y. Sun, S. Xu, Y. Wang, W. Zheng, C.Q. Sun, Coordination-resolved electron spectrometrics. Chem. Rev. **115**(14), 6746–6810 (2015)

6. C.Q. Sun, X. Zhang, J. Zhou, Y. Huang, Y. Zhou, W. Zheng, Density, elasticity, and stability anomalies of water molecules with fewer than four neighbors. J. Phys. Chem. Lett. **4**, 2565–2570 (2013)

7. Y. Peng, Y. Yang, Y. Sun, Y. Huang, C.Q. Sun, Phonon abundance-stiffness-lifetime transition from the mode of heavy water to its confinement and hydration. J. Mol. Liq. **276**, 688–693 (2019)

8. R.J. Nemanich, G. Lucovsky, S.A. Solin, Optical probes of the lattice dynamics of graphite. Mater. Sci. Eng. **31**, 157–160 (1977)

9. D. Graf, F. Molitor, K. Ensslin, C. Stampfer, A. Jungen, C. Hierold, L. Wirtz, Spatially resolved Raman spectroscopy of single- and few-layer graphene. Nano Lett. **7**(2), 238–242 (2007)

10. A.K. Gupta, T.J. Russin, H.R. Gutierrez, P.C. Eklund, Probing graphene edges via Raman scattering. ACS Nano **3**(1), 45–52 (2008)

11. C. Thomsen, S. Reich, Double resonant Raman scattering in graphite. Phys. Rev. Lett. **85**(24), 5214–5217 (2000)

12. A. Gupta, G. Chen, P. Joshi, S. Tadigadapa, P.C. Eklund, Raman scattering from high-frequency phonons in supported n-graphene layer films. Nano Lett. **6**(12), 2667–2673 (2006)

13. Z. Sun, J.R. Shi, B.K. Tay, S.P. Lau, UV Raman characteristics of nanocrystalline diamond films with different grain size. Diamond Relat. Mater. **9**(12), 1979–1983 (2000)

14. M. Yoshikawa, Y. Mori, M. Maegawa, G. Katagiri, H. Ishida, A. Ishitani, Raman scattering from diamond particles. Appl. Phys. Lett. **62**(24), 3114–3116 (1993)

15. C. Ossadnik, S. Veprek, I. Gregora, Applicability of Raman scattering for the characterization of nanocrystalline silicon. Thin Solid Films **337**(1–2), 148–151 (1999)

16. C.E. Bottani, C. Mantini, P. Milani, M. Manfredini, A. Stella, P. Tognini, P. Cheyssac, R. Kofman, Raman, optical-absorption, and transmission electron microscopy study of size effects in germanium quantum dots. Appl. Phys. Lett. **69**(16), 2409–2411 (1996)

17. M. Fujii, S. Hayashi, K. Yamamoto, Raman scattering from quantum dots of Ge embedded in SiO_2 thin films. Appl. Phys. Lett. **57**(25), 2692–2694 (1998)

18. V. Swamy, Size-dependent modifications of the first-order Raman spectra of nanostructured rutile TiO_2. Phys. Rev. B **77**(19), 195414 (2008)

19. S. Sahoo, A.K. Arora, V. Sridharan, Raman line shapes of optical phonons of different symmetries in anatase TiO_2 nanocrystals. J. Phys. Chem. C **113**(39), 16927–16933 (2009)

20. V. Swamy, A. Kuznetsov, L.S. Dubrovinsky, R.A. Caruso, D.G. Shchukin, B.C. Muddle, Finite-size and pressure effects on the Raman spectrum nanocrystalline anatase TiO_2. Phys. Rev. B **71**(18), 184302 (2005)

21. V. Swamy, D. Menzies, B.C. Muddle, A. Kuznetsov, L.S. Dubrovinsky, Q. Dai, V. Dmitriev, Nonlinear size dependence of anatase TiO_2 lattice parameters. Appl. Phys. Lett. **88**(24), 243103 (2006)

22. L. Dai, C.H. Sow, C.T. Lim, W.C.D. Cheong, V.B.C. Tan, Numerical investigations into the tensile behavior of TiO_2 nanowires: structural deformation, mechanical properties, and size effects. Nano Lett. **9**(2), 576–582 (2009)

23. X.J. Liu, L.W. Yang, Z.F. Zhou, P.K. Chu, C.Q. Sun, Inverse Hall-Petch relationship of nanostructured TiO_2: skin-depth energy pinning versus surface preferential melting. J. Appl. Phys. **108**, 073503 (2010)

24. A. Dieguez, A. Romano-Rodriguez, A. Vila, J.R. Morante, The complete Raman spectrum of nanometric SnO_2 particles. J. Appl. Phys. **90**(3), 1550–1557 (2001)

25. C.H. Shek, G.M. Lin, J.K.L. Lai, Effect of oxygen deficiency on the Raman spectra and hyperfine interactions of nanometer SnO_2. Nanostruct. Mater. **11**(7), 831–835 (1999)

26. M.J. Seong, O.I. Micic, A.J. Nozik, A. Mascarenhas, H.M. Cheong, Size-dependent raman study of InP quantum dots. Appl. Phys. Lett. **82**(2), 185–187 (2003)

27. H.-M. Cheng, K.-F. Lin, C.H. Hsu, C.-J. Lin, L.-J. Lin, W.-F. Hsieh, Enhanced resonant Raman scattering and electron−phonon coupling from self-assembled secondary ZnO nanoparticles. J. Phys. Chem. B **109**(39), 18385–18390 (2005)

28. J.W. Li, S.Z. Ma, X.J. Liu, Z.F. Zhou, C.Q. Sun, ZnO meso-mechano-thermo physical chemistry. Chem. Rev. **112**(5), 2833–2852 (2012)

29. P. Verma, L. Gupta, S.C. Abbi, K.P. Jain, Confinement effects on the electronic and vibronic properties of CdS0.65Se0.35 nanoparticles grown by thermal annealing. J. Appl. Phys. **88**(7), 4109–4116 (2000)

30. Y.N. Hwang, S.H. Shin, H.L. Park, S.H. Park, U. Kim, H.S. Jeong, E.J. Shin, D. Kim, Effect of lattice contraction on the Raman shifts of CdSe quantum dots in glass matrices. Phys. Rev. B **54**(21), 15120–15124 (1996)

31. A. Tanaka, S. Onari, T. Arai, Raman-scattering from CdSe microcrystals embedded in a germanate glass matrix. Phys. Rev. B **45**(12), 6587–6592 (1992)

32. D.S. Chuu, C.M. Dai, Quantum size effects in CdS thin films. Phys. Rev. B **45**(20), 11805–11810 (1992)

33. J. Zhang, Z. Peng, A. Soni, Y. Zhao, Y. Xiong, B. Peng, J. Wang, M.S. Dresselhaus, Q. Xiong, Raman spectroscopy of few-quintuple layer topological insulator Bi_2Se_3 nanoplatelets. Nano Lett. **11**(6), 2407–2414 (2011)

34. M. Talati, P.K. Jha, Low-frequency acoustic phonons in nanometric CeO_2 particles. Phys. E **28**(2), 171–177 (2005)

35. G. Viera, S. Huet, L. Boufendi, Crystal size and temperature measurements in nanostructured silicon using Raman spectroscopy. J. Appl. Phys. **90**(8), 4175–4183 (2001)

36. W. Cheng, S.F. Ren, Calculations on the size effects of Raman intensities of silicon quantum dots. Phys. Rev. B **65**(20), 205305 (2002)

37. J. Zi, H. Büscher, C. Falter, W. Ludwig, K. Zhang, X. Xie, Raman shifts in Si nanocrystals. Appl. Phys. Lett. **69**(2), 200–202 (1996)

38. G.X. Cheng, H. Xia, K.J. Chen, W. Zhang, X.K. Zhang, Raman measurement of the grain size for silicon crystallites. Phys. Stat. Sol. A **118**(1), K51–K54 (1990)

39. Z. Iqbal, S. Veprek, Raman-scattering from hydrogenated microcrystalline and amorphous-silicon. J. Phys. C-Solid State Phys. **15**(2), 377–392 (1982)

40. P. Gangopadhyay, R. Kesavamoorthy, K.G.M. Nair, R. Dhandapani, Raman scattering studies on silver nanoclusters in a silica matrix formed by ion-beam mixing. J. Appl. Phys. **88**(9), 4975–4979 (2000)

41. M. Fujii, T. Nagareda, S. Hayashi, K. Yamamoto, Low-frequency Raman-scattering from small silver particles embedded in SiO_2 thin-films. Phys. Rev. B **44**(12), 6243–6248 (1991)

42. B. Palpant, H. Portales, L. Saviot, J. Lerme, B. Prevel, M. Pellarin, E. Duval, A. Perez, M. Broyer, Quadrupolar vibrational mode of silver clusters from plasmon-assisted Raman scattering. Phys. Rev. B **60**(24), 17107–17111 (1999)

43. P. Gotic, M. Ivanda, A. Sekulic, S. Music, S. Popovic, A. Turkovic, K. Furic, Microstructure of nanosized TiO_2 obtained by sol-gel synthesis. Mater. Lett. **28**(1–3), 225–229 (1996)

44. L. Saviot, B. Champagnon, E. Duval, I.A. Kudriavtsev, A.I. Ekimov, Size dependence of acoustic and optical vibrational modes of CdSe nanocrystals in glasses. J. Non-Cryst. Solids **197**(2–3), 238–246 (1996)

45. P.M. Chassaing, F. Demangeot, N. Combe, L. Saint-Macary, M.L. Kahn, B. Chaudret, Raman scattering by acoustic phonons in wurtzite ZnO prismatic nanoparticles. Phys. Rev. B **79**(15), 155314-5 (2009)

46. N. Combe, P.-M. Chassaing, F. Demangeot, Surface effects in zinc oxide nanoparticles. Phys. Rev. B **79**(4), 045408-9 (2009)

47. H.K. Yadav, V. Gupta, K. Sreenivas, S.P. Singh, B. Sundarakannan, R.S. Katiyar, Low frequency raman scattering from acoustic phonons confined in ZnO nanoparticles. Phys. Rev. Lett. **97**(8), 085502 (2006)

48. C.Q. Sun, Size dependence of nanostructures: Impact of bond order deficiency. Prog. Solid State Chem. **35**(1), 1–159 (2007)

49. P. Kusch, H. Lange, M. Artemyev, C. Thomsen, Size-dependence of the anharmonicities in the vibrational potential of colloidal CdSe nanocrystals. Solid State Commun. **151**(1), 67–70 (2011)

50. X.X. Yang, Z.F. Zhou, Y. Wang, R. Jiang, W.T. Zheng, C.Q. Sun, Raman spectroscopy determination of the Debye temperature and atomic cohesive energy of CdS, CdSe, Bi_2Se_3, and Sb_2Te_3 nanostructures. J. Appl. Phys. **112**(8), 4759207 (2012)

51. F.G. Shi, Size-dependent thermal vibrations and melting in nanocrystals. J. Mater. Res. **9**(5), 1307–1313 (1994)

52. Q. Jiang, Z. Zhang, J.C. Li, Superheating of nanocrystals embedded in matrix. Chem. Phys. Lett. **322**(6), 549–552 (2000)

53. M.X. Gu, Y.C. Zhou, L.K. Pan, Z. Sun, S.Z. Wang, C.Q. Sun, Temperature dependence of the elastic and vibronic behavior of Si, Ge, and diamond crystals. J. Appl. Phys. **102**(8), 083524 (2007)

54. M.S. Liu, L.A. Bursill, S. Prawer, R. Beserman, Temperature dependence of the first-order Raman phonon lime of diamond. Phys. Rev. B **61**(5), 3391–3395 (2000)

55. J.B. Cui, K. Amtmann, J. Ristein, L. Ley, Noncontact temperature measurements of diamond by Raman scattering spectroscopy. J. Appl. Phys. **83**(12), 7929–7933 (1998)

56. H. Herchen, M.A. Cappelli, 1st-order Raman-spectrum of diamond at high-temperatures. Phys. Rev. B **43**(14), 11740–11744 (1991)

57. E.S. Zouboulis, M. Grimsditch, Raman-scattering in diamond up to 1900-K. Phys. Rev. B **43**(15), 12490–12493 (1991)

58. D.A. Czaplewski, J.P. Sullivan, T.A. Friedmann, J.R. Wendt, Temperature dependence of the mechanical properties of tetrahedrally coordinated amorphous carbon thin films. Appl. Phys. Lett. **87**(16), 2108132 (2005)

59. J. Menendez, M. Cardona, Temperature-dependence of the 1st-order raman-scattering by phonons in Si, Ge, and α-Sn: anharmonic effects. Phys. Rev. B **29**(4), 2051–2059 (1984)

60. M.E. Fine, Elasticity and thermal expansion of germinium between −195 and 275 °C. J. Appl. Phys. **24**(3), 338–340 (1953)

61. T.R. Hart, R.L. Aggarwal, B. Lax, Temperature dependence of Raman scattering in silicon. Phys. Rev. B **1**(2), 638–642 (1970)

62. M. Balkanski, R.F. Wallis, E. Haro, Anharmonic effects in light-scattering due to optical phonons in silicon. Phys. Rev. B **28**(4), 1928–1934 (1983)

63. U. Gysin, S. Rast, P. Ruff, E. Meyer, D.W. Lee, P. Vettiger, C. Gerber, Temperature dependence of the force sensitivity of silicon cantilevers. Phys. Rev. B **69**(4), 045403 (2004)

64. M.X. Gu, Y.C. Zhou, C.Q. Sun, Local bond average for the thermally induced lattice expansion. J. Phys. Chem. B **112**(27), 7992–7995 (2008)

65. C. Kittel, *Intrduction to Solid State Physics*, 8th edn. (Wiley, New York, 2005)

66. P. Perlin, A. Polian, T. Suski, Raman-scattering studies of aluminum nitride at high-pressure. Phys. Rev. B **47**(5), 2874–2877 (1993)

67. M. Kuball, J.M. Hayes, A.D. Prins, N.W.A. van Uden, D.J. Dunstan, Y. Shi, J.H. Edgar, Raman scattering studies on single-crystalline bulk AlN under high pressures. Appl. Phys. Lett. **78**(6), 724–726 (2001)

68. M. Ueno, A. Onodera, O. Shimomura, K. Takemura, X-ray-observation of the structural phase-transition of aluminium nitride under high-pressure. Phys. Rev. B **45**(17), 10123–10126 (1992)

69. M.P. Halsall, P. Harmer, P.J. Parbrook, S.J. Henley, Raman scattering and absorption study of the high-pressure wurtzite to rocksalt phase transition of GaN. Phys. Rev. B **69**(23), 235207 (2004)

70. P. Perlin, C. Jauberthiecarillon, J.P. Itie, A. San Miguel, I. Grzegory, A. Polian, Raman-scattering and x-ray-absorption spectroscopy in gallium nitride under high-pressure. Phys. Rev. B **45**(1), 83–89 (1992)

71. S. Limpijumnong, W.R.L. Lambrecht, Homogeneous strain deformation path for the wurtzite to rocksalt high-pressure phase transition in GaN. Phys. Rev. Lett. **86**(1), 91–94 (2001)

72. A. Link, K. Bitzer, W. Limmer, R. Sauer, C. Kirchner, V. Schwegler, M. Kamp, D.G. Ebling, K.W. Benz, Temperature dependence of the E-2 and A(1)(LO) phonons in GaN and AlN. J. Appl. Phys. **86**(11), 6256–6260 (1999)

73. C. Pinquier, F. Demangeot, J. Frandon, J.C. Chervin, A. Polian, B. Couzinet, P. Munsch, O. Briot, S. Ruffenach, B. Gil, B. Maleyre, Raman scattering study of wurtzite and rocksalt InN under high pressure. Phys. Rev. B **73**(11), 115211 (2006)

74. X.D. Pu, J. Chen, W.Z. Shen, H. Ogawa, Q.X. Guo, Temperature dependence of Raman scattering in hexagonal indium nitride films. J. Appl. Phys. **98**(3), 2006208 (2005)

75. M.X. Gu, L.K. Pan, T.C. Au Yeung, B.K. Tay, C.Q. Sun, Atomistic origin of the thermally driven softening of Raman optical phonons in group III nitrides. J. Phys. Chem. C **111**(36), 13606–13610 (2007)

76. G. Ouyang, C.Q. Sun, W.G. Zhu, Pressure-stiffened Raman phonons in group III nitrides: a local bond average approach. J. Phys. Chem. B **112**(16), 5027–5031 (2008)

77. L.K. Pan, C.Q. Sun, C.M. Li, Elucidating Si-Si dimmer vibration from the size-dependent Raman shift of nanosolid Si. J. Phys. Chem. B **108**(11), 3404–3406 (2004)

78. R.R. Reeber, K. Wang, Lattice parameters and thermal expansion of GaN. J. Mater. Res. **15**(1), 40–44 (2000)

79. G.A. Slack, S.F. Bartram, Thermal expansion of some diamondlike crystals. J. Appl. Phys. **46**(1), 89–98 (1975)

80. J. Zhu, J.X. Yu, Y.J. Wang, X.R. Chen, F.Q. Jing, First-principles calculations for elastic properties of rutile TiO_2 under pressure. Chin. Phys. B **17**(6), 2216 (2008)

81. Y.L. Du, Y. Deng, M.S. Zhang, Variable-temperature Raman scattering study on anatase titanium dioxide nanocrystals. J. Phys. Chem. Solids **67**(11), 2405–2408 (2006)

82. V. Swamy, A.Y. Kuznetsov, L.S. Dubrovinsky, A. Kurnosov, V.B. Prakapenka, Unusual compression behavior of anatase TiO_2 nanocrystals. Phys. Rev. Lett. **103**(7), 75505 (2009)

83. A.Y. Kuznetsov, R. Machado, L.S. Gomes, C.A. Achete, V. Swamy, Size dependence of rutile TiO_2 lattice parameters determined via simultaneous size, strain, and shape modeling. Appl. Phys. Lett. **94**, 193117 (2009)

84. A.Y. Wu, R.J. Sladek, Elastic Debye temperatures in tetragonal crystals: their determination and use. Phys. Rev. B **25**(8), 5230 (1982)

85. X.J. Liu, L.K. Pan, Z. Sun, Y.M. Chen, X.X. Yang, L.W. Yang, Z.F. Zhou, C.Q. Sun, Strain engineering of the elasticity and the Raman shift of nanostructured TiO(2). J. Appl. Phys. **110**(4), 044322 (2011)

86. F. Birch, Finite elastic strain of cubic crystals. Phys. Rev. **71**(11), 809–824 (1947)

87. F.D. Murnaghan, The compressibility of media under extreme pressures. Proc. Natl. Acad. Sci. U.S.A. **30**(9), 244–247 (1944)

88. L.W. Yang, X.L. Wu, G.S. Huang, T. Qiu, Y.M. Yang, In situ synthesis of Mn-doped ZnO multileg nanostructures and Mn-related Raman vibration. J. Appl. Phys. **97**(1), 014308-4 (2005)

89. R. Cuscó, E. Alarcón Lladó, J. Ibáñez, L. Artús, J. Jiménez, B. Wang, M.J. Callahan, Temperature dependence of Raman scattering in ZnO. Phys. Rev. B **75**(16), 165202–165213 (2007)

90. F. Decremps, J. Pellicer-Porres, A.M. Saitta, J.-C. Chervin, A. Polian, High-pressure Raman spectroscopy study of wurtzite ZnO. Phys. Rev. B **65**(9), 092101 (2002)

91. M.T. Cardona, M. L.W. Thewalt, Isotope effects on the optical spectra of semiconductors. Rev. Mod. Phys. **77**(4), 1173–1224 (2005)

92. H. Karzel, W. Potzel, M. Köfferlein, W. Schiessl, M. Steiner, U. Hiller, G.M. Kalvius, D.W. Mitchell, T.P. Das, P. Blaha, K. Schwarz, M.P. Pasternak, Lattice dynamics and hyperfine interactions in ZnO and ZnSe at high external pressures. Phys. Rev. B **53**(17), 11425 (1996)

93. K. Samanta, P. Bhattacharya, R.S. Katiyar, Temperature dependent E[sub 2] Raman modes in the ZnCoO ternary alloy. Phys. Rev. B (Condens. Matter Mater. Phys.) **75**(3), 035208-5 (2007)

94. A.A. Khan, X-ray determination of thermal expansion of zinc oxide. Acta Crystallogr. Section A **24**(3), 403-403 (1968)

95. J. Serrano, A.H. Romero, F.J. Manjon, R. Lauck, M. Cardona, A. Rubio, Pressure dependence of the lattice dynamics of ZnO: an ab initio approach. Phys. Rev. B **69**(9), 094306 (2004)

96. A.K. Swarnakar, L. Donzel, J. Vleugels, O. Van der Biest, High temperature properties of ZnO ceramics studied by the impulse excitation technique. J. Eur. Ceram. Soc. **29**(14), 2991–2998 (2009)

97. H. Alawadhi, S. Tsoi, X. Lu, A.K. Ramdas, M. Grimsditch, M. Cardona, R. Lauck, Effect of temperature on isotopic mass dependence of excitonic band gaps in semiconductors: ZnO. Phys. Rev. B **75**(20), 205207 (2007)

98. V.V. Ursaki, I.M. Tiginyanu, V.V. Zalamai, E.V. Rusu, G.A. Emelchenko, V.M. Masalov, E.N. Samarov, Multiphonon resonant Raman scattering in ZnO crystals and nanostructured layers. Phys. Rev. B **70**(15), 155204 (2004)

99. S.H. Eom, Y.M. Yu, Y.D. Choi, C.S. Kim, Optical characterization of ZnO whiskers grown without catalyst by hot wall epitaxy method. J. Crystal Growth **284**(1–2), 166–171 (2005)

100. R. Hauschild, H. Priller, M. Decker, J. Bruckner, H. Kalt, C. Klingshirn, Temperature dependent band gap and homogeneous line broadening of the exciton emission in ZnO. Phys. Status Solidi C **3**(4), 976–979 (2006)

101. Y. Fei, S. Cheng, L.B. Shi, H.K. Yuan, Phase transition, elastic property and electronic structure of wurtzite and rocksalt ZnO. J. Synth. Cryst. **38**(6), 1527–1531 (2009)

102. J.W. Li, L.W. Yang, Z.F. Zhou, X.J. Liu, G.F. Xie, Y. Pan, C.Q. Sun, Mechanically stiffened and thermally softened Raman modes of ZnO crystal. J. Phys. Chem. B **114**(4), 1648–1651 (2010)

103. B. Chen, H. Zhang, K. Dunphy-Guzman, D. Spagnoli, M. Kruger, D. Muthu, M. Kunz, S. Fakra, J. Hu, Q. Guo, J. Banfield, Size-dependent elasticity of nanocrystalline titania. Phys. Rev. B **79**(12), 125406 (2009)

104. E.S.F. Neto, N.O. Dantas, S.W. Da Silva, P.C. Morais, M.A.P. Da Silva, A.J.D. Moreno, V.L. Richard, G.E. Marques, C.T. Giner, Temperature-dependent Raman study of thermal parameters in CdS quantum dots. Nanotechnology **23**(12), 125701 (2012)

105. Y.N. Hwang, S.H. Park, D. Kim, Size-dependent surface phonon mode of CdSe quantum dots. Phys. Rev. B **59**(11), 7285 (1999)

106. N. Iles, A. Kellou, K.D. Khodja, B. Amrani, F. Lemoigno, D. Bourbie, H. Aourag, Atomistic study of structural, elastic, electronic and thermal properties of perovskites Ba(Ti, Zr, Nb)O-3. Comput. Mater. Sci. **39**(4), 896–902 (2007)

107. J. Garai, A. Laugier, The temperature dependence of the isothermal bulk modulus at 1 bar pressure. J. Appl. Phys. **101**(2), 2424535 (2007)

108. J.B. Wachtman, W.E. Tefft, D.G. Lam, C.S. Apstein, Exponential temperature dependence of Youngs modulus for several oxides. Phys. Rev. **122**(6), 1754 (1961)

109. C. Yang, Z.F. Zhou, J.W. Li, X.X. Yang, W. Qin, R. Jiang, N.G. Guo, Y. Wang, C.Q. Sun, Correlation between the band gap, elastic modulus, Raman shift and melting point of CdS, ZnS, and CdSe semiconductors and their size dependency. Nanoscale **4**, 1304–1307 (2012)

110. J. Rockenberger, L. Tröger, A. Kornowski, T. Vossmeyer, A. Eychmüller, J. Feldhaus, H. Weller, EXAFS studies on the size dependence of structural and dynamic properties of CdS nanoparticles. J. Phys. Chem. B **101**(14), 2691 (1997)

111. A.N. Goldstein, C.M. Echer, A.P. Alivisatos, Melting in semiconductor nanocrystals. Science **256**(5062), 1425 (1992)

112. S. Budak, C.I. Muntele, R.A. Minamisawa, B. Chhay, D. Ila, Effects of MeV Si ions bombardments on thermoelectric properties of sequentially deposited Bi x Te 3/Sb 2 Te 3 nano-layers. Nucl. Instrum. Methods Phys. Res. **261**(1–2), 608–611 (2007)

113. J.S. Dyck, W. Chen, C. Uher, Č. Drašar, and P. Lošt'ák, Heat transport in $Sb_{2-x} V_x$ Te$_3$ single crystals. Phys. Rev. B **66**(12), 125206 (2002)

114. S.M. Kang, S.S. Ha, W.G. Jung, M. Park, H.S. Song, B.J. Kim, J.I. Hong, Two-dimensional nanoplates of $Bi_2 Te_3$ and $Bi_2 Se_3$ with reduced thermal stability. Aip Adv. **6**(2), 801 (2016)

115. G.E. Shoemake, J.A. Rayne, R.W.J. Ure, Specific heat of n- and p-Type Bi_2 Te$_3$ from 1.4 to 90°K. Phys. Rev. **185**(3), 1046 (1969)

Chapter 24
Water and Aqueous Solutions

Abstract Phonon spectrometrics examination of the effect of pressure, temperature, molecular undercoordination, and charge injection by acid, base, and salt solvation establishes the regulations for the hydrogen bonding and electronic dynamics and the properties of the deionized water and aqueous solutions. Consistency between theory and measurements confirms the essentiality of the quasisolid phase of negative thermal expansion due to O:H–O segmental specific heat disparity, and the supersolid phase due to electrostatic polarization by ions injection or molecular undercoordination. Lewis acid and base solvation creates the H↔H anti–HB due to the excessive protons and the O:⇔:O super–HB because of the excessive lone pairs, respectively. The multifield mediation of the HB network results in anomalies of water ice and aqueous solutions such as ice friction, ice floating, regelation, superheating and supercooling, warm water speedy cooling, and critical conditions for phase transition. Extending the knowledge towards the deep engineering of liquid water would be promising.

Highlights

- Hydration of ions, lone pairs, protons, dipoles mediates the O:H–O network and solution properties.
- DPS distills the phonon abundance-order-stiffness transition of O:H–O bonds under perturbation.
- Ionic screened polarization and interanion repulsion stem the Hofmeister hydration volume.
- H↔H and O:⇔:O repulsion, ionic polarization, and solute bond contraction feature the Lewis solutions.

C. Q. Sun, *Electron and Phonon Spectrometrics*,
https://doi.org/10.1007/978-981-15-3176-7_24

24.1   Water and Aqueous Solutions

Water ice is ubiquitously important to agriculture, climate, environment, quality-of-life, and life sustainability. O:H dissociation for water harvesting and H–O bond dissociation for H_2 generating are strategies to conquering resource crisis; hydration and solvation laid foundations to ionic rejection in water desalination and protein dissolution; aqueous drug-cell and water-protein interfaces are of great importance to microbiology, disease curing, DNA regulating and signaling. Therefore, grasp with factors dictating the performance of H_2O molecules and their electrons is pivotal to deep engineering the liquid water and controlling its reaction, transition, and transport dynamics.

Water ice responds to perturbation such as compression, heating, electromagnetic radiation, and molecular undercoordination irregularly leading to anomalies such as ice friction, ice floating, ice regelation, supercooling/superheating, warm water quick cooling, etc. Acid, base and salt solvation makes the solution even amazing. The O:H–O bonds (or HB) in the hydration shells of solutes perform differently from they do in the ordinary water.

Much effort has been focused on the motion manner and dynamics of the excessive hydrated charge (proton and line-pair) and molecules with debating mechanisms. One often considers the solute motion dynamics in terms of phonon or molecular lifetime in the hydrating states, drift diffusivity, by taking the H_2O molecule as the basic structural unit. It was unclear how the H^+, HO^-, Y^+, and X^- ions functionalize collaboratively the hydration network and properties of the Lewis-Hofmeister solutions. Situations have been improved since the recent work focusing on the O:H–O bond theory for the behavior of water ice [1], aqueous solutions [2], and energetic cyclo-N_5^- molecular complexes [3].

The aim of this section is to show that a combination of the O:H–O bond cooperativity notion [4] and the DPS [5, 6] has enabled discoveries of O:H and H–O phonon relaxation under the perturbations of heating, compression, molecular undercoordination and charge injection by acid, base and salt solvation.

Figure 24.1 shows the full–frequency Raman spectra for the 0.1 malar ratio, monovalent (a) HX acid [7], (b) YHO base [8], and (c) YX salt [9] solutions collected under the ambient conditions and (d) heated water [10] (X = Cl, Br, I; Y = Li, Na, K). The Raman spectrum covers the phonon bands of O:H stretching vibration at <200 cm^{-1}, the $\angle$O:H–O bending band centered at 400 cm^{-1}, the $\angle$H–O–H bending band at 1600 cm^{-1}, and the H–O stretching band centered at 3200 cm^{-1}. The O:H stretching, molecular rotational and torsional vibrations are within the THz regime, one can hardly discriminate these contributions one from the other. The H–O stretching phonon band can be decomposed into the bulk (3200 cm^{-1}), the skin or the surface having a certain thickness (3450 cm^{-1}), and the surface dangling H–O bond or called free radical (3610 cm^{-1}) directing outwardly of the surface [4]. Likewise, the O:H stretching vibration phonon centered at 75 cm^{-1} features the undercoordination–induced skin O:H elongation and polarization, the ~200 cm^{-1}

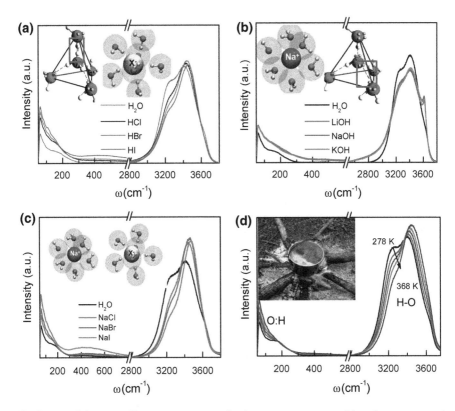

Fig. 24.1 Full-frequency Raman spectroscopy for the room-temperature, 0.1 molar concentrated **a** HX/H$_2$O [7], **b** YOH/H$_2$O [8], and **c** NaX/H$_2$O [9] solutions compared with the spectra of **d** heated water [10]. Inset a illustrates the H$_3$O$^+$:4H$_2$O unit cell with the framed H↔H point breaker and the X$^-$ point polarizer. Inset b shows the OH$^+$:4H$_2$O with the framed O:⇔:O point compressor and polarizer and the Y$^+$ polarizer. Inset c shows that the Y$^+$ and X$^-$ ions occupy eccentrically the interstitial hollow sites to form each a hydration shell through polarization and the hydrating H$_2$O dipole screen shielding. Reprinted with permission from [7–10]

peak feature the O:H vibration for the four–coordinated molecules in the ordinary bulk water.

Focusing on the evolution of the characteristic peaks for the stretching vibrations of O:H centered at <200 cm^{-1} and the H–O at >3000 cm^{-1} would suffice. Figure 24.1 insets recap the solvation reaction as follows:

$$HX + nH_2O \Rightarrow X^- + H_2O^+(H \leftrightarrow H)OH$$
$$+ (n-1)H_2O \Rightarrow X^- + (n-5)H_2O + [H_{11}O_5]^+$$
$$YHO + nH_2O \Rightarrow Y^+ + H(O^- :\Leftrightarrow: O^{2-})H_2$$
$$+ (n-1)H_2O \Rightarrow Y^+ + (n-5)H_2O + [H_9O_5]^-$$
$$YX + nH_2O \Rightarrow Y^+ + X^- + nH_2O \tag{24.1}$$

The full–frequency Raman spectra reveal that charge injection derives no vibration signatures unless the excessive feature at 3610 cm^{-1} due solute HO$^-$ contraction. Charge injection and thermal excitation do relax the H–O bond and the O:H nonbond cooperatively.

Further computational refinement [11] in Fig. 24.2 shows that Y$^+$ cation occupies the interstitial hollow site while the OH$^-$ replaces the center H$_2$O and turns the H$_2$O:4H$_2$O into an OH$^-$:4H$_2$O, which transits an O:H–O bond into the O:⇔:O super–HB. The cation forms the Y$^+$·4H$_2$O motif without any regular bond formation. The ideal Y$^+$:O^{2-} distance is $d_{Y-O} = d_{O-O} = 2.695$ Å $= \sqrt{3}a/2$ and the Y$^+$ to the six next-nearest O neighbors is $d_{Y-O} = a$ ($a = 2d_{O-O}/\sqrt{3} = 3.119$ Å at 4 °C). Being different from conventional expectation, the nearest oriented H$_2$O molecules create an electric field in the hollow sites and the Y$^+$ is subject to the Y$^+$:O^{2-} attraction in one side and Y$^+$↔H$^+$ repulsion in the opposite. The Y$^+$:O^{2-} attraction shortens itself but lengthens the H–O bond opposing to the Y$^+$:O^{2-}, while the Y$^+$↔H$^+$ repulsion does it contrastingly because of the O:H–O cooperativity. As the O:H–O configuration

Fig. 24.2 Computationally refined Y$^+$/X$^-$ (lower left) and OH$^-$/H$_3$O$^+$ (lower right) occupancy in the 4(2H$_2$O) crystal unit cell (top) of liquid water [11]. OH$^-$ or H$_3$O$^+$ replaces the central H$_2$O to form the (OH$^-$ or H$_3$O$^+$):4H$_2$O motif and turns one O:H–O bond into the H↔H anti–HB or the O:⇔:O super-HB, as framed. The Y$^+$ or X$^-$ occupies eccentrically, an offset of 0.8 Å, the interstitial hollow sites (located at the face center or corner of the unit cell) associated with four tetrahedral coordinated H$_2$O neighbors (2.6950 Å) and six next-nearest-neighbors (3.1119 Å). Reprinted with permission from [11]

and orientation conserve in the solvent, the polarization allows the oriented H_2O molecules only rotate slightly. The cation locates eccentrically by some 0.8 Å in the interstitial hollow site to polarize and relax its surroundings anisotropically [11].

This situation also applies to solvation of other solutes. The intrinsic electric field will trap the intestinal neutral molecules such as H_2 or charged ions, such as X^- and Y^+ at solvation of acid, base, and salt. Only lone pairs and protons existing in the form of OH^- and H_3O^+ can replace the H_2O to form the $(OH^-; H_3O^+):4H_2O$ associated with $H \leftrightarrow H$ anti-HB and $O: \Leftrightarrow :O$ super-HB [12].

The vibration mode features the Fourier transformation of bonds vibrating in the similar frequencies regardless of their orientations and locations in real space. The peak shape is the population function and the peak maximum the highest popularity. The peak central ω_x represents the stiffness of the vibrating bonds as a function of bond length d_x and bond energy E_x, $(\omega_x)^2 \propto E_x/d_x^2$. The peak integral is proportional to the number of bonds contributing to the phonon abundance of the vibration mode x. Thus, one can distill the fraction-stiffness transition of the O:H–O phonons by differentiating the spectrum collected from deionized water as a reference from those collected from solutions or water under perturbation such as heating and compressing upon all the spectral peak being area normalized [13]. This DPS strategy gives direct information on the O:H–O bonds transition from its vibration mode of ordinary water to the hydrating in terms of the number fraction (phonon abundance), stiffness (frequency shift), and fluctuation (line width) with much more ease and quantitative information than using the conventional spectral peak Gaussian-type decomposition.

24.2 O:H–O Bond Segmental Cooperativity

There are regulations for the O:H–O bond and the structure order of liquid water [1]. Water prefers a crystalline–like structure with well–defined lattice geometry, strong correlation, and high fluctuation. The $2H_2O$ unit cell (or $2H_2O$ motif) of C_{3v} symmetry having four HBs bridging oxygen anions. For a specimen containing N number of O^{2-} anions, there are 2 N protons H^+ and 2 N lone pairs ":" and the O:H–O bond configuration conserve regardless of structural phase [14] unless excessive H^+ or ":" is introduced [7, 8]. The $O: \Leftrightarrow :O$ inter-lone-pair repulsion couples the O:H van der Walls bond and the H–O intramolecular interactions to form the O:H–O bond that serves as the coupled oscillator pair. This oscillator features the energetics and dynamics of water ice, in terms of hydrogen bond segmental cooperative relax-ability. The segmental vibration frequency and bonding energy define the specific-heat capacity. The superposition of the segmental specific heats determine the thermodynamics of water ice and the phase structures at the ambient pressure, see Fig. 24.3. The specific-heat defines the structure phases at the ambient pressure, density anomalies, and the quasisolid (QS) phase of cooling expansion.

The motion of a H_2O molecule or the tunneling transport of a proton H^+ is subject to restriction. If the central H_2O rotates 60° or above around the C_{3v} symmetrical axis of the $2H_2O$, there will be a $H \leftrightarrow H$ anti–HB and an $O: \Leftrightarrow :O$ super–HB formed, which

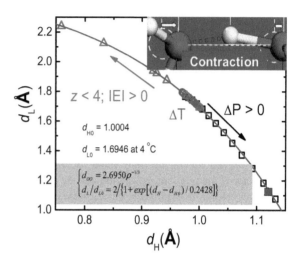

Fig. 24.3 O:H–O bond
segmental cooperative
relaxability under multifield
perturbation such as pressure
P, temperature T, molecular
CN z, and electric field E
[21]. Inset shows that O–O
length change is realized by
one segment contraction and
the other elongation and the
O:H always relaxes more
than the H–O with respect to
H at the coordination origin

is energetically forbidden. Because of the H–O bond energy of ~ 4.0 eV, translational
tunneling of the H^+ is also forbidden. Breaking the H–O bond in vapor phase needs
121.6 nm laser radiation [15], estimated 5.1 eV because the extremely low molecular
coordination numbers.

According to Einstein's relation, $\Theta_{Dx} \propto \omega_x$, and the fact that the integral of
specific heat is proportional to the bond energy E_x, the vibration frequency and the
bond energy determine the shape of the segmental specific heat $\eta_x(T/\Theta_{Dx})$ of Debye
approximation [16]. The (ω_x, E_x) is (200 cm^{-1}, ~0.1 eV) for the O:H nonbond and
(3200 cm^{-1}, ~4.0 eV) for the H–O bond. The Θ_{Dx} and the specific heat curves
are subject to the ω_x that varies with external perturbation. Figure 24.3a shows the
superposition of the specific heats η_x, which defines five phases in Fig. 24.3b showing
density oscillation over the full temperature range [16].

The hydrogen bonding thermodynamics at a certain temperature is subject to
the specific heat ratio, η_L/η_H. The segmental having a lower specific heat follows
the regular thermal expansion but the other segment responds to thermal excitation
oppositely because of the HB cooperativity by repulsion between electron pairs on
adjacent O^{2-}.

24.2.1 Physical Multifield Perturbation

As the basic structural and energy exchange unit, the O:H–O bond integrates the
intermolecular weaker O:H nonbond (or called van der Waals bond with ~0.1 eV
energy) and the intramolecular stronger H–O polar–covalent bond (~4.0 eV), rather
than either of the O:H or the H–O alone. The characteristics of the O:H–O bond is
its asymmetrical and short–range interactions and coupled with the Coulomb repulsion between electron pairs on adjacent oxygen ions [4]. O:H–O bond segmental

length and bond angle relaxation changes system energy, but structural fluctuation contributes little to the system energy on an average.

With the known H–O and O:H segmental length relaxation derived from $\rho(P)$ ($1/V(P)$) profile measured from compressed ice [18, 19] and the framework of tetrahedral-coordination for a water molecule [20], one can correlate the size d_H, separation d_{OO}, bond geometry, and mass density ρ of molecules packed in water and ice in the following manner (Fig. 24.3) [21],

$$\begin{cases} d_{OO} = 2.6950\rho^{-1/3} & (Molecular\ separation) \\ \frac{d_L}{d_{L0}} = \frac{2}{1+exp[(d_H-d_{H0})/0.2428]}; & (d_{H0} = 1.0004\ and\ d_{L0} = 1.6946\ at\ 4\ °C) \end{cases} \quad (24.2)$$

The O:H nonbond and the H–O bond segmental disparity and the O–O coupling dislocate the O^{2-} along the segmented O:H–O bond in the same direction but by different amounts under an external stimulus [16, 18, 22, 23]. The softer O:H nonbond always relaxes more than the stiffer H–O bond with respect to the H^+ as the coordination origin. The $\angle$O:H–O angle θ relaxation contributes to the geometry and mass density. The H–O relaxation absorbs or emits energy and the O:H relaxation dissipates energy caped with the 0.1 eV. The O:H–O bond segmental energy and specific heat disparity dictate the extraordinary adaptivity, cooperativity recoverability, sensitivity, properties of water and ice under external stimulus such as mechanical compression and thermal excitation.

Heating lengthens the O:H nonbond from 1.695 to 1.750 Å, but the H–O bond responds to heating contrastingly from 1.000 to 0.985 Å when heating from 273 to 377 K. The segmental disparity defines the manner of O:H–O thermal relaxation. The O:H having a lower specific heat follows the regular thermal expansion, which shortens the H–O bond by the O–O coupling [16]. Compression always shortens the O:H and lengthens the H–O regardless of the phase structures. The O:H nonbond contracts from 1.78 to 1.10 Å under compression up to 60 GPa when transits from the VII/VIII phase into phase X of identical O:H and H–O lengths. The segmental energy E_x varies with a certain power of its inverse length d_x. The phonon spectroscopy probes the length and energy as the phonon frequency shift, or segmental stiffness, in the form of $(\Delta\omega_x)^2 \propto (E/d^2)_x$. A Lagrangian-Laplacian resolution to the oscillation dynamics of the O:H–O oscillator pair has transformed the measured segmental length and vibration frequency (d_x, ω_x) int to the respective force constant and binding energy (k_x, E_x) and hence produces the potential paths of the O:H–O under continued perturbation [24, 25].

Table 24.1 summarizes the thermodynamics of O:H –O bond length and stiffness, mass density under the ambient pressure, and the molecular CN, polarization and pressure effect [16]. External perturbation changes the phase boundaries through Einstein relation $\Delta\Theta_{Dx} \propto \Delta\omega_x(P, T, z, E,...)$, referring discussions on the specific heat η_x disparity and the derives density anomalies in Chap. 11.

Numerical reproduction of the Mpemba effect—hot water cools faster [26], evidences directly the essentiality of the 0.75 gcm^{-3} mass density of the supersolid skin that promotes heat conduction outward the water of heat source. Exothermic reaction

Table 24.1 O:H–O segmental cooperative relaxation in length, vibration frequency, and surface stress with respect to $d_{L0} = 1.6946$ Å, $d_{H0} = 1.0004$ Å, $\omega_{H0} = 3200$ cm^{-1}, $\omega_{L0} = 200$ cm^{-1}, $\Theta_{DH} = 3200$ K, $\Theta_{DL} = 198$ K upon excitation by heating, compression, molecular undercoordination (skin, cluster, droplet, nanobubble), see more discussions in Chap. 11

Phase	$(T_1 - T_2)$ K	ΔT	Δd_H	Δd_L	$\Delta \omega_H$	$\Delta \omega_L$	Remark	Ref
Vapor ($\eta_L \cong$ 0)	377–	>0	–	–	–	–	H_2O monomer	
Liquid ($\eta_L/\eta_H < 1$)	277–377	>0	<0	>0	>0	<0	Liquid and solid thermal expansion	[16]
$I_c + I_h$ ($\eta_L/\eta_H < 1$)	100–258							
QS ($\eta_L/\eta_H > 1$)	258–277	>0	>0	<0	<0	>0	QS negative thermal expansion	
QS boundary ($\eta_L = \eta_H$)	258; 277	–	0	0	0	0	$\rho = 1.0$; $\rho = 0.92$ gcm^{-3}	
XI ($\eta_L \cong \eta_H \cong 0$)	0–100	<0	$\cong 0$				$\angle$O:H–O expands from 165 to 173°	
$\Delta z < 0$; $\Delta E \neq 0$ (polarization)			<0	>0	>0	<0	Polarization; supersolidity	[13, 20]
$\Delta P > 0$			>0	<0	<0	>0	d_L and d_H symmetrization	[22]

proceeds by bond elongation and dissociation while endothermic reaction proceeds by bond contraction and bond formation. The Mpemba effect integrates the O:H–O bond energy "adsorption-emission-diffusion-dissipation" cycling dynamics. The energy storage is proportional to the H–O bond heating contraction and the rate of energy emission at cooling is proportional to its first storage. The skin higher thermal conductivity due to lower mass density favors heat flow outward the solution, and the source-drain non-adiabatic dissipation ensures heat loss at cooling.

24.2.2 Acid, Base, Salt Solvation

Upon solvation, as given in Eq. (24.1), an HX acid molecule dissolves into an H$^+$ proton and an X$^-$ anion. The H$^+$ does not stay freely but bonds firmly to a H_2O molecule to form a tetrahedral H_3O^+ hydronium with one lone pair and three H$^+$, see Fig. 24.1a inset. The H_3O^+ replaces a H_2O in the center of the $2H_2O$ unit cell while its four neighbors remain their orientations because of the O:H–O regulation of interactions with their rest neighbors [1, 4]. The alteration of the ":" with a H$^+$ at the $H_2O \rightarrow H_3O^+$ transition breaks the 2N conservation with derivative of 2N +

1 number of protons and 2N-1 lone pairs in the solution. The excessive $2N + 1-(2N-1) = 2$ protons form uniquely an $H^+ \leftrightarrow H^+$ anti–HB without any other choice. The H_3O^+ remains the tetrahedron configuration having three H–O bonds and one lone pair, which is similar to the situation of $H_{2n+1}O_{2n}^+$ cluster formation with $n = 2$ and 4 [27] but no freely shuttling between oxygen ions or hopping from one site to another.

Likewise, the YOH base is dissolved into a Y^+ cation and a hydroxide (HO^- is HF-like tetrahedron with three lone pairs and one H^+). The HO^- addition transits the 2 N conservation into $2N + 3$ number of lone pairs ":" and $2N + 1$ number of protons. The excessively unbalanced $2N + 3-(2N + 1) = 2$ lone pairs can only form the $O: \Leftrightarrow :O$ point compressor between the central HO^- and one of its H_2O neighbors that reserve their molecular orientations. This $O: \Leftrightarrow :O$ compressor pertaining to each hydroxide has the same effect of but stronger mechanical compression on the network HB relaxation [22] because of the clustering of the two pairs of lone pairs and their weak binding energy to O^{2-}. The clustering of four electrons and weak bonding to O^{2-} specify the point polarizer instead of a point breaker. X^- and Y^+ in all solutions only polarize their neighboring H_2O molecules to form hydration shells.

24.3 DPS of Water and Solutions

The DPS distils only phonons transiting into their hydration states as a peak above the x–axis, which equals the abundance loss of the ordinary HBs as a valley below the axis in the DPS spectrum. This process removes the spectral areas commonly–shared by the ordinary water and the high–order hydration shells. The DPS resolves the transition of the phonon stiffness (frequency shift) and abundance (peak area) by solvation. The fraction coefficient, $f_x(C)$, being the integral of the DPS peak, is the fraction of bonds, or the number of phonons transiting from water to the hydration states at a solute concentration C.

YOH solvation broadens the main peak shifting to lower frequencies [28–30]. The solute type and concentration resolved DPS shown in Fig. 24.4 confirmed that YOH solvation indeed softens the ω_H phonon—$O: \Leftrightarrow :O$ compression lengthens the neighboring H–O bond and softens its phonon to 3100 cm^{-1} and below. Besides, a sharp peak appears at 3610 cm^{-1} that is identical to the dangling H–O bond at water surface. The sharp peaks feature the less ordered H–O bond due the HO^-. The ω_x is less sensitive to the type of the alkali cations of the same concentration. These observations confirm that the $O: \Leftrightarrow :O$ super–HB has the same but stronger effect of mechanical compression [31] and that the BOLS correlation [32] applies to aqueous solutions—intramolecular bonds of undercoordinated molecules become shorter and stronger.

Observations justify that the $O: \Leftrightarrow :O$ super–HB point compression (<3100 cm^{-1}; >220 cm^{-1}) has the same but much stronger effect of mechanically bulk compression (<3300 cm^{-1}; >200 cm^{-1}) at the critical pressure 1.33 GPa for the room-temperature water-ice transition [22], compared to the compression effect on water and ice. The

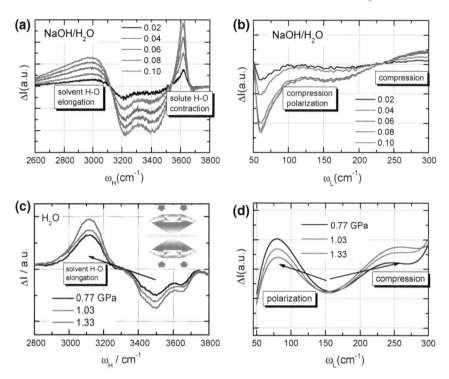

Fig. 24.4 The ω_x DPS for the **a, b** concentrated NaOH/H$_2$O solutions [8] and the **c, d** mechanically compressed water [22] under the ambient conditions. O:⇔:O compression and polarization have the same but stronger effect of pressure on the ω_H softening and ω_L stiffening. The 3610 cm^{-1} sharp feature arises from the undercoordinated solute H–O bond contraction that is the same to the dangling H–O radicals of 90% length of its bulk value

$\Delta\omega_L$ for the YOH solutions at lower concentrations duplicates the $\Delta\omega_L$ feature of the mechanically compressed water, because of the compression and polarization [18]. Compression lengthens the solvent H–O bond and softens its phonon but relaxes the O:H nonbond contrastingly. The strong effect of compression overtones the effect of Y$^+$ polarization. The excessive peak at 3610 cm^{-1} features the bond-order-deficiency induced H–O contraction of the due HO$^-$ solute, which is identical to the surface H–O dangling bond of 10% shorter, shorter and stiffer than the skin H–O bond featured at 3450 cm^{-1} for ice and water, and bulk water at 3200 cm^{-1} [33].

The broad hump at ω_H <3100 cm^{-1} shows the distant dispersion of the O:⇔:O compression forces acting on subsequent H$_2$O neighbors. The sharp feature at 3610 cm^{-1} for the YOH indicates the strong localized nature of the solute H–O bond contraction. These DPS H–O vibration peaks clarify the origin for the two ultrafast processes in terms of phonon lifetime [29, 34]. The longer 200 ± 50 fs lifetime features the slower energy dissipation of higher-frequency solute H–O bond vibrating at 3610 cm^{-1} and the other shorter time on 1–2 ps characterizes the elongated solvent H–O bond at lower-frequency of vibration <3100 cm^{-1} upon HO$^-$ solvation [35]. In the pump-probe ultrafast IR spectroscopy, the phonon population decay, or vibration energy dissipation, rate is proportional to the phonon frequency—the dynamics of higher frequency phonon is associated with a faster process.

Figure 24.5 compares the segmental ω_x DPS for the concentrated HX/H_2O solutions [7] and for the heated water [36]. The ω_L and the ω_H relax indeed cooperatively for all acidic solutions and the heated water. As the concentration increases from 0 to 0.1, X^- polarization transits the ω_L from 180 to 75 cm^{-1} and the ω_H from 3200 to 3480 cm^{-1} [37]. H$\leftrightarrow$H repulsion reverts the ω_L from 75 cm^{-1} to 110 cm^{-1}. The DPS also resolves the effect of the H$\leftrightarrow$H repulsion on the H–O bond elongation featured as a small hump below 3050 cm^{-1}. The H$\leftrightarrow$H repulsion lengthens its neighboring H–O bonds, as being the case of mechanical compression and O:$\leftrightarrow$:O point compression in basic solutions [22]. The fraction of H–O bonds elongation by H$\leftrightarrow$H repulsion decreases when the X^- turns from Cl$^-$ to Br$^-$ and I$^-$ as the stronger polarizability of I$^-$ annihilates the effect of H$\leftrightarrow$H repulsion. The small spectral valley at 3650 cm^{-1} results from the preferential skin occupation of X^- that strengthens the

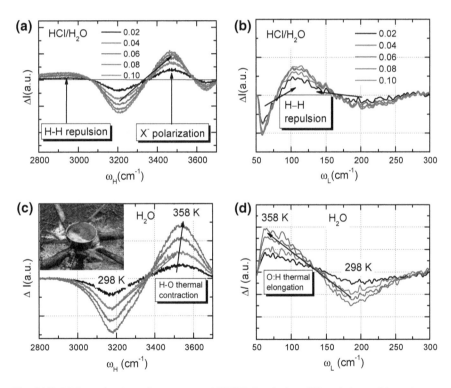

Fig. 24.5 DPS ω_x for the **a, b** concentrated HCl/H_2O solutions [7], and the **c, d** heated water [10]. The blueshift in **a** arises from X^- polarization and the humps below 3100 cm^{-1} from H$\leftrightarrow$H repulsion. The joint effect of polarization and repulsion shifts the ω_L in **b** to 110 cm^{-1}. H–O thermal contraction shifts ω_H from 3200 to 3500 cm^{-1} and the O:H elongation shifts the ω_L from 200 to 75 cm^{-1}. Reprinted with permission from [7]

local electric field. The stronger electric field stiffens the dangling H–O bond but the X^- screening weakens the signal of detection.

Acid solvation shares the same effect of segmental length and phonon stiffness relaxation and surface stress depression but different origins. X^- polarization and H–O thermal contraction stiffen the ω_H, X^- polarization and O:H thermal expansion softens the ω_L. H↔H fragilization and thermal fluctuation lowers the surface stress.

Figure 24.6 compares the ω_x DPS for the concentrated (a, b) NaCl/H$_2$O solutions [9], (c) D–O phonon for the sized (0.05D$_2$O + 0.95H$_2$O) water droplets [38, 39], and the (d) ω_H for the skins of water and ice [33, 40]. Ionic polarization and molecular undercoordination share the same effect on the H–O phonon blueshift, resulting the supersolid phase [39]. The supersolidity is characterized by the shorter and stiffer H–O bond and the longer and softer O:H nonbond, deeper O1s energy band, and longer photoelectron and phonon lifetime. The supersolid phase is less dense, viscoelastic, mechanically and thermally more stable. The O:H–O bond cooperative

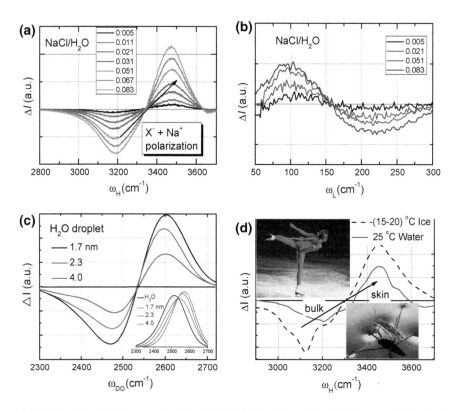

Fig. 24.6 DPS ω_x for the concentrated **a, b** NaCl/H$_2$O solutions [9], **c** D-O phonon for the sized water (0.05D$_2$O + 0.95H$_2$O) droplets [38, 39], and the **d** ω_H for the skins of water and ice [33, 40]. Molecular undercoordination transits the ω_H from 3200 for water (at 25 °C) and 3150 cm^{-1} for ice (−20 and −15 °C) to 3450 cm^{-1} that is identical to both skins of water and ice showing anomalous toughness and slipperiness

relaxation offsets boundaries of structural phases and raises the melting point and meanwhile lowers the freezing temperature of water ice—known as supercooling and superheating. The softened O:H elasticity and adaptivity and the high repulsivity of the polarized supersolid skin takes the responsibility for slipperiness of ice and the tough skin of liquid water [33]. The high thermal diffusivity of the supersolid skin ensures the Mpemba paradox—warm water cooling faster [26].

24.4 Fraction of Bond Transition and Solute-Solute Interactions

Integrating the ω_H DPS peaks for the HX, YHO, and YX yields the fraction coefficients $f_x(C)$ for the O:H–O bonds transition from its mode of ordinary water to the hydrating. The division of the fraction coefficient, $f_x(C)/C$, is proportional to the number of bonds per solute in the hydration shells, which characterizes the hydration shell size and its local electric field. A hydration shell may have one, two or more subshells, depending on the nature and size of the solute. The size and charge quantity determine its local electric field intensity that is subject to the screening by the local H_2O dipoles and the solute–solute interactions [41].

The $f_x(C)$ concentration trends show the following, see Fig. 24.7 and Fig. 24.8:

(1) The $f_H(C) \equiv 0$ means that the $H^+(H_3O^+)$ is incapable of polarizing its neighboring HBs but only breaking and slightly repulsing its neighbors [7].

(2) The $f_Y(C) \propto C$ means the constant shell size of the small Y^+ cation (radius <1.0 Å) without being interfered with by other solutes. The constant slope indicates that the number of bonds per solute conserves in the hydration shell. The electric field of a small Y^+ cation is fully screened by the H_2O dipoles in its hydration shells; thus, no cation–solute interaction is involved in the YX or YHO solutions.

(3) The $f_{OH}(C) \propto C$ (<3100, 3610 cm^{-1}) means that the numbers of the O:⇔:O compression–elongated solvent H–O bonds and the bond–order–deficiency shortened solute H–O bonds are proportional to the solute concentration. Bond order deficiency shortens and stiffens the bonds between undercoordinated atoms [4].

(4) The $f_X(C) \propto 1-\exp(-C/C_0)$ toward saturation means the number of H_2O molecules in the hydration shells is insufficient to fully screen the X^- (radius ~2.0 Å) solute local electric field because of the geometric limitation to molecules packed in the crystal–like water. This number inadequacy further evidences for the well–ordered crystal–like solvent. The solute can thus interact with their alike—only anion–anion repulsion exists in the X^-—based solutions to weaken the local electric field of X^-. Therefore, the $f_X(C)$ increases approaching saturation, the hydration shells size turns to be smaller, which limits the solute capability of O:H–O bond relaxation and polarization.

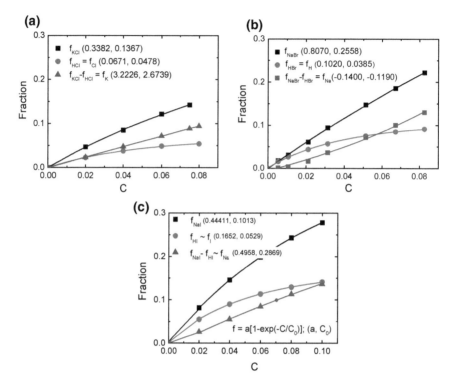

Fig. 24.7 Fraction coefficients for the O:H–O bonds transition from the mode of ordinary water to hydrating by **a** NaCl, Cl⁻, and Na⁺, **b** NaBr, Br⁻, and Na⁺, and **c** NaI, I⁻, and Na⁺. The $f_{YX}(C) = f_Y(C) + f_X(C)$ given $f_H(C) \cong 0$ and $f_{HX}(C) \cong f_X(C)$. Reprinted with permission from [7, 9]

Therefore, the $f_x(C)$ and its slope give profound information not only on the solute–solute and solute–solvent interaction but also on the relative number of bonds transiting from the referential mode of water to the hydration, by ionic polarization or O:⇔:O compression. For instance, as shown in Fig. 24.8c, the solution temperature depends linearly on the fractions of the $f_{3100}(C)$ and $f_{3600}(C)$ in the exothermic reaction. It has been clear that energy emission of H–O bond elongation by O:⇔:O compression and energy absorption of the solute H–O contraction by undercoordination heat up the solution at solvation [8].

24.5   Surface Stress, Diffusivity and Viscosity

Figure 24.9a compares the concentration dependence of the contact angle between the solutions and a glass substrate measured at 298 K. The surface stress is proportional to the contact angle. One can ignore the reaction between the glass surface and the

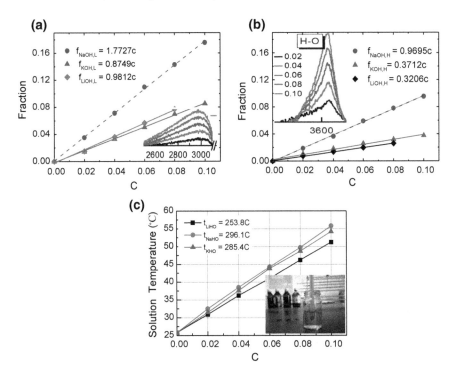

Fig. 24.8 Fraction coefficients for the **a** solvent H–O bonds elongation by O:⇔:O compression, **b** solute H–O bonds contraction by HO⁻ undercoordination, and **c** exothermal solvation of YHO solutions. Reprinted with permission from [8]

solution, as we want to know the concentration trends of the stress change at the air–solution interface of a specific solution. Ionic polarization and O:⇔:O compression and polarization enhance the stress, but the H↔H point fragilization destructs the stress. Both polarization and undercoordination form the supersolid phase; the former occurs in the hydration shell throughout the bulk, while the latter only takes place in skins. The H↔H fragmentation has the same effect of thermal fluctuation on depressing the surface stress, see Fig. 24.9b [42]. Thermal excitation weakens the individual O:H bond throughout the bulk water.

In aqueous solutions, solute molecules are taken as Brownian particles drifting randomly under thermal fluctuation by collision of the solvent molecules. The viscosity of salt solutions is one of the important macroscopic parameters often used to classify water–soluble salts into structure making or structure breaking. The drift motion diffusivity $D(\eta, R, T)$ and the solute–concentration–resolved solution viscosity $\eta(C)$ follow the Stokes–Einstein relation [43] and the Jones–Dole expression[44], respectively,

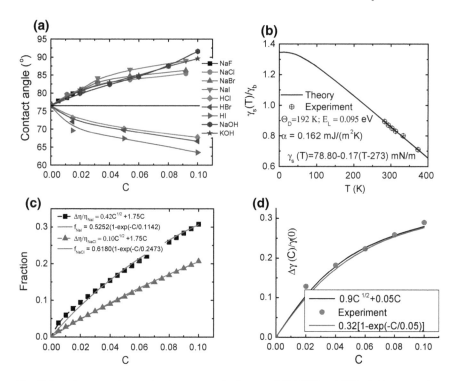

Fig. 24.9 **a** Concentration resolved contact angles between YX, YHO and HX solutions on a glass substrate [47] and **b** the thermal decay of the surface stress of liquid water [42]. Reproduction of the (**C**) $f_{NaCl}(C)$ and $f_{NaI}(C)$ for Na(Cl, I) solutions and (**d**) contact angle $\theta_{LiBr}(C)$ for LiBr solution by the Jones–Dole's expression of viscosity [44]. Reproduction of the $\gamma_S(T) \propto 1-U(T/\Theta_{DL})/E_{coh} = 1-U(T/192)/0.38$ turns out 192 K Debye temperature and the $E_L = 0/38/4 = 0.095$ eV for the O:H nonbond. $U(T/\Theta_{DL})$ is the integral of the Debye specific heat

$$\begin{cases} \dfrac{D(\eta,R,T)}{D_0} = \dfrac{k_B T}{6\pi\eta R} & (Drift) \\ \dfrac{\Delta\eta(C)}{\eta(0)} = A\sqrt{C} + BC & (Vis\,cos\,ity) \end{cases}$$

where η, R, and k_B are the viscosity, solute size, and Boltzmann constant, respectively. D_0 is the coefficient in pure water. The coefficient A and its nonlinear term is related to the solute mobility and solute–solute interaction. The coefficient B and the linear term reflects the solute–solvent molecular interactions. The $\eta(0)$ is the viscosity of neat water.

SFG measurements [45, 46] revealed that the SCN^- and CO_2 solution viscosity increases with solute concentration or solution cooling. The H–O phonon relaxation time increases with the viscosity, and results in slower molecular motion dynamics. Therefore, ionic polarization stiffens the H–O phonon and slows down the molecular motion in the semirigid or supersolid structures.

One may note that the relative viscosity and the measured surface stress due to salt solvation are in the same manner of the $f_{YX}(C)$. One can adjust the Jones–Dole viscosity coefficients A and B and fit the surface stress to match the measured $f_{YX}(C)$ curve in Fig. 24.9 c and d. The trend consistency clarifies that the linear term corresponds to Y^+ hydration shell size and the nonlinear part to the resultant of X^-–water and X^-–X^- interactions. It is clear now that both the solution viscosity and the surface stress are proportional to the extent of polarization or to the sum of O:H–O bonds in the hydration shells for the monovalent salt solutions, at least. Therefore, polarization raises the surface stress, solution viscosity and rigidity, H–O phonon frequency, and H–O phonon lifetime but decreases the molecular drift mobility, consistently by shortening the H–O bond and lengthening the O:H nonbond.

24.6 Phase Transition by Compression and Ionic Polarization

24.6.1 Spectrometric Varification

O:H–O bond segmental phonon frequencies ω_x and the critical T_{xC} for phase transition vary with the bond energy and bond length in the following manners [4]:

$$\begin{cases} w_x \propto \sqrt{E_x/\mu_x/d_x} \\ T_{xC} \propto \sum_{x=H,L} E_{xC} \end{cases} \quad (24.3)$$

The P_{Cx} and the T_C for a phase transition is correlated to the O:H–O bond energy E_{xC} ($X = L, H$):

$$T_C \propto \sum_{L,H} E_{xC} = \begin{cases} \sum_{L,H} \left(E_{x0} - s_x \int_{P_0}^{P_{C0}} p \frac{dd_x}{dp} dp \right) & (a, \text{ neat } H_2O) \\ \sum_{L,H} \left(E_x - s_x \int_{P_0}^{P_{C0}} p \frac{dd_x}{dp} dp \right) & (b, \text{solution}) \end{cases} \quad (24.4)$$

$E_{H0} < E_H$ means that the H–O bond for pure water is weaker than it is in the ionic hydration volume under the same pressure (ambient $P_0 = 100$ kPa ~ 0) [22, 23], see Figs. 24.4c and 24.6a. The integrals are energies stored into the bonds by mechanical compression. The summation is over both segments of the O:H–O bond. When the pressure increases from P_0 to P_{C0} for the neat water and to the P_C for the salted, phase transition occurs at the same temperature. At transition, the bond energy equals to the difference between the two terms in the bracket. One can assume that the change of the cross-section s_x for the specific bond of d_x length is insignificant at relaxation. To raise the critical pressure from P_{C0} to P_C at the same critical temperature ($\Delta T_C = 0$), one needs excessive energy which is the difference between the bond energy of the salted and the neat water, therefore, Eq. (24.4) yields,

$$\Delta E_x - s_x \left(\int_{P_0}^{P_c} p \frac{dd_x}{dp} dp - \int_{P_0}^{P_{c0}} p \frac{dd_x}{dp} dp \right) = 0$$

where $\Delta E_x = E_x - E_{x0}$ is the energy stored in the O:H–O bond by ionic polarization upon salt solvation. Increasing P_{C0} to P_C recovers the salt-induced bond distortion, and then phase transition occurs. This expression indicates that, if one wants to overcome the effect of ionic electrification, or molecular undercoordination, on the bond distortion for phase transition, one should increase the pressure from P_{C0} to P_C without changing the T_C. The ΔE_x varies with the solute type and concentration in different manners, as the solute-solute interaction may change with solute concentration. This principle also applies to the compressing phase transition of water nanodroplet as the ΔE_x also changes with the feature size of confinement.

Mechanical compression shortens the O:H nonbond and lengthens the H-O bond [18],

$$\frac{dd_L}{dp} < 0 \quad \frac{dd_H}{dp} > 0 \text{ and } P_C > P_{C0}$$

which derives the segmental deformation energies derived by ionic polarization,

$$\Delta E_x - s_x \left(\int_{P_0}^{P_c} p \frac{dd_x}{dp} dp - \int_{P_0}^{P_{c0}} p \frac{dd_x}{dp} dp \right) \begin{cases} > 0 (H - O) \\ < 0 (O : H) \end{cases} \quad (24.5)$$

The resultant of the ΔE_L loss and the ΔE_H gain governs the ΔP_C for the phase transition of the aqueous solutions.

24.6.2 Phonon Spectrometric Verification

24.6.2.1 Compression of Pure and Salted Ices

Figure 24.10a shows that mechanical compression softens the ω_H phonon in both ice and LiCl solutions [48]. The extent of the H–O bond softening varies with its molecular sites in the solution. The H–O bond of higher vibration frequency corresponds to those in the hydration volume, which is less sensitive to the pressure. The hydrating H–O bond relaxes less than it is in pure ice [49, 50]. The initially shorter and stiffer H–O bonds under polarization are hardly compressible and thermally more stable than they are in pure water and ice.

Figure 24.10b compares the mean O–O distance (O:H–O length) in pure ice, in LiCl- and NaCl-contained ice [51]. The O:H–O in the hydration shells of Li^+ and Na^+ is always longer than it is in pure ice. The O:H–O bond outside the hydration volume showing the same slope of variation to the pure ice. These observations confirm the

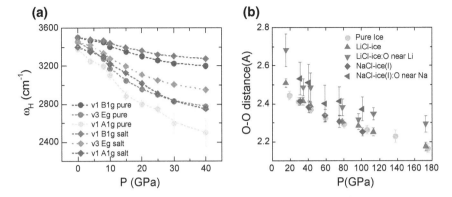

Fig. 24.10 a Raman H–O stretching mode in pure and LiCl-contained ice (0.08 mol concentration [48]) and **b** the mean O–O distance in ice (squares), LiCl-contained ice (solid circles), and in NaCl ice (solid triangles). The ν1 B1g > ν3 Eg > ν1 A1g features the Raman spectral components for the site-resolved H–O bond in the hydration volume. Open circles and triangles correspond to the mean O–O distances near the cation. Reprinted with permission from [51]

opposite effect of mechanical compression and ionic polarization on the O:H–O bond cooperative relaxation. Compression shortens and stiffens the O:H nonbond while lengthens and softens the H–O bond, but the ionic polarization does it oppositely in salt solutions.

24.6.2.2   Room Temprature NaX/H₂O Compression Icing

High-pressure Raman examination of the NaX and concentrated Na/H₂O Liquid-VI and the VI-VII phase transition confirmed consistently the above predictions [22]. The Raman spectra show abrupt changes at the P_{Cx} for the phase transition. The P_C sharp fall at the phase boundary indicates that the structural relaxation weakens the O–O repulsion, see Fig. 24.11a. Raman spectra for the room-temperature salt solution Liquid-VI and then VI-VII icing revealed the following [22]:

(1)   Three pressure zones from low to high correspond to phases of Liquid, ice VI, and ice VII, toward phase X [4].
(2)   Compression shortens the O:H nonbond and stiffens its ω_L phonon but does the opposite to the H–O bond over the full pressure range except for the P_{Cx} at phase boundaries.
(3)   At transition, the gauged pressure drops, resulting from geometric restructuring that weakens the O–O repulsion; both the O:H and the H–O contract abruptly when cross the phase boundaries.
(4)   Most strikingly, the P_{Cx} increases with the anion radius or the electronegativity difference between Na and X, following the Hofmeister series order: I > Br > Cl > F ≈ 0.

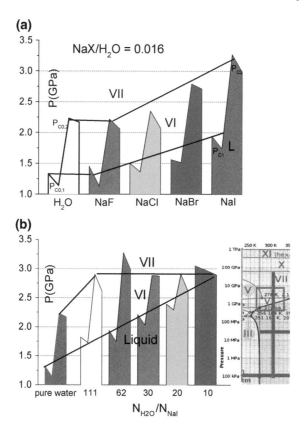

Fig. 24.11 **a** NaX type and NaI concentration dependence of the critical P_{C1} and P_{C2} for the Liquid-VI and the VI-VII phase transition at 298 K [22]. Solute capabilities of raising the critical pressures at 0.016 molar ratio follow the Hofmeister order of $I > Br > Cl > F = 0$. **b** The pressure path is along the Liquid-VI boundary for the concentrated NaI solutions to the triple phase joint at 3.3 GPa and 350 K. The involvement of the anion-anion repulsion discriminates the concentration from the type effect. The L-VI boundary is more sensitive than the VI-VII boundary to the solute concentration change [22]

24.6.2.3 Concentrated NaI/H₂O Pressure Icing: Anion-Anion Repulsion

Figure 24.11b shows the NaI/H₂O concentration dependent P_{Cx} for the Liquid-VI and VI-VII transition at 298 K [23]. Results show consistently that compression shortens the O:H nonbond and stiffens its phonons but the H–O bond responds to pressure contrastingly throughout the course unless at the phased boundaries. At higher concentrations, say 0.05 and 0.10, the skin 3450 cm⁻¹ mode are more active in responding to pressure, which evidence the preferential skin occupancy of the I⁻ anions that enhances the local electric filed.

The high-pressure Raman spectra from the concentrated NaI/H$_2$O solutions revealed the following, see Fig. 24.11b [23]:

(1) The P$_{C1}$ for the Liquid-VI transition increases faster than the P$_{C2}$ with NaI concentration till its maximum at 3.0 GPa and 0.10 concentration. The P$_{C1}$ approaches to the P$_{C2}$ and eventually meets at the 3.3 GPa and 350 K triple phase junction.
(2) The P$_{C2}$ for VI-VII transition changes insignificantly with concentration, keeping almost at the VI–VII boundary in the phase diagram, which contrasts with the trend of the solution type effect.
(3) The concentration trend of the P$_{C1}$ along the L-VI boundary is equivalent to the simultaneous compressing and heating in the phase diagram.

The discrepancy between the solute type and the concentration on the P$_{Cx}$ arises from the involvement of anion-anion repulsion that weakens the local electric field of the hydration volume at higher concentrations. Solute type determines the nature and the extent of the initial electrification; concentration increase modulates the local electric field and the extent of the initial H–O bond energy storage. The P$_{C2}$ is less sensitive than the P$_{C1}$ to the change of solute concentration because of the highly-compressed O:H–O bond is less sensitive to the local electric field of the hydration volume and the compression. The deformed O:H–O bond is harder to deform further than those less-deformed under the same pressure.

One may note that, the O:H–O bond is very sensitive to the environment such as pressure holding time, temperature, and phase precipitation. The measurement may not be readily reproducible but the trends of measurements and the physical origin always hold. The maximal P$_{C2}$, 3.27 > 3.05 > 2.23 GPa, for the 0.016 NaI/H$_2$O solution [22], the 0.10 NaI/H$_2$O, and the deionized water, demonstrates clearly the Hofmeister effect, the presence of the anion-anion repulsion, on the critical pressures for the room-temperature phase transition, as compared in Fig. 24.11.

24.7 Summary

An extension of the conventional phonon spectroscopy to the DPS has enabled resolution of the fraction and stiffness of O:H–O bonds transiting from the mode of ordinary water to their hydration and conditioned water, which amplifies solvation study from the molecular dynamics in temporal and spatial domains to hydration bonding energetic dynamics. Table 24.2 summarizes observations of the O:H–O segmental cooperative relaxation in length, vibration frequency, and water surface stress upon external excitation with respect to standard situation at 277 K.

The combination of the O:H–O bond cooperativity premise and the DPS strategy has enabled development of new knowledge, resolution to anomalies, and strategies for deep engineering of water ice in resolving the multifield effect on the HB network and properties of water and solutions. The DPS has allowed quantitative information

Table 24.2 O:H–O segmental cooperative relaxation in length, vibration frequency, and water surface stress with respect to d_{L0} = 1.6946 Å, d_{H0} = 1.0004 Å, ω_{H0} = 3200 cm^{-1}, ω_{L0} = 200 cm^{-1}, γ_s = 72.5 J·m^{-2} at 277 K upon excitation by heating, compression, molecular undercoordination (skin, cluster, droplet, nanobubble) and acid, base, salt, and organic molecular solvation [1, 4]

		Δd_H	$\Delta \omega_H$	Δd_L	$\Delta \omega_L$	$\Delta \gamma_s$	Remark	Refs.
Liquid water	Liquid heating	<0	>0	>0	<0	<0	d_L elongation; d_H contraction; thermal fluctuation	[16]
	Under-coordination					>0	d_H contraction; d_L elongation; polarization; supersolidity	[20]
	Compression	>0	<0	<0	>0	–	d_L compression; d_H elongation	[22]
Aqueous solution	YX salt	<0	>0	>0	<0	>0	Y$^+$ and X$^-$ polarization	[9]
	HX acid					<0	H↔H fragilization; X$^-$ polarization	[7]
	YOH base	>0	>0	>0	>0	>0	O:⇔:O compression; Y$^+$ polarization; solute H–O bond contraction	[35]

on the number fraction and phonon stiffness transition from the mode of ordinary HBs to hydration. The multifield meditation of the O:H–O bond in the network can thus be discriminated as follows:

(1) Surface molecular undercoordination shortens the H–O bond from 1.00 to 0.95 Å and stiffens its phonon from 3200 to 3450 cm^{-1}, meanwhile, lengthens the O:H nonbond from 1.70 to ~1.95 Å and softens the O:H phonon from 200 to 75 cm^{-1}. The dangling H–O bond is 0.90 Å long and its phonon frequency is 3610 cm^{-1}. The O:H–O bond segmental energies transit from (0.2, 4.0) to (0.1, 4.6) eV when moving from the bulk to the skin in comparison to the gaseous H–O energy of 5.10 eV. The skin mass density drops from the bulk standard of 1.0 to 0.75 g·cm^{-3}, which raised the skin thermal diffusivity by a factor of ~4/3.

Theoretical reproduction of the Mpemba paradox profiles—warm water cools more quickly, evidences the essentiality of the skin high thermal diffusivity.

(2) The nonbonding electron polarization furnishes the surface with excessive charge and dipoles. The O:H–O bond relaxation and electronic polarization endowed the aqueous hydration volume and water ice surface with supersolidity that is highly elastic, semi-rigid, electrostatic repulsive, hydrophobic, slippery, thermally more diffusive and stable with higher melting point T_m, higher surface stress, mechanical strength, but lower freezing T_N and evaporation point T_N and mass density. The degree of supersolidity increases with the surface curvature or with further lowing of the molecular coordination numbers. The supersolidity of nanodroplets, nanobubbles, and molecular clusters is higher than that of a flat surface, resulting in the super-heating/cooling/cooling of melting/freezing/evaporating. For instance, a 1.4 nm sized droplet freezes at 205 K [17] instead of the bulk at 258 K [16]. Smaller water supersolid droplet shows superfluidity, superlubricity, and superhydrophobicity, which enables its rapid flow when passing through carbon nanotubes or microchannels.

(3) Strikingly, Raman spectroscopy revealed that the skins of 25 °C water and −(15 ~ 20) °C ice share H–O bonds of identical length, energy, and vibration frequency of 3450 cm^{-1}. X-ray K edge absorption and Raman scattering unveiled that the deformed skin H–O bond is thermally much more stable than it is in the bulk. The supersolidity of the nanodroplet skin hinders vibration energy dissipation to lengthen the H–O phonon lifetime, according to the ultrafast IR spectroscopy.

(4) The O:H/H–O segmental phonon frequency softening/stiffening offset their Debye temperatures and disperses outwardly the QS phase boundaries. Hence, the supersolid skin is neither a liquid nor a solid at the ambient but rather an extended QS phase having a higher T_m (~330 K), lower T_N and T_V.

(5) Examination of the multifield effect on surface O:H–O bonds and electrons has cultivated new knowledge and strategies of regulating water evaporation and desalination, hydrogen generation, and skin supersolidity modulation for friction and hydrophobicity, etc. Besides, the theory has enabled quantitative resolutions to anomalies such as ice buoyancy, ice regelation, critical energies for phase transition, energy exchange by H–O bond relaxation, hydration shell size and droplet skin thickness determination, etc.

(6) The skin supersolidity is engineerable by O:H–O relaxation and polarization through programed stimulation. Electrostatic polarization by a parallel electric field enhances the supersolidity, which enables the water floating bridge to form and be sustained between two filled cups. An ion forms a semi-rigid supersolid hydration volume, which separates graphene oxide sheets up to 1.5 nm for the selective ion rejection. However, a superposition of the applied field and the ionic field breaks the floating bridge of acid or salt solutions; a mixture of soil grains and aqueous ions also weakens their fields than either of them alone to accelerate soil wetting by the salt solutions. Molecular undercoordination and electrostatic polarization enhances each other to lengthen and weaken the O:H nonbond whose energy dictates the T_N and T_V, being beneficial to evaporation and bio-product cryogenic reservation. Magnetization applies a Lorentz field to

flowing water dipoles, which adds a translational/rotational component to the rotational/translational motion and hence raises the mobility of the dipoles to promote the micro-circulation of human body fluid and blood.

Regarding the aqueous solutions, we have obtained quantitative information that may complement to the premises of continuum thermodynamics, molecular dynamics:

(1) H_3O^+ hydronium formation in acid solution creates the H↔H anti–HB that serves as a point breaker to disrupt the HX solution network and the surface stress. The X^- polarization dictates the O:H (from 200 partially to 110 and 300 cm^{-1}) and the H–O phonon (from 3200 to 3480 cm^{-1}) frequency cooperative shift. Acid solvation has the same effect of liquid heating on the O:H–O bond network and phonon relaxation through, respectively, H↔H fragilization and thermal fluctuation.

(2) OH^- hydroxide forms the O:⇔:O super–HB point compressor to soften the nearest solvent H–O bond (from above 3100 cm^{-1} to below), and meanwhile, the solute H–O bond shortens to its dangling radicals featured at 3610 cm^{-1}. The Y^+ polarization effect has been annihilated by the O:⇔:O compression and the solute H–O contraction. Base solvation has partly the effect of mechanical compression that lengthens and softens the H–O bond and shortens the O:H nonbond.

(3) Y^+ and X^- ions serve each as a point polarizer that aligns, stretches, and polarizes the surrounding O:H–O bonds and makes the hydration shell supersolid. The polarization transits the ω_L from 200 to 100 cm^{-1} and the ω_H from 3200 to 3480 cm^{-1}. Salt solvation has the same effect of molecular undercoordination to form the supersolid states.

(4) The solute capability of bond transition follows: $f_H(C) = 0$, $f_Y(C) \propto f_{OH}(C) \propto C$, and $f_X(C) \propto 1-\exp(-C/C_0)$ toward saturation, which indicate the nonparaxiality of protons, invariant hydration volume size of Y^+ and HO^-, and the variant X^-–X^- repulsion, respectively, and hence evidence the high structure order of H_2O molecular in the solvent matrix. The concentration trends consistent among the salt solution viscosity, surface stress, and the $f_{YX}(C)$ suggest their common origin of polarization associated with O:H–O bond transition from water to hydration shells.

(5) Solvation and compression have the opposite effect on O:H–O relaxation and the shortened H–O bond by polarization can hardly elongated than the H–O bond in pure water. At constant concentration, the P_C shows the Hofmeister sequences, but the involvement of the anion-anion repulsion alters the pattern of the critical pressure for the phase transition for the concentrated salt solutions.

(6) Compression, confinement, and heating compensate one another other on the critical pressure of phase transition at the ambient as compression always shortens the O:H nonbond while heating and confinement do oppositely in the QS phase of negative thermal expansion—H–O undergoes liquid heating and compression elongation and O:H contraction but confinement does oppositely. The Liquid-VI phase boundary is less mechanically stable than the VI-VII boundary because the involvement of the anion-anion repulsion wakens the local electric

field of the ionic hydration volume and the already deformed bond is hardly further deformable.

References

1. C.Q. Sun, Y. Sun, The attribute of water: Single notion, multiple myths. Springer Ser. Chem. Phys. **113**, 494 (2016). (Springer, Heidelberg)
2. C.Q. Sun, Solvation dynamics: A notion of charge injection. Springer Ser. Chem. Phys. **121**, 316 (2019). (Springer-Nature, German)
3. L. Zhang, C. Yao, Y. Yu, S.-L. Jiang, C.Q. Sun, J. Chen, Stabilization of the dual-aromatic Cyclo-N5⁻ Anion by acidic entrapment. J. Phys. Chem. lett. **10**, 2378–2385 (2019)
4. Y.L. Huang, X. Zhang, Z.S. Ma, Y.C. Zhou, W.T. Zheng, J. Zhou, C.Q. Sun, Hydrogen-bond relaxation dynamics: Resolving mysteries of water ice. Coord. Chem. Rev. **285**, 109–165 (2015)
5. Y. Gong, Y. Zhou, and C. Sun, Phonon spectrometrics of the hydrogen bond (O:H-O) segmental length and energy relaxation under excitation. B.o. Intelligence, Editor, China (2018)
6. C.Q. Sun, Atomic scale purification of electron spectroscopic information (US 2017 patent No. 9,625,397B2), United States (2017)
7. X. Zhang, Y. Zhou, Y. Gong, Y. Huang, C. Sun, Resolving H(Cl, Br, I) capabilities of transforming solution hydrogen-bond and surface-stress. Chem. Phys. Lett. **678**, 233–240 (2017)
8. C.Q. Sun, J. Chen, X. Liu, X. Zhang, Y. Huang, (Li, Na, K)OH hydration bondin thermodynamics: Solution self-heating. Chem. Phys. Lett. **696**, 139–143 (2018)
9. Y. Zhou, Y. Huang, Z. Ma, Y. Gong, X. Zhang, Y. Sun, C.Q. Sun, Water molecular structure-order in the NaX hydration shells (X = F, Cl, Br, I). J. Mol. Liq. **221**, 788–797 (2016)
10. Y. Zhou, Y. Zhong, Y. Gong, X. Zhang, Z. Ma, Y. Huang, C.Q. Sun, Unprecedented thermal stability of water supersolid skin. J. Mol. Liq. **220**, 865–869 (2016)
11. S. Gao, Y. Huang, X. Zhang, C.Q. Sun, Unexpected solute occupancy and anisotropic polarizability in lewis basic solutions. J. Phys. Chem. B **123**(40), 8512–8518 (2019)
12. C.Q. Sun, Unprecedented O:⇔: O compression and H↔H fragilization in Lewis solutions (Perspective). Phys. Chem. Chem. Phys. **21**, 2234–2250 (2019)
13. C.Q. Sun, J. Chen, Y. Gong, X. Zhang, Y. Huang, (H, Li)Br and LiOH solvation bonding dynamics: Molecular Nonbond Interactions and Solute Extraordinary Capabilities. J Phys. Chem. B **122**(3), 1228–1238 (2018)
14. X. Zhang, P. Sun, Y. Huang, T. Yan, Z. Ma, X. Liu, B. Zou, J. Zhou, W. Zheng, C.Q. Sun, Water's phase diagram: from the notion of thermodynamics to hydrogen-bond cooperativity. Prog. Solid State Chem. **43**, 71–81 (2015)
15. S.A. Harich, D.W.H. Hwang, X. Yang, J.J. Lin, X. Yang, R.N. Dixon, Photodissociation of H2O at 121.6 nm: A state-to-state dynamical picture. J. Chem. Phys. **113**(22), 10073–10090 (2000)
16. C.Q. Sun, X. Zhang, X. Fu, W. Zheng, J.-L. Kuo, Y. Zhou, Z. Shen, J. Zhou, Density and phonon-stiffness anomalies of water and ice in the full temperature range. J. Phys. Chem. Lett. **4**, 3238–3244 (2013)
17. F. Mallamace, C. Branca, M. Broccio, C. Corsaro, C.Y. Mou, S.H. Chen, The anomalous behavior of the density of water in the range 30 K < T < 373 K. Proc. Natl. Acad. Sci. US A **104**(47), 18387–18391 (2007)
18. C.Q. Sun, X. Zhang, W.T. Zheng, Hidden force opposing ice compression. Chem. Sci. **3**, 1455–1460 (2012)
19. Y. Yoshimura, S.T. Stewart, M. Somayazulu, H.K. Mao, R.J. Hemley, Convergent Raman features in high density amorphous Ice, Ice VII, and Ice VIII under Pressure. J. Phys. Chem. B **115**(14), 3756–3760 (2011)

20. C.Q. Sun, X. Zhang, J. Zhou, Y. Huang, Y. Zhou, W. Zheng, Density, elasticity, and stability anomalies of water molecules with fewer than four neighbors. J. Phys. Chem. Lett. **4**, 2565–2570 (2013)

21. Y. Huang, X. Zhang, Z. Ma, Y. Zhou, J. Zhou, W. Zheng, C.Q. Sun, Size, separation, structure order, and mass density of molecules packing in water and ice. Sci. Rep. **3**, 3005 (2013)

22. Q. Zeng, T. Yan, K. Wang, Y. Gong, Y. Zhou, Y. Huang, C.Q. Sun, B. Zou, Compression icing of room-temperature NaX solutions (X = F, Cl, Br, I). Phys. Chem. Chem. Phys. **18**(20), 14046–14054 (2016)

23. Q. Zeng, C. Yao, K. Wang, C.Q. Sun, B. Zou, Room-temperature NaI/H_2O compression icing: Solute–solute interactions. Phys. Chem. Chem. Phys. **19**, 26645–26650 (2017)

24. Y.L. Huang, X. Zhang, Z.S. Ma, G.H. Zhou, Y.Y. Gong, C.Q. Sun, Potential paths for the Hydrogen-bond relaxing with (H2O)(N) cluster size. J. Phys. Chem. C **119**(29), 16962–16971 (2015)

25. Y. Huang, X. Zhang, Z. Ma, Y. Zhou, G. Zhou, C.Q. Sun, Hydrogen-bond asymmetric local potentials in compressed ice. J. Phys. Chem. B **117**(43), 13639–13645 (2013)

26. X. Zhang, Y. Huang, Z. Ma, Y. Zhou, J. Zhou, W. Zheng, Q. Jiang, C.Q. Sun, Hydrogen-bond memory and water-skin supersolidity resolving the Mpemba paradox. Phys. Chem. Chem. Phys. **16**(42), 22995–23002 (2014)

27. D. Marx, M.E. Tuckerman, J. Hutter, M. Parrinello, The nature of the hydrated excess proton in water. Nature **397**(6720), 601–604 (1999)

28. Y. Crespo, A. Hassanali, Characterizing the local solvation environment of OH − in water clusters with AIMD. J. Chem. Phys. **144**(7), 074304 (2016)

29. A. Mandal, K. Ramasesha, L. De Marco, A. Tokmakoff, Collective vibrations of water-solvated hydroxide ions investigated with broadband 2DIR spectroscopy. J. Chem. Phys. **140**(20), 204508 (2014)

30. S.T. Roberts, P.B. Petersen, K. Ramasesha, A. Tokmakoff, I.S. Ufimtsev, T.J. Martinez, Observation of a Zundel-like transition state during proton transfer in aqueous hydroxide solutions. Proc. Natl. Acad. Sci. **106**(36), 15154–15159 (2009)

31. J. Chen, C. Yao, X. Liu, X. Zhang, C.Q. Sun, Y. Huang, H_2O_2 and HO^- solvation dynamics: Solute capabilities and solute-solvent molecular interactions. Chem. Select **2**, 8517–8523 (2017)

32. C.Q. Sun, Relaxation of the chemical bond. Springer Ser. Chem. Phys. **108**, 807 (2014). (Springer, Heidelberg)

33. X. Zhang, Y. Huang, Z. Ma, Y. Zhou, W. Zheng, J. Zhou, C.Q. Sun, A common supersolid skin covering both water and ice. Phys. Chem. Chem. Phys. **16**(42), 22987–22994 (2014)

34. M. Thämer, L. De Marco, K. Ramasesha, A. Mandal, A. Tokmakoff, Ultrafast 2D IR spectroscopy of the excess proton in liquid water. Science **350**(6256), 78–82 (2015)

35. Y. Zhou, D. Wu, Y. Gong, Z. Ma, Y. Huang, X. Zhang, C.Q. Sun, Base-hydration-resolved hydrogen-bond networking dynamics: Quantum point compression. J. Mol. Liq. **223**, 1277–1283 (2016)

36. X. Zhang, T. Yan, Y. Huang, Z. Ma, X. Liu, B. Zou, C.Q. Sun, Mediating relaxation and polarization of hydrogen-bonds in water by NaCl salting and heating. Phys. Chem. Chem. Phys. **16**(45), 24666–24671 (2014)

37. Y. Gong, Y. Zhou, H. Wu, D. Wu, Y. Huang, C.Q. Sun, Raman spectroscopy of alkali halide hydration: Hydrogen bond relaxation and polarization. J. Raman Spectrosc. **47**(11), 1351–1359 (2016)

38. S. Park, D.E. Moilanen, M.D. Fayer, Water Dynamics: The Effects of Ions and Nanoconfinement. J. Phys. Chem. B **112**(17), 5279–5290 (2008)

39. C.Q. Sun, Supersolidity of the undercoordinated and the hydrating water (Perspective). Phys. Chem. Chem. Phys. **20**, 30104–30119 (2018)

40. T.F. Kahan, J.P. Reid, D.J. Donaldson, Spectroscopic probes of the quasi-liquid layer on ice. J. Phys. Chem. A **111**(43), 11006–11012 (2007)

41. Y. Zhou, Y. Zhong, X. Liu, Y. Huang, X. Zhang, C.Q. Sun, NaX solvation bonding dynamics: hydrogen bond and surface stress transition (X = HSO4, NO3, ClO4, SCN). J. Mol. Liq. **248**, 432–438 (2017)

42. M. Zhao, W.T. Zheng, J.C. Li, Z. Wen, M.X. Gu, C.Q. Sun, Atomistic origin, temperature dependence, and responsibilities of surface energetics: An extended broken-bond rule. Phys. Rev. B **75**(8), 085427 (2007)

43. J.C. Araque, S.K. Yadav, M. Shadeck, M. Maroncelli, C.J. Margulis, How is diffusion of neutral and charged tracers related to the structure and dynamics of a room-temperature ionic liquid? Large deviations from Stokes-Einstein behavior explained. J. Phys. Chem. B **119**(23), 7015–7029 (2015)

44. G. Jones, M. Dole, The viscosity of aqueous solutions of strong electrolytes with special reference to barium chloride. J. Am. Chem. Soc. **51**(10), 2950–2964 (1929)

45. T. Brinzer, E.J. Berquist, Z. Ren, S. Dutta, C.A. Johnson, C.S. Krisher, D.S. Lambrecht, S. Garrett-Roe, Ultrafast vibrational spectroscopy (2D-IR) of CO2 in ionic liquids: Carbon capture from carbon dioxide's point of view. J. Chem. Phys. **142**(21), 212425 (2015)

46. Z. Ren, A.S. Ivanova, D. Couchot-Vore, S. Garrett-Roe, Ultrafast structure and dynamics in ionic liquids: 2D-IR spectroscopy probes the molecular origin of viscosity. J. Phys. Chem. lett. **5**(9), 1541–1546 (2014)

47. C.Q. Sun, Aqueous charge injection: Solvation bonding dynamics, molecular nonbond interactions, and extraordinary solute capabilities. Int. Rev. Phys. Chem. **37**(4), 358–363 (2018)

48. L.E. Bove, R. Gaal, Z. Raza, A.A. Ludl, S. Klotz, A.M. Saitta, A.F. Goncharov, P. Gillet, Effect of salt on the H-bond symmetrization in ice. Proc Natl Acad Sci U S A **112**(27), 8216–8220 (2015)

49. A.F. Goncharov, V.V. Struzhkin, M.S. Somayazulu, R.J. Hemley, H.K. Mao, Compression of ice to 210 Gigapascals: Infrared evidence for a symmetric Hydrogen-bonded phase. Science **273**(5272), 218–220 (1996)

50. A.F. Goncharov, V.V. Struzhkin, H.-K. Mao, R.J. Hemley, Raman spectroscopy of dense H 2 O and the transition to symmetric hydrogen bonds. Phys. Rev. Lett. **83**(10), 1998 (1999)

51. Y. Bronstein, P. Depondt, L.E. Bove, R. Gaal, A.M. Saitta, F. Finocchi, Quantum versus classical protons in pure and salty ice under pressure. Phys. Rev. B **93**(2), 024104 (2016)

Chapter 25
Perspectives

Abstract The notion of multifield-driven bond oscillation dynamics and the theory-driven phonon spectrometrics have enabled quantitative information on the bond relxation under perturbation and bond transformation by chemical reaction with ever-deep and consistent insight into the perturbation-relaxation-property correlation of liquid and solid crystals. Progress evidences the essentiality of bonding and electronic dynamics and the spectrometric engineering in uncovering the physics and chemisrty behind the processes of matter and life. Coupling of the intramolecular bonding and the intermolecular O:H attractive, H↔H and O:⇔:O repulsive non-bonding interactions would lead to promising breakthroughs in molecular science and engineering.

25.1 Attainments

The presented notion of multifield bonding dynamics and the theory-driven phonon spectrometrics have enabled the ever-deep and consistent understanding about perturbation-relaxation-property correlation of liquid and solid crystals. The following summarize the major progress:

(1) The LBA approach and the multifield bond oscillation have reconciled the effect of P, T, CN, and charge injection on the bond relaxation and property evolution of the liquid and solid crystals. A stimulus mediates the properties by bond relaxation from one equilibrium to another.

(2) Crystal size reduction creates three types of phonons unseen in the bulk – size-reduction induced E_{2g} (TO mode) blueshift, A_{1g} (LO) redshift, and the emerging of the LFR mode in the THz frequencies. Nanostructures prefer the core-shell configuration and the performance of bonds in the skin shell dictates the size dependency of liquid and solid crystals.

(3) The phonon frequency shift, bandgap, and elastic modulus follows a Debye thermal decay, which offers information of the atomic cohesive energy E_{coh} and the Debye temperature; mechanical compression stiffens the phonons nonlinearly and provides information of biding energy density E_{den} and elastic modulus.

C. Q. Sun, *Electron and Phonon Spectrometrics*,
https://doi.org/10.1007/978-981-15-3176-7_25

(4) Strikingly, the DPS filters the fraction-stiffness transition of bonds from the reference state to the conditioned, which has derived the thickness of skin-shell thickness of liquid and solid nanocrystals up to two atomic diameters.

(5) Charge injection in terms of anions, cations, electrons, lone pairs, protons, and molecular dipoles by acid, base, and salt solvation mediates the O:H–O bonding network and properties of a solution through O:H formation, H↔H fragilization, O:⇔:O compression, electrostatic polarization, H_2O dipolar screen shielding, solute-solute interaction, and undercoordinated H–O bond contraction.

(6) O:H–O segmental specific-heat disparity derives the quasisolid phase of negative thermal expansion. Molecular undercoordination and electrostatic polarization result in the supersolid phase that disperses the QS boundary outwardly. Excessive protons and lone pairs form the H_3O^+ hydronium and HO^- hydroxide, which turns an O:H–O bond into the H↔H anti–HB breaker upon acid solvation and O:⇔:O super–HB compressor and polarizer on base solvation. Ions polarize and stretches the O:H–O bond and shifts the segmental frequencies.

25.2 Perspectives

Progress described in this part recommend the following

(1) The notion of the multifield bonding dynamics may overcome limitations of the Gibbs free energy, the molecular dynamics premise, and the Grüneisen parameter, $\partial\omega/\partial x_i$, for materials properties under perturbation. External stimulus mediates the properties by relaxing the interatomic bonding and the associated energetics, densification, localization, entrapment and polarization of electrons. Gibbs energy, $dG(P, T, N, ...) = SdT + VdP + \mu dN +...$, is valid for macroscopic statistic systems and gaseous phases by taking the degrees of freedom as plain variables with S, V, μ being the entropy, volume, and chemical potentials. Molecular dynamics takes the molecules as the basic structure unit with little attention to the coupling of the intramolecular and intermolecular interactions, such as water and molecular crystals.

(2) It is important to note that nonbonding weak O:H interactions, and even the H↔H and O:⇔:O repulsions are the keys to the performance of substance though they contribute insignificantly to the Hamiltonian and disobey the dispersion of the Schrödinger equations. Water forms the simplest, ordered, fluctuating molecular crystal because of its equal numbers of lone pairs and protons compared with other molecular crystals.

(3) The phonon spectrometrics and the DPS is powerful yet convenient to probe the bond relaxation and transition dynamics without needing the critical high-vacuum conditions. The phonon spectrometrics is suitable for all sorts substance,

liquid and solid, varying from conductors to insulators under any applied perturbations, and provides quantitative information about bonding dynamics and consistent insight into the physics behind observations.

(4) As an independent degree of freedom that laid foundation of defect physics, surface chemistry, and nanoscience and nanotechnology, atomic and molecular undercoordination has yet received deserved attention. Atomic CN reduction shortens and stiffens bonds between the undercoordinated atoms, which modify electrons in various energy bands, and dictate the performance of substance.

(5) It is necessary to take liquid water as crystal with numbers of protons and lone pairs and the O:H–O configuration conservation and molecular and proton motion restriction. Charge injection by solvation does not form new bonds between the solute and solvent unless acid and base solvation occurs, which breaks the conservation rule of pure water by forming the $(H_3O^+;$ $HO^-):4H_2O$ motifs. New kinds of intermolecular $H \leftrightarrow H$ and $O: \Leftrightarrow :O$ repulsive interactions need particular attention.

(6) The O:H–O segmental specific heat disparity derived QS phase may apply to negative thermal expansion of substance with multiple interaction potentials being involved, such as $ZrWO_8$ and graphite. Each potential corresponds to a specific heat and the superposition of such specific heat curves specifies the structural phases of different specific heat ratios.

Understanding and strategies may extend to wherever the regular chemical bonds and the segmented hydrogen bonds are involved. It would be very promising for one to keep arm and mind open, and always move and look forward to developing experimental strategies and innovating ideas toward resolution to the wonders.

Appendix A
Advantages and Limitations of the Electron and Phonon Spectrometrics

C. Q. Sun, *Electron and Phonon Spectrometrics*,
https://doi.org/10.1007/978-981-15-3176-7

Table A.1 Advantages, capabilities, and limitations of the electron and phonon spectrometrics. Excises have opened a subject paving the path toward conventionally-unexpected information and consistent insight into the intrinsic bond-electron-property cooperativity of substance

Spectrometrics	Experimentally-orientated spectroscopy data processing		Electron diffraction (VLEED) [6, 7] (1992–2003)
	Phonon relaxation (DPS) [1–3] (2007–2020)	Electron emission (ZPS) [4, 5] (2004–15)	
Conventional spectroscopy	Peak decomposition needing constraints, peak maximal normalization, artificial/extrinsic factors involvement; empirical description with freely adjustable parameters		Hard sphere dislocation scheme incapable of reproducing measurements
Objective	Ever-unknown information and comprehension		
Strategic principles	• The nature of the chemical bond bridges the structures and properties of crystals and molecules [8] • Bond and nonbond relaxation and the associated localization, entrapment, and polarization of energetic electrons mediate the macroscopic performance of substance [6] • Fourier real-energy-domain transition enables the average-bond oscillating dynamics		Multi-beam interference by crystal geometry and by elastic and damping surface potential barrier
Key strategy	$\Delta\omega \propto (E_z/d^2)^{1/2}$ (Lagrangian oscillation [9]) Phonon frequency shift and repopulation	$\Delta E_v \propto \Delta E_z$ (Tight binding [10]) Binding energy shift and electronic states	Bond geometry—atomic position conversion; SPB least parameterization; computer auto-optimization
Generlization	Logic correct, concise, and consistence; free from artificial assumption; least calculations A detectable quantity $\Delta Q(x_i)/Q(x_0) = f[x_i, d(x_i), E(x_i)]$ functionalization on stimuli x_i (atomic site-resolved $Y_z \propto E_z/d_z^3$; $T_{Cz} \propto zE_z$; $\Delta E_{Gz} \propto \Delta E_z$; etc.)		Chemisorption bond-band-barrier dynamics; SPB-geometry superposition; massive computation
Detecitng sources	Light of various wave lengths; Phonon reflection and absorption	X-ray, UV light, electron, electric bias; electron emission	LEED beam (6–16 eV) (00) pattern I–V curve
Experimental conditions	Unnecessary	Ultrahigh vacuum; conductor or semiconductor	Ultrahigh vacuum; dedicated and sophisticated data collection
Artefacts	Nonlinear contribution at equilibrium is out of instrument resolution		—

(continued)

Table A.1 (continued)

	Multiple phonon resonant scattering; Optical phonon thermal degeneration	Charging effect; Initial- and final-state recombination	
Sample conditions	Solid conductors and semiconductors		Ideally pure crystal with programable controlled dose of chemisorption gas
	Liquids and insulators	Hardly possible	
External stimuli	Pressure, strain		
	Temperature, irregular coordination, charge injection, electrification		Annealing; aging; dosing; azimuth angle change
Basic information	Atomistic, local, dynamic, quantitative		Unit cell sized information on
	Bond and nonbond relaxation and oscillation—phonon frequency and abundance transition	Bond relaxation derived behavior of electrons in various bands/levels—entrapment and polarization	bond-band-barrier transition by chemisorption; unification of morphology; crystallography; energy spectroscopy of outside second atomic layer
New information	Bond length and energy; binding energy density, atomic cohesive energy		
	(1) Vibration frequency excited shift Intergrain interaction–THz wave; bond nature index	(1) Core level of an isolated atom and shift	(1) Surface relaxation and reconstruction; bond geometry, angle, length
	(2) Collective and single dimer vibration; nanograin shell thickness (skin/core volume ratio)	(2) Registry and layer order resolved binding energy	(2) sp^3-orbital hybridization and de-hybridization; nonbonding lone pairs, missing-row, and dipole formation
	(3) Debye decay; Debye temperature; thermal expansion; elastic modulus	(3) Monolayer skin and point defect bond energy and electronic states; energy states of terrace edges, adatoms	(3) Four-stage Cu_3O_2 on Cu(001) surface
	(4) Compressibility; bond force constant	(4) Interfaces, adsorbate site; nonbonding states; antibonding polarization	(4) 2D Brillouin zones; electron effective mass; inner potential constant
	(5) Transition of bond number, stiffness, fluctuation by perturbation; etc.	(5) Screening effect; potential splitting; charge redistribution in reaction; etc.	(5) Potential barrier rippling; work function modulation; four valance states generation, etc.

(continued)

Table A.1 (continued)

Conceptual development	Bond-phonon-property correlation O:H-O bond cooperativity and potentials H ↔ H anti-HB; O:⇔:O super-HB; Supersolidity; quasisolidity; THz sources; etc.	Bond-electron-energy correlation Irregular atomic CN; Entrapment and polarization n/p-type catalyst; Dirac-Fermion; hydrophobicity; lubricity; etc.	Bond-band-barrier correlation; sp^3-orbital hybridization; surface bond contraction; lone pairs; polarization; host electronegativity, atomic radius, and crystal geometry resolve 4-stage bonding kinetics, etc.
Impact	**Performance of electrons, bonds, nonbonds and molecules in the energetic-spatial-temporal domians** Molecular science; aqueous solutions; foods and drugs; life science and biochemistry; porous and nanostructures; functional materials and devices; geometric multifield insulators; sensors; catalyst design; energetic materials; energy management; and environment science; etc.		Chemisorption bonding dynamics; Basic chemical and physical sciences

Appendix B
Introduction of the "Spectral Studio" Analytical Package

This book is associated with an analytical package, "Spectral Studio", developed in collaboration with the Hongzhiwei Technology (Shanghai) Co., Ltd. This package is focused on extracting the un-expected genomic information of bonding, electronic, and molecular dynamics and energetics of liquids and solids from the energy shift of electrons emitted from various energy bands and frequency shift of phonons of different modes by physical perturbation and chemical reaction.

The spectroscopies include electronic AES, STS, UPS, XPS, XAS, APECS and phononic Raman and RTIR. Physical perturbation includes atomic and molecular undercoordination (adatoms, point defects, terrace edges, surfaces, and nanostructures of various shapes), mechanical and thermal excitation, electric and magnetic fields; chemical reaction includes hetero-coordination (interfaces, alloys, compounds, etc.), chemisorption, aqueous solvation.

Gained information includes the energy level of an isolated atom, effective atomic CN, local bond length and energy, binding energy density, skin thickness of nanocrystals, elastic modulus, Debye temperature, single bond force constant, charge transfer in reaction, orbital screening coefficient.

The package use visualized interface for input, output, display and data processing. It is easy to operate (Fig. B.1).

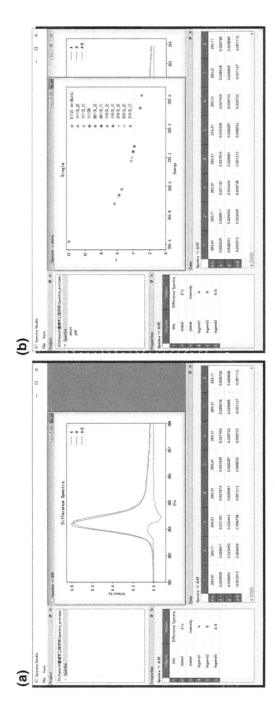

Fig. B.1 "Spectral studio" Interface display. Exemplary of **a** two electronic spectra and their difference **b** registry and layer-order resolved atomic coordination number

References

1. X.X. Yang, C.Q. Sun, *Raman detection of temperature CN 106908170A* (2017)
2. Y. Gong, Y. Zhou, Y. Huang, C.Q. Sun, *Spectrometrics of the O:H-O bond segmental length and energy relaxation (CN 105403515A)* (China, 2018)
3. Y.L. Huang, X.X. Yang, C.Q. Sun, *Spectrometric evaluation of the force constant, elastic modulues, and Debye temp[erature of sized matter disclosure at evaluation* (2018)
4. X.J. Liu, M.L. Bo, X. Zhang, L. Li, Y.G. Nie, H. Tian, Y. Sun, S. Xu, Y. Wang, W. Zheng, C.Q. Sun, Coordination-resolved electron spectrometrics. Chem. Rev. **115**(14), 6746–6810 (2015)
5. C.Q. Sun, Atomic scale purification of electron spectroscopic information (US 2017 patent No. 9,625,397B2) (United States, 2017)
6. C.Q. Sun, Relaxation of the Chemical Bond. Spr. Ser. Chem. Phys. **108**, 807 (Springer-Verlag, Heidelberg, 2014)
7. C.Q. Sun, Oxidation electronics: bond-band-barrier correlation and its applications. Progr. Mater. Sci. **48**(6), 521–685 (2003)
8. L. Pauling, *The Nature of the Chemical Bond*, 3rd edn. (Cornell University press, Ithaca, NY, 1960)
9. Y. Huang, X. Zhang, Z. Ma, Y. Zhou, G. Zhou, C.Q. Sun, Hydrogen-bond asymmetric local potentials in compressed ice. J. Phys. Chem. B **117**(43), 13639–13645 (2013)
10. M.A. Omar, *Elementary solid state physics: principles and applications* (Addison-Wesley, New York, 1993)

Index

A

Acceptor-like catalyst, 14, 15, 81, 85, 87, 96, 167, 245

Adatom, 14, 15, 81, 85, 87, 96, 245

Adsorption, 4, 6, 11, 83, 86, 93, 94, 96, 164, 187, 189, 191–193, 198, 199, 227, 247, 252, 256, 278, 289, 306, 337, 341, 348, 476

Antibonding, 9, 15, 26, 39, 40, 46, 47, 49, 110, 181–183, 186, 187, 190, 192, 193, 197, 199, 225, 244, 246, 257, 272, 273, 279, 280, 282, 288, 338, 343, 362, 503

Atomic cohesive energy, 7, 13, 15, 28, 31, 35, 39, 41, 42, 45, 55, 59, 63, 67, 74, 75, 123, 147, 156, 163, 165, 172, 174, 243, 245, 247, 370, 371, 373, 387, 396, 402, 415, 428, 429, 431, 434–436, 456-458, 461–463, 497, 503

Atomic undercoordination, 4–6, 8, 10, 25, 59, 63, 68, 74, 86, 114, 118, 123, 131, 135, 145, 146, 181, 194, 196–199, 209, 222, 245, 258, 271, 280, 370, 372, 374, 378, 384, 402, 415, 434

Auger parameter, 36, 73, 74, 75, 78, 108, 128, 132, 134, 264

Auger Photoelectron Coincidence Spectroscopy (APECS), 13–15, 45, 46, 50–52, 55, 101, 102, 108, 110–117, 119–121, 243–247, 505

Auger process, 14

Azimuth angle, 48

B

Binding energy density, 7, 15, 45, 55, 59, 67, 74, 172, 174, 243, 371, 373, 387, 393, 402, 410, 415, 428, 430, 432, 438, 447, 459, 461, 503, 505

BOLS-NEP theory, 25, 31, 33–35, 38, 40, 81, 87, 89, 91, 102, 108, 135, 143, 152, 156, 211, 212, 244, 246

Bond stiffness, 396, 409, 449

Brillouin zone, 48, 251, 253, 258, 259, 311, 317, 331, 332, 334–336, 339, 343, 382, 384, 396, 397, 419, 459, 503

C

Charge injection, 217, 236, 245, 371, 372, 409, 410, 469, 470, 472, 497, 498, 503

Charge transfer, 12, 103, 109, 110, 117, 175, 269, 331, 358, 359, 505

Cu_3O_2 bonding, 343

D

Dangling bond, 143

Debye temperature, 505

Debye thermal decay, 370, 371, 378, 379, 384, 405, 407, 415, 419, 429, 434, 437, 447, 453, 456, 457, 460–462, 497

Donor-like catalyst, 15, 85, 163

Drift mobility, 485

Droplet, 11, 216, 222, 223, 371, 447–449, 476, 480, 490, 491

CPSIA information can be obtained
at www.ICGtesting.com
Printed in the USA
LVHW082121120420
653141LV00002B/105